Springer Geography

The Springer Geography series seeks to publish a broad portfolio of scientific books, aiming at researchers, students, and everyone interested in geographical research.

The series includes peer-reviewed monographs, edited volumes, textbooks, and conference proceedings. It covers the major topics in geography and geographical sciences including, but not limited to; Economic Geography, Landscape and Urban Planning, Urban Geography, Physical Geography and Environmental Geography.

Springer Geography—now indexed in Scopus

Yile Chen · Liang Zheng · Jianyi Zheng
Editors

City-Sea Symbiosis

Settlement Morphology Types in Zhuhai and Macau

Springer

Editors
Yile Chen
Faculty of Humanities and Arts
Macau University of Science
and Technology
Taipa, Macao

Liang Zheng
Faculty of Humanities and Arts
Macau University of Science
and Technology
Taipa, Macao

Jianyi Zheng
Faculty of Humanities and Arts
Macau University of Science
and Technology
Taipa, Macao

ISSN 2194-315X ISSN 2194-3168 (electronic)
Springer Geography
ISBN 978-3-032-13724-1 ISBN 978-3-032-13725-8 (eBook)
https://doi.org/10.1007/978-3-032-13725-8

This book was funded by (1) the Macau Science and Technology Development Fund (FDCT)'s Funding for Innovation and Technology Promotion (0080/2023/ITP2) and (2) Faculty Research Grants by Macau University of Science and Technology (FRG-MUST) (grant number: FRG-25-067-FA). The funders had no role in study conceptualization, data curation, formal analysis, methodology, software, decision to publish, or preparation of the manuscript. There was no additional external funding received for this study.

This Springer imprint is published by the registered company Springer Nature Switzerland AG
The registered company address is: Gewerbestrasse 11, 6330 Cham, Switzerland

Contents

The Development of Settlements and Morphological Space in the Xiangshan Area

Geographical and Cultural Background of Modern Xiangshan Urban Form

Yile Chen

Abstract This article reviews and summarizes the development history of Xiangshan County, along with the core concepts of urban form analysis and analysis of settlement spatial morphology. It also provides exemplary analyzes of the unique case studies of Zhuhai and Macau—Tangjiawan Ancient Town, Guishan Islands, Hengqin Islands, Beishan Village, and Macau. Finally, it discusses Prospects and Challenges and provides an overview of each chapter in the book.

Keywords Geographical and cultural · Xiangshan (Heungsan) · Urban form

1 Introduction

1.1 Development History of Xiangshan County

Xiangshan (Heungshan) County, also spelled Hsiangshan, Siangshan, and Heungsan, was a former county in Southern China [1, 2]. It was an administrative region that existed between 1152 and 1925 [3, 4]. According to the *Taiping Huanyuzhi* (太平寰宇志) of the Song Dynasty, the name Xiangshan came from a place called Wugui Mountain, where there were many exotic flowers and the fragrance spread for dozens of miles. Deng Guangjian, a patriotic poet of the Southern Song Dynasty who lived in seclusion in Xiangshan, described this island isolated in the ancient Lingdingyang in his *Brief Record of Fuxu Mountain* (浮虛山記略). The county consists of the entirety of Xiangshan Island, an island in a bay where three rivers emptied into the sea, part of the Pearl River Delta [5].

Y. Chen (✉)
Faculty of Humanities and Arts, Macau University of Science and Technology, Avenida Wai Long, Tapai, Macau, China
e-mail: chenyile@must.edu.mo

Y. Chen et al. (eds.), *City-Sea Symbiosis*, Springer Geography,
https://doi.org/10.1007/978-3-032-13725-8_1

It was originally separated from the continent by the distributaries in the delta, until it became a peninsula sometime during the seventeenth century [6–11], but remained separated from the mainland by waterways. The county covered the modern-day Zhuhai City, Zhongshan City and a part of Nansha District of Guangzhou City in the Guangdong Province of the People's Republic of China (PRC), and the Macau Special Administrative Region of the People's Republic of China, formerly known as Portuguese Macau until 1999.

Because the salt industry was a monopoly throughout many dynasties, it not only supported the economy but also drove administrative development in Xiangshan County's early history. Salt inextricably links the establishment and evolution of Xiangshan County [12]. During the Qin and Han Dynasties, it was under the jurisdiction of Panyu County, Nanhai Commandery, and during the Eastern Jin Dynasty, it was under the jurisdiction of Dongguan County. It wasn't until the second year of the Tang Dynasty's Zhide reign (A.D. 757) that the government established a military camp called Xiangshan Town in present-day Shanchang Village (山場), Zhuhai City, due to its abundant production of sea salt [13]. Town was a military organization in the Tang Dynasty rather than an administrative system and was only set up in important military locations. Xiangshan Township was carved out of the Bao'an County and Dongguan County. Xiangshan County is in the subtropics. Since the Han and Tang Dynasties, extensive mudflats and shallow bays have emerged, providing the area with abundant sea salt resources. Especially during the Tang and Song Dynasties, the diversion of the Xijiang River and Beijiang River accelerated the deposition of silt between the islands at the Pearl River Estuary [14, 15], gradually connecting the land and islands, forming natural salt pans.

In the fifth year of Yuanfeng in the Northern Song Dynasty (A.D. 1082), Xiangshan Fort was established and still belonged to Dongguan County. It was not until the 22nd year of Shaoxing in the Southern Song Dynasty (A.D. 1152) that Xiangshan Village was upgraded to Xiangshan County due to its development, and the coastal areas of Nanhai, Panyu, Dongguan, and Xinhui counties were placed under the jurisdiction of Xiangshan County. When Xiangshan was established in 1152, Shiqi (Shekki) was selected to be the administrative center [5]. At that time, Xiangshan County was under the jurisdiction of Guangzhou Prefecture. Around the twelfth century (during the Northern Song Dynasty), this township became a separate county. In the third year of the Hongwu reign of the Ming Dynasty (A.D. 1370), the Ming government established the Guangzhou Maritime Customs Office to conduct tribute trade with foreign countries and designated Lampacau (浪白澳) (now Nanshui Town, Zhuhai City) in the south of Xiangshan as a port for foreign merchant ships to dock and trade [16, 17].

According to the "*Da De Nan Hai Zhi*" (大德南海志) of the Yuan Dynasty, the areas under the jurisdiction of Xiangshan County include Wenshun Township (文順鄉), Yanfuli (延福里), Huangziwei (皇字圍), Sziwei (四字圍), Daziwei (大字圍), Wuchangziwei (五常字圍), Gongziwei (恭字圍), Liangziwei (良字圍), Denengziwei (得能字圍), Yongle Township (永樂鄉), Yongning Township (永甯鄉), Tanzhou (譚洲), Huangqijiao (黄旗角), Dalan (大欖), Xiaolan (小欖),

and Haizhou (海洲). From the Yuan Dynasty to the Ming Dynasty, sedimentation at the Pearl River Estuary intensified, forming the "Eighteen Sandbars of the West Sea" in the northwest of Xiangshan and the "Sixteen Sandbars of the East Sea" in the northeast. Numerous sandbars transformed into land, which increased the amount of agricultural land available. The geographical landscape gradually evolved from isolated islands in the sea to an estuarine plain with alternating islands and lagoons. The continuous siltation at the mouth continuously created new land resources, driving the development of sandy land to become the core production activity during the Ming and Qing Dynasties. From the Ming and Qing Dynasties to modern times, the jurisdiction of Xiangshan was gradually divided. Topographically, low mountains and hills, exemplified by Wugui Mountain and Phoenix Mountain, form the regional framework, surrounding the central and coastal alluvial plains, forming a complex geographical pattern of "mountains-hills-fields-water." This topography, combined with hydrological characteristics, determined the spatial distribution of settlements along river mouths, sandbars, and transportation routes, laying the natural foundation for the subsequent "core city + satellite town" urban system.

During the mid-sixteenth century, in the Ming Dynasty, the Portuguese settled in Macau. This was permitted by the Ming Dynasty, but despite the Portuguese presence, Macau was still under the jurisdiction of Xiangshan (Fig. 1.1). Macau

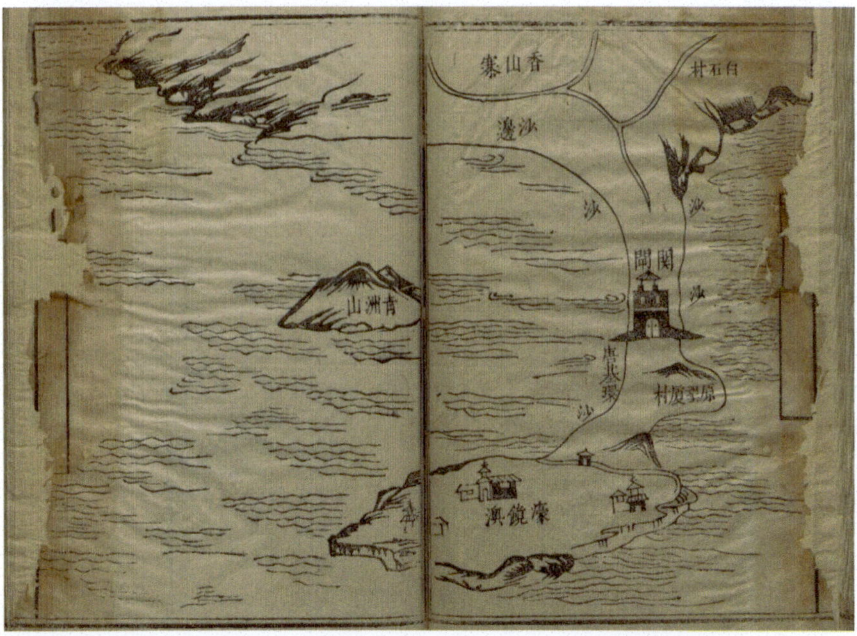

Fig. 1.1 Maps of Xiangshan County and Macau, selected from *Xiangshan County Local Gazetteers* of the Kangxi Era, 1673. (*Image source* Originally collected in the National Library of China, accessed online from the Macau University of Science and Technology Library)

became a full Portuguese-administered areas in 1847 [18]. This colony later expanded to include Ilha Verde, Taipa, and Coloane. During the early Qing Dynasty (Fig. 1.2), Xiangshan completed its transformation from a "salt industry economy" to an "agriculture-commerce complex economy." Sand field development technology matured, and clans became deeply involved in resource management, formulating clan rules to ensure production order, demonstrating the unique survival wisdom of coastal areas. After the Qianlong reign of the Qing Dynasty, the salt industry in Xiangshan Salt Field began to decline. The cause was the continuous sedimentation carried by the Pearl River. Over time, vast alluvial plains gradually formed in many areas. The sea gradually receded from the Xiangshan Salt Field, and the salt pans became tidal fields [19]. By the Daoguang reign of the Qing Dynasty, the Xiangshan Salt Field had completely emerged from the sea onto land, becoming a vast expanse of sand. With the passage of time, the Xiangshan Salt Field has become a historical memory.

After the nineteenth century, Xiangshan County, leveraging its strategic proximity to Hong Kong and Macau, became a pioneering region in modern China. On one hand, the culture of overseas Chinese communities flourished, with large numbers of people seeking livelihoods in Southeast Asia, Europe, and the United States. The capital and technology they brought back fueled the emergence of

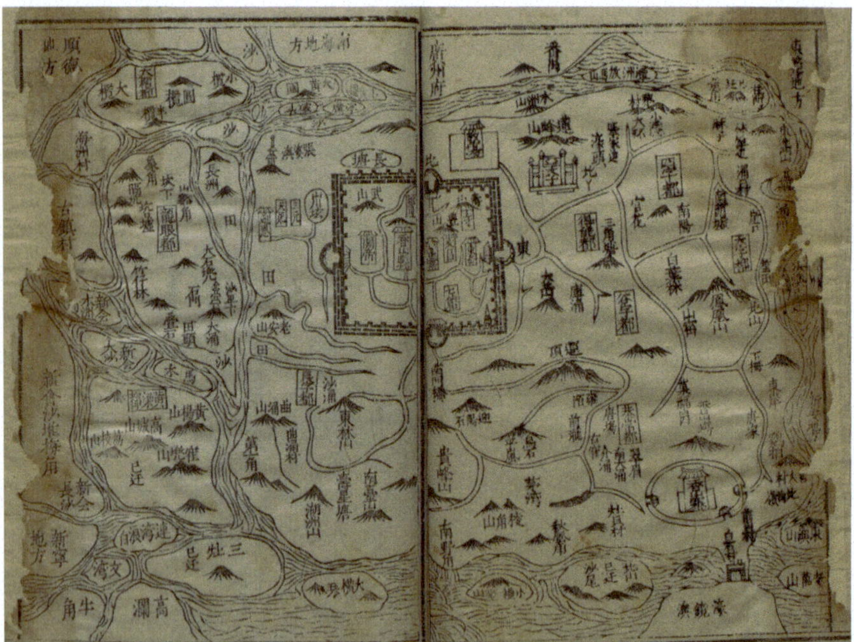

Fig. 1.2 Maps of Xiangshan County, selected from *Xiangshan County Local Gazetteers* of the Kangxi Era, 1673. (*Image source* Originally collected in the National Library of China, accessed online from the Macau University of Science and Technology Library)

local commerce and modern industry. On the other hand, as the hometown of Dr. Sun Yat-sen, it became a crucial birthplace of the democratic revolution, a scene of frequent intellectual enlightenment and social change.

In the first year of the Republic of China (1912), it was a county in Guangdong Province, in the Republic of China. Then, Guangdong Province abolished Guangzhou Prefecture, and Xiangshan County was directly under the jurisdiction of the Guangdong Provincial Governor's Office. It was renamed Zhongshan (then usually romanized "Chungshan") in April 1925, in honor of the founder of the Republic of China [20], Sun Yat Sen, a Xiangshan native. In December 1952, Yumin County (漁民縣) was created out of islands from Hsiang-shan County, Pao-an County, and Tung-kuan County. This county was renamed Chu-hai County (珠海縣) in April 1953, and was expanded [7–11]. Chu-hai County was briefly merged with Hsiang-shan County in March 1959 and was reconstituted in April 1961.

From 1961 onwards, the territory of Chung-shan County was reduced by 43%, with areas ceded to the counties of Tou-men, Pan-yu, Shun-te, and Hsin-huei. Its romanization was changed to Zhuhai following the introduction of the Pinyin orthography for romanizing Chinese characters [7–11].

In 1979, Zhuhai County (珠海縣) became Zhuhai City (珠海市). In December 1983, Zhongshan County (中山縣) became Zhongshan City (中山市). Originally a county-level city, Zhongshan was elevated to prefecture level in January 1988 [21].

A 6.81 km^2 special economic zone was established within Zhuhai City in August 1980 [22]. The special economic zone was expanded to 15.16 km^2 in June 1983 and 121 km^2 in April 1989. In May 1983, Doumen County (斗門縣) became administrated under Zhuhai City. Although Xiangshan County is no longer within its original geographical administrative scope, this area has gradually developed into an important economic node on the west bank of the Pearl River Estuary. Its historical and geographical foundation and development context have profoundly influenced the spatial pattern of the west bank of the Guangdong-Hong Kong-Macau Greater Bay Area today.

1.2 Research and Analysis Methods

This book (*City-Sea Symbiosis: Settlement Morphology Types in Zhuhai and Macau*) is based on the interdisciplinary perspective of urban and rural planning and urban morphology research in European human geography. It uses the following research methods:

(1) Historical Data Analysis Method. By consulting various versions of ancient books such as "Xiangshan County Chronicles" (香山縣志), "Guangdong General Chronicles" (廣東通志), "Tangjiawan Town Chronicles" (唐家灣鎮志), and "Quanhai Tuzhu" (全海圖註), we went to archives, museums, urban

planning exhibition halls, libraries, and other institutions in Macau, Zhuhai, and Shunde to collect Chinese and foreign maps from various periods for research. The main ones are ancient Chinese maps, navigation maps drawn by the West during the era of great navigation, and maps of the Macau surrounding area drawn by the Portuguese Macau government.

(2) M.R.G. Conzen's Urban Plan Pattern Analysis Method. By studying the plan maps of various periods, we extracted and summarized the three morphological elements, namely plan units, architectural forms, and land use patterns. These three elements were then analyzed and superimposed to form urban morphological regions.

(3) Field Research and Interviews. Field investigations, interviews, and records were conducted on the status of cities, villages, and reclaimed sites. For Macau, we investigated the historical city area, dock buildings, street spaces, etc., of the Macau Peninsula. The research also included selecting the best and most representative modern gardens, landscapes, and buildings in Tangjiawan, Zhuhai, for on-site research and interviews with residents and relevant officials, such as Director Guan of Gongle Garden, Professor Yang of the Zhuhai Museum, and Director Li of the Lu Family Ancestral Hall. Furthermore, on-site mapping and photographic footage were collected to obtain authentic and effective survey data, providing a solid foundation for subsequent research and analysis.

(4) ArcGIS Analysis. This mainly includes kernel density analysis and network analysis [23]. ArcGIS density analysis is a commonly used spatial analysis method. It can analyze the clustering of features in a limited area based on the input feature dataset and generate a raster density surface. Kernel density analysis is used to calculate the unit density of a given feature in the surrounding area. For example, this book uses the kernel density analysis tool in ArcGIS to calculate the clustering of heritage attraction points in urban space in each historical period. The analysis results can, to a certain extent, reflect the changes in the focus of urban development in different historical periods. The network analysis tool in ArcGIS can perform service area analysis, nearest facility point analysis, and optimal path solutions by constructing a network dataset. The service area function is used to perform accessibility analysis on the location of existing world heritage sites, thereby screening the range of attraction points for path optimization. The nearest facility point analysis can find points closer to the core area from near to far and further screen the attraction points. The optimal path analysis can find the optimal path between multiple stop points and is used to solve an optimized path between all selected attraction points.

2 Literature Review

2.1 Core Concepts of Urban Form Analysis

Morphology is a discipline that specifically studies the nature of biological forms. Unlike anatomy, which breaks down organisms into individual units, it requires that life forms be viewed as organic systems, focusing on the microscopic analysis of parts and their connection to the whole [24]. As cities evolve and change, geographers and historians have introduced morphology, the study of the structure of objects and the relationships between their components, into urban research, creating a crucial tool for understanding the physical form and historical development of cities—urban morphology [25]. Originating in biology, urban morphology officially became a discipline when German geographer Otto Schlüter published his paper *Über Den Grundriss Der Städte* in the late nineteenth century [26].

Conzen's town-plan analysis method is a highly influential school of thought [27]. Influenced by Schlüter's view that urban form is composed of elements such as land, settlements, transportation routes, and buildings, Conzen began using large-scale maps to conduct detailed analysis and research on town plans. In 1960, Conzen published *Alnwick, Northumberland: a study in town-plan analysis* [28]. Through a detailed plan analysis of the small town of Alnwick in Northumberland, he established a preliminary framework for the study of urban morphology, thus giving rise to the Conzen school of urban morphology [29]. Conzen believed that urban landscapes are the result of historical geography, that the formation of urban plan patterns is closely related to the social and economic development of the period, and that their evolution is constrained by planar characteristics. Historical maps can examine and analyze the planar characteristics of towns in different historical periods [30]. Conzen's core idea is to analyze the structure and evolution of cities from a dynamic perspective of historical development. The fundamental concept of Conzen's urban morphology theory is to define the urban landscape as the research object. Conzen categorizes the urban landscape as a combination of three elements: town plan, building form, and land use. The town plan is composed of streets, plots, and building blocks. Finally, these three elements of the urban landscape are analyzed and superimposed to form different hierarchical urban morphological zones (Fig. 1.3). Management and protection recommendations are then proposed for each zone, thereby forming a dynamic approach to urban landscape conservation, management, and research. Currently, some domestic scholars are combining Conzen's urban morphology theory with the architectural typology of the Italian Caniggia School to explore urban morphology analysis methods suitable for China [31]. Research related to Conzen's urban morphology theory still primarily focuses on large cities or old urban areas with a long history. Most of the research focuses on the preservation of historic cities, old urban areas, and ancient villages. While there are some studies on the evolution of urban morphology in new towns and districts that emerged after the founding of the People's Republic of China, they are limited. When studying old or historical

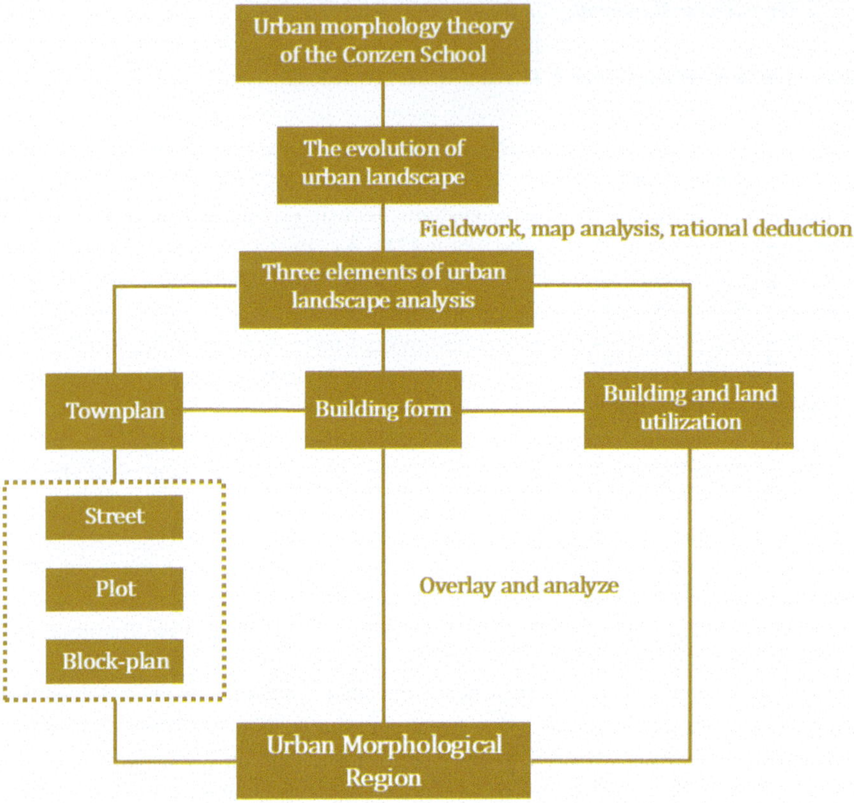

Fig. 1.3 Analytical framework. (*Image source* drawn by the author)

urban areas, researchers often rely on old or planning maps for data collection. With the advancement of technology and the faster update speed of maps, the urban morphology research of new towns or new districts increasingly relies on high-definition historical satellite images and uses GIS-related software to make more quantitative analysis [32].

M. R. G. Conzen's school of urban morphology and its town-plan analysis method are relatively mature and effective, having achieved some success in the study of urban morphology in both classical and modern European and American cities. In recent years, numerous studies in China have incorporated Conzen's urban morphology methods. However, the localized application of Conzen's urban morphology in China has primarily focused on old cities or ancient villages. In the context of China's large-scale and rapid urbanization, the morphological development of new urban areas, which plays a crucial role, has received relatively little attention. China's reform and opening-up policy has had a unique impact on the pace of urban expansion and construction compared to Western urban development. The Xiangshan area, situated in the Guangdong-Hong Kong-Macau Greater

Bay Area and endowed with unique historical and cultural landscapes, has also been influenced by the local culture of Xiangshan. Applying Conzen's urban morphology theory to empirically examine the evolution of urban morphology in the Xiangshan area has significant implications for further expanding the research and application of Conzen's urban morphology theory in China, particularly within the Guangdong-Hong Kong-Macau Greater Bay Area.

2.2 *Analysis of Settlement Spatial Morphology*

2.2.1 Island Settlements

Island settlements refer to residential and living areas formed by human settlement on islands. Combining the geographical concept of "island" with the human activity of "settlement," they are villages, towns, or other settlements on islands and form an integral part of human geography research. At present, there are few studies specifically on island village architecture, but the exploration of architecture is involved in the study of the spatial pattern and planning of island towns. Huan et al. [33] used the argumentation method of comparing today and tomorrow to conclude that the island's unique resources have caused changes in settlement nature, environmental transportation infrastructure, the urbanization process, soft conditions, and other aspects of the island's human settlement environment. Weiwei [34] used the Xiazhi Cha'ao Village in Zhoushan as an example to analyze the regional characteristics of island village site selection, residential building layout, architectural form, and building materials under the influence of special geographical environment and climate, and explored the ecological design methods, survival concepts, and cultural connotations of island residential buildings in adapting to the environment. Rui [35] proposed the need to establish standard data to study ancient island villages and create a traditional village spatial pattern. Weiqiu [36] pointed out that in the process of island city development, we should learn from excellent cases and formulate the development direction of the island in the early stage of development as early as possible. Rong et al. [37] took the Flower and Bird Community in Si County as an example, took the construction of its fishing culture as the core goal, sorted out and analyzed the current situation and characteristics of the rural landscape, and explored the design path of revitalizing the four aspects of village cultural resources, architectural environment, architectural space, and building materials under the perspective of island tourism. Jue et al. [38] studied the design method of island environments under the guidance of texture renewal theory and took the stone house renovation of Dashuwan Village in Dongfushan as an example to carry out the application practice of texture renewal in environmental design, hoping to provide a feasible texture renewal design path for the texture protection of remote islands in eastern Zhejiang through design theory and design practice. Jue et al. [39] took the digital protection and construction of traditional villages in Dongji Island, Zhoushan, Zhejiang, as a practical case to further verify the feasibility of the theoretical framework and technical path of digital dynamic protection of cultural genes of traditional villages in remote islands. Including 3D cultural heritage maps, 3D animations, and other digital protection results, the construction of the remote island digital platform was completed. Haidong [40] mainly studied the architectural form and construction technology of island dwellings in Dongtou, southern Zhejiang, and summarized the regional characteristics of the architectural form and construction of island dwellings under special conditions, such as marine characteristics, cultural roots, appropriate technology, and ecological strategies.

So far, there are four aspects of research results on traditional villages in Guishan Town. Shougan et al. [41] based on the summary of the characteristics of the eastern archipelago of Zhuhai, combined with the preparation of the Guishan Island control plan, focused on its population composition, scale, and supporting service facilities. Lihao et al. [42], through the investigation of the current living environment of Guishan Island, as well as the village and building space, and existing village space and residential buildings based on one of the factors affecting the living environment, revealed that the buildings and village spaces of the Wanshan Islands and the living environment are mutually influential. Ge [43], through the analysis of the local architectural style characteristics, explored the architectural style design mechanism that can inherit the traditional architectural features and adapt to the changes of time. Guangxi [44] conducted a field investigation on the residential buildings in the Guishan Island area, combined with the unique architectural features of Lingnan traditional residential buildings, and put forward some views on the new Lingnan coastal residential buildings. In recent years, an increasing number of studies have focused on the traditional villages of Guishan Town, indirectly indicating their explorability and cultural value; conversely, these four existing studies cover three areas: urban planning, human settlement environment, and new residential buildings, demonstrating a certain breadth. However, the depth of exploration of the traditional village architecture in Guishan Town is insufficient. They are all explaining the existing phenomena, but no constructive opinions have been given on the practices. At the same time, there is also the problem of insufficient systematic research.

2.2.2 Tangjiawan Ancient Town

Despite Zhuhai's Tangjiawan Town belonging to the Cantonese region, systematic research on Zhuhai and the Tangjiawan area is relatively limited. Existing studies primarily focus on village histories, news reports, and periodicals. This is largely due to the relatively recent designation of Tangjiawan as a Famous Historical and Cultural Town in China. When Tangjiawan applied for the designation, Professor Ruan Yisan of Tongji University and his team conducted preliminary research on the town in 2006 [45]. This study primarily traced Tangjiawan's history and defined the scope of its designation but did not go further than the forms and decorations of traditional architecture. This document served as a foundational resource for this study. Following Tangjiawan's successful application for historical and cultural town designation, Yuan Hao (2012) of South China University of Technology conducted a systematic review of its historical features [46]. Through multiple field trips, the author summarized the characteristics of Tangjiawan's historical architecture, from its overall planning and street interface characteristics to the characteristics of individual buildings, from a macroscopic to a microscopic perspective. Finally, the article provides a preliminary analysis of Tangjiawan's status and offers relevant recommendations. This study incorporates these recommendations and further explores the study of Tangjiawan's ancestral halls. Liu Zhenhuan and Tian Zhongxun (2012) analyzed the cultural phenomenon of the integration of Chinese and Western cultures in Tangjiawan in modern times and explored the historical value and modern significance of Tangjiawan culture [47]. Feng Jiang et al. (2014) reviewed the evolution of villages in Tangjiawan and analyzed the settlement forms and distribution of ancestral temple buildings in Tangjiawan Town [48], providing important basic data support for the relationship between Tangjiawan settlements and ancestral halls. Furthermore, a research team from the School of Geography and Planning at Sun Yat-sen University (2014) conducted a comprehensive survey of traditional residential buildings and their communities in Zhuhai, exploring their current spatial distribution [49]. They identified the diffusion and dissemination of traditional residential culture in Zhuhai, as well as the main types and distinctive features of architectural spaces. Finally, they proposed strategies for the preservation and sustainable development of traditional residential buildings. In recent years, Zhuhai High-tech Zone has combined new rural construction to compile the "*Huitong Traditional Village Protection and Development Plan*" (會同傳統村落保護發展規劃) and the "*Qiaoao Traditional Village Protection and Development Plan*" (淇澳傳統村落保護發展規劃) to inherit the long history and culture of Tangjiawan and has successively published books such as "*History of Jishan Village*" (雞山村史), "*Tangjiawan Town Chronicles*" (唐家灣鎮志), and "*History of Guantang Village*" (官塘村史話).

3 Uniqueness of the Case Study

3.1 Zhuhai City

3.1.1 Tangjiawan Ancient Town

Tangjiawan is in northern Zhuhai, Guangdong Province. Bordering Zhongshan City to the north and the Pearl River Estuary to the east, it was formed based on the former Tangjia Town and Jinding Town. The town stretches approximately 12 km from north to south and 17 km from east to west, with a coastline of approximately 20 km, covering a total area of approximately 139 km^2

(including reclaimed land). It is located between longitudes 130° 07′ and 114° 18′ east and latitudes 21° 04′ and 227′ north. To the north and west, it borders the towns of Nanmeng and Sanxiang in Zhongshan City; to the south and southwest, it shares borders with Jiejiu Village and Dongkeng Village in Qianshan Subdistrict, Xiangzhou District, and Shenqian Village in Cuixiang Subdistrict. To the east, it faces the Pearl River Estuary and Lantau Island in Hong Kong. The Jinxingmen Waterway flows through Jinxingmen into Lingdingyang. Rich in historical and cultural resources, it is a key development area within Zhuhai's urban master plan. Directly administered by the Zhuhai Municipal Government, it encompasses 16 communities: Tangjia, Tangle, Nazhou, Yongfeng, Qi'ao, Beisha, Huitong, and Dong'an.

During the Northern Song Dynasty, Fuchong (Fuchung) (釜涌), over a hundred miles south of the prefecture, had a silver mine on the seashore, and common people rushed to mine it. Fuchong, the predecessor of Tangjiawan, was once one of the 11 districts under Xiangshan County. "Tangjia Village" first appears in the preface to the *Liang Family Genealogy* (梁氏家乘), dated 1524 during the Ming Dynasty. Because the village faced the bay, it was renamed "Tangjiawan." Tangjiawan Town was part of Xiangshan County from the Northern Song Dynasty to the Republic of China, so the people and clans of Tangjiawan must be viewed within Xiangshan County. Xiangshan County was originally a community of immigrants, with Xiangyi clans primarily originating from Jiangxi and Zhuzhuxiang in Nanxiong, with some also coming from the Cantonese region. The *Continuation of the Xiangshan County Chronicle* (香山縣誌續編), published during the Republican era, records that the major surnames in present-day Zhuhai, including Tang, Liang, Rong, Yang, Wu, Lu, Bao, and Wei, all migrated from Nanxiong and other areas between the first year of the Kaixi reign of Emperor Ningzong of the Southern Song Dynasty and the Xianchun reign of Emperor Duzong. People from various regions often settled together in the same villages, based on their surnames. Immigration, however, signifies cultural integration. As immigrants settled, the culture of the Central Plains merged with the local environment of Tangjiawan. This cultural exchange fostered the courageous, pioneering, and receptive character of the people of Tangjiawan, laying the foundation for modern trading culture. During the Republican era, Xiangshan County, Sun Yat-sen's hometown, was renamed Zhongshan County in his honor. After the founding of the People's Republic of China, the State Council approved the establishment of Zhuhai County on May 1, 1953, with the county government initially located in Tangjiawan Town. In July 2006, the Zhuhai Municipal Government established the High-Tech Zone, merging it with Tangjiawan Town. During this period, Xiangshan County, where Tangjiawan is located, was one of the birthplaces of modern Chinese history. Immersed in the spirit of modernization, its location in southern Lingnan, connected to Hong Kong and Macau, and deeply influenced by Western cultural exchange, it became a hub for transportation and trade between China and the rest of the world. Furthermore, it served as a model county for Zhongshan during the Republican era. This period also saw the emergence of several historical figures who made outstanding contributions to modern Chinese politics, culture, diplomacy, and education. It also saw the rise of a renowned group of real estate agents, creating a historically rare Xiangshan-Zhuhai real estate agent group. Since modern times, Tangjiawan has produced numerous pioneers and influential figures in Chinese industry and commerce. Influenced by traditional values, the people of Tangjiawan, despite their struggles abroad, still valued the construction of ancestral halls in their hometown. An ancestral hall is a public building shared by a branch, family, or lineage, built for ancestor worship and to honor blood-related ancestors. This includes large ancestral halls, ancestral halls, public ancestral halls, and family temples. Ancestral halls are structures used to worship and honor ancestors according to the patriarchal system and can be seen as the materialized symbol of this system. Their spatial structure reflects traditional Chinese ethical values. Currently, over forty ancestral halls remain in the Tangjiawan area. The houses they built upon their return not only incorporate the aesthetics of Lingnan architecture but also incorporate Western elements, creating a unique style. This demonstrates the significant influence of Western culture on much of Tangjiawan's residential architecture and garden landscapes in modern times. In summary, this unique cultural phenomenon emerged within a unique historical context, enriching the historical and cultural spirit of Zhuhai and shaping Tangjiawan's distinctive regional culture.

3.1.2 Guishan Island

Guishan Island is the political, economic, and cultural center of the Wanshan Marine Development Pilot Zone. It is the closest island in the Wanshan Archipelago to Zhuhai city, nestled between Hong Kong, Shenzhen, Macau, and the mainland of Zhuhai. It is approximately 17 nautical miles west of Macau and Xiangzhou, with Xiangzhou Pier a mere 45-min sailing distance. It is only three nautical miles north of Lantau Island in Hong Kong. Connected primarily with neighboring islands, it serves as the maritime hub of the Wanshan District. The villages studied in this book are Guishan Village and Guihai Village, located on the eastern side of the central section of Guishan Island. Guishan and Guihai Village are two natural villages within Guishan Town, connected from north to south. The protected area of the traditional villages of Guishan and Guihai extends from Jianshe Road in the north to the mountain boundary east of the Guishan Cultural Center and Binhai Avenue in the west. The village boundaries on the south and west sides encompass approximately 13.03 ha. Guishan Village, located at a higher elevation in the north, covers an area of approximately 5.12 ha. Guihai Village, located at a lower elevation in the south, covers an area of approximately 2.28 ha. This study primarily focused on Guishan and Guihai villages. It is noteworthy that the housing in the border area between the two villages resembles the existing conditions of Guishan and Guihai villages. The structures of Guishan and Guihai villages also influenced the town's housing, leading to its inclusion in the research for this book.

Liang Sicheng once said, "Architecture originated from practical needs and was constrained by natural physics… Its structural systems and forms are shaped by the material environment." Island villages, shaped by the diversity of their environments, have developed unique cultural connotations: First, their unique geographical environment means that islands lack convenient transportation. Compared to inland areas, interaction, connection, and integration with the culture and art of other regions are much less common. This closed and rigid nature has become a unique cultural characteristic of islands, a culture that has continuously retreated and concentrated over time. Second, the land on islands is mostly composed of rock, making it difficult to cultivate crops, thus distinguishing them from the small-scale peasant economy that emerged under the agricultural civilization of inland areas. Life and production in island regions largely rely on the ocean, fostering a distinctive maritime and fishing culture. It's within this broader context that settlement location and architectural layout have gradually shaped their unique "spatial perspective."

Guishan Village and Guihai Village are located on an island. The residents of Guishan Village are engaged in both agriculture and fishing and are registered as agricultural households. The residents of Guihai Village make their living by fishing and are registered as fishermen. The two villages share a common way of living and production, which is quite different from the inland areas where agriculture is the main industry. The characteristics of the island space pattern are mainly the following four points:

(1) The Guihai Village group is in the valley of Jiawei Main Mountain, with Guihai Village Square as the center. The buildings are arranged around the square, which is relatively introverted and has a very distinct cohesiveness. The buildings are all built in a small valley and arranged in sequence along the contour lines. Therefore, the road shape of the village is mostly perpendicular to the contour lines, with a prominent fishbone shape, and the roads are mostly terminal roads. There are also "Z"-shaped roads, which are winding and graceful, organically enclosing the village together. The village buildings are built from the bottom up. The higher the building goes, the lower the density and the sparser the layout. Looking up from the center of the village, the whole village is in a radial shape.
(2) The complex terrain of the island mountain is not suitable for building complex forms. The floor plans of the houses are usually simple rectangles, L shapes, and "concave" shapes. Since Guihai Village was built by the government, it adopted a distribution method, forming some houses with a "two rows and one road" pattern. Among them is the collective housing built by the Guihai Village government and distributed to the villagers. There are five "I"-shaped multi-bay houses. One row can accommodate multiple households. The building's shape is very simple and without any decoration. The roofs were sloping in the early days and flat in later years. The houses are arranged relatively densely to resist frequent typhoons. There are gaps or walkways between each building, or even village roads. There is less flat

land on the island, so it is convenient to build according to the contour lines, which requires less earthwork. At the same time, the roof of the lower floor can be used as the front yard of the upper floor or as a public road, thereby improving land utilization. The sea is severely affected by tides, rain, and storms. Building houses on higher slopes can block the impact of tidal waves. In addition, residents at higher altitudes can use the slopes and valleys to grow vegetables and fruits to meet their daily needs.

(3) In other parts of China, the first floor of mountain buildings is usually made of natural cliffs that are slightly leveled as the internal walls of the building, and the second floor of the house is built on the mountain terrain. However, Guishan Town is different. Generally, the ground is leveled (excavated, leveled, or processed at the same time) before ground construction is carried out. The flattened base is the main body of the mountain rock, which has a strong bearing capacity and is safer for building houses. The filled base is usually used as a road or courtyard. The house is separated from the mountain, which prevents the house from being damaged by the mountain in the event of a disaster. The back of the mountain in the flattened situation is usually built with a retaining wall to prevent the house from being damaged by rolling rocks. The main visual effect is when walking on the road perpendicular to the contour line, people can see the stone retaining wall; when looking down from a high place, they can see the whole building instead of just a part of it.

(4) The building is built according to the terrain of the mountain. The building is built from the lower position of the coastal avenue near the sea to the upper layer of the hillside. The flat roofs of Guishan Town are staggered, and the characteristics of the island mountain settlement are typical. The overall spatial form reflects the integration of "mountain-city-sea." The overall architectural form follows the mountain's terrain, stacked layer by layer, facing the sea, and the building is intertwined with the mountain and the sea.

In addition, as a typical island village, the residential buildings also present different characteristics.

(1) To adapt to the marine climate and environment, the buildings are relatively small and low. Most of the buildings in Guishan Village are three-bay, while those in Guihai Village are mainly "I"-shaped with multiple bays. Most buildings are made of local materials because transport between the island and mainland is hard. To resist wind, moisture, and salt, the buildings are all built with stones mined on the island. The color of the buildings matches that of the island soil, making them appear as if they have naturally grown from the island and blend in with their surroundings.

(2) The distribution of resources is mostly mountainous with little vegetation and tall trees. Stone walls are very resistant to typhoons and have excellent thermal insulation properties. Therefore, the primary building material used in the structure is stone. In the early years, some buildings also used stone strips for door and window lintels. Now, most lintels and beams are made of reinforced concrete.

(3) The mountainous terrain of the island is complex and not suitable for building complex forms. The floor plans of residential houses are usually simple rectangles, L-shaped, and "concave" shapes. The government built Guihai Village, adopting a distribution method that resulted in some houses featuring a "two rows and one row" pattern.
(4) The resources in the island area are limited, the economic development is relatively slow, and there is a lack of the ability to build large buildings. The scale and volume of houses are small, and most of them are built for families. Currently, there are five "I"-shaped multi-bay houses among the collective housing built by the Guihai Village Government and distributed to the villagers, which can accommodate multiple households in a row. The building form is basic, without any decoration. The early roofs were sloping, while the later ones are mostly flat.
(5) The houses are arranged relatively closely to resist frequent typhoons. There are gaps or walkways between each building, or even village roads. Consequently, fires have never damaged the village buildings, and only typhoons have caused damage.
(6) The buildings are built on the hillside, from the lower part of Binhai Avenue near the sea to the upper part of the hillside. The flat roofs of Guihai Village are staggered and arranged in an orderly manner, which is a typical feature of island mountain settlements.

Therefore, the villages of Guishan Island embody the efforts and wisdom of island residents in adapting to their living environment. Traditional architecture bears witness to the traditional lifestyle, craftsmanship, and values of Chinese fishermen. Exploring the cultural value of Guishan Island villages will provide strong support for tourism development and serve as an excellent way for the public to understand China's maritime and fishing culture. Concurrently, constructing a historical landscape map has far-reaching practical significance for enhancing the public's intangible cultural heritage.

3.1.3 Hengqin Island

Hengqin is located on the west side of the Pearl River Estuary, facing the Taipa and Coloane of Macau to the east across the water, facing Zhuhai Free Trade and Wan Chai across the Maliuzhou Waterway to the north, the Modaomen Waterway to the west, and the South China Sea to the south. Hengqin Island was originally two separate islands, Great Hengqin Island and Little Hengqin Island. The topography of the two islands resembled two guqins lying on the sea, hence their name. Historically, Little Hengqin Island, the opposite mountain (today's high mountain in Wan Chai), the Macau Peninsula, and Taipa formed the inner Shaptsz Moon waterway; the Great Hengqin Island and Little Hengqin Island Taipa, and Coloane formed the outer Shaptsz Moon waterway. The inner and outer Shaptsz Moon were also called large and small Shaptsz Moon or upper and lower Shaptsz Moon.

The terrain of Little Hengqin Island is high in the east and low in the middle, with a long and narrow shape running east–west. Hengqin Island is characterized by high mountains, with the highest peak, Naobei Mountain, at 457.7 m, the third highest in the Wanshan Archipelago, offering commanding views of the entire island and the surrounding sea. The mountain slopes extend outward in four directions, with deep bays on the east and west sides, forming a shape resembling the letter X. Reclamation in the 1970s connected the two islands into a single entity, collectively known as Hengqin Island.

3.1.4 Beishan Village

Beishan Village, located in the eastern part of Nanping Town, Xiangzhou District, Zhuhai City, is named after the North Mountain upon which it sits. Historically, a river sank along the village's northern edge. With its back to the mountain and its front to the water, the village, surrounded by mountains and water, embodies a vibrant and lively atmosphere. Its historical landscape closely preserves the overall layout and style of clan villages in the Pearl River Estuary region during the Qing Dynasty. Buildings in the village face both south and north, their distinct and well-defined forms creating a dense, irregular pattern. Within the village, buildings are arranged in rows, oriented north–south, with narrow spaces between them, forming a comb-like pattern. A two-meter-wide lane runs between each row of buildings, serving as the primary transportation route for Beishan villagers. The area within the core conservation area of Beishan Village boasts the most complete historical features, the highest concentration of historical buildings, and the best-preserved conditions, maintaining a clear and integrated structure. The settlement's outer areas are primarily lanes and courtyards, with shops flanking some accessible lanes. Within the core area, existing cultural relics, historical buildings, traditional architecture, and historical environmental elements occupy 2.23 ha., or 25%, of the total area of 8.97 ha. These include eight immovable cultural relics from the Qing Dynasty, covering approximately 6,796 m^2, or 7.58% of the core protected area; historical buildings, approximately 8,503 m^2, or 9.48% of the core protected area; and traditional-style buildings, approximately 13,790 m^2, or 15.37%. Among them, Chengchuan Yang's Ancestral Hall and Baoxia Yang's Ancestral Hall have been put into use again in the form of art exhibitions and art centers through integration with the cultural and creative industries, attracting a large number of people.

3.1.5 Macau

Macau is in the Pearl River Delta on the southeast coast of China, at 113° 35′ east longitude and 22°14′ north latitude, about 60 km from Hong Kong in the northeast. Macau consists of Macau Peninsula (9.3 km^2), Taipa Island (7.9 km^2), Coloane Island (7.6 km^2), Cotai Reclamation Area (6.1 km^2), Zone A of the New District (1.4 km^2), Zone C of the New District (0.3 km^2), and Macau Administration Zone on the Artificial Frontier Island of Zhuhai-Macau of the Hong Kong-Zhuhai-Macau Bridge (0.7 km^2) [50]. The Macau Peninsula pier district (mainly the Macau Inner Harbour) is one of the earliest urban areas established in Macau since its opening as a port. However, regarding the delineation of the Inner Harbour's boundaries, Macau's official historical records do not provide a clearly defined scope. The term "Inner Harbour" first appeared on maps in the *1962 Macau map*, marked as "Port Interior," where the geographic location corresponds to the western side of the Macau Peninsula. In 1834, a city and harbor map of Macau labeled the area as "Inner Harbour," referring mainly to the coastal area south of Patane (沙梨頭) and along the western side of the peninsula. Historical maps document the continuous expansion of the Inner Harbour area over time. The Inner Harbour Revitalization Plan received approval in 1990 and underwent continuous revisions thereafter. Based on available data, the Macau Peninsula pier district can be defined as the coastal port and harbor areas, as well as the mixed-use commercial and residential zones between Macau and Port City.

Furthermore, the Historic Centre of Macau, inscribed on the World Heritage List in 2005, is the most complete and largest extant complex of historical buildings in China, showcasing a blend of Eastern and Western elements. It perfectly preserves the spatial imprint of Macau from the sixteenth to the early twentieth centuries as a crucial hub on the Maritime Silk Road and a frontier for the intersection of Chinese and Western cultures. It serves as a prime example for studying the evolution of modern maritime trade, religious dissemination, architectural integration, and urban form. Spatially, the district encompasses not a single, contiguous area but rather, centered around the Macau Peninsula, connects 22 historic buildings and eight plazas, forming a "connected pattern." Culturally, its core characteristic is a "living expression of the fusion of East and West": architecturally, it boasts Gothic and Baroque Western churches, Lingnan-style Chinese temples, and innovative architectural decorative forms that blend Eastern and Western influences. Functionally, Macau served as both an administrative and religious center under Portuguese rule and a commercial hub for Chinese settlement and Sino-Western trade. It continues to this day to encompass a complex mix of residential, commercial, cultural, and touristic functions, where residents' daily lives intertwine with visitors' experiences, perpetuating the vitality of this "living heritage." The Historic Center of Macau is not merely a static architectural relic but a dynamic witness to the evolution of urban form. Its transformation from a "small fishing village" to a "commercial port" and finally to a "multicultural district" reflects the impact of maritime trade on urban settlements since modern times. Its unique characteristics, such as the preservation of historical textures amidst high-density development and the spatial negotiation of diverse cultural contexts, offer critical perspectives on the preservation of historical heritage in contemporary coastal cities and the study of cross-cultural settlement patterns. They hold irreplaceable empirical value in exploring topics such as "cultural identity of heritage" and "sustainable development through the symbiosis of city and sea."

4 Prospects and Challenges

4.1 New Development Opportunities Under the Policy

The concept of the "Greater Bay Area" was first proposed in March 2015 in the "Belt and Road Initiative" document jointly released by the National Development and Reform Commission, the Ministry of Foreign Affairs, and the Ministry of Commerce under the State Council of the People's Republic of China. The Guangdong-Hong Kong-Macau Greater Bay Area (GBA) is an urban cluster encompassing the Pearl River Delta and Lingdingyang in the People's Republic of China. It encompasses nine contiguous cities in Guangdong Province: the two first-tier cities and sub-provincial-level cities of Guangzhou and Shenzhen; the seven prefecture-level cities of Zhuhai, Foshan, Dongguan, Zhongshan, Jiangmen, Huizhou, and Zhaoqing; and the two special administrative regions of Hong Kong and Macau. Spanning approximately 56,000 km^2, it accounts for approximately 30% of Guangdong Province's land area. The *Outline Development Plan for the Guangdong-Hong Kong-Macau Greater Bay Area* was released by the General Office of the State Council on February 21, 2019. It consists of 11 chapters, including planning background, overall requirements, spatial layout, and accelerating infrastructure connectivity. It proposes four central cities—Hong Kong, Macau, Guangzhou, and Shenzhen—as the core engines of development. With a population of 86 million and a GDP exceeding RMB 13 trillion as of 2022, the GBA boasts the highest per capita GDP and one of the strongest economies in China. Within the framework of the Guangdong-Hong Kong-Macau Greater Bay Area, Macau is positioned as a world center of tourism and leisure and a service platform for trade and economic cooperation between China and Portuguese-speaking countries. Its role is to strengthen cooperation with the Chinese Mainland, promote moderate economic diversification, and serve as a link for openness and cooperation in the Belt and Road Initiative. To achieve these goals, Macau actively promotes economic, social, and cultural exchanges, strengthens infrastructure connectivity, and supports youth innovation and entrepreneurship. These policy drivers also provide new opportunities for urban development.

Furthermore, driven by national policies, Hengqin has become a key development area within the Guangdong-Hong Kong-Macau Greater Bay Area, ushering in a critical period of urban transformation and rapid growth. In 1992, the Guangdong Provincial Party Committee approved the establishment of the Hengqin Economic Development Zone. Hengqin's geographical proximity to Macau and its advantageous ecological environment have gradually revealed its enormous development potential. In 2009, the central government officially announced the development of Hengqin, and the State Council formally approved the *Hengqin Overall Development Plan*, incorporating Hengqin into the Zhuhai Special Economic Zone and establishing the Hengqin New Area. This became China's third national-level new area, following the Tianjin Binhai New Area and the Shanghai Pudong New Area. The plan explicitly aimed to develop Hengqin into a model area for close cooperation between Guangdong, Hong Kong, and Macau, driving the Pearl River Delta, serving Hong Kong and Macau, and spearheading development. On March 11, 2011, Hengqin's development was incorporated into *China's 12th Five-Year Plan*, elevating the development of the Hengqin New Area to national strategic status. On March 5, 2012, the then-Premier of the State Council explicitly stated in the *Government Work Report*: "Support Macau in developing itself into a world leisure and tourism center, advance the development of the Hengqin New Area, and promote moderate and diversified economic development." The *Outline Development Plan for the Guangdong-Hong Kong-Macau Greater Bay Area*, issued and implemented by the State Council on February 18, 2019, explicitly calls for "building on the strong foundation of cooperation between Guangdong, Hong Kong, and Macau, fully leveraging the role of major cooperation platforms such as Shenzhen Qianhai, Guangzhou Nansha, and Zhuhai Hengqin, and exploring new models of coordinated and collaborative development." On September 5, 2021, the Central Committee of the Communist Party of China and the State Council issued the *Overall Plan for the Construction of the Hengqin Guangdong-Macau Deep Cooperation Zone*. This plan clearly states that the Guangdong-Macau In-Depth Cooperation Zone in Hengqin will serve as a new way to help Macau's economy grow in a balanced way, improve the lives and job opportunities for Macau residents, support the "One Country, Two Systems" policy, and boost the development of the Guangdong-Hong Kong-Macau Greater Bay Area. On the morning of September 17, 2021, the Guangdong-Macau In-Depth Cooperation Zone in Hengqin Management Office was officially unveiled, ushering in a new phase of full implementation and accelerated advancement for the Guangdong-Macau In-Depth Cooperation Zone in Hengqin.

4.2 Development Prospects

"Bay Area Collaboration" promotes cross-city integration of city-sea symbiosis. Within the context of the Guangdong-Hong Kong-Macau Greater Bay Area's "coordinated development of land and sea and regional linkage," the city-sea symbiosis of Zhuhai and Macau holds the potential for cross-administrative boundary integration. Zhuhai boasts vast coastal space and island resources (such as the Hengqin and Wanshan Archipelago), while Macau possesses mature experience in high-density city-sea interaction. Leveraging the Guangdong-Macau In-Depth Cooperation Zone in Hengqin, both can promote a shift in settlement patterns from "independent development of a single city" to "cross-city collaborative symbiosis." For example, they could jointly plan a "coastal settlement belt around the west bank of the Pearl River Estuary." Zhuhai would focus on island ecological settlements and coastal science and technology innovation spaces, while Macau would leverage its advantages in coastal commercial and cultural clusters to form a complementary city-sea system combining ecological protection, industrial innovation, and cultural experience. On the other hand, the improvement of cross-city transportation networks (such as the Hong Kong–Zhuhai–Macau Bridge (HZMB) and the Zhuhai–Zhuhai Airport intercity railway) will further strengthen the spatial connection between Zhuhai Hengqin and the Macau Peninsula and Taipa and promote the city-sea symbiosis of the two places from "physical proximity" to "functional integration." For example, the coordinated development of Macau's coastal tourism facilities and Zhuhai's island resort resources will form a "one-hour city-sea leisure circle" and promote the upgrading of settlement forms to "networked and complex."

Prioritizing ecology guides green innovation in city-sea symbiosis. The global goal of carbon neutrality and China's strategy of building an ecological civilization are injecting green development momentum into the city-sea symbiosis model in Zhuhai and Macau. In the future, Zhuhai can leverage its natural landscape of "mountains embracing the sea and islands facing each other" to explore ecologically sensitive city-sea settlements. For example, in the Tangjiawan coastal area, a "low-impact development" model could be adopted, preserving native mangrove wetlands and creating a "coastal green corridor + sponge community" settlement structure, fostering a mutually beneficial symbiosis between urban development and the marine ecosystem. In the Wanshan Archipelago, a "zero-carbon island settlement" could be developed, utilizing clean energy such as wind and solar power, combined with seawater desalination technology, to create an island city that is both ecologically livable and a model for low-carbon development. Macau can explore "vertical ecological" innovations in high-density urban-sea spaces. For example, it can promote the "three-dimensional greening + coastal ventilation corridor" design in reclaimed new areas (such as Zona A dos Novos Aterros Urbanos) to alleviate the urban heat island effect. At the same time, it can optimize coastal wetland restoration technology, enhance the ecological resilience of artificial reclamation areas, and achieve a balance between "high-density development" and "marine ecological protection," providing a green paradigm for the city-sea symbiosis of high-density coastal cities around the world.

Cultural empowerment can activate the distinctive expression of the city-sea symbiosis. Zhuhai and Macau's unique resources of "maritime culture, overseas Chinese culture, and a fusion of Chinese and Western cultures" support the differentiated expression of this symbiotic city-sea model. For example, Zhuhai can tap into local cultural genes, such as Water-Faring Community (Dan Jia) culture and coastal settlements of overseas Chinese communities. In coastal areas, they can preserve the traditional waterfront settlement structure while incorporating maritime cultural displays and overseas Chinese memory museums, transforming the city-sea model into a spatial vehicle for the "transmission of maritime culture." Macau, on the other hand, can leverage its cultural advantages of melding East and West to deepen the cultural connection between its historic city district and coastal space. For example, this can involve restoring coastal settlement remains and preserving the diverse combination of Chinese temples, Portuguese houses, and coastal piers. Furthermore, in conjunction with the "Historic Center of Macau" World Heritage status, a coastal cultural promenade can be developed. This symbiotic city-sea model serves residential and commercial functions and becomes a living space showcasing the fusion of Chinese and Western cultures, enhancing the cultural identity and appeal of the city-sea models of both cities.

4.3 Common Problems Exist in Rapidly Developing Urban New Areas

China's rich history boasts numerous historical and cultural villages and towns. These villages and towns record a wide range of information, from local politics and economy, social life, customs and habits to landscapes and architectural styles, representing an invaluable legacy left by our ancestors. Following the reform and opening-up policy, national, provincial, and municipal governments have placed considerable emphasis on the preservation of historical and cultural heritage. Starting in the 1980s, ancient buildings in some villages and towns were gradually designated as national key cultural relics protection sites, leading to the development and emphasis on relevant conservation measures and regulations. The newly revised *Cultural Relics Protection Law of the People's Republic of China*, promulgated in 2003, explicitly mandates the protection of "towns, subdistricts, and villages that are particularly rich in cultural relics and possess significant historical value or revolutionary significance." Since 2003, the country has announced 119 National Historical and Cultural Cities in three batches, both through centralized announcements and subsequent additions. Furthermore, the country has announced 181 Chinese Historical and Cultural Towns and 197 Chinese Historical and Cultural Villages in five batches. In 2007, Tangjiawan Town in Zhuhai City received the designation of a National Historical and Cultural Town due to its rich historical and cultural heritage. Soon after, the Guangdong Federation of Literary and Art Circles and the Guangdong Folk Artists Association jointly launched the "Guangdong Ancient Village Preservation Project," mobilizing a large number of experts to conduct academic research and compilation, among other rescue efforts. To date, over 100 ancient villages and towns in Guangdong have been designated, including Tangjiawan Town in Zhuhai. In recent years, the Zhuhai Municipal Government has conducted comprehensive research on the planning and design for preserving and developing Tangjiawan Ancient Town, developing a preservation and development strategy for the town. Numerous experts and scholars have contributed to the project's conservation and development.

Historically, urban residents comprised only a small portion of humanity. However, over the past two centuries, the world has experienced rapid globalization, with half of the world's population now living in cities, profoundly impacting global sustainability. According to data from the National Bureau of Statistics, urbanization was very slow in the three decades between the founding of the People's Republic of China in 1949 and the beginning of reform and opening up, with China's urbanization rate increasing only from 11 to 18%. However, since the beginning of reform and opening up in 1979, China's urbanization has entered a period of rapid development. The national urbanization rate has risen from 18.96% in 1979 to 45.76% in just over four decades, reaching a new high of 64.72% in 2021.

In this state of rapid development, new urban areas have emerged as a widespread model of urbanization to alleviate the pressures of overcrowded old cities and resource shortages. However, this period of urban construction has also revealed some problems. For example, while blindly pursuing modernization, many cities have neglected consideration of the natural and cultural environments in urban planning and design, resulting in a "similar" phenomenon, with a high degree of similarity in urban form, structure, and texture. Faced with rapid urban development, how can cities cultivate their own unique urban landscapes and characteristics and develop distinctive planning and architectural languages? Therefore, during the development process, it is necessary to periodically review and summarize the actual development process of cities to facilitate future development.

5 Overview of the Book

This book is divided into three parts:

Part I: The Development of Settlements and Morphological Space in the Xiangshan Area; Part II: Traditional Architecture Types in the Xiangshan Area; and Part III: Evaluation and Optimization of Modern Urban Space in the Xiangshan Area. Based in the Xiangshan area of South China, these three parts explore spatial and architectural forms from the perspectives of urban form, past, present, and future.

5.1 Part I: The Development of Settlements and Morphological Space in the Xiangshan Area

The first section covers four important settlement cases: commercial settlements (Macau Inner Harbor), clan settlements (Tangjiawan Ancient Town), island settlements (Guishan Island), and cross-border settlements (Hengqin). These four chapters discuss in detail their urban form characteristics, or island morphological evolution.

The chapter, *Commercial Settlements: Macau Inner Harbor Urban Form Characteristics*, not only analyzes the spatial morphological development of Macau's Inner Harbor but also compares it to changes in parts of the core area of Newcastle upon Tyne, UK. This section examines the process of land infill in and around the market area, the expansion of commercial functions into adjacent areas, and adaptive and augmentative redevelopment (including the creation of breakout streets). This section highlights the Portuguese commercial district in Macau, which was significantly influenced by colonial planning, local culture, and maritime trade.

The chapter, *Clan Settlements: Tangjiawan Ancient Town Diversified Residential Form Characteristics*, divides the existing urban landscape into units through the superposition of morphological units, building types, and land use, thereby reflecting the morphological development, structure, and composition characteristics of Tangjiawan Ancient Town and its residential settlements.

The chapter, *Island Settlements: The Traditional Villages of Guishan and Guihai on Guishan Island*, analyzes two traditional villages on Guishan Island in the Wanshan Archipelago of Zhuhai, exploring their settlement forms and cultural evolution within the unique island environment. By examining the natural environment, cultural background, architectural features, settlement evolution, cultural characteristics, and spatial structure of these two villages, it reveals the interplay between nature and culture in the formation of island settlements and the mechanisms of their adaptation.

The chapter, *Cross-border Settlements: Island Morphological Evolution and Urban Development Characteristics in Hengqin*, shows in detail four important periods of the evolution of Hengqin Island. The first is the Double Shaptsz Moon period (before 1949—natural alluvial deformation)—Hengqin Island was originally divided into Great Hengqin Island and Little Hengqin Island, known as Jing'ao and Xie Nu Gorge during the Song Dynasty, respectively. In the modern period of foreign trade, the waters of Shaptsz Moon around Hengqin Island were once anchorages for foreign merchant ships. During the Portuguese Macau government, the Portuguese made several attempts to occupy Hengqin Island but ultimately failed due to strong resistance. After the Opium War, trade activity decreased, and there was no longer a sight of large ocean-going ships docked at Shaptsz Moon. With the reclamation activities carried out in Hengqin and Macau beginning in the 1960s–70s, the outer cross gate east to west was blocked, ending the double cross gate period. The second is the period of encirclement of the central ditch (1949–1989—the first large-scale development). Shortly after the founding of the People's Republic of China, China experienced a difficult time. In the 1970s, Zhuhai and Shunde jointly cooperated to develop the central ditch of Hengqin. After building two embankments in the east and west, the Great Hengqin Island and Little Hengqin Island were connected into one. Until 2010, the Zhuhai Municipal Government reclaimed all 14 km^2 of state-owned land use rights in the Shunde Encirclement Area of the Hengqin Center Valley for nearly 3 billion yuan. The third is the period of marginal development (1990–2009—the early stage of urbanization)—with the abolition of villages and towns in 1989, the establishment of the Zhuhai Hengqin Economic Development Zone in 1992, The Lord opened the second large-scale encirclement campaign of Hengqin, the entire Hengqin reclamation expansion to the east, west, and north, and established the Hengqin Bridge and Lotus Bridge and built several highways, strengthening the transportation connections between Hengqin and the surrounding areas. At the same time, in 1995, for the first time, Hengqin was included in the scope of the *Zhuhai*

City Master Plan, marking the beginning of Hengqin's urbanization. The fourth is the period of integrated development (2009 to present—stable form, entering the stage of integrated development and construction). In 2009, Zhuhai Hengqin New Area was officially established, and the State Council officially approved the *Hengqin Overall Development Plan*. In 2021, it was upgraded to the Guangdong-Macau In-Depth Cooperation Zone in Hengqin under the joint administration of Guangdong and Macau. At this stage, the island form zone of Hengqin is stable, and with the promotion of policies, Hengqin gradually enters the development stage of development and construction of large urban areas. At present, the country has agreed to reclaim another 27.9 km^2 of area on the southern edge of Hengqin Island to develop tourism.

5.2 Part II: Traditional Architecture Types in Xiangshan Area

The second part focuses on the unique architectural space types in the Xiangshan area, covering ancestral hall architecture, garden landscape, wharf architecture, and the "Pátio" and "Beco" architectural forms.

The chapter, *Ancestral Hall Architecture: The Clan Culture in Tangjiawan Ancient Town*, explores the relationship between Tangjiawan village form and the distribution of ancestral halls, the form and components of ancestral halls, the materials and craftsmanship used, the decorative themes, and the major influencing factors. Tangjiawan villages exhibit three types of layouts: comb-shaped, water-network, and fort-style. The comb-shaped layout serves as the fundamental model for all three types, from which the water-network and fort-style villages derive. Based on these three village forms, ancestral halls in Tangjiawan are primarily distributed within and around the village. The form of Tangjiawan ancestral halls, influenced by the patriarchal system, is divided into two main types: interior and exterior spaces. Combining these elements, ancestral halls in Tangjiawan present six types of configurations. These six types consistently adhere to the spatial composition principles of "respect for hierarchy, with the most respected at the center," "circular" worship routes, "alternating void and solid" spatial variations, and "enclosed exterior and open interior" relationships. The components of the Tangjiawan ancestral halls are primarily distributed across six areas: foundation, columns, walls, doors and windows, beams, and roof. Different materials are used for each component, each yielding a variety of craftsmanship based on its characteristics. The primary materials include wood, stone, brick, tile, oyster shell, rammed earth, plaster sculpture, and painted paintings. These materials are adorned with a rich variety of themes, which can be categorized into four main types: auspicious birds and beasts, fruit trees, flowers and birds, traditional ornamentation, and cultural artifacts. Overall, the factors influencing the Tangjiawan ancestral halls can be categorized into three levels: macro, meso, and micro. The macro factor is the patriarchal system, which influences the form and decorative features of the Tangjiawan ancestral halls. The meso level is the clan policies of the Cantonese region, which influence the distribution of the Tangjiawan ancestral halls. The micro level is influenced by the natural environment and Xiangshan culture of Tangjiawan. The natural environment dictates the characteristics of the ancestral hall components and the choice of materials. Xiangshan culture has enabled the Tangjiawan Ancestral Hall to exhibit strong adaptability and accommodation.

The chapter, *Garden Landscape: A Fusion of Chinese and Western Cultures in Tangjiawan*, explores the selection of relatively well-preserved typical garden landscapes in Tangjiawan for field research, summarization, and organization. Through analysis and research into the spatial layout, landscape elements, and cultural connotations of Tangjiawan's modern garden landscapes, the chapter aims to summarize the characteristics of Tangjiawan's modern garden landscapes and the design connotations with local characteristics, providing a reference and basis for the protection and utilization of Tangjiawan's modern garden landscapes.

The chapter, "*Wharf Architecture in Macau: The Typological Transformation from Timber Structures to Reinforced Concrete*" takes the 17 wharves along the Inner Harbour of the Macau Peninsula from Barra (媽閣) to Lam Mau Tong (林茂塘) as the objects of study, systematically revealing the coupled evolution of port–city relations, industrial organization, architectural types, and governance mechanisms.

The chapter, "*Pátio*" *and* "*Beco*" *Architectural Forms: Chinese Living Spaces in Macau*, focuses on analyzing the interpretation and distribution of Macau "Pátio" and "Beco," the origin and cause of "Pátio" and "Beco," the spatial characteristics of traditional Chinese style "Pátio" and "Beco," and the development and evolution of "Pátio" and "Beco" living form in the early twentieth century.

5.3 Part III: Evaluation and Optimization of Modern Urban Space in Xiangshan Area

The third part focuses on the present of traditional urban space, using evaluation methods such as AVC, spatial syntax, GIS, POI, and other big urban data analysis methods to evaluate and reflect.

The chapter, *Carrying Capacity Assessment of Traditional Village Revitalization: Zhuhai Beishan*, boldly introduces the AVC (Attraction, Vitality, Capacity) evaluation method, establishes an evaluation system and model for Beishan Village, and determines indicator weights. AVC stands for attractiveness, vitality, and carrying capacity. Attractiveness refers to the factors that satisfy tourists' aesthetic tastes and entertainment and leisure needs; vitality refers to the potential for long-term development of a tourist destination; and carrying capacity refers to the carrying capacity of the ecological and resource environment. Finally, the chapter summarizes and analyzes the comprehensive AVC evaluation results of Beishan's cultural and creative transformation and local culture and points out the interplay between cultural and creative transformation and the continuation of local culture.

The chapter, *Potential Assessment of Traditional Residential Area Revitalization: "Pátio" and "Beco" in Macau Peninsula*, combines the integration, traversability, and intelligibility of space syntax with GIS to explore the current characteristics of this low-level street space. These characteristics include accessibility and openness at different scales within each Pátio and Beco street, as well as the cognition-related characteristic of neighborhood harmony. The resulting quantitative results are then combined with previous cultural value assessments to derive a comprehensive revitalization potential index for each Pátio and Beco street on the Macau Peninsula. This index provides a planning reference for the orderly revitalization of Pátio and Beco streets.

Finally, the chapter, *Route Optimization for the Activation of Historical Urban Areas: Macau Route*, systematically analyzes the historical stratification of the Macau Peninsula from the perspective of its historic urban landscape and constructs a spatiotemporal model of heritage. Based on the concept of the stratification of urban heritage, it then develops a stratification-richness evaluation system for urban cultural heritage trails. It analyzes and evaluates existing trails in the core area of Macau's Historic Center and utilizes Geographic Information System (GIS) tools to optimize the evaluation results. Finally, the optimized trails are compared with existing trails in the core area to verify the effectiveness of the optimization process. This provides new insights into route selection for the development of heritage-themed trails and hopefully enhances public understanding of the stratification of urban heritage and strengthens public awareness of its protection.

References

1. Zhang, H., Fang, X., & Qi, J. (2012). Study on Zhongshan's urban construction in modern era under the influence of Xiangshan commercial ideologies. In *2nd International Conference on Electronic & Mechanical Engineering and Information Technology* (pp. 717–723). Atlantis Press. https://doi.org/10.2991/emeit.2012.153
2. Xie, H., & Tho-ard, M. (2025). Folk culture patterns and dissemination in Xiangshan cultural area of Guangdong Province. วารสาร ปราชญ์ ประชาคม, *3*(4), 198–207. https://so12.tci-thaijo.org/index.php/watmahasawat_jsc/article/view/3537
3. Wang, Y., & Yan, Z. (2006). *Centuries and millennia: Tracing and interpreting the origins of Xiangshan culture*. Guangdong People's Publishing.
4. Feng, J., Yin, J., & Chen, Y. (2024). Spatial evolution and influencing factors of water-town settlements in Southern China: Evidence from 942 villages in Zhongshan City. *Journal of Asian Architecture and Building Engineering, 24*(5), 4053–4072. https://doi.org/10.1080/13467581.2024.2396611
5. Lai, H. M. (2007). *Zhongshan, Zhuhai, and Macau: Geographical and Historical Notes*.
6. Lai, H. M. History of Guangdong. https://www.summaryplanet.com/summary/History-of-Guangdong.html

7. *Guangdong Sheng Zhuhai Shi diming zhi* [Gazetteer of geographical names of Zhuhai City, Guangdong Province] (Guangzhou: Guangdong Keji Chubanshe, 1989).
8. *Guangdong Sheng Zhongshan Shi dimingzhi* [Gazetteer of geographical names of Zhongshan Municipality, Guangdong Province] (Guangzhou: Guangdong Keji Chubanshe, 1989).
9. *Zhongshan Shi diming zhi* [Gazetteer of geographical names of Zhongshan City] (Guangzhou: Guangdong Keji Chubanshe, 1989).
10. *Zhongshan Shi zhi* [Gazetteer of Zhongshan City] (Guangzhou: Guangdong Renmin Chubanshe, 1996).
11. *Zhuhai Shi zhi* [Gazetteer of Zhuhai City] (Zhuhai: Zhuhai Chubanshe, 2001).
12. Li, S. (2025). Exploration, discovery, and foresight: Salt production archaeology in China. In P. N. Eubanks, A.A. Dumas, H. McKillop & M. Alexianu (Eds.) *Meridians of Salt. Interdisciplinary Contributions to Archaeology*. Springer, Cham. https://doi.org/10.1007/978-3-031-96692-7_15
13. Guiqi, L., & Shengbo, G. (2015). Changes and attributes of the basic administrative divisions in Xiangshan County during the Qing Dynasty. *Journal of Guangdong Second Normal University (廣東第二師範學院學報), 35*(1), 95–104.
14. Ge, Q., Liu, J. P., Xue, Z., et al. (2014). Dispersal of the Zhujiang River (Pearl River) derived sediment in the Holocene. *Acta Oceanologica Sinica, 33*, 1–9. https://doi.org/10.1007/s13131-014-0407-8
15. Tang, Y., Zheng, Z., Huang, K., Chen, C., Chen, Z., Lu, H., Wu, W., Lin, X., Zhang, X., & Li, H. (2023). Holocene evolution of the pearl river delta: Mapping integral isobaths and delta progradation. *Journal of Marine Science and Engineering, 11*(10), 1986. https://doi.org/10.3390/jmse11101986
16. Ptak, R. (2020). Chinese navigation near the Coast of Central Guangdong: Nantingshan and nantingmen in ming times. *Monumenta Serica, 68*(2), 339–367. https://doi.org/10.1080/02549948.2020.1831209
17. Gebhardt, J. (2014). Negotiating barriers: Cross-cultural communication and the Portuguese mercantile community in Macau, 1550–1640. *Itinerario, 38*(2), 27–50. https://doi.org/10.1017/S0165115314000345
18. Adonso, R., & Pereira, F. G. (1986). The political status and government institutions of Macau. *Hong Kong LJ, 16*, 28.
19. Zhongshan Municipal People's Government Office, December 26, 2022, the past of Xiangshan Salt Field, https://www.zs.gov.cn/zjzs/zsgs/content/post_219575.html. Last accessed 04 October 2025.
20. Williams, M. (2023). Villages of the fragrant hills. *Heritage and History in the China-Australia Migration Corridor, 5*, 25.
21. Liu, Y., Li, Z., & Jin, J. (2014). Pseudo-urbanization or real urbanization? Urban China's Mergence of administrative regions and its effects: A case study of Zhongshan City, Guangdong Province. *The China Review*, 37–59.
22. Sheng, N., & Tang, U. W. (2013). Zhuhai. *Cities, 32*, 70–79. https://doi.org/10.1016/j.cities.2013.02.006
23. Portnov, B. A., & Zusman, M. (2014). Spatial data analysis using kernel density tools. In *Encyclopedia of business analytics and optimization* (pp. 2252–2264). IGI Global. https://doi.org/10.4018/978-1-4666-5202-6.ch203
24. Lauder, G. V. (1981). Form and function: Structural analysis in evolutionary morphology. *Paleobiology, 7*(4), 430–442. https://doi.org/10.1017/S0094837300025495
25. Kassim, S. J., & Kamaruddin, K. (2023). City-Form: Morphologies. *Eco-Urbanism and the South East Asian City: Climate, Urban-Architectural Form and Heritage, 3*.
26. Schlüter, O. (1899). *Über den Grundriss der Städte*. (Vol. 34). WH Kühl.
27. Min, W. (2012). *Research on the evolution and driving forces of urban morphology in the Overseas Chinese New Village area of Guangzhou*. Ph.D. thesis, South China University of Technology, Guangzhou.

28. Conzen, M. R. G. (1960). Alnwick, Northumberland: A study in town-plan analysis. *Transactions and Papers (Institute of British Geographers), 27*, iii–122.
29. Li, S., & Jialu, G. (2016). A review of the research on the theory of Kang Zeen urban morphology in China. *Architecture and Culture, 12*, 212–213.
30. Song, F., Xu, L., Hou, A., Zhang, J., & Wang, J. (2011). *Chengzhen pingmian geju fenxi: Nuosenbolan Annike anli yanjiu (Alnwick, Northumberland: A study in town-plan analysis) by MRG Conzen*. China Architecture and Building Press.
31. Yinsheng, T., Kai, G., & Wei, T. (2010). Urban morphology research and urban historical protection planning. *Urban Planning, 04*, 21–26.
32. Shuo, G. (2021). *Research on the morphological evolution of urban new districts in China (Master's degree thesis)*. Nanjing University.
33. Huan, Z., Zhu, W., & Yuliang, Z. (2011). Changes in human settlement environment under the influence of island-specific resources: A case study of Zhoushan Archipelago. *Hua Zhong Jian Zhu, 29*(12), 98–102.
34. Weiwei, L. (2015). *A preliminary study on the regional construction of residential buildings on Zhoushan islands: A case study of Cha'ao Village on Xiazhi Island*. Doctoral dissertation, Hangzhou: China Academy of Art.
35. Rui, T. (2016). *A brief discussion on the protection and development of ancient island villages: A case study of Zhoushan area*. Master's thesis, Tianjin: Tianjin University.
36. Weiqiu, Y. (2017). Protection and improvement of historical buildings in the urbanization development of Zhoushan islands. *Wen Hua Xue Kan, 11*, 142–144.
37. Rong, Z., Jijun, D., & Wei, T. (2019). An analysis of the path to revitalize historical and cultural villages on islands: A case study of the Flower and Bird Community in Shengsi County. *Popular Literature and Art, 16*, 260–261.
38. Jue, Y., Han, Z., & Yani, W. (2019). Research on environmental design of traditional island villages based on texture renewal—Taking the stone house renovation of Dashuwan Village in Dongfushan as an example. *Art Education, 2019*(12), 187–188.
39. Jue, Y., Han, Z., Rong, Y., Qun, Z., & Yani, W. (2019). Digital living protection of cultural genes of traditional villages in remote islands in east-ern Zhejiang—Taking the digital protection and construction of Dongji Island as an example. *Research on Natural and Cultural Heritage, 12*, 6–9.
40. Donghai, L. (2020). *Research on architectural form and construction of stone dwellings in southern Zhejiang islands*. Master's thesis, Huaqiao University.
41. Shougan, X., & Wenjun, L. (2011). Research on the scale control meth-od of tourism-oriented island town planning—Taking Guishan Island in Zhuhai as an example. In China Urban Planning Society (Eds.), *Transformation and Reconstruc-tion—Proceedings of the 2011 China Urban Planning Annual Conference* (pp. 2694–2704). Southeast University Press.
42. Lihao, Y., & Siru, G. (2018). The impact of the spatial form of villages in the Wanshan Islands on the human settlement environment—Taking Guishan Village and Guihai Village on Guishan Island as an example. *Architecture and Culture, 3*, 162–163.
43. Ge, W. (2020). Research on the local style characteristics and control guidelines of island architecture—Taking Guishan Island as an example. *Decoration, 01*, 136–137.
44. Guangxi, G., & Wenfeng, B. (2021). Research on new coastal residential buildings in Lingnan—Taking Guishan and Guihai Villages on Guishan Island in Zhuhai as an example. *Urban Architecture, 18*(8), 74–76.
45. Tangjiawan: Timely protection and rational planning restore the century-old town to its former glory. http://www.heritagewuhan.cn/page?detail&type=zhuanjiadstr&aid=f8cc1c706eedc478017158a969720237
46. Hao, Y. (2012). *Research on the historical architectural features of Tangjiawan Town, Zhuhai City*, Master's thesis, South China University of Technology.
47. Zhenhuan, L., & Zhongxun, T. (2012). A preliminary study on the historical and cultural value of Tangjiawan. *Reform and Opening Up, 09X*, 189–190.

48. Jiang, F., & Zexuan, P. (2014). From clan village to pastoral city: Tangjiawan experiment in Zhongshan model county during the Republic of China. *Journal of Architecture, 9*, 117–122.
49. Research Group, School of Geography and Planning, Sun Yat-sen University. (2014). A survey on the spatial distribution, main types, and typical characteristics of traditional dwellings in Zhuhai. *Journal of Zhuhai Economic and Social Development, 1*, 591–640.
50. Huang, Y., Chen, Y., Song, J., Zheng, L., Yang, S., Gao, Y., Li, R., & Huang, L. (2025). Explore the ultra-high density urban waterfront space form: An investigation of Macau Peninsula Pier district via Point of Interest (POI) and space syntax. *Buildings, 15*(10), 1735. https://doi.org/10.3390/buildings15101735

The Development of Settlements and Morphological Space in the Xiangshan Area

The first part covers four important settlement cases: commercial settlements (Macau Inner Harbor), clan settlements (Tangjiawan Ancient Town), island settlements (Guishan Island), and cross-border settlements (Hengqin). These four chapters discuss in detail their urban form characteristics, or island morphological evolution.

Commercial Settlements: Macau Inner Harbor Urban Form Characteristics

Jianyi Zheng

Abstract Detailed research on long-term changes in the physical form of the cores of traditional urban areas has hitherto been largely concerned with Western, mainly European, cities. The findings and concepts of M. R. G. Conzen in England, mainly published in the second half of the twentieth century, have been especially influential. In this paper some of his conclusions about part of the core of Newcastle upon Tyne, England are examined in relation to findings in the corresponding area of central Macao, China. Macao is one of the earliest settlements in Eastern Asia established by Europeans, though primarily influenced over much of its existence by its predominantly Chinese population. Comparisons are made of the two cities in respect of (1) the processes of repletion in and around the market areas; (2) the expansion of commercial functions into adjoining areas; and (3) adaptive and augmentative redevelopment (including the creation of breakthrough streets). The evidence suggests that a number of Conzen's concepts of morphological development formulated in the West are applicable in Eastern Asia. However, in the case of the cities studied, there are marked differences in the timing of developments and outcomes in the urban landscape. Central Newcastle has been heavily reshaped relatively early by the Industrial Revolution and neoclassical planning. In the commercial area of Macao, colonial planning, local culture, and sea trading have been major influences.

Keywords Macao · Newcastle upon Tyne · Market centers · M. R. G. Conzen · Redevelopment

1 Introduction

Cities in different cultural environments are mostly characterized by their traditional urban cores. The physical changes of traditional urban cores in western cities had been widely concerned since M.R.G. Conzen published his iconic study on Alnwick [1] and Newcastle upon Tyne [2]. Recently, comparisons between the East, especially

J. Zheng (✉)
Faculty of Humanities and Arts, Macau University of Science and Technology, Macau, China
e-mail: jyzheng@must.edu.mo

Y. Chen et al. (eds.), *City-Sea Symbiosis*, Springer Geography,
https://doi.org/10.1007/978-3-032-13725-8_2

China, and the West have increasingly received more attention from scholars (for example, J.W.R. Whitehand, K. Gu, M.P. Conzen, and S. Yao). Comparative works on fringe-belt and residential areas have been especially fruitful [3–5].

Market areas within urban cores are important in both the West and China. British medieval towns and cities had their specialized markets and they were greatly changed in the course of the Industrial Revolution [6]. Similarly, ancient Chinese cities have been greatly transformed by the development of market streets in the urban cores after Song Dynasty [7]. It is a cross-cultural perspective, from which we can explore the different landscapes in east–west cities.

What is the difference of the evolutions in the traditional urban core between the West and East with the developments of commerce? In this paper, some of Conzen's conclusions about part of the core of Newcastle upon Tyne, England are examined in relation to findings in the corresponding area of central Macao, China.

2 Cases and Methodology

Newcastle upon Tyne, established in the Norman period, is the regional capital of north-eastern England with a population in 2011 of 280,200.[1] In its development over more than 800 years, its strategic situation proved to be a major economic advantage [2].

Macao, located west of the Pearl River estuary, is one of the earliest settlements in Eastern Asia established by Europeans. It has been primarily influenced over much of its existence by its predominantly Chinese population since Portuguese occupied in 1557. In the course of its development, Macao was an ancient Chinese free-trading port [8] and now is a tourism and leisure city with a population in 2011 of 552,503.

The urban core of each city was around a market which became the nucleus for development of an early CBD. Inspired by the concepts employed by M.R.G. Conzen in the study on Newcastle, the morphological evolution of traditional urban core of both cities has broadly speaking been transformed in three ways: repletion and plot metamorphosis within existing street blocks, commercial expansion into new areas, and commercial redevelopment in existing areas. By town-plan analyses, the morphogenetic plan units of both cities will present the intricate landscapes.

[1] http://webarchive.nationalarchives.gov.uk/20160105160709/http://www.ons.gov.uk/ons/rel/mro/news-release/census-result-shows-increase-in-population-of-the-north-east/censusnortheastnr0712.html.

3 Early Market Area and Developing Directions

The urban core of each city was developed from early market area characterized by their geographical environment. The pre-urban nucleus in Newcastle was on the riverside. The urban core subsequently extended to the north over the plateau which was some 24 m above the riverside (Fig. 1). In 1723, the urban core had retained the salient features of the medieval market area (Fig. 1), surrounded by burgages and extended along the arterial ribbon. The triangular old market place was defined by the frontages of the burgages [2].

The Portuguese settlement of Macao, arranged by the Central Street along the ridge within the walled town, was limited at the middle of Macao peninsular. Because

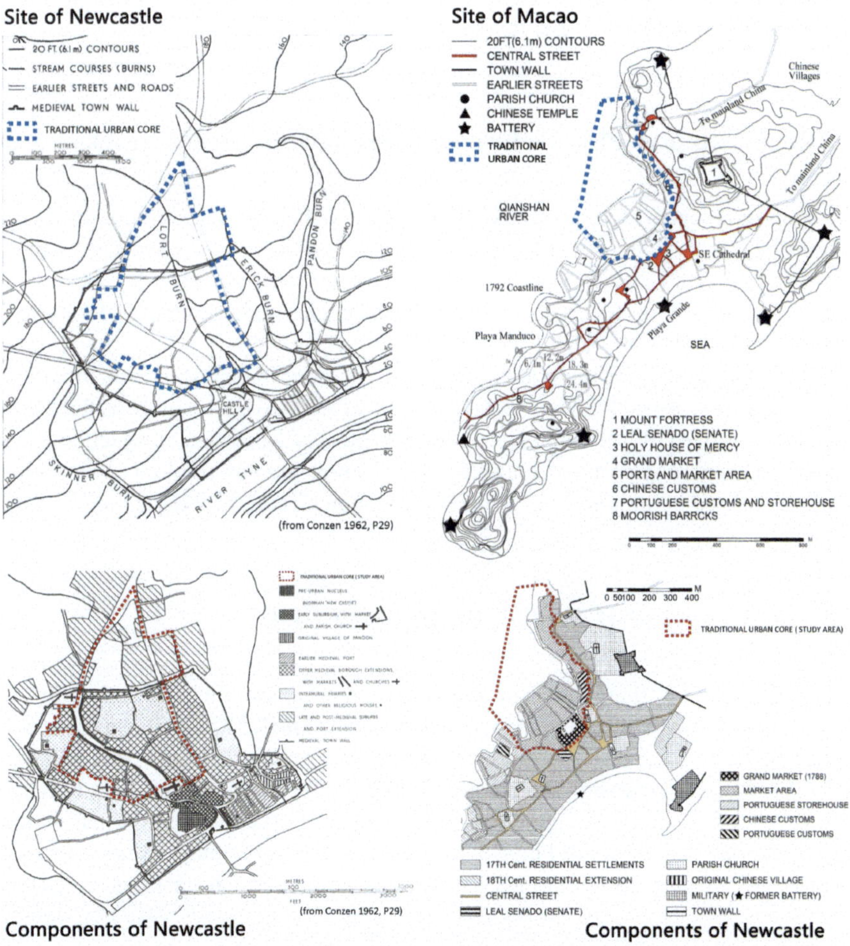

Fig. 1 Comparison of sites and components (*Source* created by the author)

of the narrowness of the territory and defense works in north, east, and south, the market area is mainly centralized on the lower land of the north-western part of Central Street, which is less than 6m in elevation (Fig. 1). It resembled early Portuguese coastal cities in the Roman period, in that upper class residential areas, social and administrative buildings, and defensive works are at a higher elevation and trade is in the downtown, as in the case of Lisbon [9].

The market area was the most important commercial center in Macao[2] However, Macao's commercial areas were to develop toward the coastline. The earliest market place temporarily located at the Largo do Senado was convenient for the Portuguese residents living along the Central Street to trade with Chinese before 1785. In 1788, the military camp of Qing Dynasty, located between Largo do Senado and coastline, was rebuilt as a large market place named Grand Bazar [10, 11] so as to form a unified market area near the riverside (Fig. 1). In 1831, the original market area had been composed of different functions [12].

4 Repletion and Plot Metamorphosis Within Street Blocks

The concept of repletion, employed by M.R.G Conzen [1], was the most common way of transformation. In the street blocks of both cities, repletion, associated with plots metamorphosis, began early and has continued through to the present day. Two street blocks (NCL01-02) in Newcastle and three street blocks in Macao (MAC01-03) were chosen for special study. These were in the immediate vicinity of or fronting their respective markets. Their repletion and metamorphosis can be traced from historical maps. Important aspects of the extent of their changes over time are reflected in building coverage (BC) and plot concentration (PC) [2]. In particular repletion tends to result in increasing plot concentration. The average amount of plot concentration can be measured by plot areas per meter of plot boundaries (M^2/M).

4.1 Development of NCL01-02 and MAC01-03

Long, narrow plots (burgages) are the basic elements fronting the streets in the urban core of Newcastle. Originated in the medieval period, NCL01-02 were cut and shaped by three breakthrough streets, Grainger St., Market St., and Grey St. after 1839 [6]. Initially, buildings were concentrated on the burgage heads, the burgage tails being left as yards and gardens. In map 1830,[3] repletion was limited within the boundaries of individual burgages, where accessory buildings were gradually constructed. In map 1862, burgages in NCL02 had almost been filled up, while

[2] Boletim Da Provincia De Macau E Timor, 1867 No.18, pp. 97.

[3] Map1830 is Oliver's Map. Source: British Library. Ordnance survey maps of 1862, 1896, 1951, 1967, 1984, 2015. The map date is the publication date. *Source* http://digimap.edina.ac.uk/

those in NCL01 hadn't. It showed in early stages of the Industrial Revolution, the commercial activities were mainly around the early market place. However, burgages in NCL01 had subsequently been hugely transformed because of the move of the commercial gravity.

MAC01-03 are composed of Wei-Li and peripheral plots. In map 1831 [12], MAC01 was a large street block enclosed by individual plots along the streets. It contained an open space in the middle used for temporary warehouses. MAC02-03 was at an early stage divided into large plots. MAC01-03 had already been filled by Wei and Li before 1869.[4] Wei (Patio in Portuguese) and Li (Beco in Portuguese) are distinct fabric patterns within the street blocks. As a neighborhood unit, it refers to a series of terraced houses built in uniform style and forming a small spatial unit. As the lowest level access where two sides of buildings open their front or back doors, Wei-Li are alleyways of less than three meters in width, providing access to dwellings and an array of plots along the two sides of the street [13]. Basically, Wei is a lane and Li is a cul-de-sac, which provides a public right access to the main street. They are similar to a "lane," "court," or "yard" in a burgage, which are means of providing access where buildings have been added to an existing plot [1]. In this way, MAC01-03 gained increasing numbers of plots and buildings for dwellers. Despite of the absence of a detailed survey map before 1900, it could be conjectured that the repletion continued within MAC01-03 according to the amounts of buildings recorded in official registers of public streets.[5] After 1914, the plan of MAC01 had been obviously changed after the redevelopment of Sun-Ma-Lo Street. MAC02 is almost no change, but MAC03 is in massive urban fallows after 1999 (Fig. 2).

4.2 *Changes of BC and PC in NCL01-02 and MAC01-03*

The BC cycles of NCL01 and MAC01 over times were more obvious than the others. The climax phase of NCL01 was 1896–1984, while that of MAC01 was 1900–1954. A new repletive process of NCL01 and MAC01 began respectively in 2002 and 1980. Meanwhile, PC in both was increasing, but the extent of NCL01 was larger, particularly in 1896–1968. Until 2015, it was almost twice of that in NCL02 and MAC01. But, PC in MAC01 decreased during 1831–1900, while slightly increased again afterwards.

The BC in MAC02 and NCL02 were in slightly recessive phase after 1830. MAC03 kept relatively stable during 1900–1999, but obviously declined after 1999. As for PC, the progress of NCL02 can be easily found after 1984, but MAC02-03 underwent almost no change. Two differences of recessive phenomenon can be seen. In NCL02, small plots were amalgamated into big plots with low building

[4] Idem 2, 1869 No.30, pp.141–143.

[5] Idem 2, 1880 No.52, pp. 19–31; Boletim Official Do Governo Da Provincia De Macau, 1897 No.6, pp. 95–105; Cadastro das vias publicas de Macau 1906.

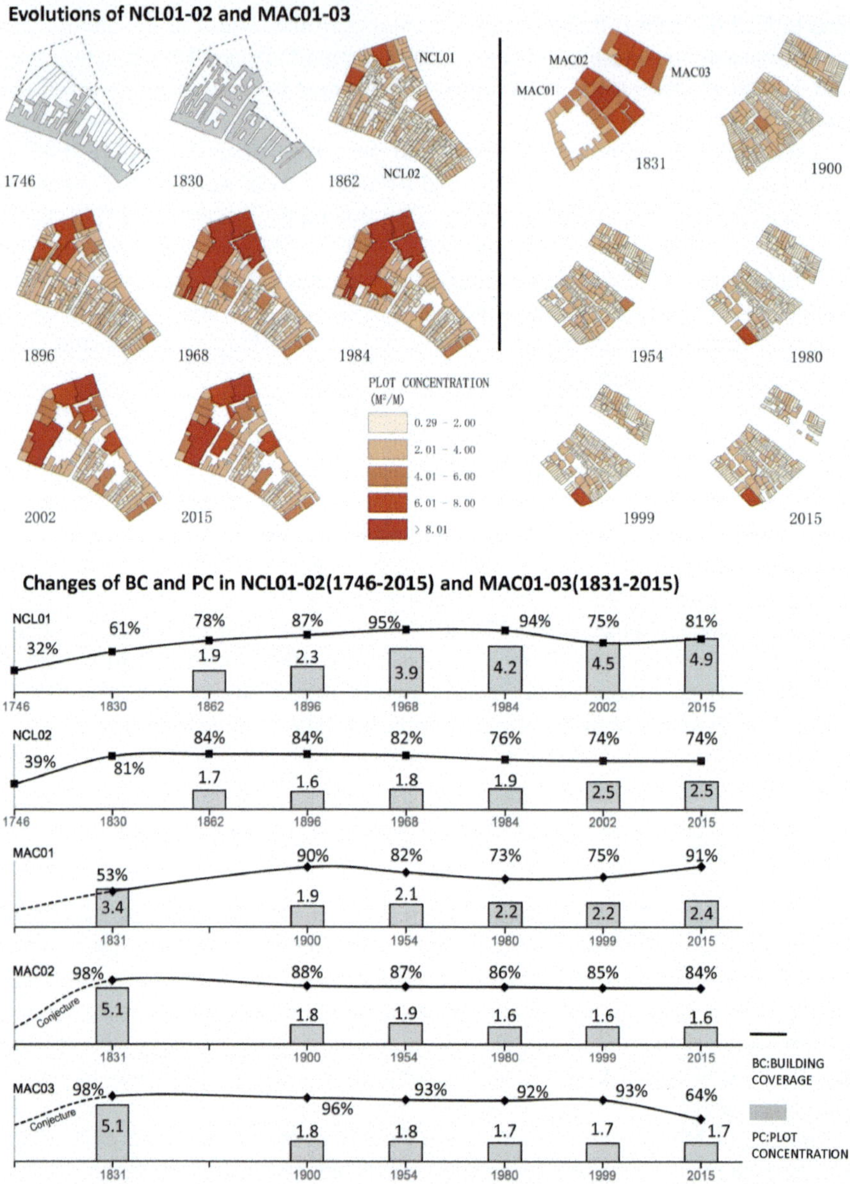

Fig. 2 Comparison of the cycles within street blocks (*Source* created by the author)

coverage, which were waiting for new developments. In MAC02-03, plot patterns didn't change, but their decays led to urban fallows within individual plots.

BC and PC clearly reflects that repletion and metamorphosis of street blocks in Newcastle is more intensive than those in Macao. MAC01 and NCL01 are relatively more active than the others, because both were closer to the main commercial streets. MAC02-03 still keep the small scale of plot patterns, similar to NCL02 in 1896.

4.3 Metamorphosis in Residual Burgage Complexes (RBC) and Residual Wei-Li (RWL)

As early plot patterns, burgage and Wei-Li are basic elements of traditional urban cores in both cities. With long-term evolutions, they had undergone considerable changes. According to the types of RBC employed by Conzen [14], the Fig. 3 shows their distributions.

After Conzen's study in 1962, RBC had hugely changed. Case far from the early market place, 17W, 17E, 18C, 14C, were replaced by new shopping center. Case near the early market place, 11, 10S, 21N, 21S, were replaced by modern commercial buildings. Only those around the early market place still kept some features of burgage series, which were gradually changing with the introduction of new buildings. About 61.1% of RBC were metamorphic types or demolished.

The residual Wei-Li had survived widely in Macao's urban core. Compared to RBC, only about 18.7% of RWL were metamorphic types, but RWL's PC was relatively smaller. It helps to preserve the historical town plan of Macao's early market area, but about 62.9% of RWL were of the recessive type. Furthermore, 9.4% of RWL are in urban fallow. They are mainly located in the middle of traditional urban core.

So, the conservation of RBC was being threatened by commercial redevelopments, while RWL was facing the extensive decline.

5 Commercial Expansion

The expansion area in Newcastle is part of today's commercial center located north of the old market area, including the triangular area at the traffic junction of two arterial streets where there were high commercial values. Macao's expansion area was to the west of the reclaimed land of the early market area.

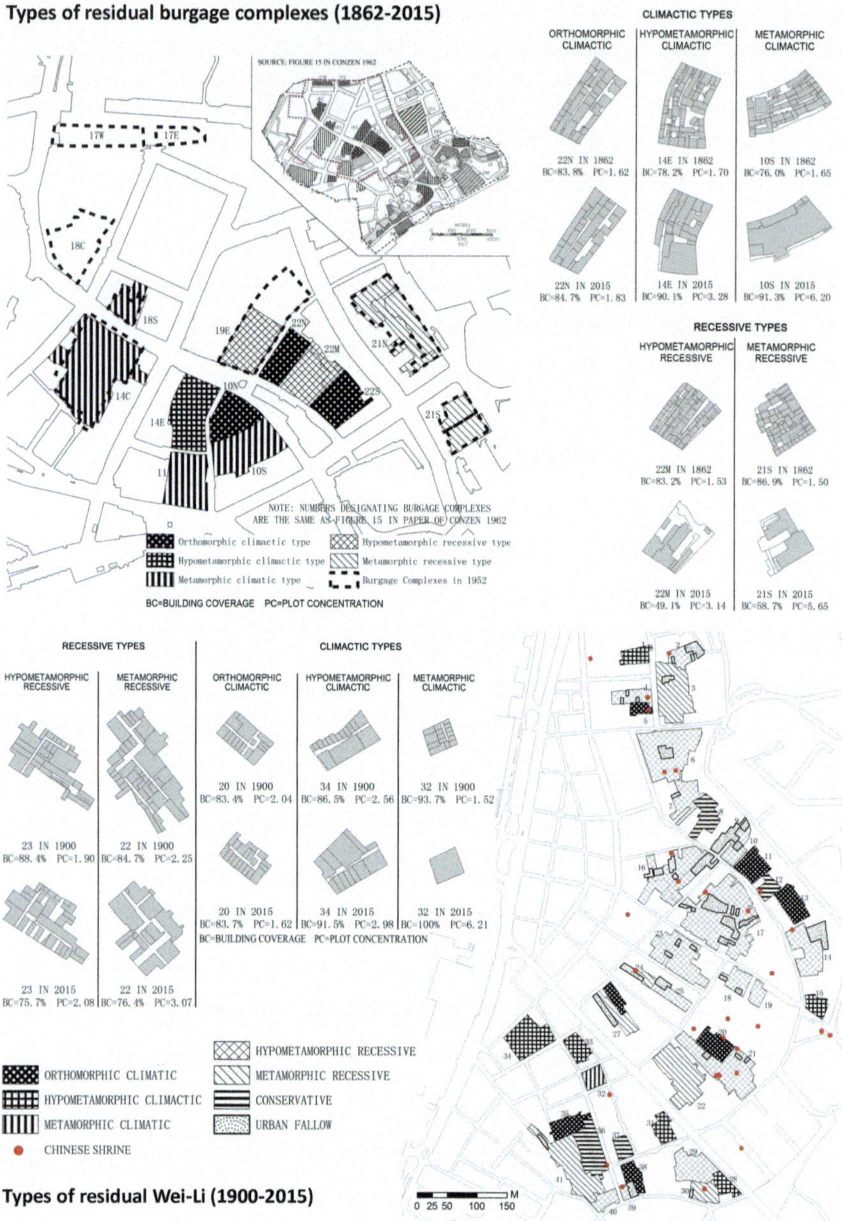

Fig. 3 Comparison of RBC and RWL (*Source* created by the author)

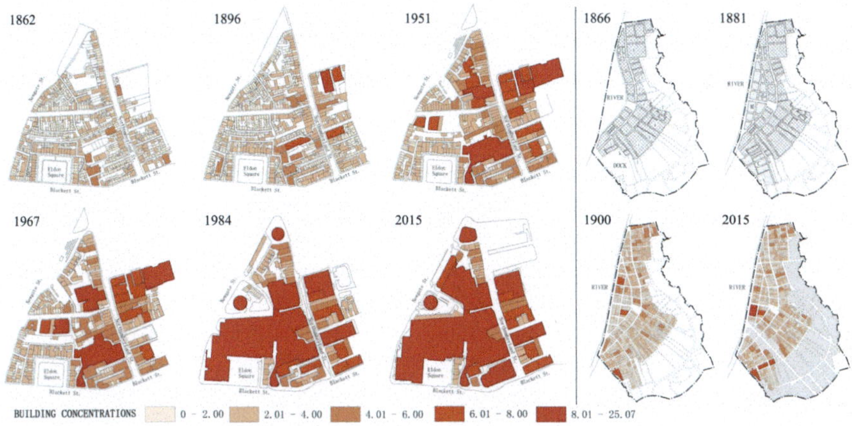

Fig. 4 Comparison of building concentration in expansion areas (*Source* created by the author)

5.1 Newcastle's Expansion from 1824

The expansion in Newcastle was in 1770–1830 [2], earlier than that in Macao. The developments of Blackett Street and Eldon Square during 1824–1826 were symbolic. In map 1830, residential accretions, which was laid out as small plots and blocks, were evident, as well as commerce, community services, a hay market, and chapels (Fig. 4).

In map 1862 and map 1896, the gravity center of the growing CBD had been pulled toward the street junction at Grey's Monument. The expansion area proceeded with several site successions along the Northumberland Street. Plots, for example, Elswick Cycle Works and Cement Works, yielded to commercial expansion, the alienation being small but foreshadowing more extensive redevelopments.

In map 1951, big department stores began to occupy along Northumberland Street, which had emphasized the move of the commercial center of the town away from Grainger Street [15]. After the 1970s, the shopping center occupied most of this area, the small plot patterns almost being wiped off.

5.2 Macao's Expansion from 1832

Macao's expansion was evident in the maps of 1866 and 1881.[6] In 1832–1866, new street system was laid out as a series of small grid blocks with back-to-back plots except some blocks were irregular and somewhat larger so as to adapt to the 1831 coastline (Fig. 4). The new areas provided shophouses, Wei-Li, several new ports,

[6] Macau com as ilhas e costas adjacentes (1865–1866), from Macao Archives: MNL.10.18d.CART. Planta do Palácio da Justiça e Fazenda (1880–1881), from Macao Archives: MNL.05.09.CART.

and Chinese temples. A new commercial street parallel to the coastline, named Rua Nova d'El Rei, extended the early commercial streets. The gravity of commercial center was moved to the junction at Largo do Pagode do Bazar, one side facing to coastline. Then, the new and old were reorganized in an H-type commercial street system.

During 1867–1881, private firms led the constructions of two expansion areas (Fig. 4). Miguel area, named after its Macao-born Portuguese developer, developed into a quayside with shophouses. It straightened the coastline of 1866 and was laid out as the small orthogonal grid street blocks. Caldeira area, developed by Chinese developer, reclaimed from an old dock and its street system and plot patterns were flexible and mixed to adapt the surroundings. After 1881, the features were generally preserved, except for the area of Sun-Ma-Lo redevelopment.

6 Commercial Redevelopment

Following a distinction made by Conzen, redevelopment can be divided into two types: augmentative and adaptive [14]. Augmentative redevelopment entails augmentation of an existing street system. In contrast, in the case of adaptive redevelopment no new streets are created, redevelopment being adapted to the existing street pattern.

6.1 Newcastle from 1834

The earliest augmentative redevelopment in Newcastle that can be ascertained directly from map evidence was along the breakthrough street of Dean Street, Mosley Street, and Collingwood Street in 1784–1810. But Newcastle's extensive redevelopment scheme was proposed to the government in 1834 by Richard Grainger.

Initially, the system of eleven new streets was instituted in 1835–1839. Part of Clayton Street, Grainger Street, and Grey Street was breakthrough in character. These three streets became principal streets by cutting through former burgages, and taking over the sites of Anderson Place and the Nun's Fields. Grey Street and Grainger Street joined at a newly created focal point, Grey's Monument. Afterwards, the former two were continuously extended to the south-west in 1855–1870 (Fig. 5). The plots and buildings along associated new streets were contemporaneously redeveloped. The scheme created an enormous extent of the Victorian integuments and improved sanitations. Although the building replacements continued after 1896, the street system was well preserved.

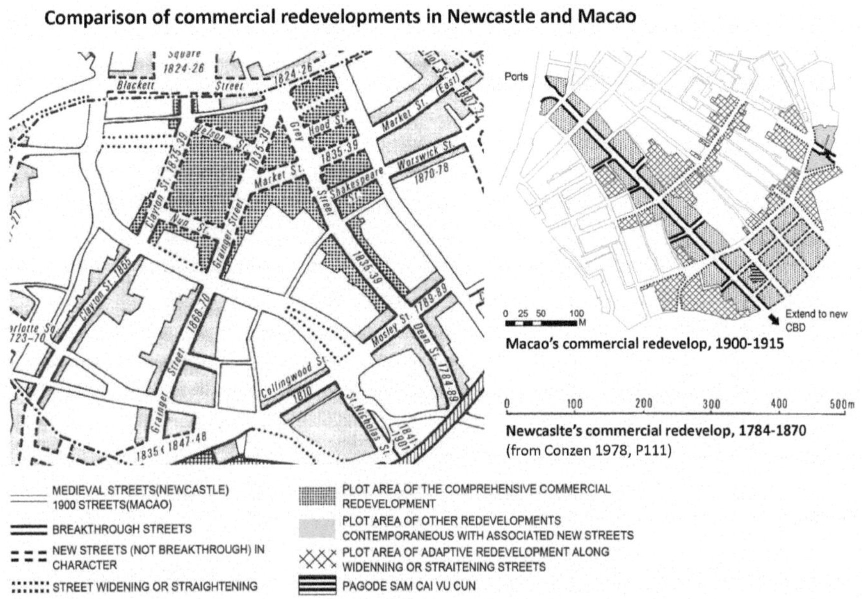

Fig. 5 Comparison of commercial redevelopment in Newcastle 1870 and Macao 1915 (*Source* created by the author)

6.2 *Macao from 1897*

The augmentative redevelopment of Grand Bazar was first proposed by Chinese developers in 1897[7] and had been rebuilt before 1905.[8] Six small rectangular street blocks were divided by five new streets, two of which in the middle were for two market buildings. To improve the sanitations and transform the congested streets, Macao's redevelopment scheme covered the whole of the urban core as it existed in 1900.[9] However, the revised plan of 1903[10] only retained a breakthrough street, Sun-Ma-Lo, and the widening of associated streets was proposed. Sun-Ma-Lo was built in 1908–1915 by cutting the existing urban core, providing a thoroughfare to the new CBD in the south (Fig. 5). Commercial activities began to gather in the new shophouse plots along Sun-Ma-Lo.

Afterwards, a number of plots were amalgamated or divided, but they were controlled in a small scale. In the cases of two market buildings, the two street blocks were amalgamated to build a new market complex in 1950.

[7] Macao Archives: MO/AH/AC/SA/01/01862.

[8] Idem: MO/AH/LS/0712.

[9] Idem: MNL.03.09.Cart.

[10] Idem: MNL.05.27.Cart.

6.3 *Breakthrough Streets: Clayton Street and Sun-Ma-Lo Street*

Breakthrough streets, adapting the increasing needs for communications, inevitably entailed demolishing existing buildings and the erasure of plots. Clayton Street is about 436 m long from Blackett Street to Fenkle Street and about 15 m in width. Before 1862, except for Grainger Market, most of the plots along Clayton Street were small and of various widths and depths. Initially, to make land acquisitions as few as possible, the new plots associated with breakthrough streets were irregular and isolated from the existing plots. Then adjoining plots were amalgamated to form larger plots. However, driven by commercial interests, some plots were greatly transformed to adapt the new shopping malls after 1950s (Fig. 6).

Sun-Ma-Lo Street is about 437 m from the Senate to the western coastline, with a 10-m wide road and 3-m colonnaded veranda (Qilou) on both sides. The route ran through the congested plots and streets of the old urban core. Similar to the land acquisitions of Clayton Street, Sun-Ma-Lo was mainly composed of small plots with 3–5 m in width but varying in depth.

Contrast to advanced metamorphosis of plots in Clayton Street, the changes in Sun-Ma-Lo were slight in the way of plot amalgamations or division. However, the changes of facades and owners were very active in Sun-Ma-Lo, especially in 1910s–1970s (Fig. 6). The styles of the facades were mixed with four styles: eclectic style in 1910s–1920s, Luso-Chinese style in 1920s–1930s, art-deco style in 1938s–1950s, and modern style. Art-deco style was predominant during 1930–1950. In the cases of 75 available documents of buildings, 50 of them had been art-deco style during 1930–1950. Thirty-five of these art-deco buildings were designed by the same Portuguese architect. Sun-Ma-Lo Street was strongly affected by global trends of architectural styles. But, in the case of the reconstructions of Luk Kok in Sun-Ma-Lo [16], big commercial centers were withheld because of contrary to the conservation legislation after 1970s.

7 Comparison and Interpretation

The complex results of repletion, expansion, and redevelopment in each city have created the present plan of the urban core, presenting an intricate pattern of plan units (Fig. 7).

In both cities, the transformations, especially expansion and redevelopment, have close relations to the traffics, which can bring huge commercial values. Newcastle benefited from the technological progress of land communication during the Industrial Revolution, while Macao depended on the maritime trades. In the case of breakthrough street, Sun-Ma-Lo acts as an important connection to the harbors in new CBD. Clayton Street and Grainger Street connected the northern commercial center with the railway station in the south.

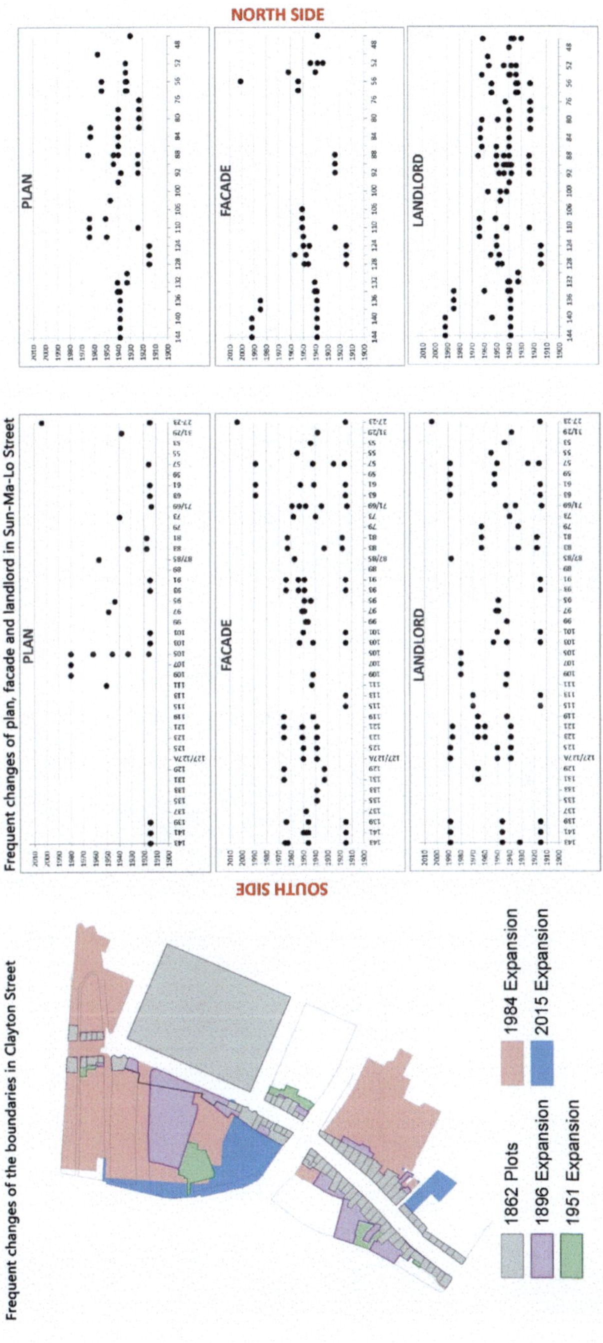

Fig. 6 Comparison of the changes along breakthrough streets (*Source* created by the author)

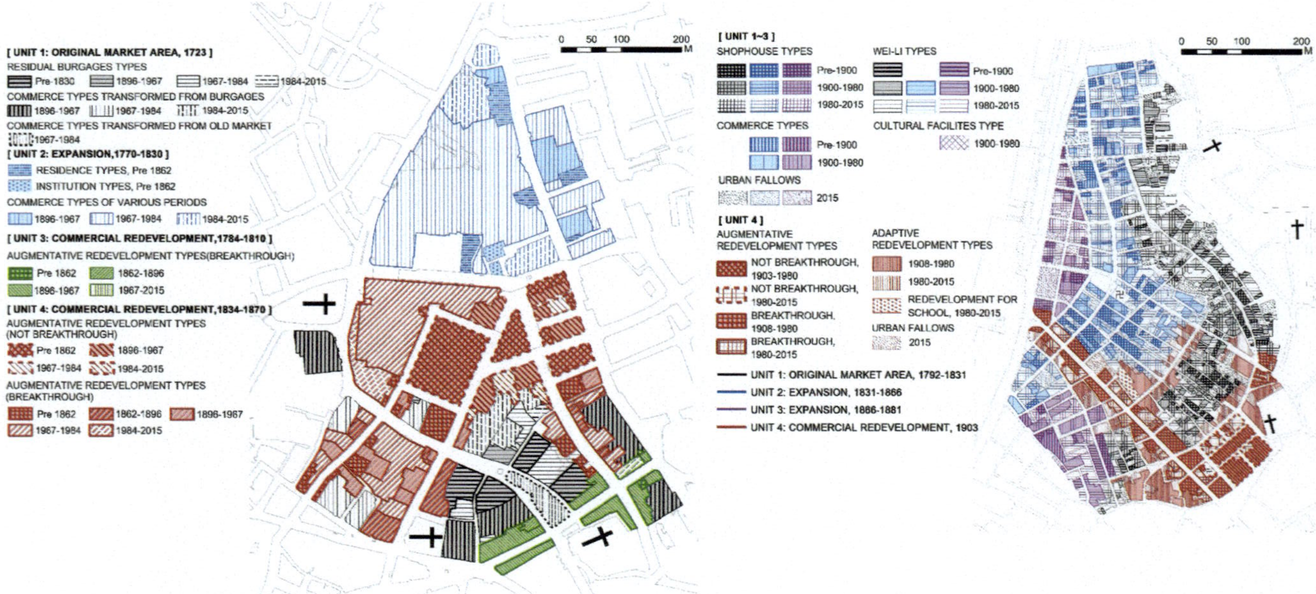

Fig. 7 Comparison of the morphogenetic plan units (*Source* created by the author)

Political environments are another impetus of transformations. The disappearance of a threat from Scotland had earlier been an important factor in the growth of the town beyond its medieval walls. However, Portugal was forced to open Macao and expand the new commercial areas with new ports so as to compete with Chinese cities opened after Opium War, especially the rise of Hong Kong. During the inter-war period, as not involved in the two world wars, a great deal of immigrants and money gathered in Macao. As an important way of modernization in the early twentieth century, Qilou commercial streets were prevalent in southern China during the 1920s and 1930s. The construction of Hong Kong's and Guangzhou's Qilou was earlier than Macao's [17], Sun-Ma-Lo being unusual for its breakthrough character as early as 1915. However, Newcastle underwent whole regenerations after the post-war recovery.

Macao has more different cultural impacts than Newcastle. The early cultural impacts came from Portuguese and Chinese. The grid street system in expansion areas, well influenced by the Pombal's downtown planning of Lisbon in 1755, contrasts with the early market area composed of Wei-Li with Chinese shrines (Fig. 3). After twentieth century, in the case of Qilou redevelopment, impacts from British south-Asia colonies [18] and Europe reflect in the plot patterns and building styles.

After 1960s–70s, contrasting strongly with the large-scale replacements of buildings in Newcastle was the extremely piecemeal plan-types in Macao. Newcastle was greatly impacted by "clean sweep" modernist planning of regenerations which was in full swing within UK [19]. Although the disillusion on "conservation-led regeneration" had been promoted after 1970s [20], the accelerated commercial functions stimulated high replacement rates of inherited plot patterns and buildings in Newcastle, of which about 59.24% had been changed tremendously in 1967–2015. However, Macao's large-scale redevelopments were controlled by conservation rules after 1970s. Moreover, a number of commercial lands were scattered in new CBD, which was evident in the map of land use 1972 [21]. On the other hand, increasingly residential needs caused the spontaneous building replacements in 1980–2015, the amount being about 55.9% of the totality. Eventually, it prejudiced the historical characters and leads to lose the economic vitalities.

8 Conclusion

First of all, the study demonstrates a number of Conzen's concepts of morphological development formulated in the West that are applicable in Eastern Asia. Conzen's concepts established a basis for examining different cities in different cultural backgrounds. Newcastle and Macao underwent the same ways of transformation but marked differences in the timing of developments and outcomes in urban landscapes.

Secondly, the study reveals two morphological renovations in west–east cities. The first transformation of western cities in the Industrial Revolution is earlier than that of Chinese cities in early twentieth century. Although west–east cities underwent the second transformation in 1970s–2010s, Newcastle and Macao respectively represent constructive destructions and preservative recessions.

Finally, the fragmented townscapes in traditional urban core of Chinese cities are the morphological combination with the relatively complicated influences on culture and socio-economy, especially after mid-nineteenth century, while the western urban cores tend toward uniformity integuments reshaped by the modern commercial complexes, which further develop the Conzen's study on Newcastle in 1960s. It could be predicted that the method can be tested elsewhere in other cases.

Acknowledgements I am very grateful to Prof. J.W.R. Whitehand and UMRG, University of Birmingham for their helps.

Funding The study was funded by China National Natural Science Foundation of China (NSFC No.51308232) and China Scholarship Council (201407540005).

Institutional Review Board Statement: Not applicable for studies not involving humans or animals.

Informed Consent Statement: Not applicable for studies not involving humans.

Conflicts of Interest The authors declare that the research was conducted in the absence of any commercial or financial relationships that could be construed as a potential conflict of interest.

Data Availability Statement The datasets used and analyzed during the current study are available from the corresponding author on reasonable request.

References

1. Conzen, M. R. G. (1960) *Alnwick, Northumberland: A study in town-plan analysis*. Institute of British Geographers Publication 27 (George Philip, London).
2. Conzen, M. R. G. (1962). The plan analysis of an English city centre. In J. W. R. Whitehand (ed.) *The Urban Landscape: Historical Development and Management*. Institute of British Geographers Special Publication 13 (Academic Press, London) 25–53.
3. Whitehand, J. W. R., Gu, K., & Whitehand, S. M. (2011). Fringe belts and socioeconomic change in China. *Environment and Planning B: Planning and Design, 38*, 41–60.
4. Conzen, M. P., Gu, K., & Whitehand, J. W. R. (2012). Comparing traditional urban form in China and Europe: A fringe-belt approach. *Urban Geography, 33*, 22–45.
5. Yao, S. (2013). Comparative study between urban morphology of the historical districts in guangzhou and birmingham, unpublished Ph.D. thesis, South China University of Technology.
6. Conzen, M. R. G. (1978). The morphology of towns in Britain during the industrial era. In J. W. R. Whitehand (Ed.), *The urban landscape: Historical development and management Institute of British Geographers Special Publication 13* (pp. 87–126). Academic Press.
7. Tian, Y. S. (2011). *Towards the opening city: Study on the market streets of Dongjing Song Dynasty*. SDX Joint Publishing Company.
8. Wu, Z. (2009). The establishment of Macao as a special port city and ensuing debates. *Social Sciences in China, 30*(2), 116–133.
9. Oliveira, V., & Pinho, P. (2010). Lisbon. *Cities, 27*(5), 405–419.
10. Ljungstedt, A. (1836). *An historical sketch of the Portuguese settlements in China*. Jame's Munroe & Co.
11. Liu, F. (1999). Colecção de documentos sinicos do IAN/TT referentes a macau durante a dinastia qing (Macao Foundation, Macao).

12. Amaro, A. M. (1998). *Das cabanas de palha às torres de betão: assim cresceu Macau* (Instituto Superior de Ciencias Sociais e Politicas, Macau).
13. Wang, W. J., & Cheong, C. K. (2011). *Regenerating pátio: Study of Macao's historical urban fabric* (Culture Institute of Macao, Macao).
14. Conzen, M. R. G. (2004) Thinking about urbanform: Papers on urban morphology, 1932–1998(Peter Lang, Oxford).
15. Ayris, I. (1997). A City of Palaces: Richard Grainger and the Making of Newcastle (Newcastle Libraries & Information Service, Newcastle upon Tyne).
16. Prescott, J. A. (1993). *Macaensis Momentum: A fragment of architecture: A moment in the history of the development of Macau.* Hewell Publications.
17. Lin, C. (2001). The initial study on the developmental reason of Guangdong modern arcade building. *Huazhong Architecture, 5*, 89–91.
18. Huang, S. J. (2013). Regulations and Practice of Arcades in Guangzhou during the Republic of China. *Academic Research, 3*, 123–133.
19. Pendlebury, J. (2001). Alas Smith and Burns? Conservation in Newcastle upon Tyne city centre 1959–68. *Planning Perspectives, 16*(2), 115–141. https://doi.org/10.1080/02665430010018239
20. Pendlebury, J. (2002). Conservation and regeneration: Complementary or conflicting processes? The case of Grainger Town, Newcastle upon Tyne. *Planning Practice and Research, 17*(2), 145–158. https://doi.org/10.1080/02697450220145913
21. Victor, F. S. S. (2012). *Macau through 500 Years: Emergence and Development of an Untypical Chinese City*. Enrich Professional Publishing.

Clan Settlements: Tangjiawan Ancient Town Diversified Residential Form Characteristics

Junxin Song

Abstract This paper uses the urban morphology method of the Conzen School to analyze the development process of the urban morphology of Tangjiawan Ancient Town and provides theoretical and methodological guidance for the protection of historical landscapes based on understanding the evolution of urban morphology. Researchers studied the layout and changes of traditional ancient towns in South China by conducting field research, looking at historical records, analyzing urban shapes, and interviewing people. Morphological units, building types, and land use are included. The analysis of the morphological process in Tangjiawan Ancient Town was conducted based on the urban morphology theory of the Conzen School and the town's actual situation. From 1522 to the present, Tangjiawan Ancient Town experienced five typical development stages: the Fuchongjing period (1522–1750), the five fortress pattern periods in the late Qing Dynasty (1750–1925), the Zhongshan Model County period (1925–1934), the slow development period (1934–1967), and the rapid development period (1967–present). The result is a visualization of the process of urban form evolution. The results of the division also reflect the morphological development, structure, and compositional characteristics of the ancient city of Tangjiawan. The morphological map of the Tangjiawan Ancient Town is inconsistent with the current economic-oriented development of ancient town planning. This shows that current planning is less concerned with the physical form of architectural heritage, which may lead to further damage to the landscape of the ancient town. Unlike previous studies, this chapter discusses in depth the relationship between residential plan units and plots, revealing the characteristics of clan settlements in South China. This study provides a strong reference for the follow-up planning of ancient town protection of the same type from the perspective of morphology and area and offers ideas for the subsequent preparation of protection plans.

Keywords Historic urban landscape · Urban morphology · Conzen school · Morphological map

J. Song (✉)
Institute for Research On Portuguese Speaking Countries, City University of Macau, Avenida Padre Tomás Pereira, Taipa, Macau, China
e-mail: a23092100150@cityu.edu.mo

Y. Chen et al. (eds.), *City-Sea Symbiosis*, Springer Geography,
https://doi.org/10.1007/978-3-032-13725-8_3

1 Introduction

1.1 Research Background

In 2011, China's urbanization rate surpassed 50% for the first time, reaching 51.27%. This means that China's urban population surpassed its rural population for the first time, signaling the transformation of Chinese society from "rural China" to "urban China," a pivotal point in China's urbanization. China's urbanization rate reached 58.52% in 2017 [1]. According to the "2013 China Human Development Report," published by the United Nations Development Program [2], China's urbanization rate will reach 70% by 2030, and its urban population will surpass 1 billion. The rapid urbanization of China has caused significant harm to urban cultural heritage [3]. The hot topic of the National Two Sessions in 2022 was rural revitalization, and the highest-rated-related search terms are smart agriculture and digital villages [4]. The use of the information platform has also received positive responses from the national urban management department and local governments. To support the development of the platform and the development of the historical context, various relevant departments have issued a series of relevant documents and policies. The efforts to protect Chinese urban historical and cultural heritage have reached a new starting point. The heritage of digital technology to support cultural heritage protection is also at the core of UNESCO's work this year [5]. The digital management of cultural heritage is a mainstream trend. The "Quebec Declaration," adopted at the 17th ICOMOS Conference in 2008 [6], also pointed out the importance of digital protection. However, how can we visualize the evolution of historically important ancient towns to urban areas? This is helpful for the digital protection of cultural heritage and subsequent heritage space planning. It can also combine visual results with publicity and education to build awareness of cultural protection for future generations.

1.2 Literature Review

In May 2005, the World Heritage Center and other professional institutions jointly held an international conference on "World Heritage and Contemporary Architecture: Managing Historic Urban Landscapes" in Vienna [7–9]. The formation of the "Vienna Memorandum," which used the term "historic urban landscape" for the first time to discuss the contemporary development pressures faced by historical cities, is discussed. There are currently two research directions for historic urban landscapes abroad: the Italian School and the Conzen School [4, 10–12]. The Italian School mainly analyzes from the perspective of typology, establishes a set of typology processes and methods, and applies them to architectural design. Typology was transformed from the proposals of Durand and Quincy to the development of neorationalist scholars such as Rossi, the Creel Brothers, Viedler, and Mudu [13, 14]. Although they have different views on city status, they all admit that the type expresses the urban culture and collective memory and can continue throughout the continuous evolution of time. However, the disadvantage is that although this school tries to construct a whole system research structure from city to architecture, it still only remains at the architectural scale in the end; thus, applying it to the historical landscape of cities and towns is difficult. Conzen mainly analyzed the urban form from the perspectives of structure and composition and the coordination of various parts of the urban fabric at different scales during the process of change [15–23]. Conzen's method focuses on analytical and conceptual information rather than pure narrative to study the city and determine how it is built. Scholars such as Whitehand, J. W. R., Slater, T. R., and Kropf, K. S. developed Conzen's theory and finally formed the Conzen School [24].

From the perspective of heritage protection, Professors RON and Zhou Jian noted in 2013 that the "historic urban landscape is an updated approach to heritage management" [25]. The starting point is to recognize and confirm that any historical town has accumulated value, and it is necessary to combine different disciplines to analyze and plan the urban protection process to avoid the separation of these values in the planning and development of modern cities. From the perspective of value recognition, in 2011, Professor Lu Zhou's postgraduate student proposed in his master's thesis *Cohesion and Continuity: Tangible Elements of Urban Historic Landscape and Related Issues* that parade elements are an important part of the historic urban landscape. From the perspective of value cognition, in 2010, some scholars proposed the idea and method of identifying historic urban landscapes based on the understanding of urban form [26]. He sorted and discussed the urban landscape protection and management methods and policy frameworks proposed by the school of urban psychology represented by Conzen, providing a substantive research framework for historic urban landscapes and clarifying the analysis elements of urban landscapes: physical form, social groups, urban history, and the development process [27, 28]. In other parts of the world, urban morphology developed earlier. After experiencing various problems with urbanization, Western scholars engaged in in-depth discussions on urban morphology. However, the focus of the research is constantly changing, and no research methods or research frameworks applicable to all cities have been established. Considering China's national conditions, many scholars have achieved important results with different theories and practices as China's urbanization progresses. However, the application of Conzen's urban morphology theory to ancient towns is still in the exploratory stage, and its theoretical methods should be further strengthened and valued.

According to Conzen School theory, an urban landscape is composed of three elements: plane pattern, architectural texture, and land use. At present, the characteristic unit map of historic urban landscapes is mainly delineated according to the landscape characteristics of historical regions and has the following four parts: topography, land use, construction characteristics, and regional characteristics. This approach insufficiently considers the structural and historical context of the landscape. The Conzen School reflects the heritage value carried by the urban landscape, which can provide a more scientific basis for the division of spatial units in historic urban landscapes. When the morphological unit area is divided according to the theory of historic urban landscapes combined with Conzen's urban morphology theory, the morphological unit division is adjusted. The urban landscape research of the Conzen School uses multilevel morphological region boundaries to divide morphological regions, which can provide more detail when dividing urban landscape forms and reflect the morphological characteristics of historical evolution.

1.3 Problem Statement and Objectives

Although many ancient villages and towns in China have distinguished histories, changes in their geographical locations and social systems impact them today. However, Tangjiawan, an ancient town with a relatively short history, has been able to continue to develop. Especially under the special circumstances that most of the rural areas were sluggish after the late Qing Dynasty, Tangjiawan did not decline but became a model county in Zhongshan and strengthened its historical traces. After 1979, the reform and opening seized Tangjiawan. The local government took the opportunity to make full use of the superior conditions in the region and vigorously improved the investment environment, introduced capital, equipment, and technology, and established joint ventures. In 2004, the town's GDP reached 2.39065 billion yuan.

Tangjia Ancient Town has a long history. As early as the Tang Dynasty, primitive villages appeared in some places in Tangjiawan. By the Northern and Southern Song Dynasties, many people from the Central Plains had moved in and built their own villages. Due to its geographical proximity to Macau, people from Tangjiawan worked and studied in Macau earlier than those from other villages, where they met advanced Western technology, ideology, and a religious culture. Under this influence, a group of influential pioneers in various industries in modern China gradually emerged. In the field of modern industry and commerce, Tong King-Sing is representative of the pioneers of modern industry and commerce. These include the China Merchants Steamship Company, the first civilian enterprise in China; the Kaiping Mining Bureau, the first coal mine; the Dragon Locomotive, the first train; the Tangxu Railway; and the Tangshan Fine Cotton Soil (Cement) Factory, the first cement factory. Tong King-Sing played a pivotal role in the development of the modern Chinese economy. The unique cultural environment has made Tangjiawan a unique Lingnan culture and folklore that combines the ancient and the modern, the intersection of the north and the south, and the integration of Chinese and foreign cultures. As an ancient town with such a rich historical and cultural heritage, the protection of this heritage is facing great challenges.

Current research on China's historic urban landscapes mostly reorganizes the composition system of urban historical spaces through value accumulation, which has a certain adverse impact on its value recognition and protection management. The Conzen School developed a research method oriented toward spatial development, which provides good support for the understanding and conservation management of historic urban landscapes. Under the guidance of this theory, the heritage value of Tangjia Ancient Town was identified, aiming to explore management methods suitable for the protection and management of historical urban landscapes in China and more widespread urban landscapes. According to the research framework of the Conzen School, this research aims to obtain the division results of urban landscape units. Among the three elements of the landscape unit, the plane unit is relatively stable and affects changes in building types and land use properties. Therefore, the focus of this research is mainly on the division of plane units and in-depth analysis and summary of the street system, plot combination, and building base of the ancient town is provided. Regarding the research on building types, since horizontal and vertical sections are required and the workload of basic data collection is relatively large, this study uses drones to take bird's eye views for direct observation and classifies buildings based on age and style. Regarding the land use method, since the Tangjiawan area has experienced several wars and has a long history, there is a lack of historical maps, and the land use cannot be verified. Therefore, the current land use is summarized.

2 Research Methods and Analytical Framework

2.1 Research Methods

At present, domestic scholars have seldom analyzed the historical urban landscapes of ancient towns. This paper attempts to apply historical urban landscape and Conzen School methods to ancient towns in China and proposes a morphological analysis framework suitable for the Tangjiawan Ancient Town. The morphological element framework, evolutionary process, and morphological area of the Tangjiawan Ancient Town are analyzed, and constructive suggestions for its protection are proposed. At the same time, in the Tangjiawan Ancient Town area, there currently only exist surveys and maps of individual public buildings, and the research has not yet been promoted to the whole village. Streets, plots, and land use are not included, which are of great significance for the study of form.

(1) Onsite Field Investigation: Onsite investigations, surveys, and map records of ancient towns should be conducted, current problems in ancient towns should be understood, and relevant information should be collected.
(2) Conzen Urban Morphological Analysis Method: Taking the historical urban landscape characteristics of Tangjiawan Ancient Town in Zhuhai city as the research object, under the guidance of this method, the development and renewal work are discussed. In addition, Conzen's urban morphology method was used to further divide the management units.
(3) Geographical Information System (GIS) Analysis Method: Based on GIS tool functions, the angles of land parcels in the parcel combination are analyzed.
(4) Combined with the above steps, the evolution of the urban form of the ancient town is finally visualized.

2.2 Framework of the Morphological Elements

Tangjiawan Ancient Town is in the northern part of Xiangzhou District, Zhuhai City, Guangdong Province (Fig. 1). It is backed by the lush green Phoenix Mountain and faces the vast blue waves of the Pearl River Estuary. It is a typical settlement in the Pearl River Delta region with both the landform characteristics of mountains and seas and a profound historical heritage. It is connected to Lingdingyang (Lingding Channel) in the east, Zhongshan City in the west, and the Pearl River Estuary in the north. Its unique geographical location made it a coastal defense stronghold and commercial node in the Ming and Qing Dynasties [29, 30]. It still retains the spatial texture that integrates the mountain and sea pattern and the defense system.

To understand the specific role played by each morphological element in linking the spatial relationships of the ancient town, it is necessary to decompose the complex spatial form of the ancient town into different morphological element frameworks. Based on this research, the typical characteristic factors that play a key role in the formation of the spatial intention of the ancient town and the spatial structure—organization relationship of the ancient town are selected. These characteristic factors can be used as evidence to guide the protection of the overall style and appearance of the ancient town. Under the guidance of the above ideas, combined with the analysis of morphological elements in Conzen's morphology and the spatial morphological characteristics and actual conditions of the Tangjiawan Ancient Town (Fig. 2), the ancient town wall, five fortress patterns, temples, farmland, and public space were analyzed.

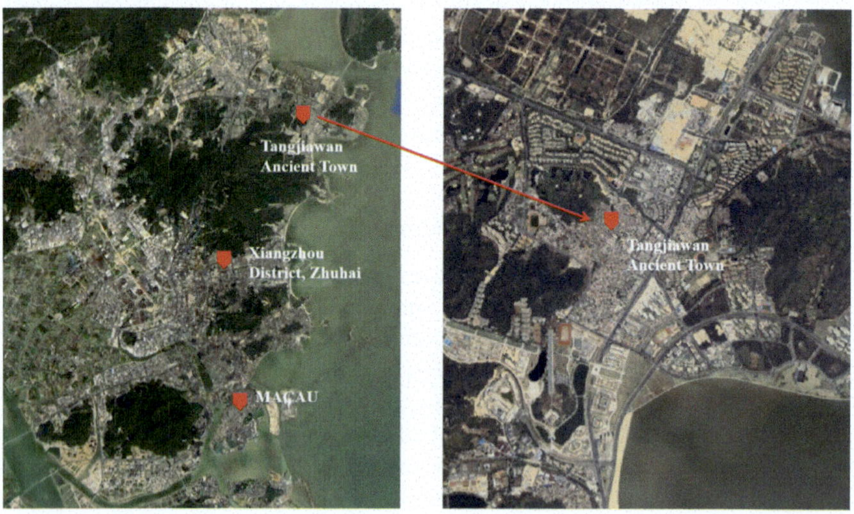

Fig. 1 Tangjiawan Ancient town location map. (*Image source* The author revised and drew it based on Google Earth)

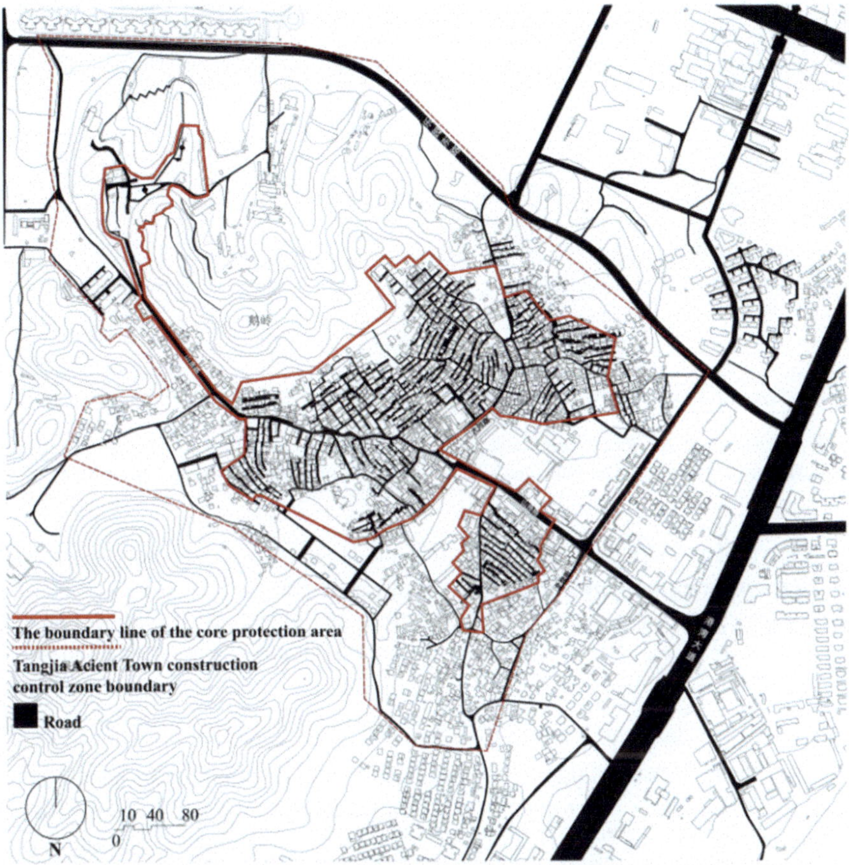

Fig. 2 The spatial range of Tangjiawan ancient town. (*Image source* In the 2016 publication titled "Conservation planning of historic town, Tangjiawan ancient town, China" (p. 50), the author made changes and created drawings based on this work)

(1) Ancient City Walls: The earthen walls and gates of Tangjiawan Ancient Town form a complete and closed defense system to protect against the outside. Tangjiawan Ancient Town also built an earthen wall to prevent entry by bandits. It was first built in the late Ming Dynasty, and it was rebuilt many times from 1815 to 1852 and in the early years of the Republic of China [31, 32]. The main body was completed in 1854, and there are still remnants of the ancient city wall near Gongle Garden (Fig. 3). The circumference of the city wall is 260 ft. The five gates of Wan'an, Longqing, Yingxun, Qiming, and Hengqu provide entry through the wall at various locations. In the late period of the Republic of China (1912–1949), the five gates were successively demolished, and most of the walls were also demolished. Only the sections from Longqing Gate to Lingyuanpu,

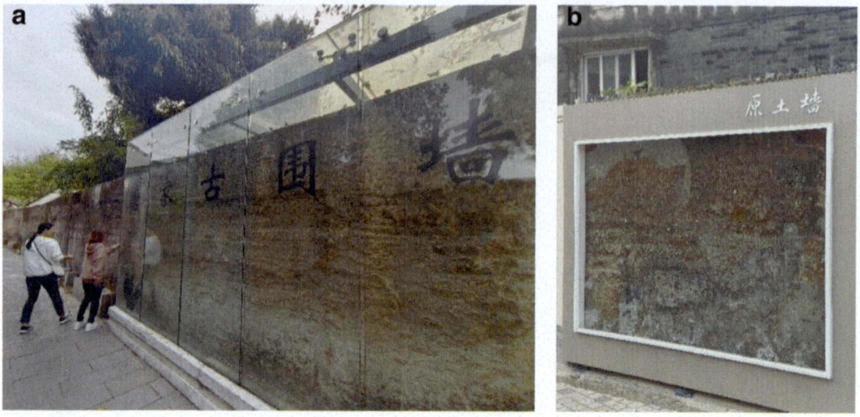

Fig. 3 The status of the remaining walls in Tangjiawan ancient town. **a** Ancient city walls in Tangjiawan ancient town. **b** Rammed earth wall. (*Image source* Photographed by the author)

at the foot of Yifeng Mountain in Ealing and Houlou Lane, Houmen Forest, and Hengqu Lane, remained, totaling approximately 580 m.

(2) Fortress-style settlements were a common type of defensive clan settlement during the Ming and Qing Dynasties. Tangjia Village is the largest and best-preserved fortress settlement in the Tangjiawan Ancient Town area. According to records, Tangjia Village began to form a basic pattern of "five fortresses" with "dividing houses" of clans in the late Qing Dynasty. Among them, those with the surname Tang lived in the first to third forts, and those with the surname Liang lived in the fourth and fifth forts. Generally, there were approximately 70–80 households in each fort, which is difficult to verify accurately at present. From the analysis of the morphological characteristics of the five existing forts, the following characteristics are presented. First, they conform to the traditional Feng Shui pattern. The five fortresses in Tangjiawan Ancient Town face the sea and are backed by the Efeng Ridge, with Longgang Mountain on the left and Ma Mountain on the right facing southeast from the northwest. Second, the five forts are adjacent to each other without a clear boundary, but each has a definite spatial order. The five forts were located on different platforms, and the low-lying areas between the platforms formed a "Y"-shaped main road connecting the five forts, namely, Shanfang Road and Datong Road. On the other hand, the spatial orientation axis of the buildings of the five forts was chosen according to their respective micro topographies. The five forts of the Tang family are built of rammed earth with a total length of approximately 5 km, a height of approximately 3 m, and a width of approximately 0.5 m. They are built on rammed earth, and only fragments remain (Fig. 3). There were five gates, namely, the Wan'an Gate, Longqing Gate, Yingzhao Gate, Qiming Gate, and Hengqu Gate, but they no longer exist.

(3) In the coastal areas of the Lingnan area, there are temples in every village, and the objects of worship are very practical. It was common to have multiple sacrifices

in one temple. The common idols are Mazu, Guan Gong, and Empress Guanyin. The most famous temple in Tangjiawan Ancient Town is the Tang Family Three Temples, which is the general name for the Shengtang Temple, Wenwu Emperor Temple, and Jinhua Temple (Fig. 4). It is located at the intersection of Datong Road and Xindizhi Street and has an area of approximately 1,500 m^2. It was rebuilt in the 40th year of Qianlong in the Qing Dynasty, and it was completely overhauled in the second year of Tongzhi. The current three temples were developed on this basis. From an architectural point of view, the three temples all have two entrances, with a courtyard inside, an incense pavilion in the middle, and a hard mountain top in the entrance hall. The temples are connected by water alleys and corridors. Lingnan has a hot and rainy climate. This feature must be considered in the layout of the building complex. There is a one-meter-wide roadway between the courtyards, which is not only used for transportation but also serves drainage and fire prevention functions. The building adopts a mixed structure of piercing buckets, lifting beams, and brick walls. Facade brick walls were built using grinding bricks to seam technology. Due to the additions during successive dynasties, stone columns and wooden columns were used together. The three temples show people's good wishes and sustenance for a harmonious life, and they are also the materialized form of expectations in life. The township office was once located here, and it housed the Tang Family Training Bureau in 1886 and the Tang Township Public School in 1907, and disputes among the people were often resolved there. Under the shackles of feudalism, it is quite fair and enlightening to use folk temples instead of ancestral halls and patriarchal clans to address issues such as belief, administration, education, and public security.

(4) Public Space: The main spaces for the public activities of villagers in Tangjiawan Ancient Town are squares, which are generally located in the open spaces in front of public buildings, for example, in front of the large ancestral hall of the Liang family and in front of the vast ancestral hall. There is also a square in front of the market at the intersection of Shanfang Road and Datong Road. In the past, every second day of the second month of the lunar calendar during the "Land Festival,"

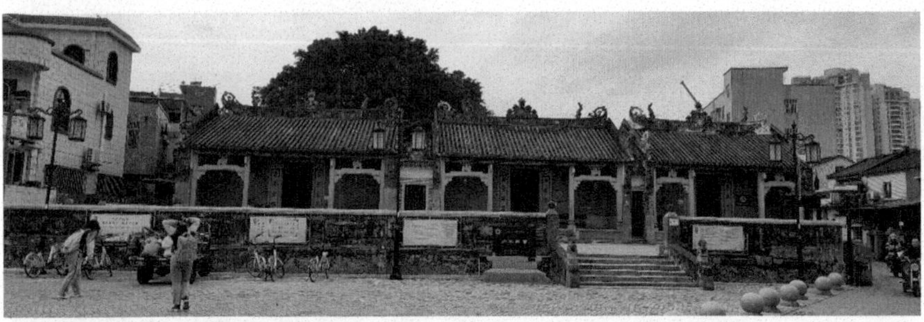

Fig. 4 The status of the Tang family three temples in Tangjiawan ancient town. (*Image source* Photographed by the author)

villagers held mass activities such as "grabbing firecrackers" and dragon and lion dances in the square in front of the temple. Gong Le Garden, located at No. 234 Shanfang Road, is also an important public space in Tangjiawan Ancient Town (Figs. 5 and 6). As the name suggests, Gong Le Garden means having fun with the people, and it is better to have fun alone than to have fun together. Therefore, Tang Shaoyi, the first prime minister of the Republic of China (1912–1949), opened the private garden he built in 1915 and named it Gongle Garden for use by fellow villagers and even by people of the whole country [33, 34]. The garden is surrounded by mountains and faces the sea, with beautiful scenery, covering an area of more than 500,000 mu and planted with rare trees, flowers, and lychees that are mostly imported from abroad.

(5) Ancient wells and ancient trees together constitute important public space nodes in the Tangjiawan Ancient Town (Figs. 7 and 8). People gather by wells and under old trees. There are two types of ancient wells in Tangjiawan Ancient Town: a single well with one borehole and a double well with two boreholes. There are also two forms of twin wells: one is two eyes side by side, and the other is two eyes tangent to each other. These ancient wells generally have a diameter of approximately 0.5–1 m and a depth of approximately 1–5 m. The well platform is generally made of granite, most of which is square with a small round part. Due to the terrain limitations, the shape of the well platform will also change accordingly.

(6) Farmland: The Tangjiawan Ancient Town region was an isolated island in the Lingding Ocean outside the Pearl River Estuary in ancient times. As the sea level

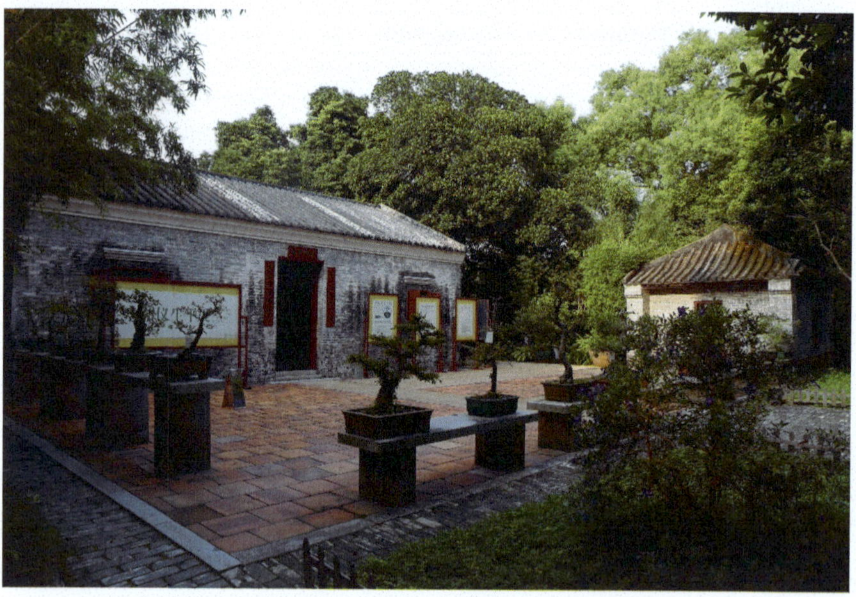

Fig. 5 Gongle Garden in Tangjiawan ancient town. (*Image source* Photographed by the author)

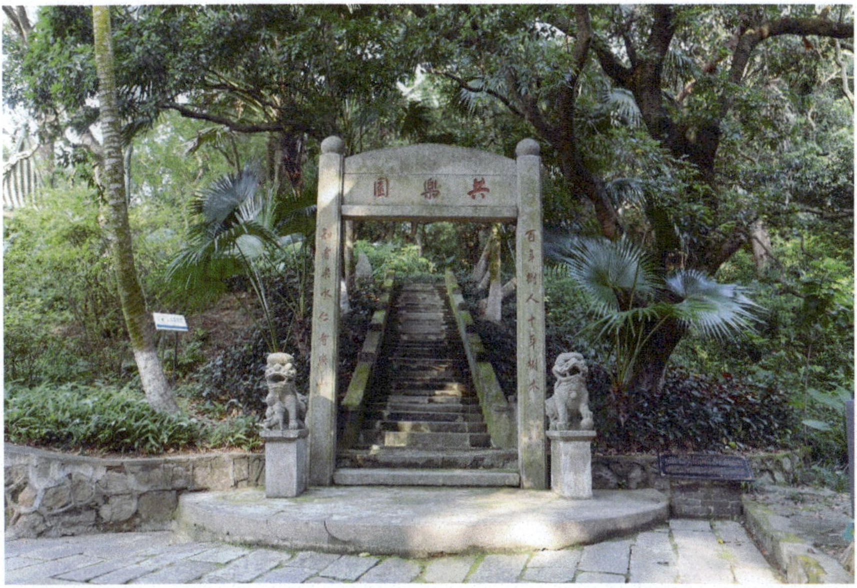

Fig. 6 Gongle Garden in Tangjiawan ancient town. (*Image source* Photographed by the author)

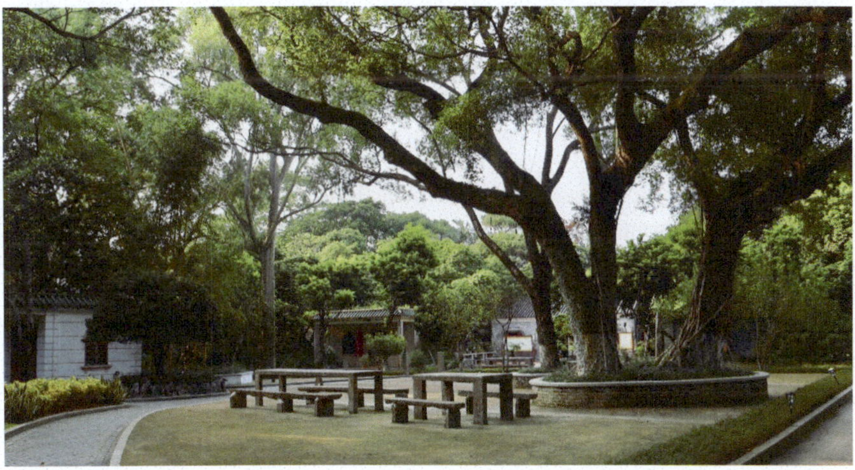

Fig. 7 Ancient trees in Tangjiawan ancient town. (*Image source* Photographed by the author)

gradually decreased, the land area continued to expand. By the start of the nineteenth century, Tangjiawan Ancient Town had completely emerged from the sea and turned into land. Since then, the upper reaches of the Pearl River have been continuously depositing a significant amount of sediment in the bay of Tangjiawan Ancient Town, causing it to rapidly build up. The beaches of Shanghai,

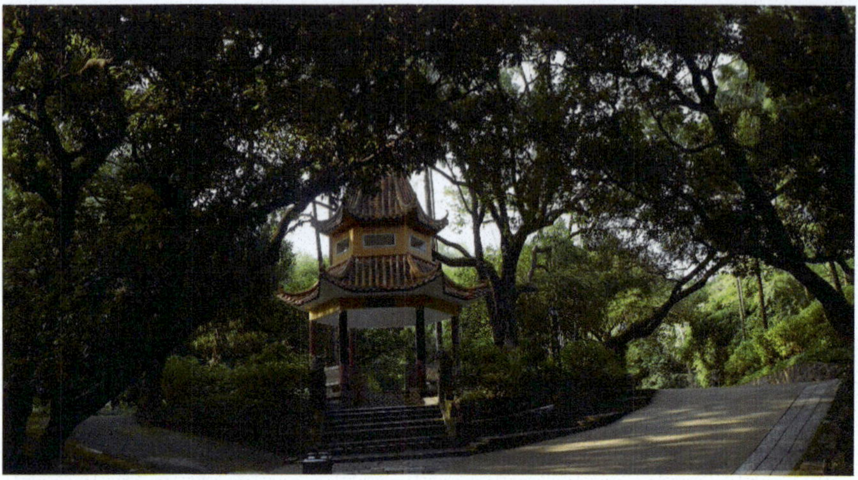

Fig. 8 Ancient trees in Tangjiawan ancient town. (*Image source* Photographed by the author)

Kwun Tong, and Longtoujiao in Houhuan led to Xiangzhou, Nanyang, and San Francisco in the United States at the end of the Qing Dynasty [35, 36]. In the 1980s, they evolved into shoals. In conjunction with the artificial reclamation that occurred at the end of the nineteenth century and the start of the twenty-first century, most of the shoals became land. The Municipal Bureau of Land and Resources began expropriating all rural collective land in the city in 1991. Since 1997, all collective land in the Tangjiawan Ancient Town area has been requisitioned for municipal government use, except for some land set aside for the production and housing of villagers in accordance with local policies. As a result, the current pattern of ancient towns has developed, while the farmland has essentially vanished.

3 Tangjiawan Ancient Town Townplan Analysis

3.1 Historical Development Stages of Tangjiawan Ancient Town

From 1522 to the present, Tangjiawan Ancient Town experienced five typical development stages: the Fuchongjing period (1522–1750), the five fortress pattern periods in the late Qing Dynasty (1750–1925), the Zhongshan Model County period (1925–1934), the slow development period (1934–1967), and the rapid development period (1967–present).

(1) Fuchongjing Period (1522–1750): According to an old map from 1522, the most famous name of the place where Tangjiawan was located at that time was "Fuchongjing," and it is generally inferred that this place is related to water and gold. Silver mines were discovered in Fuchongjing, and the center of the map was Xiangshan Town at that time, which was very impressive. The earliest people living at the foot of Phoenix Mountain were families with the three surnames of Jiang, Cheng, and Feng. The area was originally named "Sanjia Village" (a village where people from families with three surnames live together) and later changed to Tangjia Village.

(2) Five Fortress Pattern Period in the late Qing Dynasty (1750–1925): Fortress-style settlements were a common type of defensive clan settlement during the Ming and Qing Dynasties in China. Tangjia Village was the largest and best-preserved fortress settlement in the Tangjiawan area at that time. According to records, Tangjia Village began to form a basic pattern of "five castles" with "dividing houses" belonging to clans in the late Qing Dynasty. Moreover, the area around each fort basically remained unchanged for the next six hundred years.

(3) Zhongshan Model County Period (1925–1934): During the period of the Republic of China (1912–1949), the design of model counties was based on or aimed at local autonomy, and it was promoted throughout the country [37, 38]. Therefore, the model county became a comprehensive honor to praise local achievements in autonomy, education, and industry. There were fewer new buildings during this period. In 1930, the county government of Zhongshan County moved to Tangjia Town, which has an urban area of less than two square kilometers.

    As usual, large-scale construction projects were needed, but Tang Shaoyi (chairperson of the Zhongshan County Political Training Committee at the time) did not do so [39]. In contrast, the construction of county government embodied more economical and practical characteristics, fully respected the original living pattern, and used it to transform it into a political, economic, and cultural center at the county level. Ruizhi Ancestral Hall was built in the Guangxu period (1875–1908) and was an important government office in Zhongshan County. Tang Shaoyi's private garden (Gong Le Garden) became a high-level reception center that often received government officials and celebrities from all walks of life [1, 40]. The Wangci Mountain House at the beginning of Shanfang Road was also the office of the county magistrate. However, at this time, the five-fort structure was destroyed, and the city gates were demolished.

(4) Slow Development Period (1934–1967): During the Anti-Japanese War, Tangjia Ancient Town experienced several major military incidents. In 1934, soldiers from Zhongshan County caused problems in Tangjiawan. In 1938, the Japanese army attacked Tangjiawan. The anti-Japanese troops in Zhongshan County dug four air raid shelters and trenches on Liushi Mountain to prevent Japanese attacks. Unfortunately, the Tangjia Dongshan blockhouse, the Tang Family Ancestral Hall, and the Bianshan Maiyuan were destroyed by shells. After several battles, the layout of Tangjia Ancient Town changed, and the original

layout of the ancient town was no longer recognizable. After the founding of New China in 1953, to strengthen the construction of coastal defenses and develop fishery and agricultural production, the Government Council approved the establishment of Zhuhai County, and the county government was in Tangjiawan. So, Tangjiawan became the political, economic, and cultural center of the county. In 1959, Zhuhai County merged with Zhongshan County. When Zhuhai County was restored in 1961, the People's Government of Zhuhai County was in Xiangzhou, and the economic center shifted. Especially during the "Cultural Revolution," the town's economic construction developed relatively slowly.

(5) Rapid Development Period (1967–present): After 1979, the reform and opening swept through Tangjiawan. The local government seized the opportunity, made full use of the superior conditions in the region, vigorously improved the investment environment, introduced capital, equipment, and technology, established joint ventures, cooperated, developed "three in one subsidy" enterprises, and accelerated the development of foreign India inner linked and self-run enterprises. Overseas Chinese, Hong Kong, and Macau compatriots actively donated commemorative facilities and souvenirs. In 1994, individuals from overseas, Hong Kong, and Macau donated 500,000 RMB to renovate the 500-year-old "Three Temples of the Tang Family." After the establishment of Tangjiawan Town in 2002, the economic development of the region accelerated. The town government actively built the Tangjiawan Cultural Ancient Town brand and vigorously developed the tertiary industry, integrating tourism, catering, and commerce, with an annual growth rate of more than 20%. In 2007, Tangjiawan Ancient Town was rated as a famous historical and cultural town.

3.2 *Street System Analysis*

In the composition of plan units, the street system is the most stable form element. We can divide the main roads in Tangjiawan Ancient Town into main roads and secondary roads (Fig. 9). Shanfang Road and Datong Road have been the two most important roads in Tangjiawan in modern times, and they are also the main streets of Tangjiawan Ancient Town. In 1932, these two roads served as branch lines connecting Xiazha to Tangjiawan, located east of Qiguanche Road. Both sides of the road host a concentration of important buildings, including ancestral halls, temples, former celebrity residences, hotels, and stations. The Macau Green Island Cement Factory originally produced cement for the ground. The secondary roads are mainly alleys extending from Shanfang Road and Datong Road in the ancient town. Inside Tangjiawan Ancient Town, there are many ancestral halls arranged along Shanfang Road and Datong Road. These ancestral halls control the internal spatial structure of Tangjiawan Ancient Town and the street interface of each main road, forming a continuous ancient town control line. It is also connected with external ancestral halls and temples through Y-shaped streets, which together constitute the traditional road system of Tangjiawan Ancient Town.

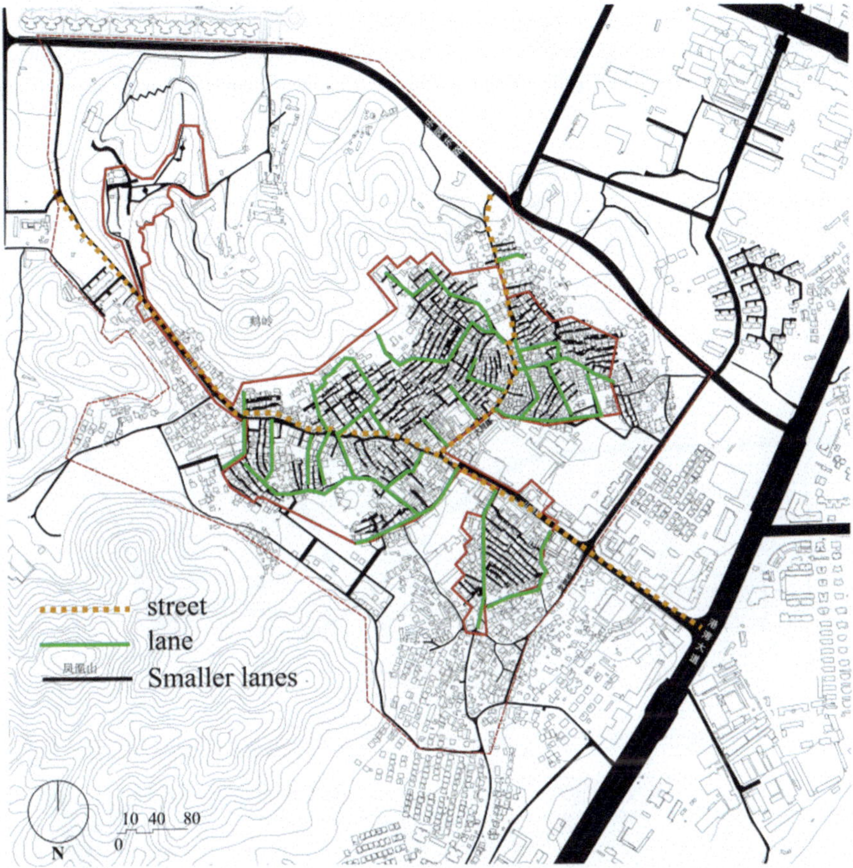

Fig. 9 Internal street system of Tangjiawan ancient town. (*Image source* Drawn by the author)

The lanes in the ancient town can be roughly divided into three categories. In the first category, the width of the roadway is approximately ten meters, such as on Shanfang Road, the street at the entrance of Yuwo Tang's Ancestral Hall, D/H > 1. In the second category, the width of the laneway is approximately 5–8 m, and it is the main laneway inside the historic district. The heights of the buildings on both sides are mainly 1 to 2 floors; for example, at the Datong Road Xiaoqing Family School entrance, D/H $\approx$ 1. In the third category, the width of the roadway is approximately 2 m, mainly used by pedestrians. This is the main auxiliary roadway in the ancient town, and the building heights on both sides of the roadway are mainly 1–2 floors, and D/H < 1. An example is the street behind Yuwo Tang's Ancestral Hall.

Furthermore, the outline of the block can be divided into several descriptive units for later analysis. Since the historical map depicts the internal streets of the ancient town rather vaguely, the drawing of the basic unit of the street is based on the current roads of the ancient town (Figs. 10 and 11).

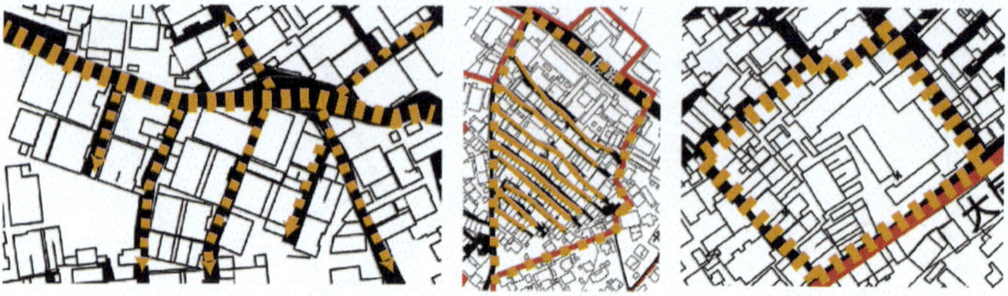

Fig. 10 Classification of street unit types. (1) Bifurcation type. (2) Comb-shaped. (3) Street-along type. (*Image source* In the 2016 publication titled "Conservation planning of historic Town, Tangjiawan ancient town, China" (p. 50), the author made changes and created drawings based on this work)

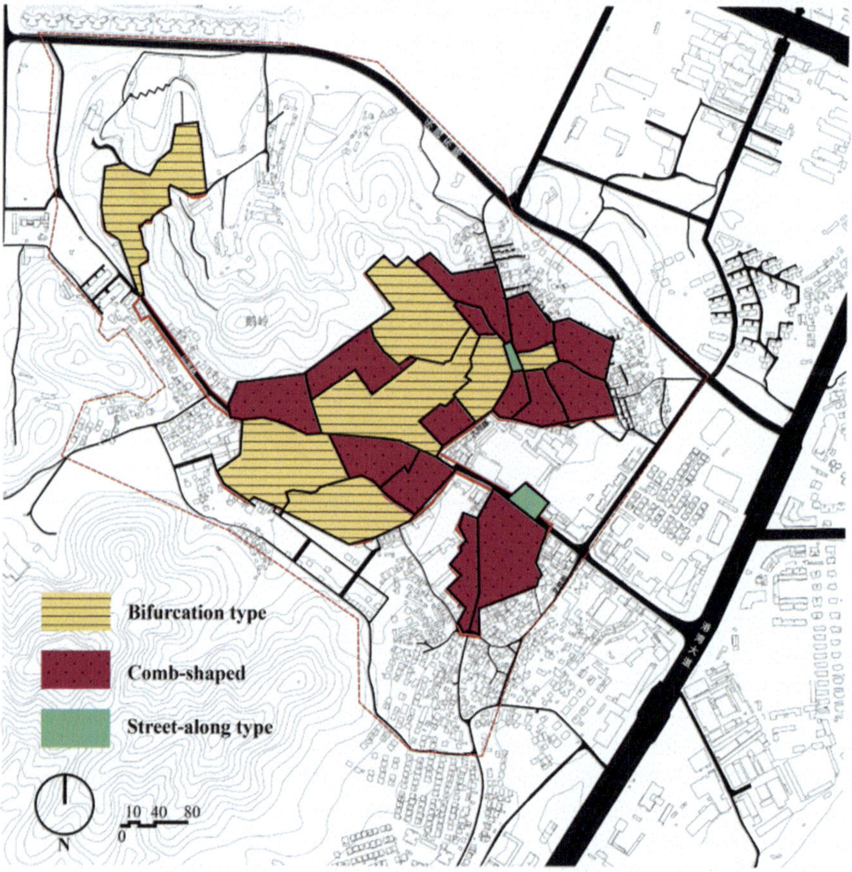

Fig. 11 Combination unit map of the Tangjiawan ancient town plot. (*Image source* In the 2016 publication titled "Conservation planning of historic Town, Tangjiawan ancient town, China" (p. 50), the author made changes and created drawings based on this work)

(1) Bifurcation Type: This form is the most common road form in ancient towns. It often branches from the main road of the ancient town and connects various building groups in series (Fig. 10).
(2) Comb-Shaped: As the name suggests, this road network consists of a horizontal line and several roads perpendicular to the horizontal line. This form effectively responds to the sloping terrain, connects roads to different elevation lines, and facilitates the internal traffic management of each building group. At the same time, the roadways of this form are usually at the end of section.
(3) Street-Along Type: In this type, there is no roadway inside the building group, and the building group directly connects to the main road of the village and town space.

Based on the unit analysis of the street and lane system, further research is carried out on the land combination mode of the inner group of the ancient town. Since Tangjiawan Ancient Town experienced several battles during the Anti-Japanese War, the original structure of the ancient town and the area of property rights are no longer available. Plots can be classified into independent plots, large ancestral hall plots, single-row plots, irregular plots, urban fringe belt plots, multi-row plots, irregular plots at the same angle, and single-row four-sided street plots (Fig. 12).

(1) Independent Plots: There is only a single building on the plot; the number of plots is small, and the plots are located along the edge of the ancient town.
(2) Large Ancestral Hall Plots: The square in front of the ancestral hall often serves other functions and hosts various activities, and the plot of the ancestral hall is often large.
(3) Single-Row Plots: This type of plot is widely distributed throughout ancient towns. The internal building structure is relatively regular, and the slope topography and roadway direction are the determining factors in the formation of these plots.
(4) Irregular Plots: The internal building directions of this type of land plot vary, and the land area is relatively large.
(5) Urban Fringe Belt Plots: Larger plots with scattered buildings near the walls of the ancient city.
(6) Multi-row Plot: The terrain restricts the building combination and follows the contour line.
(7) Irregular Plots at the Same Angle: The distribution of buildings in this plot type is relatively disordered, but they all face the same direction and follow a contour line distribution. The GIS calculations also included four regular distributions of angles of 320, 30, 80, and 170.
(8) Single-row Four-Sided Street Plots: This type of land is rare in Tangjiawan Ancient Town and is in the depths of the ancient town and seems to be wedged within and occluded by the adjacent land.

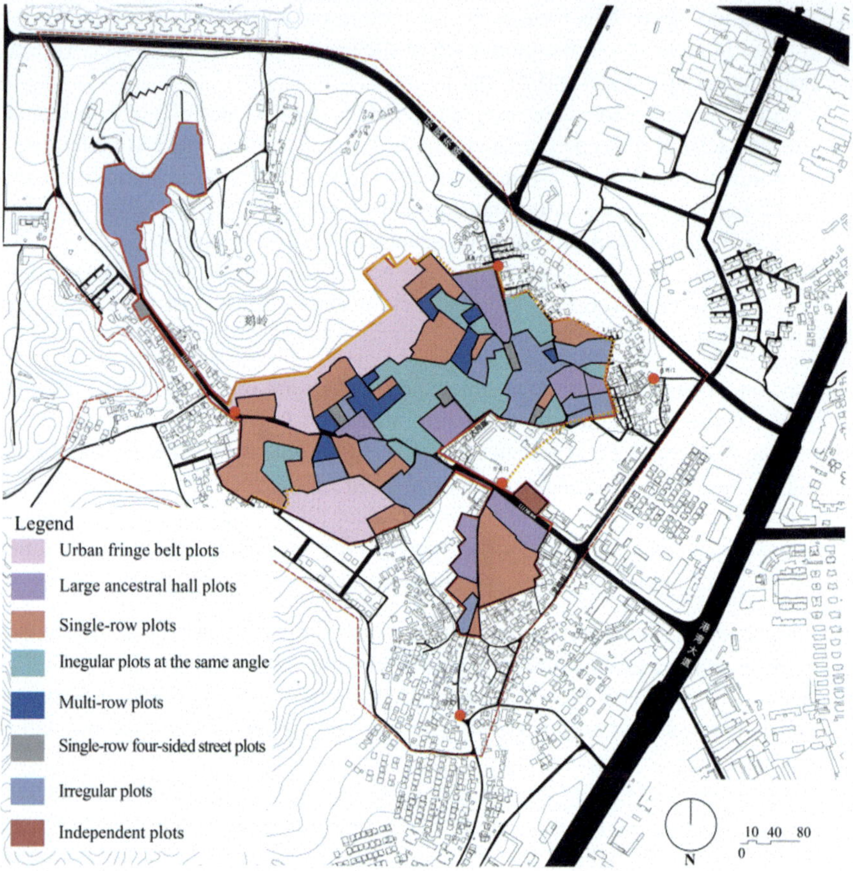

Fig. 12 Combination unit map of the Tangjiawan ancient town plot. (*Image source* In the 2016 publication titled "Conservation planning of historic town, Tangjiawan Ancient Town, China" (p. 50), the author made changes and created drawings based on this work)

3.3 Block Plan of Buildings

Tangjiawan Ancient Town is located on the coast of China and is surrounded by mountains in the southeast and northwest and flats in the northeast, facing the Pearl River Estuary and Venus Gate. Affected by the western Pacific monsoon climate, the dominant wind direction is southeasterly. The climate is hot, humid, and rainy, with typhoons and heavy rains in the summer, strong solar radiation, and long periods of high temperatures. Therefore, ventilation, heat protection, typhoon protection, and rainstorm protection are problems that must be solved in residential buildings, and they are reflected in the architectural layout and space combination. Generally, traditional houses are mainly located from northwest to southeast to cater to the dominant wind direction in summer. The architectural layout is typical of traditional

Lingnan dwellings [41–44]: a "three-room, two-corridor" courtyard house with one bright, two dark, and three bays or a "three-room, two-corridor" courtyard house and other variations (Fig. 13). The courtyard style is the most common; the former is two rooms deep, and the latter is an expansion of the former, three rooms deep (Fig. 14). There is a hall in the middle of the main room and rooms on the left and right, forming a bright hall and a dark room. The building is balanced on the left and right; it is closed to the outside and open to the inside. Additionally, a double house also existed in the ancient town where two families shared a front yard [45, 46]. In this case, two brothers usually shared the house, each with his own room. However, due to the large area and small depth, this kind of architectural form is not conducive to saving land, so it is rarely distributed throughout ancient towns.

With the development of urban society and the economy in the Ming and Qing Dynasties, the living standards of residents continued to rise, and the pursuit of an urban-living culture became more intense. However, due to the development and

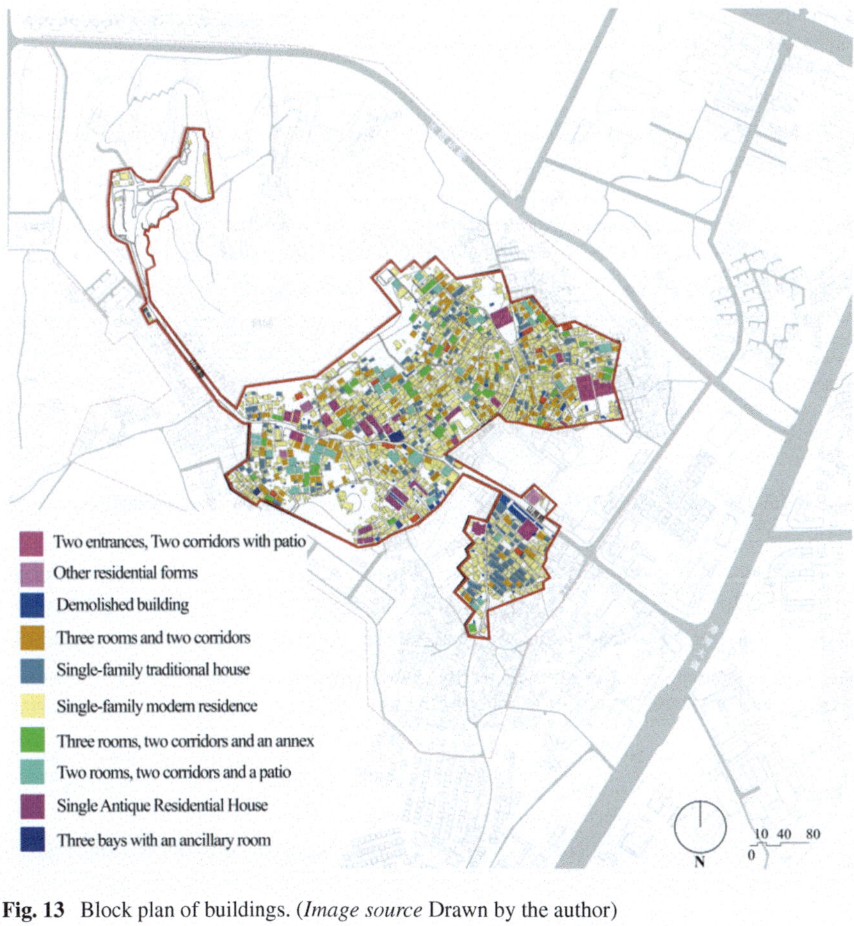

Fig. 13 Block plan of buildings. (*Image source* Drawn by the author)

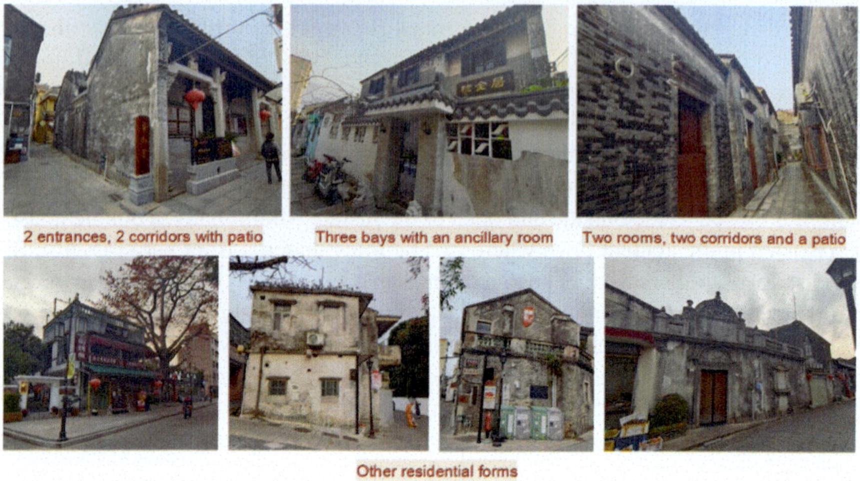

Fig. 14 The typical block plan of a building type in Tangjiawan Ancient town. (*Image source* Drawn by the author)

prosperity of commerce and the growth of families, the land supply in the ancient town was tight, making multibay, shallow, and deep traditional houses unable to adapt to the new city, and a new architectural form of single-bay houses emerged. However, this type of building is still very different from the bamboo tube house in Guangzhou in terms of nature and base. Since the layout of the Tangjiawan Ancient Town was relatively fragmented and the roads of the ancient town were relatively tortuous, some irregular plots were divided. Therefore, in the forms of the above several buildings, ancillary rooms are added according to the terrain, forming buildings with different shapes that adapt to the base (Figs. 15 and 16).

3.4 Plan Unit Division

The division of plan units is based on the superposition of the above three plan units. Since the street base description unit is the most stable factor, it is used as the basemap. The parcel composition unit provides the framework for the building block plan and is therefore the secondary basemap. Finally, the building block-plan map is superimposed to form a plan unit (Fig. 17).

(1) Traditional and Modern Mixed Residential Plots: Traditional and modern dwellings coexist in this type of flat unit, and the plots have variable shapes, so they are classified as one category.
(2) Public Building Plots: There are public spaces such as squares in this type of planar unit; most of these planar units are located at the intersections of streets and alleys, and the plots vary in size.

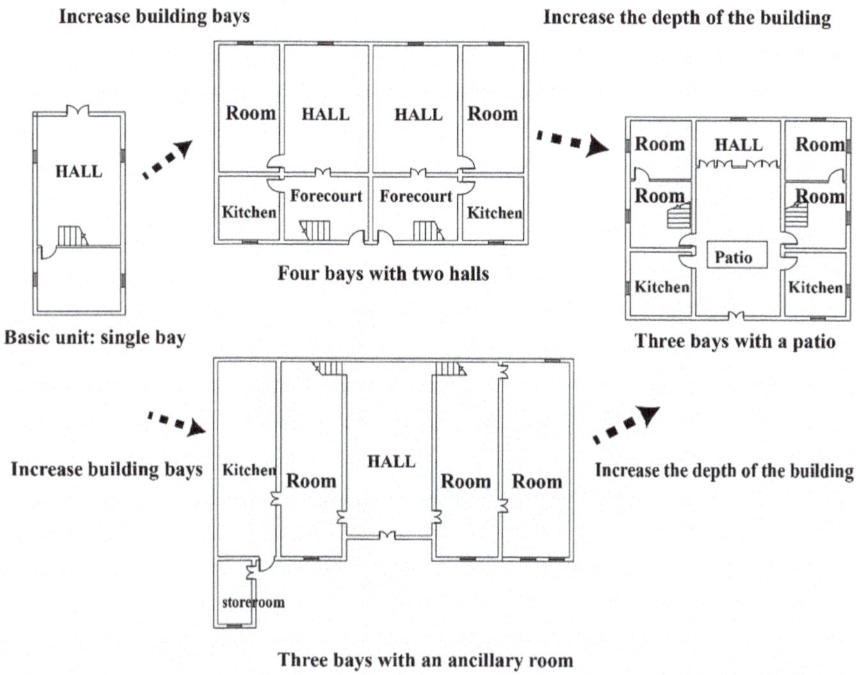

Fig. 15 Changes in the form of residential buildings in Tangjiawan ancient town. (*Image source* Drawn by the author)

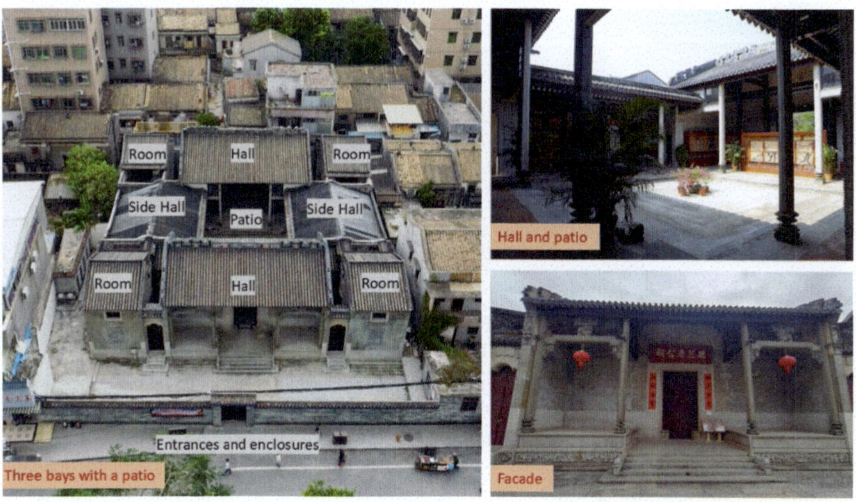

Fig. 16 Analysis of typical three-bay and two-corridor building layouts in Tangjiawan ancient town. (*Image source* Drawn by the author)

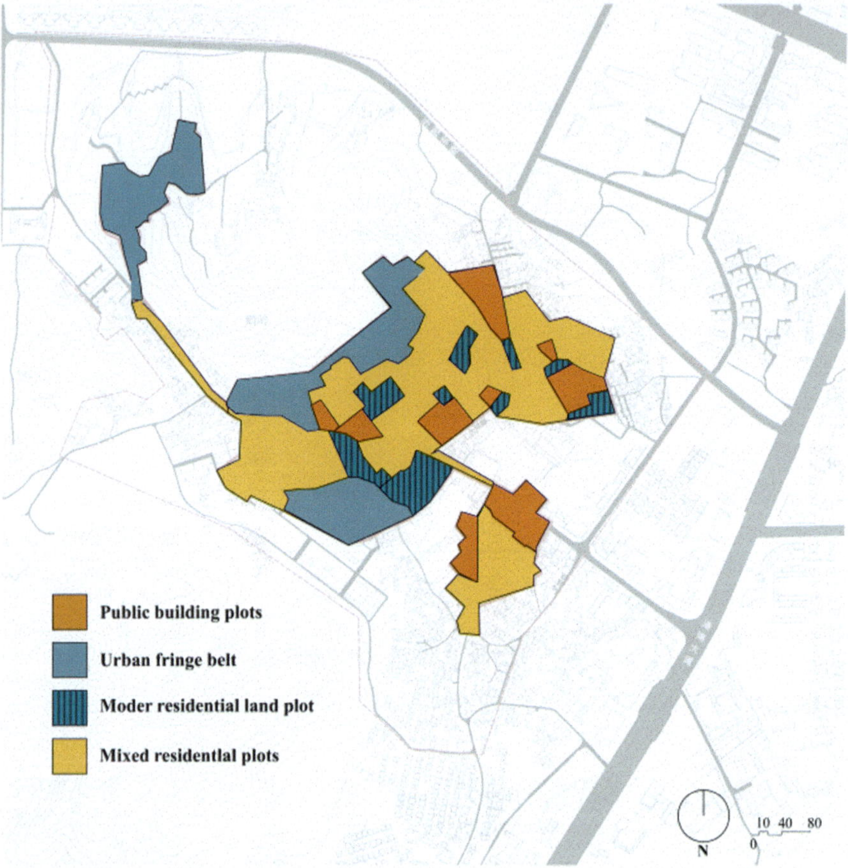

Fig. 17 Plan unit division of Tangjiawan ancient town. (*Image source* Drawn by the author)

(3) Urban Fringe Belt: The fringe belt of Tangjiawan Ancient Town is a belt-shaped area with an ancient wall as the axis. This type of flat unit building density is low and relatively loose, so it is classified as one category.

(4) Modern Residential Land Plot: The plot unit is dominated by modern buildings, the land area is small, and the distribution is relatively loose.

4 Building Type and Land Use Type Unit Division

4.1 Classification of Building Types

(1) Public Buildings: There are 28 public buildings in Tangjiawan Ancient Town, mainly family ancestral halls, private schools, temples, and former residences of celebrities. These include Cuiping Tang's Ancestral Hall, Longgang Ancient Temple, Ruizhi Ancestral Hall, Juchuan Tang's Ancestral Hall, Tang Shaoyi's Former Residence, Pingli Temple, Tangjia Old Station, Fengtai Liang's Ancestral Hall, Xiaoqing Family School, Tangle Majing, Peiyu Liang's Ancestral Hall, Wenwu Emperor Temple, Shengtang Temple, Jinhua Temple, Tang Bao'e Former Residence, Tangjia Ancient Wall, Junxu Liang's Ancestral Hall, Shoushan Liang's Ancestral Hall, Bujiu Liang's Ancestral Hall, Yueji Liang's Ancestral Hall, Yuwo Tang's Ancestral Hall, Wangci Shanfang, Gongle Garden, Xuejingtang, Juzhuang Tang's Ancestral Hall, Quanda Tang's Ancestral Hall, Xianyao Tang's Ancestral Hall, and Tangjia Ancient Well (Fig. 18).

(1) Modern Buildings in the Period of the Republic of China (1912–1949): Most buildings were brick and wood structures or mixed structures with flat roofs. The overall architectural style was varied and uneven, and the building quality was mostly average.
(2) Traditional Buildings in the Period of the Republic of China (1912–1949): Most of them were built with blue bricks, and a few were built with stone or oyster shells. The ground pavement was an all-lime soil foundation or block stone foundation. There were various forms of roofs, mainly hard mountain tops with gray tiles with exquisite decorations on roof ridges, which have typical Lingnan characteristics [47, 48]. The houses were dominated by wooden doors. The thickness of the stone beams and stone door frames were combined with the elegance of gray bricks. There were also some hollow wrought-iron doors. The windows were mostly ordinary casement windows and glazed windows; the sashes were made of wood; and the upper part was basically decorated with plaster. Most railings were made of concrete, and hollow, decorated concrete grids or vase rails were made of ceramic.
(3) During the Period of the Republic of China (1912–1949): Some of the traditional structures were rebuilt into modern dwellings; the old and the new ones were intertwined, and the old and new residential buildings had different styles, which greatly damaged the texture and style of the ancient town.
(4) Residential Houses in the 1950s to 1970s: These buildings, which were primarily steel–concrete structures influenced by foreign residential culture, were of Western style.
(5) Residential Houses After the 1980s: Most of these buildings were multistory buildings without traditional features.

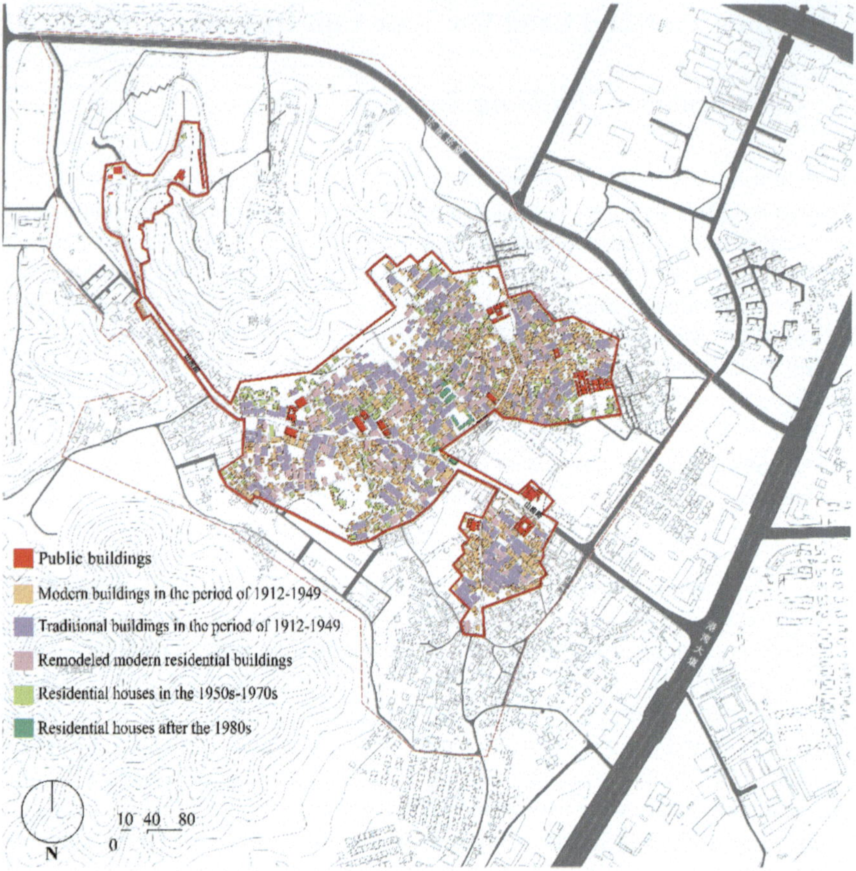

Fig. 18 Building type map of Tangjiawan Ancient town. (*Image source* Drawn by the author)

4.2 Residential Building Types

(1) Ming and Qing Dynasties to the Late Nineteenth Century: The Formation of Traditional Lingnan Residential Buildings

In terms of morphological characteristics, the residential buildings in this stage are based on the "three rooms and two corridors" courtyard as the basic unit, with a plan of "one bright and two dark" three bay, a hall in the middle, and living rooms on both sides, connected to the courtyard through a courtyard. Due to their coastal location, they are often invaded by pirates, and high walls and narrow windows are added to the residential buildings to form a ventilation system of "high walls and narrow alleys." The building materials are mostly green bricks and oyster shell walls. The roof is made of gray tiles on the gable roof, and the ridge decoration is mainly made of gray sculptures and brick carvings, reflecting the ritual order of clan settlement. In terms of spatial layout, influenced by the terrain, some residential buildings are arranged linearly along Shanfang Road and Datong Road, or form a "comb style layout" between terraces, such as the row of residential buildings in Wubao, Tangjia Ancient Town. The representative architecture preserved in this period is No. 5, Yatang Lane, Tangjiawan. No. 5, Yatang Lane, Tangjiawan Ancient Town is a typical Lingnan residence in the late Qing Dynasty. The blue brick wall is matched with western style arch doors and windows. The layout of patio and corridor continues the traditional "three rooms and two corridors" shape, and the roof is a round ridge roof.

(2) From the end of the Qing Dynasty to the Republic of China (late nineteenth century to 1949)—the "opening period" of combining Chinese and Western cultures

As the birthplace of modern study abroad culture, the overseas Chinese in Tangjiawan Ancient Town introduced Western architectural elements after returning home, forming a combination of Chinese and Western architectural styles such as "arcade buildings" and "Western style buildings" [49–51]. For example, the expansion of Tang Shaoyi's former residence combines Western style arcades with Lingnan courtyards, and the doors and windows are decorated with colored glass and Roman columns, reflecting the fusion of Eastern and Western aesthetics. At the same time, political and business celebrities (such as Tang Shaoyi and Tong Kwo On) have built mansions here, which combine the grandeur and practical functions of official residences. For example, Tang Shaoyi gifted his private garden "Gongle Garden" to the villagers, making it a public cultural space. In terms of architectural innovation, some residential buildings have begun to use new materials such as cement and arches. The facade decoration has simplified traditional carving and added European details such as Baroque moldings, but still retains traditional layouts such as courtyards and halls, showcasing the clever fusion of old and new architectural styles.

(3) From the Reform and Opening Up to the Present (1980s–present): Balancing Modernization and Heritage Protection

Under the promotion of urbanization, newly built residential buildings are mostly 2–5 story self-built buildings, which lack coordination with the traditional style of the ancient town and are scattered irregularly throughout the town. The building mainly adopts a simple reinforced concrete structure, with a compact layout and a lack of traditional decoration. In 2007, Tangjiawan Ancient Town was awarded the title of "National Historical and Cultural Town." The government initiated renovation projects, such as the renovation of a 400-m demonstration block on Shanfang Road in 2024, restoring the old to its original state and introducing new formats such as intangible cultural heritage markets and Live houses to attract young tourists.

4.3 Classification of Land Use Types

The current land is mainly used for village construction, public management and public service facilities, commercial service facilities, public facilities, green land, and squares. The village construction land refers to the current villagers' residences, with a land area of 55.98 ha, including Tangle Community and Tangjiawan Community. The land area for public management and public service facilities, including middle schools, primary schools, administrative offices, and other facilities, is 14.03 ha (Fig. 19). The current land use situation reflects the development status of the ancient town as the mainstay. Unfortunately, its historical and cultural value is not well utilized at present.

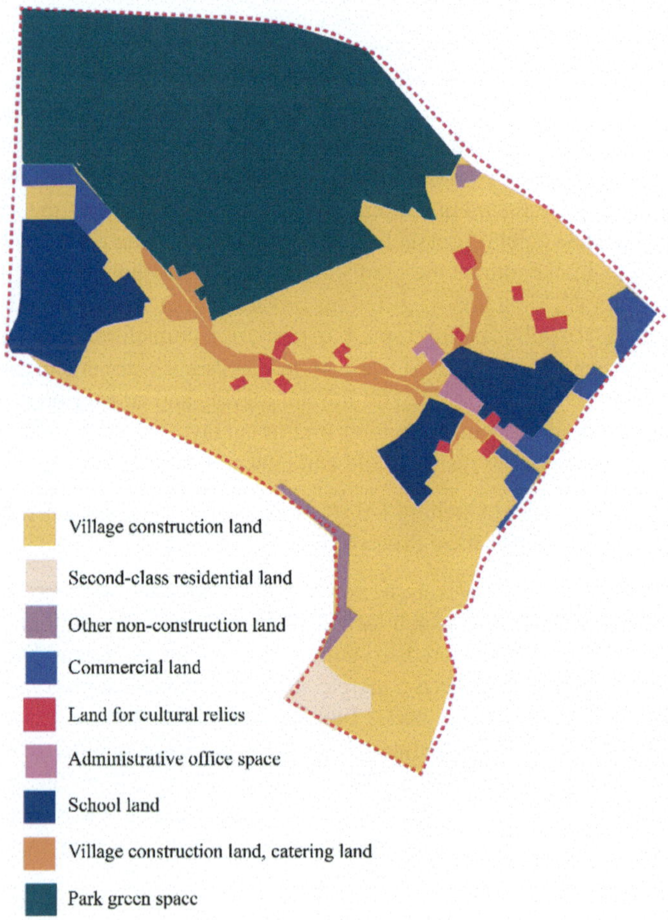

Fig. 19 Land use map of Tangjiawan Ancient Town. (*Image source* Drawn by the author)

5 Urban Morphological Unit Division and Characteristics

The street system unit, plot combination unit, building type unit, and land use unit division diagrams above simply superimpose the division of morphological regions. In the process of superposition, overlapping areas appear, indicating that there are many similar elements in this area, and they can be divided into the same unit area.

5.1 Morphological Area Division

(1) The First-Level Dividing Line: The area is divided into planar units formed by superimposing the three plans of the street and alley plan, the plot combination map, and the building block-plan map.
(2) Second-Level Dividing Line: Based on first-level zoning, the building type distribution map and the land use type map are superimposed, and the morphological units are further subdivided (Fig. 20).

According to the figure, based on the morphological regions initially divided by the plane pattern (Fig. 20), nine morphological regions are superimposed:

(1) Traditional Residential Areas: Most of these areas are row-by-row arrangements in the plane and are different from other subareas in terms of building function and plot size. Morphological subregions with uniform characteristics are thus formed.
(2) Green Space: There is a large area of green space in a park area, and some buildings are scattered among it.
(3) Modern Residential Areas During the Republic of China (1912–1949): These areas were formed earlier, but their architectural style is relatively modern, which conflicts with the style of the ancient town.
(4) Modern Residential Areas in the 1950s to 1970s: These areas were formed relatively late and can be classified as subareas according to their row and multistory forms.
(5) Unused Land: This land is located at the edge of the city, and currently, no buildings are built on it.
(6) Mixed Commercial and Residential Area: The first floor or the entire building of the residential area in this area is transformed into a commercial space, which is generally arranged facing the street, and the land use is villagers' catering and commercial land.
(7) Reconstruction of Traditional Residential Buildings into Modern Districts: Second floors will be added to traditional residential buildings or modern ancillary buildings, and the architectural style will be chaotic.
(8) Public Building Area: Most of the buildings in this area were built in the late Qing Dynasty and had obvious architectural styles and large plots. It is an area with independent shape characteristics in Tangjiawan Ancient Town.

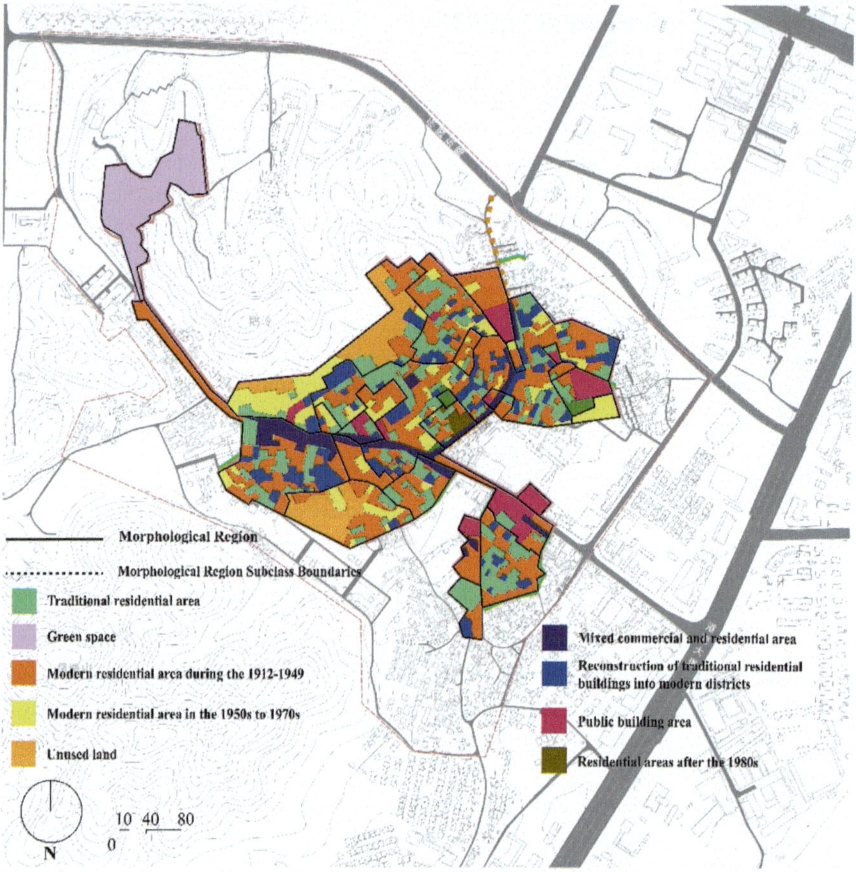

Fig. 20 Morphological region division of the Tangjiawan Ancient Town. (Image source: drawn by the author)

(9) Residential Areas After the 1980s: This is the latest form, with irregular plots and inconsistent architectural styles.

5.2 *Overall Characteristics of the Urban Form*

The morphological evolution process starts with traditional villages and focuses on modernized villages. Before 1925, the main construction body was the various families, the morphological feature was the five-fort pattern, and the evolution direction was southeast. In 1934, the main body of Tangjiawan Ancient Town became the national government, and the morphological characteristics at this time were large land units with uniform types and efficient uses. The ancient town of Tangjia in the middle has experienced several battles, but due to the lack of historical maps, it is speculated that some morphological areas were destroyed based on subsequent changes in the pattern. After the reform and opening, the rural system changed, and after the land reform movement, the government's restructuring and private participation in development made the ancient town a collage-style unit of "public–private blur." After the land use of the ancient town reached the saturation level in terms of construction, renovation activities began, changing residential buildings into commercial use or modernizing parts, forming a pattern of overlapping and mixing forms in various historical periods.

The stable primary and secondary arterial road systems constitute the morphological skeleton of Tangjiawan Ancient Town. The main and secondary arterial road systems that formed in the late Qing Dynasty are the morphological skeleton of Tangjiawan Ancient Town, which has guided the spatial form evolution in historical development. The ancestors of the Tangjiawan Ancient Town experienced collective migration. There were ancestral halls and housing ties. Thus far, the residential structure of the five fortresses has remained stable. The Mult surname settlements dominated by Liang and Tang carried a large amount of original historical information. From the end of the Qing Dynasty to the Republic of China, the size of the land was determined by the division of houses and the needs of life, and the architectural form of three rooms and two corridors formed the early small land plots. After the reform and opening, with the change of the construction subject, the large amount of land that originally belonged to a unit was further subdivided and planned.

6 Conclusions

This paper uses the urban morphology method of the Conzen School to analyze the development process of the urban morphology of Tangjiawan Ancient Town and provides theoretical and methodological guidance for the protection of historical landscapes based on understanding the evolution of urban morphology. The first is the understanding of the historical urban landscape level, especially a richer and more diverse understanding of the authenticity and integrity of heritage. It emphasizes local values, responds to bottom-up needs, and maintains the integrity and continuity of a city rather than simply dividing it into different types of "units." Therefore, the practice of delineating isolated "protection areas" and confining protection within the delimited space often results in the destruction of the integrity and continuity of the city.

Despite the increasing application of UNESCO's recommendations on historic urban landscapes in heritage conservation communities, research and practice to date have been limited. A historic townscape can observe and experience the urban landscape through urban form and an urban interior. Tangjiawan Ancient Town is in a state of rapid development. During intensifying pressures for change, the dynamic identity and changing character of the city are expressed through the identification of structures, places, and other traditional cultural elements of the historic urban landscape and the analysis of their historical context and evolution. With the flow of capital and the participation of experts from other regions, recent changes in Tangjiawan Ancient Town have been influenced by the development experience of large cities. The preparation of the master plan involved academics and professionals from Guangzhou and other larger cities, and the planning approach was largely influenced by national guidelines. Therefore, a historic urban landscape strategy needs to be integrated into the guidelines and regulations of the national strategic planning process. This will require support from policymakers and better coordination among government agencies. For example, more effective collaboration between the Ministry of Culture and Tourism, the Ministry of Housing and Urban–Rural Development, and their corresponding local offices is essential to ensuring good integration of urban planning and heritage management. In the process of industrial and tourism development, the development of Tangjiawan Ancient Town tends to prioritize material and economic goals over meeting the basic cultural and spiritual needs of the local community. Therefore, historically sensitive and community-based urban landscape management, including urban design and place-making, is a particularly important endeavor. Treating historic urban landscapes as infrastructures for livable cities is critical for ensuring sustainable development, the continuation of local cultures, and the re-connection of people and places. Urban design and place-making have become important tools for driving cities to meet the needs of local communities and visitors. They utilize existing elements and create others as needed to recreate spatial unity and continuity. Employing inclusive urban design and place-making practices grounded in an understanding of the processes by which places evolve can strengthen connections between people and places and between old and new structures. More strategic, holistic urban design and place-making strategies that involve all stakeholders can enhance local identity and individuality and help avoid repeating the development experiences of large cities. Unfortunately, the potential of morphological urban design and place-making has not yet been clearly defined in the planning of the Tangjiawan Ancient Town. The protection and planning of such ancient towns need more attention in the future.

Note: This chapter adapts and deepens the existing research results (*The evolution of traditional ancient towns in South China: Tangjiawan Ancient Town in Zhuhai city*, https://doi.org/https://doi.org/10.1080/13467581.2024.2349754) and adds a discussion on residential space form. The author of this article is one of its main researchers.

Acknowledgements The evolution of traditional ancient towns in South China: Tangjiawan Ancient Town in Zhuhai city, © 2024 The Author(s), taken from Journal of Asian Architecture and Building Engineering, © 2024 Architectural Institute of Japan, Architectural Institute of Korea and Architectural Society of China, reprinted by permission of Informa UK Limited, trading as Taylor & Francis Ltd, https://www.tandfonline.com. *This is an Open Access article distributed under the terms of the Creative Commons Attribution-Non Commercial-No Derivatives License (*http://creativecommons.org/licenses/by-nc-nd/4.0/*), which permits non-commercial re-use, distribution, and reproduction in any medium, provided the original work is properly cited, and is not altered, transformed, or built upon in any way.*

Funding This research received no external funding.

Institutional Review Board Statement: Not applicable for studies not involving humans or animals.
Informed Consent Statement: Not applicable for studies not involving humans.

Conflicts of Interest The authors declare that the research was conducted in the absence of any commercial or financial relationships that could be construed as a potential conflict of interest.

References

1. Song, J., Chen, Y., Zheng, L., & Tang, Q. (2025). The evolution of traditional ancient towns in South China: Tangjiawan Ancient Town in Zhuhai city. *Journal of Asian Architecture and Building Engineering, 24*(3), 2114–2132. https://doi.org/10.1080/13467581.2024.2349754
2. World Bank. (2012). *World development report 2013: Jobs*. The World Bank.
3. Jigyasu, R. (2014). The intangible dimension of urban heritage. In *Reconnecting the City: The historic urban landscape approach and the future of urban heritage* (pp. 129–159). https://doi.org/10.1002/9781118383940.ch5
4. Li, W., Zhang, P., Zhao, K., Chen, H., & Zhao, S. (2023). The evolution model of and factors influencing digital villages: Evidence from Guangxi. *China. Agriculture, 13*(3), 659. https://doi.org/10.3390/agriculture13030659
5. Xiao, W., Mills, J., Guidi, G., Rodríguez-Gonzálvez, P., Barsanti, S. G., & González-Aguilera, D. (2018). Geoinformatics for the conservation and promotion of cultural heritage in support of the UN sustainable development goals. *ISPRS Journal of Photogrammetry and Remote Sensing, 142*, 389–406. https://doi.org/10.1016/j.isprsjprs.2018.01.001
6. Yen, Y. N. (2012). A CH based integrative management framework on the value priority aspect. In *Progress in cultural heritage preservation: 4th international conference, EuroMed 2012, Limassol, Cyprus. Proceedings 4* (pp. 841–848). Springer. https://doi.org/10.1007/978-3-642-34234-9_90
7. Alberts, H. C., & Hazen, H. D. (2010). Maintaining authenticity and integrity at cultural world heritage sites. *Geographical Review, 100*(1), 56–73. https://doi.org/10.1111/j.1931-0846.2010.00006.x
8. Rössler, M. (2014). World heritage cultural landscapes: 1992–2012. In *Conserving cultural landscapes* (pp. 29–46). Routledge.
9. Rössler, M., & Lin, R. C. H. (2018). Cultural landscape in world heritage conservation and cultural landscape conservation challenges in Asia. *Built Heritage, 2*(3), 3–26. https://doi.org/10.1186/BF03545707
10. Cataldi, G. (2013). Translating 'Alnwick'into Italian: A tribute to MRG Conzen G. Cataldi. *Urban Morphology, 17*(1), 56–57. https://doi.org/10.51347/jum.v17i1.4556
11. Gauthiez, B. (2004). The history of urban morphology. *Urban morphology, 8*(2), 71–89. https://doi.org/10.51347/jum.v8i2.3910
12. Mosharraf, H. (2021). Urban morphology schools: A review of the English, Italian, and French schools of thought. *Gazi University Journal of Science Part B: Art Humanities Design and Planning, 11*(3), 399–407.
13. Hong, X. (2022). *Invisible intervention in public space: the vernacularization of western architectural typology in the urban public space of eastern Quanzhou* (Master's thesis, Universitat Politècnica de Catalunya). http://hdl.handle.net/2117/375915
14. Matsui, K. (2016). Does typology analyze the city? Or does the city define the typology? Typologies in Italy and Japan. In *11th International Symposium on Architectural Interchanges in Asia (ISAIA 2016), Sendai, Japan*
15. Whitehand, J. W., & Conzen, M. R. G. (1981). *The urban landscape: Historical development and management*. Academic Press.

16. Moudon, A. V. (1997). Urban morphology as an emerging interdisciplinary field. *Urban Morphology*, 3–10.
17. Whitehand, J. W. (2007). Conezenian urban morphology and urban landscapes. In *Proceedings, 6th internathional space syntax symposium.*
18. Mirmoghtadayi, M. (2006). In Farsi: Pishnehad raveshi baraye tahlil shakhsiat shahr (Suggest a way to analyze the "personality" of the city). *MOHIT SHENASI, 39*, 129–140.
19. Oliveira, V. (2019). J.W.R. Whitehand and the Historico-geographical approach to urban morphology. Springer International Publishing.
20. Larkham, P. J. (1998). *Urban morphology and typology in the United Kingdom.* Typological Process, Ed. Attilio petruccioli, Agha khan program for Islamic architecture.
21. Pourmohammadi, M., Sadr Mousavi, M., & Jamali, S. (2011). In farsi: Vakavi maktab haye morphology shahri, A review of urban morphology schools. *Arid Regions Geographic Studies, 5*(2), 1–16.
22. MahmeliAbyaneh, H. (2012). In Farsi: Moghayese tatbighi makateb morphology shahri be manzoor takmil an bar asas dastgah tahlili maktab eghtesad siasi faza (Comparative study on schools of urban morphology studies in order to complete them, based on analytical system of political economy school). *ARMANSHAHR, 4*(7), 159–171.
23. Moudon, A. V. (1998). The changing morphology of suburban neighborhoods. In Attilio Petruccioli (Ed.). Typological process and design theory. Aga Khan Program for Islamic Architecture.
24. Oliveira, V. (2019). An historico-geographical theory of urban form. *Journal of Urbanism: International Research on Placemaking and Urban Sustainability, 12*(4), 412–432. https://doi.org/10.1080/17549175.2019.1626266
25. Kong, P., & Z. Jian. (2015). Historic urban landscape, a management approach for cities of heritage-values. In *Revisiting Kathmandu: Safeguarding living urban heritage: Proceeding of an international symposium* (Vol. 12). UNESCO Publishing.
26. Wang, S., & Gu, K. (2020). Pingyao: The historic urban landscape and planning for heritage-led urban changes. *Cities, 97*, Article 102489. https://doi.org/10.1016/j.cities.2019.102489
27. Bandarin, F., & Van Oers, R. (2012). *The historic urban landscape: Managing heritage in an urban century.* John Wiley & Sons.
28. Zeayter, H., & Mansour, A. M. H. (2018). Heritage conservation ideologies analysis–Historic urban Landscape approach for a Mediterranean historic city case study. *HBRC journal, 14*(3), 345–356. https://doi.org/10.1016/j.hbrcj.2017.06.001
29. Chen, H. (2024). Zhuhai's craft-based intangible cultural heritage. In *Proceedings of the 2024 5th International Conference on Big Data and Social Sciences (ICBDSS 2024).* (p. 446). Springer Nature. https://doi.org/10.2991/978-94-6463-562-1_40
30. Yan, W., Jun, L., & Li, Y. (2024). The Monuments and cultural spaces among South China Historical Trail in Zhuhai. *Journal of Sustainable Urbanization, Planning and Progress, 8*(1), 39–46.
31. Wang, Y., Song, J., Zhang, J., Huang, Y., & Yang, S. (2025). Analysis of Rammed earth wall erosion in traditional village dwellings in Zhuhai City. *Coatings, 15*(5), 526. https://doi.org/10.3390/coatings15050526
32. Chen, Y., & Chen, J. (2022). Analysis and research on spatial nodes of historical villages based on spatial syntax theory: Cuiwei Village in Qianshan area of Zhuhai as an example. *Journal of Computer Science and Technology Studies, 4*(1), 07–34. https://doi.org/10.32996/jcsts.2022.4.1.2
33. Xiaoming, Z., Bo, Z., & Feng, D. (2005). Research on historical blocks of the National Research center for historical and cultural cities: Tangjiawan Ancient Town, Zhuhai. *Guangdong. Urban Planning, 29*(5), 97–97.
34. Luo, Y. (2020). Protection and utilization of historical gardens from the perspective of cultural and tourism integration: A case study of Tangjia Gongle Garden in Zhuhai. *Cultural Relics Appraisal and Appreciation.*
35. Liu, M., Chen, G., Li, G., Huang, Y., Luo, K., & Zhan, C. (2023). Landscape evolution and its driving forces in the rapidly urbanized Guangdong-Hong Kong–Macao Greater Bay Area, a

case study in Zhuhai City. *South China. Sustainability, 15*(17), 13045. https://doi.org/10.3390/su151713045
36. Yaping, L., Chao-Yu, W., Jiyu, C. (2002). Reclamation and Regulation in the Pearl River Delta, China. In: Chen, J., Eisma, D., Hotta, K., Walker, H.J. (Eds.), Engineered coasts. Coastal systems and continental margins (Vol. 6). Springer. https://doi.org/10.1007/978-94-017-0099-3_10
37. Feng, J., Yin, J., & Chen, Y. (2024). Spatial evolution and influencing factors of water-town settlements in Southern China: evidence from 942 villages in Zhongshan City. *Journal of Asian Architecture and Building Engineering*, 1–20. https://doi.org/10.1080/13467581.2024.2396611
38. Feng, J., Chen, Y., Lou, S., & Yin, J. (2023). Research on land use pattern change and its influencing factors based on the coupling model: Evidence from Zhongshan, China. *Geocarto International, 38*(1), 2280559. https://doi.org/10.1080/10106049.2023.2280559
39. Zhang, H., Fang, X. S., & He, X. W. (2012). The contribution of overseas Chinese towards the development of modern Zhongshan city. *Advanced Materials Research, 368*, 3292–3299. https://doi.org/10.4028/www.scientific.net/AMR.368-373.3292
40. Lam, J. F., Ong, C. E., Wu, T., He, R., & Li, S. (2022). The right to entrepreneurial expression over heritage significance: The micropolitics of an adapted cafe in a historic Chinese town. *Built Heritage, 6*(1), 26. https://doi.org/10.1186/s43238-022-00076-8
41. Yan, L., & Chen, Y. (2021). Stained glass in Lingnan historic buildings and its application in Manchuria windows. In *Journal of Physics: Conference Series* (Vol. 1885, No. 3, p. 032052). IOP Publishing.
42. Chen, Y. (2019). The development of decorative art in Lingnan house garden in Macau—taking Lingnan Oyster shell window and Manchuria window as examples. In *3rd international conference on art studies: Science, experience, education (ICASSEE 2019)* (pp. 479–484). Atlantis Press.
43. Chen, Y. (2019). Research on the development of the style of Lingnan Homestead in Macau—taking mandarin's house as an example. In *2019 International conference on architecture: Heritage, traditions and innovations (AHTI 2019)* (pp. 251–257). Atlantis Press.
44. Zheng, L., Chen, Y., Yan, L., & Zhang, Y. (2024). Automatic detection and recognition method of Chinese clay tiles based on YOLOv4: A case study in Macau. *International Journal of Architectural Heritage, 18*(10), 1551–1570. https://doi.org/10.1080/15583058.2023.2246029
45. Ng, J. K. P. (2022). Traditional ventilation skills of Lingnan Chinese architecture–a case study of Macau Mandarin's house. In *International civil engineering and architecture conference* (pp. 556–563). Springer Nature Singapore, Singapore. https://doi.org/10.1007/978-981-19-4293-8_57
46. Song, Y., & Liao, C. (2022). Structural materials, ventilation design and architectural art of traditional buildings in Guangdong, China. *Buildings, 12*(7), 900. https://doi.org/10.3390/buildings12070900
47. Yan, L., & Chen, Y. (2021). Study of roof tiles in Lingnan traditional buildings and roof drainage technologies. In *IOP conference series: Earth and environmental science* (Vol. 768, No. 1, p. 012151). IOP Publishing.
48. Zhang, W., Xiao, S., Wang, W., & Seong, D. (2024). A comparative study of traditional decorative patterns in the Lingnan Culture of China. 문화산업연구, *24*(3), 67–77. https://doi.org/10.35174/JKCI.2024.09.24.3.67
49. Tang, Q., Zheng, L., Chen, Y., Chen, J., & Yang, S. (2025). Innovative design method for Lingnan region veranda architectural heritage (Qi-Lou) facades based on computer vision. *Buildings, 15*(3), 368. https://doi.org/10.3390/buildings15030368
50. Wu, Y., Chen, Z., Lin, X., & Tu, X. (2025). A visual analysis and review of Chinese Qilou architectural heritage based on citespace. *Buildings, 15*(10), 1638. https://doi.org/10.3390/buildings15101638
51. Ji, H., Wu, S., Ye, B., Wang, S., Chen, Y., & Deng, J.-Y. (2023). Exploring the implementation path of passive heat-protection design heritage in Lingnan buildings. *Buildings, 13*(12), 2954. https://doi.org/10.3390/buildings13122954

Island Settlements: The Traditional Villages—Guishan and Guihai—on Guishan Island

Tao Xia

Abstract This chapter focuses on two traditional villages—Guishan Village and Guihai Village—on Guishan Island, part of the Wanshan Archipelago in Zhuhai, and examines the settlement patterns and cultural evolution shaped by the island's unique environmental context. As a strategic node at the Pearl River Estuary, Guishan Island possesses distinctive geographical and ecological resources. From the perspectives of Hakka and Tanka cultures, the chapter analyzes the spatial organization, architectural characteristics, and cultural expressions of the two villages under environmental constraints. Guishan Village, built along the mountain slopes, exhibits typical features of Hakka agrarian culture, characterized by a defensive and agriculture-oriented spatial order. In contrast, Guihai Village, organized around the harbor, reflects Tanka culture's fishery-oriented economy and its open, adaptive spatial layout. By investigating the natural setting, cultural background, settlement evolution, and spatial structure of the two villages, the chapter reveals the interplay and adaptive mechanisms between nature and culture in the formation of island settlements. Finally, from the perspective of tourism development and cultural heritage conservation, the chapter discusses sustainable development strategies for Guishan and Guihai villages, aiming to provide theoretical insights and practical implications for the study and preservation of similar island settlements.

Keywords Traditional villages · Settlement morphology · Spatial structure · Environmental adaptation

1 Introduction

Guishan Island's traditional settlements represent significant cultural heritage within the Wanshan Archipelago of Zhuhai [1]. Their unique natural geography and the fusion of diverse cultures have given rise to two prototypical villages: Guishan Village and Guihai Village. This chapter centers on the spatial evolution of Hakka agrarian

T. Xia (✉)
Zhuhai Dechuang Construction Engineering Consulting Co., Ltd., Zhuhai, China
e-mail: so19929013@126.com

Y. Chen et al. (eds.), *City-Sea Symbiosis*, Springer Geography,
https://doi.org/10.1007/978-3-032-13725-8_4

culture and Tanka (Dan) fishing culture on the island [2, 3], exploring the mechanisms of settlement formation, spatial characteristics, and pattern logic under environmental constraints.

Strategically positioned at the estuary of the Pearl River, Guishan Island is a core island within the Wanshan Archipelago [4], offering both strategic significance and rich ecological resources. The island's topography, ascending from north to south, provides favorable conditions for settlement placement. Nevertheless, its tropical maritime climate—with high temperatures, humidity, salinity, frequent typhoons, and relative scarcity of freshwater—has profoundly shaped architectural forms, land-use patterns, and resource utilization strategies. Guishan Village, settled by migrating Hakka communities, and Guihai Village, established by the locally indigenous Tanka people, exhibit striking cultural contrasts and spatial layout differences: the former is nestled into the hillside and reflects agrarian cohesion and defensiveness; the latter clusters around a harbor and demonstrates openness with a fishing-oriented spatial logic.

This chapter systematically examines the natural environment, cultural foundations, evolutionary trajectories, and spatial pattern features of both villages. The structure is organized as follows:

- Section 2 analyzes the island's geography and climate, elucidating impacts on site selection and spatial composition.
- Section 3 traces the migration history and cultural characteristics of the Hakka and Tanka groups.
- Sections 4 and 5 detail the evolutionary processes and morphological traits of Guishan Village and Guihai Village, respectively.
- Section 6 conducts a comparative analysis of the two settlements to investigate adaptive strategies and cultural divergence.
- Section 7 explores the pattern logic and the cultural expression of the "mountain–village–sea" spatial sequence.
- Section 8 synthesizes commonalities and differences and evaluates each village's contemporary value in the contexts of tourism development and cultural heritage conservation.

By presenting the Guishan Island case, this chapter aims to unveil the logic underlying island settlement formation in multicultural contexts and to propose sustainable development pathways. It is intended to furnish theoretical insights and practical examples for related studies and field applications.

2 Natural Environment and Settlement Formation Conditions on Guishan Island

The distinctive natural environment of Guishan Island has provided a multifaceted foundation—geographical, geomorphological, and climatic—for the development of the settlements of Guishan and Guihai Villages. Strategically positioned at the mouth of the Pearl River, the island boasts not only a superior geographic location but also abundant ecological resources. Its terrain, maritime climate, and natural endowments have significantly influenced the site selection and spatial configuration of the two villages, while also shaping local characteristics in architectural materials and livelihood strategies.

The subtropical maritime climate—characterized by high temperatures, humidity, salinity, and frequent typhoons—combined with the scarcity of freshwater, has driven the islanders to devise a series of adaptive strategies [5, 6]. This section analyzes the environmental conditions of Guishan Island, highlighting how these factors have served as key drivers in the spatial evolution of the Hakka agrarian culture and the Tanka fishing culture.

2.1 Geographical and Geomorphological Features

Guishan Island is located at the core of Zhuhai's Wanshan Archipelago and occupies a strategic position along a major navigation channel at the Pearl River Estuary (Fig. 1). Situated just 17 nautical miles from both Macau and Zhuhai's Xiangzhou District, the island serves as both a maritime transportation hub and a mediator for cultural exchange. The island's terrain is generally higher in the south and lower in the north, with a stepped mountain topography. The highest point, He Peak (or He-shan-ding), reaches an elevation of 232 m. This mountainous landscape has endowed Guishan Village with natural defensive advantages and supports the development of terraced agricultural practices on mountain slopes.

The island's coastline is divided into a naturally preserved pebble shoreline on the east and a modified artificial shoreline on the west (Fig. 2). The eastern coast retains rich pebble deposits and a stable coastal ecosystem, providing favorable conditions for the long-term development of Guishan Village. In contrast, the engineered shoreline on the western coast offers infrastructural support for Guihai Village's fishing port, catalyzing the formation and expansion of its coastal settlement pattern.

Granite is widely distributed across the island and serves as the principal construction material for both villages. Its superior resistance to wind erosion and salt damage makes it highly suited for the typhoon-prone maritime environment, and it is extensively used in wall construction [7, 8], courtyard enclosures, and building foundations.

The island's vegetation is predominantly tropical and subtropical, including species such as Podocarpus, masson pine, acacia groves, and banyan trees. These

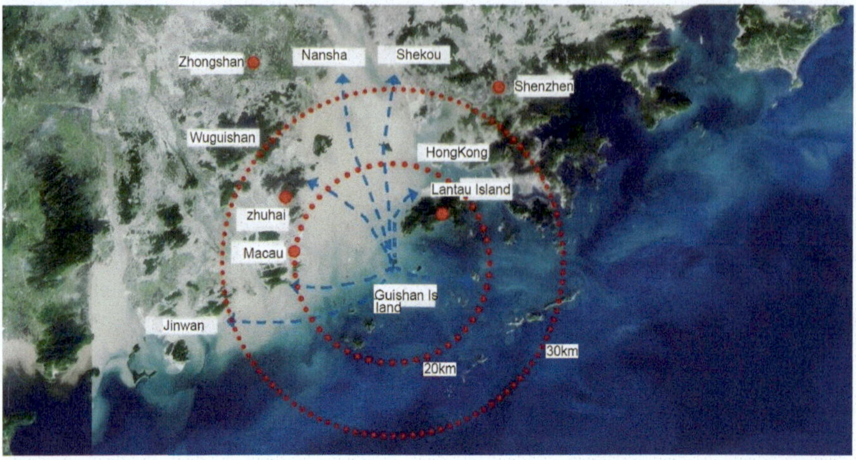

Fig. 1 Location of Guishan Island

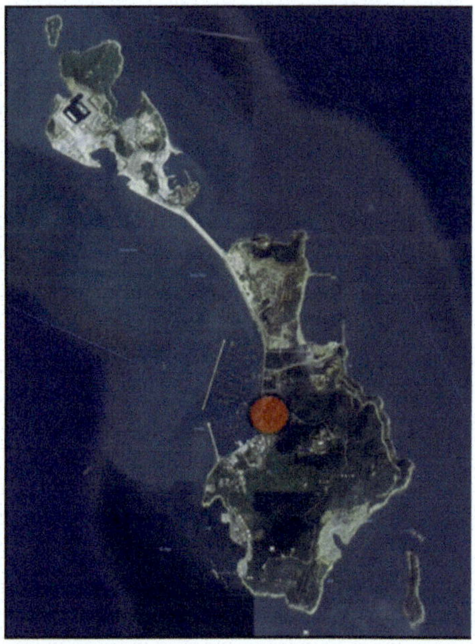

Fig. 2 Locations of Guishan and Guihai villages

not only sustain local biodiversity but also shape a landscape image that fuses mountainous and coastal features. This natural backdrop enriches the agricultural context of Guishan Village and the maritime culture of Guihai Village.

2.2 Climatic Conditions and Environmental Challenges

Guishan Island experiences a typical subtropical maritime climate, with an average annual temperature of approximately 23 °C. Summers are hot and rainy, while winters are mild and relatively dry. The main precipitation period spans from April to September, accounting for over 80% of the annual rainfall, with monthly averages reaching 200–350 mm. The dry season lasts from November to February, during which precipitation drops sharply to just 10–20 mm per month. Frequent typhoons, along with high humidity and salinity, impose strict demands on the structural resilience and material durability of local buildings.

To enhance wind resistance, buildings in Guishan Village typically feature thick granite walls and steeply pitched roofs, which facilitate rainwater runoff and improve structural stability. In contrast, Guihai Village, located along the coast, adopts strategies such as reduced building density and optimized building orientation to mitigate wind pressure.

The scarcity of freshwater is a persistent challenge on the island. Both villages have implemented site-specific rainwater harvesting and storage systems. Leveraging its elevated terrain, Guishan Village utilizes a system of stone wells and hillside diversion channels to maximize rainwater collection. Guihai Village, on the other hand, has developed centralized storage pools and channel networks to enhance water distribution during the rainy season, supporting both fishing operations and daily domestic needs.

In terms of agriculture, Guishan Village primarily cultivates drought-resistant crops such as sweet potatoes to improve soil quality and enhance the viability of hillside farming. Guihai Village, meanwhile, engages in the cultivation of salt-tolerant fruits and vegetables, forming an integrated "agriculture–fisheries" dual-use system. These adaptive strategies reveal the islanders' ecological wisdom in confronting environmental pressures and underscore the foundational role of the natural environment in shaping local culture.

2.3 Summary

The natural environment of Guishan Island has shaped the development trajectories of Guishan and Guihai Villages, each with distinct characteristics. The former, supported by terraced terrain and granite resources, evolved into a typical mountainous agricultural settlement; the latter, leveraging port infrastructure and artificial shoreline conditions, developed as a coastal fishing village. Climatic challenges and resource constraints have driven differentiated architectural and agricultural strategies, reflecting cultural responses to nature and offering practical insights for future settlement conservation and sustainable development.

3 Cultural Backgrounds of Guishan and Guihai Villages

Guishan Village was established by the Hakka people, while Guihai Village was founded by the Tanka (Dan) community. These two groups exhibit marked differences in cultural traditions, social organization, and spatial structure. The Hakka are primarily agrarian, emphasizing kinship-based lineage systems and self-sufficiency, whereas the Tanka rely on fishing, with settlement patterns closely tied to aquatic environments. This section examines how these distinct cultural features shaped the spatial configurations of their respective villages.

3.1 Hakka Culture and the Settlement Pattern of Guishan Village

The Hakka people migrated southward from China's Central Plains and developed a culture characterized by agricultural practices and clan-based patriarchal systems [9–11]. Their traditional settlements, such as weilongwu (enclosed dragon-shaped houses) and circular tulou structures, are known for their defensive qualities, mirroring a self-reliant economic model and the need for protection in mountainous environments.

Guishan Village demonstrates typical characteristics of Hakka cultural landscapes. The village is arranged along the natural terrain, with houses built from granite and topped with pitched roofs, enhancing wind resistance and drainage capacity. Terraced fields surround the village, forming an integrated "residence–agriculture–water" spatial structure. This layout not only accommodates agricultural production but is also deeply interwoven with the clan-based system, reflecting the strong connection between kinship and land ownership.

3.2 Tanka Culture and the Settlement Pattern of Guihai Village

The Tanka, traditionally residing along China's southern coastline, have a way of life closely intertwined with the marine environment [12–14]. Tanka culture emphasizes a fusion between fisheries and aquatic ecosystems, often reflected in their water-based dwellings such as boat houses and stilted shoreline residences.

The layout of Guihai Village embodies these cultural traits. The village is densely clustered along the harbor, with primarily single-story houses featuring low, flat roofs designed to minimize wind exposure. Fishing boats and gear are frequently moored near homes, resulting in a compact "boat–dwelling–sea" spatial formation [15]. This configuration reveals Tanka's heavy reliance on marine resources and significantly shapes their social structure and modes of production.

3.3 *Cultural Influences on Spatial Organization*

From a spatial perspective, the differences between Guishan and Guihai Villages stem from their distinct cultural underpinnings. Under the influence of Hakka culture, Guishan's spatial structure exhibits traits of defensiveness, lineage orientation, and agricultural functionality. In contrast, Guihai's layout—shaped by Tanka culture—is characterized by openness, functionality, and a strong orientation toward marine livelihoods.

These divergences manifest not only in architectural forms but also in modes of production, social organization, and resource utilization. Guishan Village forms a harmonious settlement model through the integration of family and agricultural systems, while Guihai Village establishes a water-adapted system that fuses residential and fishery functions. This interaction between culture and space illustrates the adaptive strategies of both communities on Guishan Island, offering critical insights into how traditional societies construct cultural identity through spatial design.

3.4 *Summary*

In conclusion, the spatial forms of Guishan and Guihai Villages are deeply shaped by Hakka and Tanka cultural traditions. Hakka culture reinforces mountainous settlements through kin-based social systems and defensive architecture, whereas Tanka culture gives rise to coastal settlements rooted in fishery and aquatic adaptation. The interweaving of these two cultures not only demonstrates the diverse settlement patterns on Guishan Island but also provides compelling evidence for the co-constitutive relationship between culture and environment.

4 Hakka Agricultural Settlements: The Formation and Transformation of Guishan Village

The spatial evolution of Guishan Village reflects how Hakka agrarian culture adapted to an island environment. Early settlers, driven by defense imperatives, established their homes atop the mountain, resulting in a terraced settlement pattern. As social conditions stabilized and population increased, residents gradually relocated to the slopes, incorporating fishing into their livelihoods and forming a hybrid agri-fishing economic space. This section divides the settlement's development into two key phases—initial reclamation and later stabilization—and explores how Hakka culture shaped residential forms under specific environmental and socio-political conditions.

4.1 *Initial Reclamation Phase: Characteristics of Hilltop Settlements*

The formation of Guishan Village dates to the Kangxi era of the Qing Dynasty (1662–1722), when Hakka immigrants arrived under the constraints of the "Migration Ban" (Qianjie Ling) and "Maritime Prohibition" (Jinhai Ling) policies. In response to threats from piracy and external incursions, settlers chose to reside atop Jiawei Mountain, leveraging its elevated terrain for defensive advantage. The village layout unfolded along the mountain's contour lines, forming a well-ordered, step-like configuration that exemplifies the Hakka emphasis on spatial order and hierarchy.

Architectural design retained traditional defensive features. Granite—abundantly available on the island—was used for walls, while gray-tiled roofs were adopted to cope with the island's humid and windy climate. Some dwellings clustered around a central ancestral hall (Shihe Tang), forming the core of the settlement. The compact arrangement of houses enhanced defensibility and promoted internal cohesion. This "enclosed aggregation" spatial model not only facilitated efficient resource integration but also reinforced kinship-based spatial organization, supporting both agricultural production and collective defense.

This stage exemplifies how Hakka communities responded to environmental and social pressures by organizing space through collectivist mechanisms.

4.2 *Stabilization Phase: Relocation and Agri-Fishing Integration*

From the late 1970s to early 1980s (1972–1983), loosening policy restrictions, improved social stability, and population growth prompted residents of Guishan Village to relocate from mountaintop areas to the slopes. This migration improved agricultural conditions and allowed for more efficient land use. The new settlement pattern became more open and adaptable. While agriculture remained a core livelihood, residents increasingly engaged in fishing, establishing a diversified agri-fishing economy.

The newly constructed houses were distributed along the terrain, still aligned with contour lines, preserving the traditional emphasis on spatial order. Fishing-related facilities were typically located at the periphery of the settlement, indicating clearer functional zoning and enhanced spatial flexibility. Building materials and forms largely followed traditional practices—granite walls and pitched roofs were retained to improve resistance to wind and moisture. Additionally, some newer structures incorporated stone cisterns for water storage, addressing the island's freshwater scarcity.

This period marked both a physical relocation from hilltops to slopes and a shift in economic structure—from a mono-agricultural model to a hybrid agri-fishing system. The spatial organization thus began to reflect a balance between collectivist order and functional adaptation.

4.3 Summary

The development trajectory of Guishan Village illustrates the adaptive refinement of Hakka agrarian culture in the face of complex environmental and policy contexts. During the initial reclamation phase, a defensive hilltop settlement formed a highly ordered, collective living environment. In the stabilization phase, relocation and industrial integration expanded spatial boundaries and diversified economic structures, achieving ecological synergy between agriculture and fishery.

Throughout both phases, core Hakka cultural principles—especially familial collectivism and spatial order—shaped settlement layout, construction materials, and spatial organization (Fig. 3). These enduring influences form the analytical foundation for comparative settlement studies and spatial pattern analysis in subsequent sections.

5 Tanka Fishing Settlements: The Formation and Evolution of Guihai Village

The formation and transformation of Guihai Village illustrate the adaptive processes of Tanka fishing culture in a maritime island environment. The initial waterborne lifestyle, with clustered boats along the harbor, gradually gave way to a land-based settlement. This transition was largely driven by government intervention, resulting in a stable village structure centered around fishermen's dormitories and Guihai Plaza. This evolution highlights the flexibility and openness characteristics of Tanka

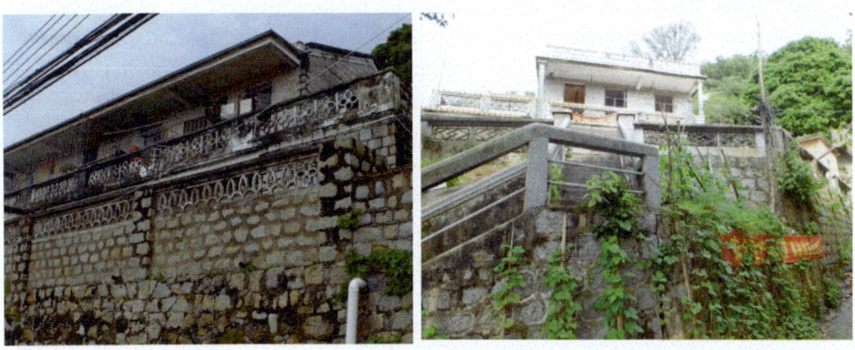

Fig. 3 Individual buildings in Guishan village. (*Image source* Fieldwork by the author)

culture. By analyzing the formation, development, and stabilization phases of Guihai Village, this section explores how Tanka traditions shaped the settlement's spatial morphology, offering a compelling case for understanding adaptive strategies in island fishing communities.

5.1 Formation Phase: Harbor-Based Waterborne Settlement

The Tanka people of Guihai Village trace their origins to a waterborne ethnic group from the Pearl River Delta, whose livelihoods depended heavily on fishing and maritime transport. During the late Qing and Republican periods, abundant fishery resources in the Pearl River estuary led to the congregation of Tanka communities near the harbor of Guishan Island [16, 17]. This gave rise to a distinctive boat-dwelling settlement in which residents lived aboard their vessels. These boats, typically constructed of timber and bamboo, were designed with functional versatility in mind, allowing for fishing, residence, and storage, thus reflecting the Tanka values of freedom and practicality.

The settlement emerged concentrically around the harbor, with boats anchored along the shoreline in arrangements that fluctuated with tidal changes and fishing cycles. The spatial organization of this waterborne settlement embodied the Tanka cultural principles of "free movement" and "flexibility," creating a unique community form and laying the spatial foundation for subsequent land-based habitation.

5.2 Development Phase: Amphibious Transition and Adaptive Architecture

After the founding of the People's Republic of China, government policies promoted the relocation of the Tanka people from boats to land [18], initiating a transitional stage in Guihai Village during the early 1950s. Tanka gradually began constructing temporary land-based dwellings near the harbor, accommodating tidal shifts and the demands of fishing production. These transitional structures were primarily made of timber and bamboo, often elevated to mitigate seawater intrusion and humidity. Lightweight materials such as wooden planks and reeds were used for the walls, while roofs were covered with tiles or thatch to ensure ventilation and moisture resistance.

The architectural layouts of this stage retained the dispersive features of the boat-dwelling settlement, with houses freely arranged along the coastline to facilitate boat docking and fish processing. The transitional design highlights the adaptive capacity of Tanka culture, accommodating the physical and functional requirements of the shift from a marine to a terrestrial lifestyle, and providing a structural basis for later permanent constructions.

5.3 Stabilization Phase: Fishermen's Dormitories and Guihai Plaza

Since the 1960s, the government initiated a housing program for the Tanka population, leading to the emergence of a stable land-based settlement in Guihai Village. The fishermen's dormitories were built with granite walls and sloped roofs to withstand typhoons and saline-alkali conditions. The spatial layout was oriented around the harbor and the newly established Guihai Plaza, continuing the dispersive spatial character of earlier phases (Fig. 4). The plaza became the central space for both communal and economic activities, serving not only as a venue for fish trading and village gatherings but also as a site for cultural events and social interaction. This significantly enhanced the settlement's cohesiveness and openness.

The spatial synergy between the harbor and the plaza optimized the functional efficiency of fishery operations. Villagers engaged in both fishing and the cultivation of salt-tolerant crops, creating a hybrid economy that integrated marine and agricultural practices. The spatial organization of this phase consolidated Guihai Village's economic base while diversifying its social structure and spatial functions, again reinforcing the Tanka values of freedom and utility. Guihai Plaza emerged as a cultural symbol of the Tanka community, playing a critical role in shaping local social life and sustaining economic vitality.

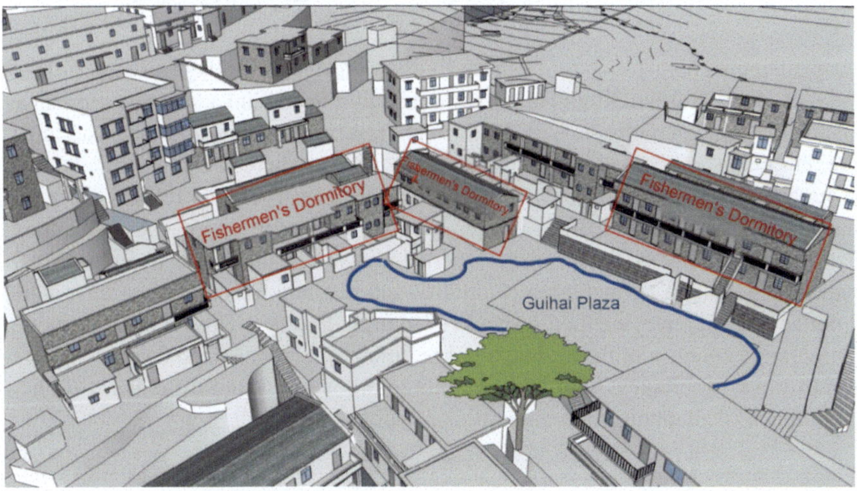

Fig. 4 Fishermen's Dormitories and Guihai Plaza in Guihai village. (*Image source* Drawn by the author)

5.4 Summary

The formation and evolution of Guihai Village exemplifies the adaptive and progressive character of Tanka fishing culture within the island context. The harbor-based waterborne settlement of the formation phase provided the spatial groundwork for subsequent land-based habitation. The development phase, marked by amphibious transitions and adaptive architecture, enabled a smooth shift from boat to land. In the stabilization phase, the construction of fishermen's dormitories and Guihai Plaza established a practical, dispersive settlement structure that balanced fishery production with communal functions.

Throughout all phases, the Tanka ethos of freedom and pragmatism guided the transformation of the settlement, resulting in an open, harbor- and plaza-oriented spatial layout coupled with environment-responsive architectural solutions. These characteristics form a strong foundation for the comparative analysis in Sect. 6 and the spatial order study in Sect. 7. Furthermore, Guihai Village offers meaningful practical insights for the conservation and sustainable development of traditional settlements.

6 Settlement Characteristics of the Two Villages: Environmental Adaptation and Cultural Expression

The settlement characteristics of Guishan Village and Guihai Village reveal the differentiated adaptive strategies of Hakka agrarian culture and Tanka fishing culture in an island environment. Guishan Village, through its terraced layout and robust architecture, embodies the cohesiveness of Hakka traditions, whereas Guihai Village, with its harbor-oriented dispersive layout and flexible constructions, reflects the pragmatism of Tanka culture. Both villages adopted distinctive approaches in building materials, structural systems, site selection, and resource utilization to address challenges such as high temperatures, humidity, salinity, and freshwater scarcity, while simultaneously conveying their respective cultural identities. This section conducts a comparative analysis of the architectural features, site layouts, and resource utilization of the two villages, highlighting both commonalities and differences in their environmental adaptation and cultural expression.

6.1 Architectural Features: Materials, Structure, and Function

The architectural practices in both villages demonstrate careful material selection and structural design to adapt to typhoons, high temperatures, humidity, and saline conditions, while also reflecting divergent production demands (see Table 1).

Table 1 Comparative summary of architectural features

Component	Feature	Guishan village	Guihai village
Roof	Form	Double-pitched gabled roofs, some flat	Predominantly flat, with tented roofs
	Structure	Pan tiles with layered insulation, drainage channels	2% slope, cross-ribbed ceramic tiles
	Function	Ventilation and drainage, agriculture-oriented	Drying fish, fishery-oriented
Walls	Form	Courtyard walls with carvings, enclosing garden space	Simple stone enclosures, mainly retaining walls
	Structure	Granite, 360–450 mm thick, well maintained	Granite, 300–400 mm thick, poorly maintained
	Function	Supporting agricultural life, planting	Moisture control and fishery storage
Doors and windows	Form	Single/double wooden doors, lattice glass windows	Wooden doors, white frames with fish or wave motifs
	Size/Color	Doors $\leq$ 1200 mm wide, windows $\leq$ 1000 mm, wood/gray tones	Doors $\leq$ 1200 mm, windows $\leq$ 1000 mm, blue/white marine paint
	Function	Agricultural ventilation and wind resistance	Ventilation and moisture control for fishery

Source by the author

Shared Features and Modifications: Both villages employ granite walls (300–450 mm thick) and small openings for doors and windows, enhancing their resilience against typhoons. In recent years, government-led renovations have introduced glass windows and organized drainage systems. Guihai Village, during its transitional phase, utilized elevated timber or bamboo structures, later shifting to granite walls with a focus on ventilation and resistance to saline–alkali conditions.

6.2 *Site Selection and Spatial Layout: Adaptation to Topography and Function*

The site selection and spatial layout of the two villages are shaped by both topographical conditions and production needs, resulting in distinct spatial patterns (Fig. 5).

Guishan Village: Located on the slopes of Jiawei Mountain, the dwellings are arranged along contour lines. The alleys are narrow (1–2 m wide), with a regular layout that facilitates agricultural cultivation and the transport of farming tools. Courtyards are typically located at the rear of the buildings (approximately 10–20 m^2), serving as spaces for vegetable planting or tool storage, with an emphasis on privacy.

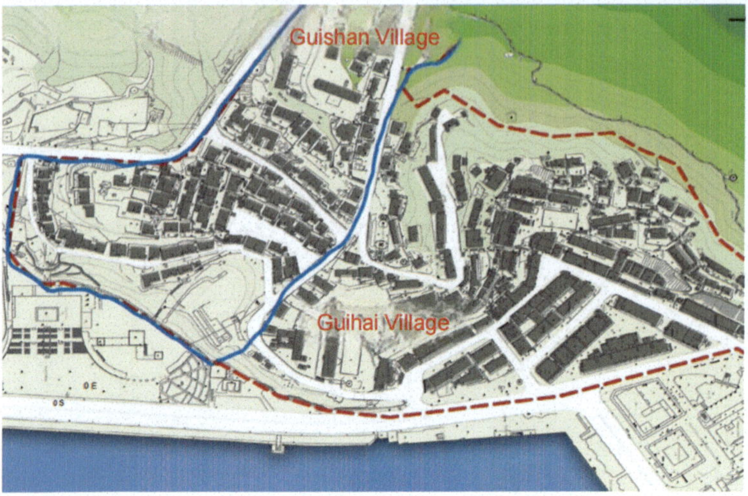

Fig. 5 Spatial layout of the two villages. Guishan village arranged along contour lines (left); Guihai village distributed along the coastline and around a central square (right). (*Image source* Drawn by the author)

Guihai Village: Situated near the harbor, houses are distributed along the coastline and around a central square. The alleys are relatively wider (2–4 m), with a more winding circulation path. Courtyards are positioned in front of the houses (approximately 15–35 m^2), partially used for drying fish catches or storing fishing equipment, integrating both residential and productive functions.

Commonalities: Due to topographical constraints, both villages exhibit relatively small courtyard areas ($\leq$40 m^2). The alleys not only serve as access routes but also promote community interaction. While Guihai Village retains a more complete traditional character, recent years have seen some unauthorized constructions, slightly disrupting the original spatial order.

The data presented in this section lays the groundwork for the subsequent analysis of spatial order in Sect. 7, illustrating how topography and production requirements shape spatial form and providing a basis for the conservation strategies discussed in Sect. 8.

6.3 *Resource Utilization: Optimizing Survival and Production*

Due to freshwater scarcity and the demands of production activities, the two villages have developed distinct strategies for resource utilization.

Guishan Village: Rainwater is collected through stone wells and diversion channels to meet the needs of agricultural irrigation and domestic use. Drought-resistant crops such as sweet potatoes are cultivated, making effective use of the mountainous terrain.

Guihai Village: In addition to using stone wells for water storage, rainwater is also collected from low-lying areas. Salt- and alkali-tolerant fruit trees and vegetables are planted, and a fish–farming–agriculture integrated model has emerged, emphasizing the harbor's economic role—particularly in fish trading.

Commonalities: Both villages rely on stone wells and crop selection to address the scarcity of freshwater, demonstrating adaptive strategies for island survival. While Guishan Village is more agriculture-oriented, Guihai Village highlights its fishing-oriented functions.

7 Spatial Patterns of the Two Villages: Spatial Articulation of Nature and Culture

Building upon the comparative analysis of architectural features, site layouts, and resource utilization of Guishan Village and Guihai Village in Sect. 6, this section delves into the spatial order of the two villages, exploring how the interaction between natural and cultural forces shapes their spatial expression. Section 6 provided empirical data on alley dimensions and courtyard distributions, which this section synthesizes through spatial syntax and morphological analysis to distill the "mountain–village–sea" spatial sequence. This reveals how the cohesiveness of Hakka culture and the openness of Tanka culture have molded Guishan Village's grid-like order and Guihai Village's dispersive texture. These insights lay the spatial theoretical groundwork for the protection and sustainable development strategies discussed in Sect. 8.

7.1 Spatial Order Under the Mountain–Water Environment

The mountain and water environment of Guishan Island underpins the spatial configurations of the two villages, generating a characteristic "mountain–village–sea" sequence (Fig. 6). Guishan Village, situated on the terraced slopes of Jiawei Mountain, organizes its dwellings along contour lines (with alleys 1–2 m wide as detailed in Sect. 6.2), resulting in a vertical grid order descending from the summit to the foothill. Alleys connect seamlessly with terraces, optimizing both agricultural productivity and community management. By contrast, Guihai Village is located near the harbor, where dwellings line the coastline and Guihai Square (alleys 2–4 m wide), forming a horizontally dispersive texture centered on the harbor, well-suited to fishing activities and maritime connectivity (Fig. 7). The spatial structures of both villages exhibit

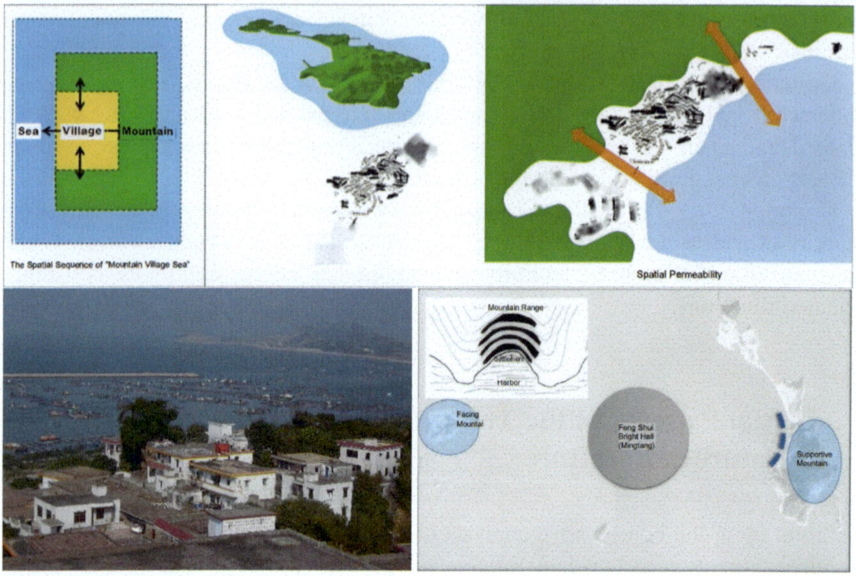

Fig. 6 Spatial sequence of "mountain–village–sea" on Guishan Island. (*Image source* Drawn by the author)

adaptive characteristics shaped by topographical differences. Buildings extend naturally along the terrain, generating vibrant figure-ground patterns that reflect the diversity of island settlements (Fig. 8). The spatial arrangements manifest four principal modes—"house–courtyard," "house–alley," "house–courtyard–alley," and "house–courtyard–yard–alley"—which reveal the constraints imposed by topography and the daily needs of villagers.

7.2 Village Texture of Tanka Fishing Culture

The village texture of Guihai Village centers around its harbor and Guihai Square, forming a distinct dispersive spatial order (Fig. 9). Based on layout data from Sect. 6.2 (alleys 2–4 m wide, courtyards 20–40 m^2) (Fig. 10), dwellings are distributed freely along the coastline and around the square, with winding alleys that optimize the functioning of the fishing port and provide convenient thoroughfares for trade and daily life. The flexible alley dimensions balance private spaces with the requirements for public activities, thereby enhancing community interactions and visitor experience. Guihai Square serves as the economic and social nucleus of the settlement, hosting fish trade and festive events, which further strengthen the openness and solidarity of the village. Through this adaptable spatial organization, the dispersive texture accommodates tidal fluctuations and integrates closely with the marine

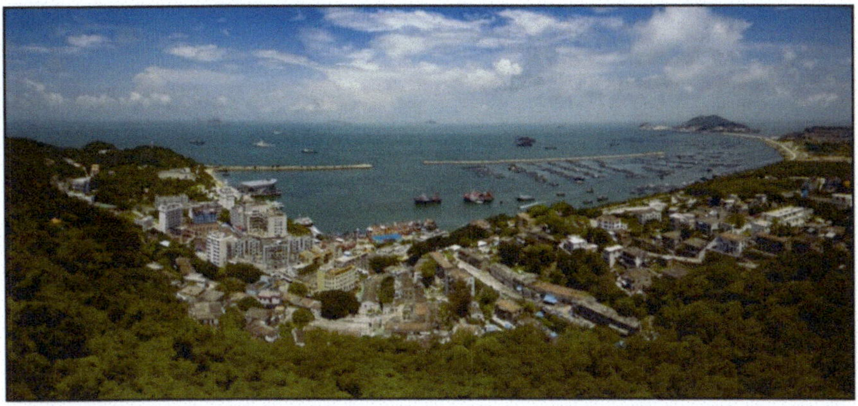

Fig. 7 Architectural spatial morphology of natural villages in Guishan Town. Comparative figure-ground textures showing Guishan Village's regular grid and Guihai Village's flexible dispersive pattern. (*Image source* photographed by the author)

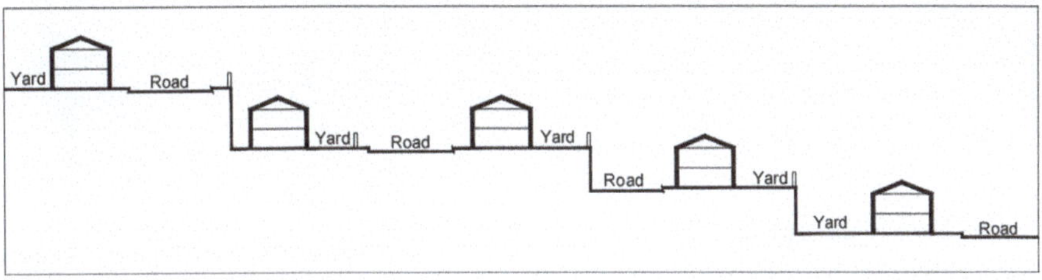

Fig. 8 Spatial sequence of "mountain–village–sea" on Guishan Island spatial arrangement patterns. Illustrative diagrams of the four spatial modes ("house–courtyard," etc.), highlighting topographic and functional constraints. (*Image source* Drawn by the author)

environment, embodying the practicality and profound maritime reliance of Tanka culture.

7.3 Spatial Order of Hakka Agrarian Culture

The spatial order of Guishan Village encapsulates the cohesiveness and meticulous hierarchy of Hakka culture (Fig. 11). Based on the layout data from Sect. 6.2 (alleys 1–2 m wide, courtyards 10–20 m^2), dwellings are arrayed along contour lines on the hillside (Fig. 12), with alleys forming a grid structure that intricately links terraced fields to residential spaces, thus enhancing both agricultural production and community management. Initially, the design focused around the ancestral "Shihetang" hall, reinforcing clan solidarity. Today, the village continues to follow this grid layout with clear and orderly boundaries, exemplifying collective traditions. The stepped spatial

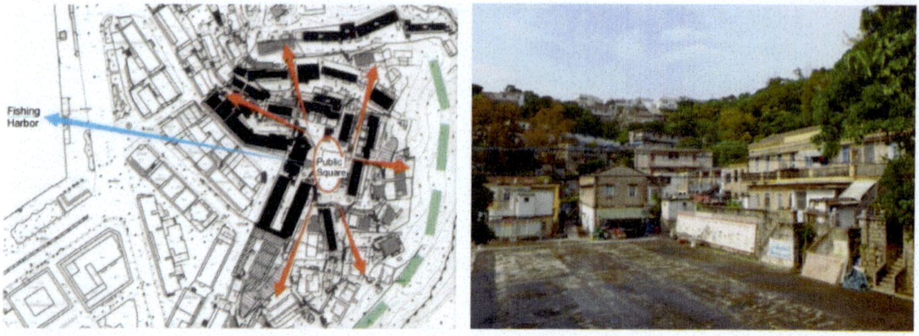

Fig. 9 Dispersive village texture of Guihai village. Dispersive spatial organization centered on the harbor and square, adapted to fishery functions. (*Image source* Drawn by the author)

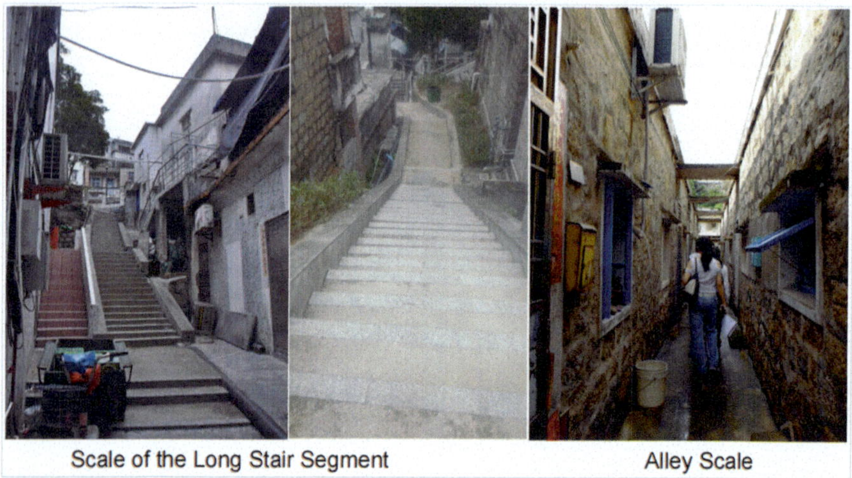

Fig. 10 Current conditions of historical alleys in the village. Guihai village alleys combine narrow and wide segments, balancing privacy and public use. (*Image source* Photographed by the author)

order adeptly conforms to mountainous terrain and agricultural practices, producing a defensive and cultivation-centered spatial arrangement that vividly reflects the inward orientation of Hakka culture and its rigorous social organization.

7.4 Summary

The spatial layouts of Guishan Village and Guihai Village, forged by the interplay of natural environments and cultural traditions, display pronounced commonalities and differences. Both settlements rely on a "mountain–village–sea" spatial framework, skillfully adapting to the island's topographical and climatic constraints. Guishan

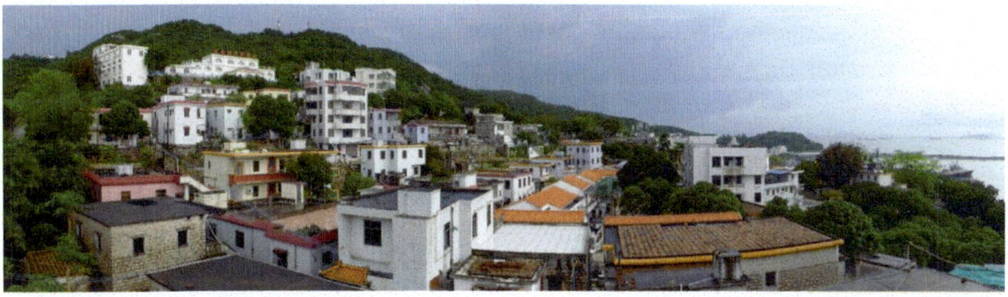

Fig. 11 Stepped spatial layout of Guishan village. (*Image source* Photographed by the author)

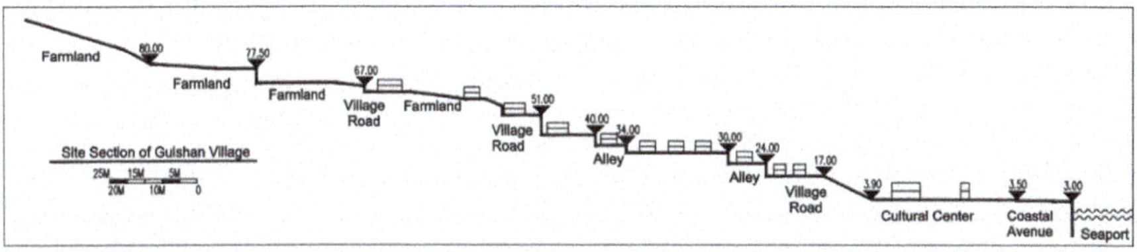

Fig. 12 Grid order along contour lines, adapted to hillside agriculture and clan management. (*Image source* Drawn by the author)

Village exhibits a regular grid and stepped arrangement that accentuates the cohesion of Hakka culture and its agrarian orientation, thereby reinforcing clan solidarity. In contrast, Guihai Village employs a more dispersive layout integrated with its harbor square, showcasing the openness and fishery-centered character of Tanka culture, while enhancing marine production and community interaction. This analysis highlights the use of spatial modes such as "house–courtyard" and "house–alley," which not only enrich street scales but also elevate the functionality and residential experience of the villages. These insights provide theoretical support for the protection and sustainable development strategies presented in Sect. 8, underscoring the need to balance traditional spatial order with contemporary growth in island settlements.

8 Conclusion and Outlook: Conservation of Traditional Spatial Order and Sustainable Renewal

This section synthesizes the findings from Sects. 1 through 7, systematically reviewing the morphological evolution and spatial order characteristics of Guishan Village and Guihai Village within the island context. It reveals the differentiated spatial expressions arising from the Hakka agrarian culture and Tanka maritime culture. Against the backdrop of increasingly weakened traditional landscapes and declining cultural functions, this study further proposes targeted strategies for conservation and renewal, aiming to provide theoretical support and practical references for the preservation and sustainable development of island settlements in Zhuhai and the wider Greater Bay Area.

9 Summary of Spatial Expressions Under Nature–Culture Interactions

In this chapter, Sects. 1 and 2 highlighted the profound influence of Guishan Island's mountainous terrain and marine climate on settlement sitting and architectural models, underscoring the fundamental constraints imposed by the natural environment on spatial form. Sections 3 and 5 examined the cultural dimensions, analyzing the inward-looking mountainous logic of Hakka agrarian culture and the outward-oriented marine logic of Tanka fishing culture. Guishan Village, inhabited by Hakka people, exhibits a grid layout along contour lines that emphasizes clan structure and agricultural efficiency, while Guihai Village, shaped by Tanka culture, develops a dispersive pattern around the harbor, enhancing fishery accessibility and community openness.

Section 6 further compared differences between the two villages in building materials, roof types, architectural details, and spatial layouts, revealing their proactive adaptations to the environment and spatial projections of cultural identity. Building on this, Sect. 7 distilled the "mountain–village–sea" spatial sequence, clarifying the defensive and orderly layout logic of Hakka settlements versus the fluid and adaptive spatial strategies of Tanka settlements. Although these two settlement styles each possess distinctive features, they face common challenges, such as the continuous spread of unauthorized constructions, fragmentation of cultural spaces, and gradual loss of original appearances. These pressing issues highlight the urgent need for organic conservation and systematic renewal measures.

9.1 Strategies for Conservation and Sustainable Development

Drawing on the empirical analyses from Sects. 6 and 7, this study proposes the following multidimensional pathways to address the weakening of traditional spatial order and disruptions to cultural functions:

- Restoring Identifiability of Traditional Landscapes: New or renovated projects should be encouraged to incorporate traditional architectural elements such as granite walls and double-pitched or flat roofs, while avoiding extensive use of modern materials like glass curtain walls. This approach can effectively mitigate the visual impact of contemporary construction on historical landscapes and preserve the cultural distinctiveness of settlements.

- Maintaining Continuity of Spatial Patterns and Morphology: This entails protecting the stepped grid layout unique to Guishan Village and the radial structure centered on the harbor in Guihai Village. Achieving this requires strict regulation of illegal constructions and refined spatial guidance to ensure continuity of historical patterns and prevent damage to traditional appearances.
- Revitalizing Cultural Functions through Local Practices: Supporting traditional festivals, clan ancestral worship, and daily fishing and farming activities can help re-embed these cultural practices into the spatial environment, thereby sustaining the interactive dynamics among living practices, culture, and space. This ensures the continued vitality of traditional lifestyles within contemporary contexts.
- Leveraging Harbor Resources and Tourism Potential in Guihai Village: The goal is to achieve a dual integration of culture and tourism—enhancing economic vitality while safeguarding the authenticity and adaptability of cultural spaces. This requires adjustments to infrastructure and the promotion of relevant developments to achieve a harmonious coexistence between culture and economy, maintaining original cultural appearances while strengthening modern adaptability.

9.2 Research Significance and Future Prospects

Through field surveys and spatial interpretation, this study elucidated the spatial structural characteristics of island settlements formed under the constraints of natural topography and guidance of cultural systems. It proposed a conservation and development approach centered on "cultural embeddedness–pattern continuity–functional revitalization." This research not only provides theoretical underpinnings for spatial planning and cultural revitalization of traditional villages in the Guangdong–Hong Kong–Macau Greater Bay Area but also offers references for the conservation of cultural landscapes in other peripheral regions.

Future research can focus on several key areas: first, comprehensively assessing the potential impacts of tourism development on settlement morphology and cultural continuity, particularly in analyzing the extent of such disturbances; second, employing digital modeling and participatory planning tools to progressively realize the renewal of traditional island settlements under the premise of preserving cultural characteristics, thereby ensuring their long-term sustainable development.

Funding This research received no external funding.

Conflicts of Interest The authors declare that the research was conducted in the absence of any commercial or financial relationships that could be construed as a potential conflict of interest.

Institutional Review Board Statement Not applicable for studies not involving humans or animals.

Informed Consent Statement Not applicable for studies not involving humans.

References

1. Xu, Y., & Dou, W. (2024). Research on the development of culture and art in Zhuhai Island under the strategy of rural revitalization. In *2024 International Conference on Humanities, Arts, and Cultural Industry Development (HACID 2024)* (pp. 66–74). Atlantis Press. https://doi.org/10.2991/978-2-38476-281-1_10
2. Li, Y. (2022). Research report of Island tourism and well-being cultural ecological construction strategy of Zhuhai City. In *Innovation Economics and Management Research (IEMR), Volume 2, ISSN: 2949–1304. The 3rd International Conference on Management, Economy and Law (ICMEL 2022).* http://journals.isccac.org/d/file/lunwen/2023-02-09/40a62c0a6c512c7815d98a27313f9821.pdf
3. Kan, X., & Li, H. (2018). Study on tourism cooperation between Zhuhai Island and Macau in the times of Greater Bay Area of Guangdong, Hong Kong and Macau. In *3rd international conference on contemporary education, social sciences and humanities (ICCESSH 2018)* (pp. 1061–1064). Atlantis Press. https://doi.org/10.2991/iccessh-18.2018.239
4. Yao, L. H. (2022). Research on the conservation and utilization of Subtropical Island stone dwellings in China. *Journal of the Science of Design, 6*(2), 41–50. https://doi.org/10.11247/jsd.6.2_2_41
5. Tang, Y., Meng, Q., & Ren, P. (2023). Spatial distribution and concentrations of salt fogs in a coastal urban environment: A case study in Zhuhai city. *Building and Environment, 234*, Article 110156. https://doi.org/10.1016/j.buildenv.2023.110156
6. Zhang, N., & Zhang, S. Y. (2014). Zhuhai traditional residence in the site selection and planning for natural ventilation. *Applied Mechanics and Materials, 584*, 859–862. https://doi.org/10.4028/www.scientific.net/AMM.584-586.859
7. Song, J., Chen, Y., Zheng, L., & Tang, Q. (2025). The evolution of traditional ancient towns in South China: Tangjiawan Ancient Town in Zhuhai city. *Journal of Asian Architecture and Building Engineering, 24*(3), 2114–2132. https://doi.org/10.1080/13467581.2024.2349754
8. Huang, W., & Liu, H. (2022). Floristic analysis of Vascular plants in Sanzhou Island, Zhuhai, China. In *IOP conference series: Earth and environmental science* (Vol. 1011, No. 1, p. 012049). IOP Publishing. https://doi.org/10.1088/1755-1315/1011/1/012049
9. Campbell, G. (1912). Origin and Migration of the Hakkas. *Chinese Recorder, 43*(8), 473–480.
10. Leong, S. T., Wright, T., & Skinner, G. W. (1997). *Migration and ethnicity in Chinese history: Hakkas, Pengmin, and their neighbors.* Stanford University Press.
11. Chen, Y., Peng, H., Zheng, H., Luo, Y., & Guan, R. (2025). Exploring the spatial distribution characteristics and formation mechanisms of Hakka folk settlements: A case study of Hakka traditional architecture in southeastern China. *Humanities and Social Sciences Communications, 12*(1), 1–19. https://doi.org/10.1057/s41599-025-04561-0
12. He, G., Zhang, Y., Wei, L. H., Wang, M., Yang, X., Guo, J., & Zhang, X. (2022). The genomic formation of Tanka people, an isolated "Gypsies in water" in the coastal region of Southeast China. *American Journal of Biological Anthropology, 178*(1), 154–170. https://doi.org/10.1002/ajpa.24495
13. Hong, S. Y., & Han, D. S. (2023). A study on the water-faring community and architectural forms of the "Tanka people" in Macau from the Ming and Qing dynasties to the modern period. *Journal of Architectural History, 32*(3), 7–20. https://doi.org/10.7738/JAH.2023.32.3.007
14. Lin, T., & Su, P. (2024). The myth of the Others: Western representations of the Dan people and boat clusters in the island city of Guangzhou, China (1842–1900). *International Journal of Asian Studies, 21*(2), 238–256. https://doi.org/10.1017/S1479591423000566
15. Chen, Y. (2022). Research on the protection of Macau water-faring community (Dan Jia) from the perspective of urban heritage. In P. Mendonça, & N. D. Cortiços (Eds.), *Proceedings of the 7th international conference on architecture, materials and construction. ICAMC 2021. Lecture notes in civil engineering* (Vol. 226). Springer. https://doi.org/10.1007/978-3-030-94514-5_49
16. Fung, H. H. (1998). Thousand-year oppression and thousand-year resistance. *Chinese Sociology & Anthropology, 30*(3), 75–99. https://doi.org/10.2753/CSA0009-4625300375

17. Daming, Z., & Yingqiang, Z. (1995). Rural urbanization in Guangdong's Pearl River Delta. *Chinese Sociology & Anthropology, 28*(2), 47–102. https://doi.org/10.2753/CSA0009-4625280247
18. Anderson, E. N. (1970). The boat people of South China. *Anthropos, 65*(1/2), 248–256. http://www.jstor.org/stable/40457623

Cross-Border Settlements: Island Morphological Evolution and Urban Development Characteristics in Hengqin

Tong Ling

Abstract Hengqin, which is in rapid development, is an important area for strengthening of links between Guangdong, Hong Kong, and Macau in the Greater Bay Area. Since the earliest record of Hengqin's place name in the Jiajing-Xiangshan County Record over 400 years ago, Hengqin has developed from a group of frontier islands into a strategic area. Its topography and urban morphology have undergone tremendous changes, yet there is a gap in the field of research on urban morphology in Hengqin. In this paper, M. R. G. Conzen's Urban Morphology methodology, Town-Plan Analysis, is applied to the study of Hengqin. First, Hengqin's urban development is divided into four morphological periods based on the history and morphological characteristics. Then, the relationship between the urban morphological characteristics of different periods and the contributions (e.g., social, political, cultural, historical, etc.) was analyzed in terms of town plan, building and land utilization, and building forms. The types of plan units of each period also be shown. Finally, the connections with Macau are analyzed during different morphological periods.

Keywords Urban morphology · Greater Bay Area · Types of plan units

1 Introduction

1.1 Historical Progression and Population of Hengqin

Hengqin is located on the west side of the Pearl River Estuary, west of Cotai Island and south of Wan Zai (灣仔) in Zhuhai City. Its total area is about 106 km^2 now. Originally, the main land of Hengqin were two separated islands, named the Greater Hengqin Island and the Lesser Hengqin Island, forming the famous Shizimen (十字門, literally "Cross Gate") Waterways with the mountains around Cotai Island in Macau and Wan Zai in Zhuhai. As the original shape of Shizimen Waterways is like

T. Ling (✉)
Guangdong Shunkong Zihua Technology Co., Ltd, Foshan, Guangdong, China
e-mail: corrinaling@163.com

Y. Chen et al. (eds.), *City-Sea Symbiosis*, Springer Geography,
https://doi.org/10.1007/978-3-032-13725-8_5

two crosses, they are usually named Inner Shizimen and Outer Shizimen (Fig. 1). Since ancient times, Hengqin has belonged to Xiangshan (香山) County, and after the founding of the People's Republic of China it belonged to Zhongshan County, and later to Zhuhai. With the official implementation of the *Master Plan of Hengqin (2009)*, the Hengqin New Area was established. On September 17, 2021, it was changed to Guangdong-Macau In-Depth Cooperation Zone in Hengqin (Fig. 2).

The population of Hengqin was less than 1,000 in the 1960s, and only 956 were recorded in the early 1970s at the beginning of the land reclamation [1]. The population on the island increased slowly with the development of the reclamation, which was 2,406 in 1989 (Compilation Committee of *Gazetteer of Geographical Names of Zhuhai City, Guangdong Province*, 1989) [2]. In 1998 it rose to 3060 (The Editorial Committee of *The Atlas of Zhuhai City*, 1998) [3]. Following the establishment of the Hengqin New Area in 2009, the population surged. In 2010, the total population was 6,914, and it had reached 33,450 by 2015. According to the latest data from the seventh national census, the permanent population of Hengqin in 2022 was 43,618 (Fig. 3).

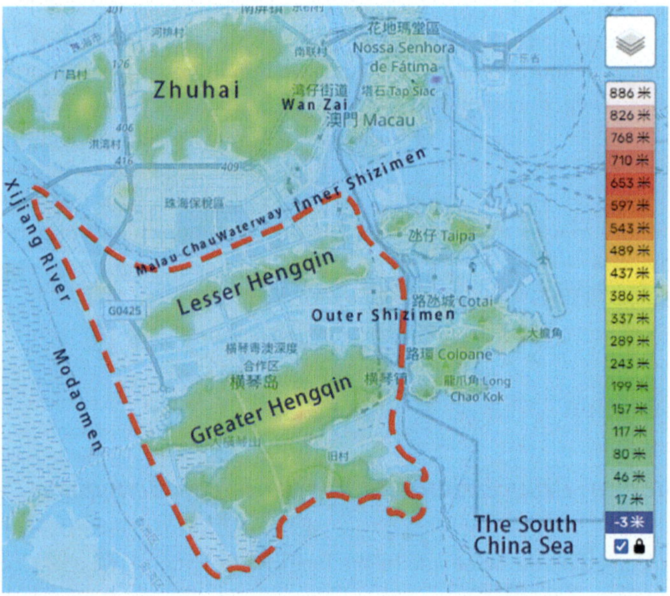

Fig. 1 Geographical analysis map of Hengqin (*Source* Modified from OpenStreetMap by the author)

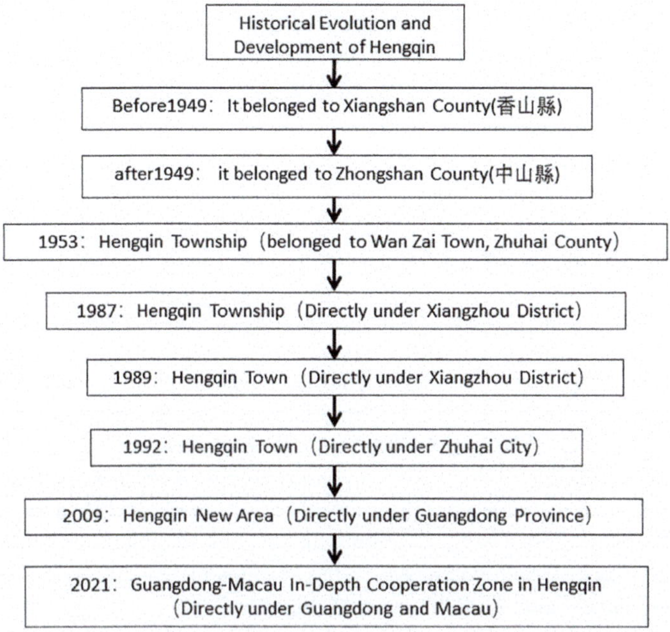

Fig. 2 Historical evolution and development of Hengqin (*Source* Created by the author)

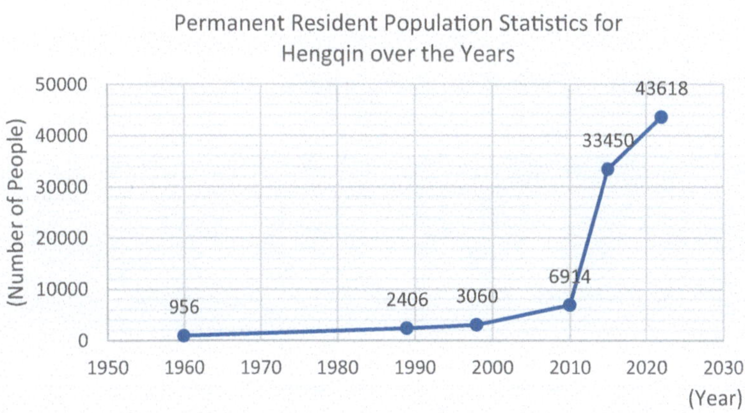

Fig. 3 Permanent resident population statistics for Hengqin over the years (*Source* Created by the author. The original data comes from Stathetics Bureau of Guangdong-Macao In-Deep Cooperation Zone in Hengqin)

1.2 Periodization of Urban Morphology in Hengqin

Before 1949, Hengqin was sparsely populated with low development intensity, and the island form evolved mainly naturally. During the difficult period in the early times of China's foundation, the reclamation boom led to the first large-scale development of the island. Greater Hengqin and Lesser Hengqin were connected after the reclamation of the Zhongxin Channel (the waterway between the greater and lesser Hengqin Islands). In the early 1990s, Hengqin began the early stages of urbanization under the influence of reform and opening-up, and the area of Hengqin was expanded again by land reclamation along the edges close to Zhuhai and Macau. In twenty-first century, the shape of the island became relatively stable, and in 2009 the *State Council enacted the Master Plan of Hengqin (2009)*, marking the official development of Hengqin by the Central Government. Subsequently, influenced by the official policies, Hengqin started the phase of integrated development and construction. According to the history and morphological characteristic, Hengqin's urban development is divided into four morphological periods: (1) Inner and Outer Shizimen Period (before 1949); (2) Zhongxin Channel Reclamation Period (1949–1989); (3) Period of Development Along the Margins (1990–2009); and (4) Period of Integrated Development (2009–present) (Fig. 4).

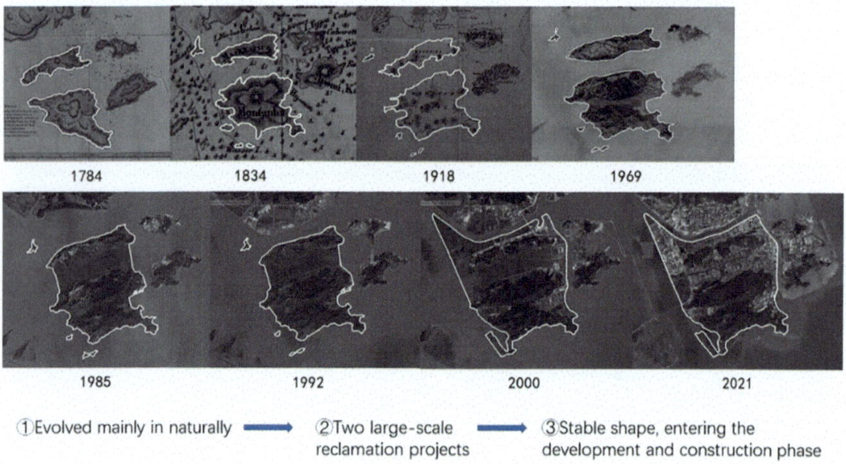

Fig. 4 Evolution of Hengqin Island's topography (*Source* Created by the author based on old maps and satellite images)

2 Inner and Outer Shizimen Period (Before 1949)

2.1 An Important Route for Maritime Trade

2.1.1 Location and Historical Role of Hengqin

To the east of Hengqin Island is the famous Shizimen Waterway located on the west side of the Pearl River Estuary. Historically, the Hengqin Islands, Taipa, Coloane, Mangzhou (Mong-chau) Islets, Malau Zhou (馬騮洲, also Malau-chau, literally means Monkey Island in Cantonese) constituted an archipelago network interconnected by the Inner and Outer Shizimen Waterways as its pivotal hub. They were often called the "Shizimen Islands" (十字門諸島) and were also once labeled as the "Shizimen Islets" (十字門洲) (Fig. 5). The Shizimen Waterway area was a strategic node for maritime trade and defense.

Hengqin has been on an important corridor of the ancient Maritime Silk Road since ancient times, and it has been recorded that as early as the Tang Dynasty, Hengqin was on the route from Guangzhou to Dashi (Arab/Persian) [4]. The Guangdong Provincial Gazetteer (Jiajing edition) of the Ming Dynasty records that merchant ships from Siam (Thailand), Champa (Vietnam), and other foreign states anchored at coastal havens such as Langbai Ao and Shizimen, noting that "Shizimen was the most convenient anchorage" [5].

Shizimen Waterways held critical maritime defensive significance, given their strategic position at the oceanic gateway facing the open sea. There was Shizimen

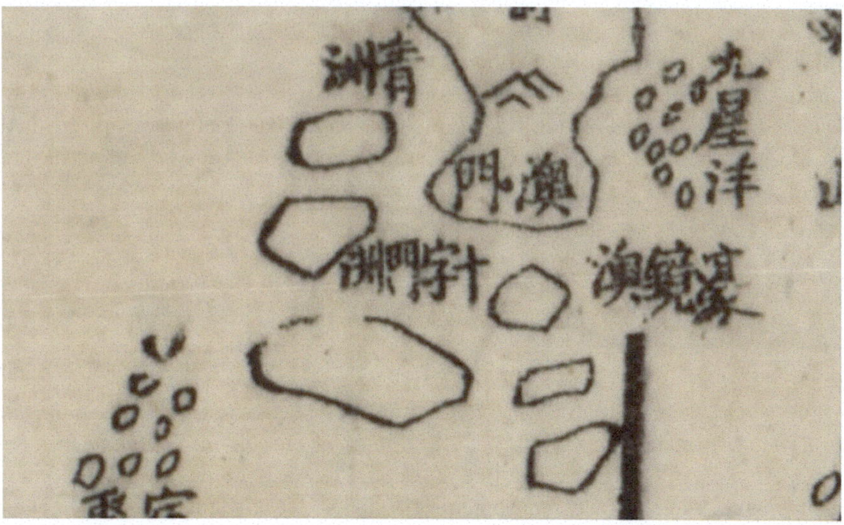

Fig. 5 Partial map of Guangdong Province in 1864, during the reign of Emperor Tongzhi of the Qing Dynasty (*Source* From Academia Sinica Digital Culture Center website)

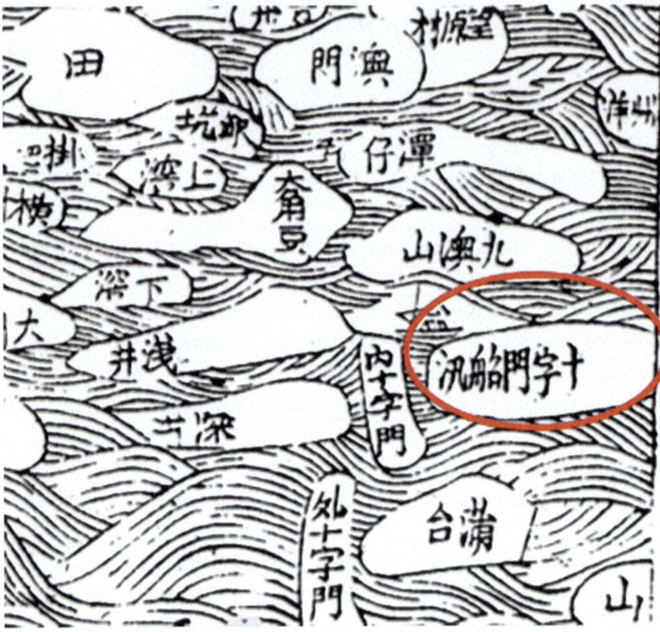

Fig. 6 Map of Shizimen Naval Patrol Station (*Source* Yin Guangren, Zhang Rulin, 1751, Ref. [7])

Naval Patrol Station (十字門船汛) for defense and inspection during the period of Qianlong Dynasty (Fig. 6). After Macau was completely occupied by the Portuguese, the central government strengthened its control over the frontier islands around Hengqin. Since the 13th Year of the Guangxu Reign (1887), troops were garrisoned on the islands [7]. Prior to 1949, the Gongbei Customs established branch offices on Mangzhou and Malau Zhou near Lesser Hengqin (Fig. 7), with a fort additionally constructed on Mangzhou (Fig. 8). In 2011, an ancient cannon was discovered near Ma Liu Zhou within the Shizimen Central Business District, providing further evidence of the strategic defensive importance of the Shizimen waterways.

2.1.2 Historical Toponymy of Hengqin: Sino-Foreign Designations and Key Place Names

During this period, Greater Hengqin and Lesser Hengqin were not connected, and they had a lot of different historical names, respectively (Appendix Table 2). Greater Hengqin Island, historically known by various toponyms including Jing Ao (井澳), Xiannu Ao (仙女澳), Shenjing (深井 or Shenjing Mountain 深井山), Qianjing (淺井), Greater and Lesster Hengqin (大小横琴), and Hengqin Village (横琴村). In ancient Chinese, "澳(Ao)" often referred to a bay. Lesser Hengqin Island was named Tudiwan (土地灣), Duowei (舵尾), Toi-ko-ke-tou (大角豆), Qingjiao (青角), Shizimen Mountain (十字門山), Macarira (馬格裡勒), Chi-sion-long (長沙

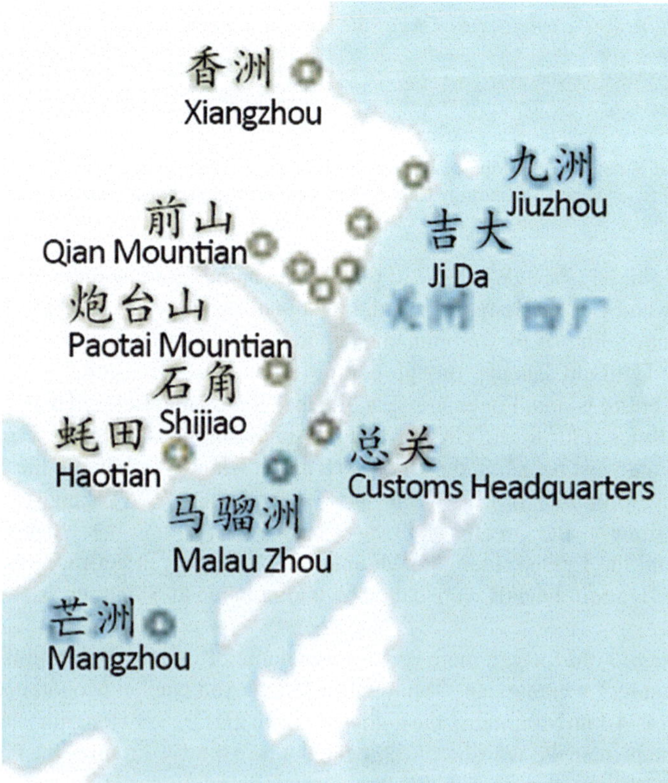

Fig. 7 Partial diagram showing the locations of the Gongbei Customs Headquarters and its branches before 1949 (*Source* Compilation committee of *The History of Zhuhai Customs* (1991–2010) (Chief Editor), 2016, Ref. [8])

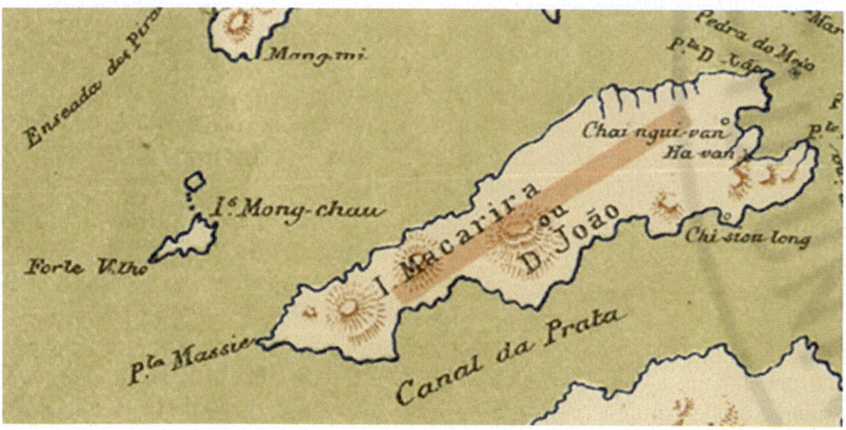

Fig. 8 Partial map of Macau Province, 1909 (*Source* From Macau University of Science and Technology Library)

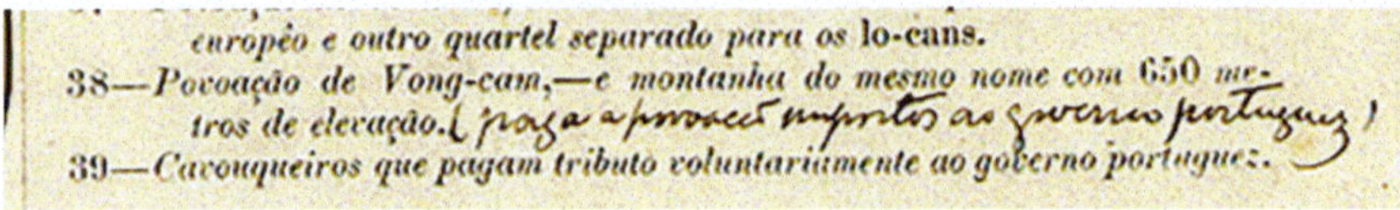
europêo e outro quartel separado para os lo-cans.
38—Povoação de Vong-cam,—e montanha do mesmo nome com 650 metros de elevação.
39—Cavouqueiros que pagam tributo voluntariamente ao governo portuguez.

Fig. 9 Partial annotations of Map of Portuguese Colonies, 1901 (*Source* Collected from Macau University of Science and Technology Library, Provenance: Geographical Society of Lisbon)

欄), and Moluobu (嚤囉埠). Some of these place names are just one place name on the Greater and Lesser Hengqin Island, but sometimes they are used to refer to the whole island. Some of these toponyms denoted particular sites across the Greater and Lesser Hengqin Islands, though certain terms were occasionally extended to refer to the entire island. For example, while the Greater and Lesser Hengqin Mountains sometimes refer to the Greater and Lesser Hengqin Islands, sometimes refer to the east and west peaks on the Greater Hengqin Island. And due to the similarity in Cantonese pronunciation between "黃 (Wong4)" and "橫 (Waang4)," it is also wrongly written as the Greater and Lesser Huangqin (大小黃琴), Huangqinzhou (黃琴洲), and the Lesser Huangqin Mountain (小黃琴山). In addition, the Greater and Lesser Hengqin Islands were called by a joint name as Shizimen Islets (十字門洲) [9–11].

According to old foreign maps, the foreign name of Greater Hengqin Island was "Ilha Montanha," which means "Mountain Island." And Naobei Mountain (腦背山, also known as Adam Mountain), the highest peak on the Greater Hengqin Island, is the third highest peak in the Wanshan Archipelago. From the 1901 map of the Portuguese colony (Fig. 9), it is recorded in the 38th note: "The mountain of the same name on the Greater Hengqin Island, is about 650 m high." Although there is a discrepancy between this and the present measurement, it shows that the mountain on the Greater Hengqin Island was so majestic that impressed the Portuguese. Meanwhile, there are Portuguese names transliterated in Cantonese, such as Ilha Tai-vong-cam (大橫琴山) or vong-can (橫琴).

The name "Macarira" or "D. João" is commonly found on foreign maps, while "Macarira" has many similar forms in foreign maps, such as "Maquareira," "Mackkarera," "Mackkarira," and "Macarira" (Appendix Table 3). Since the use of "Macareira" in Portuguese literature in 1685, "Macareira" has become the relatively common foreign name of Lesser Hengqin Island [10], and "馬格裡勒" is its Chinese transliteration. It is possible that "Macareira" originates from the Portuguese word for cassava "Macaxeira," which is from the Portuguese translation of the word "薯" in the place name Shuliang Gang (薯莨崗) on the Lesser Hengqin Island [10]. D. João I was one of the great monarchs who ushered in the era of Portuguese maritime geographic exploration. According to historical records, the Portuguese army under King D. João I (1357–1433) conquered Ceuta in 1415, which is the beginning of the Portuguese great geographic discovery [12]. It is assumed that this was the reason for naming the then newly discovered island. In addition to this, there are also Portuguese names transliterated in Cantonese, such as Sio-Vong-Cam (小橫琴) and Toi-ko-ke-tou (大角豆).

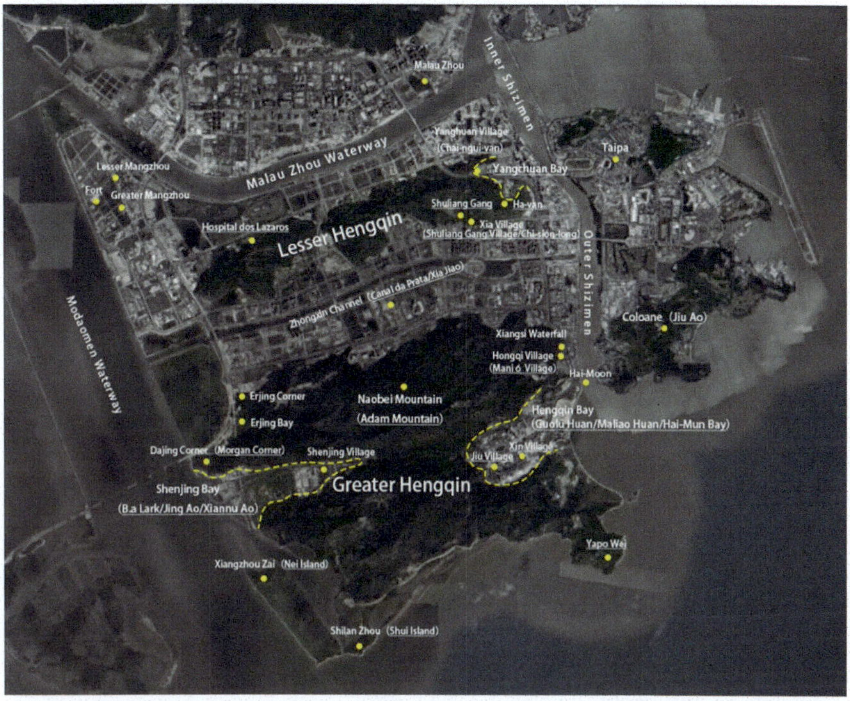

Fig. 10 The locations of important place names on the island on a modern map (*Source* Self-drawn, old place names are underlined)

There were three important bays in the earliest stage of the form development in Hengqin, namely, Shenjing Bay (深井灣), Hengqin Bay (橫琴灣), and Yangchuan Bay (洋船灣) (Fig. 10).

Shenjing Bay, named for being the deepest bay on the island, was also known as Jing Ao (井澳) and Xiannu Ao (仙女澳) [11]. The name "Shenjing" is found earlier in the *Complete Maritime Atlas: Guangdong Coastal Map* (Fig. 11), and later there was a village named Shenjing (深井村) in this area. Jing'ao is the oldest place name on the Great Hengqin, which was recorded in the Jiajing-Xiangshan County Record and the Wanli-Guangdong General Records. Shenjing Bay is labeled as "B.a Lark" in foreign maps (Fig. 12), and the full name is "Bay Lark."

Hengqin Bay is the largest bay on the Great Hengqin Island, where the Jiu Village (舊村), the first village in Hengqin, was later formed (Compilation Committee of *Gazetteer of Geographical Names of Zhuhai City, Guangdong Province*, 1989) [2]. At the beginning of the twentieth century, it was known as Manió Huan (馬料環), Guolu Huan (過路環), and B.a de Hai-Moon (海門灣). It is assumed that Manió Huan (馬料環) was named after a village called "Manió (馬尿)" to the north of it, and "尿" and "料" in Cantonese are the same pronunciation. Hengqin Bay faces the busy Shizimen Waterway, and the closest distance between Hengqin Bay and Coloane is only 200 m, which is known as Hai-Moon (海門, means the door to open sea),

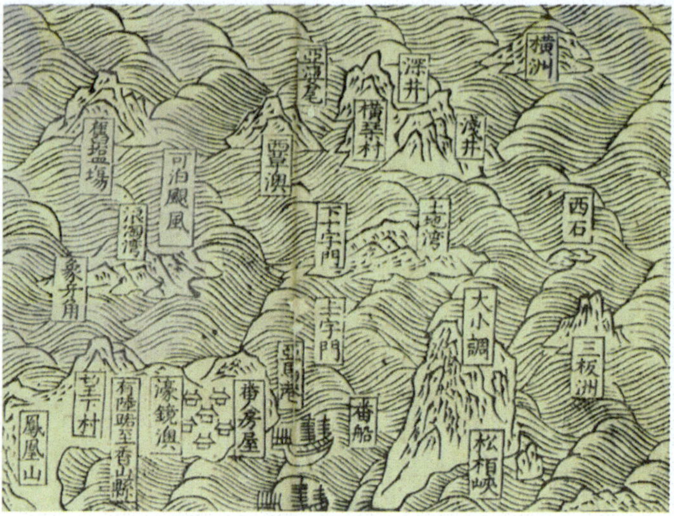

Fig. 11 Complete maritime atlas: Guangdong coastal map (*Source* Tang Kaijian and Zhou Xiaolei, 2020, Ref. [11])

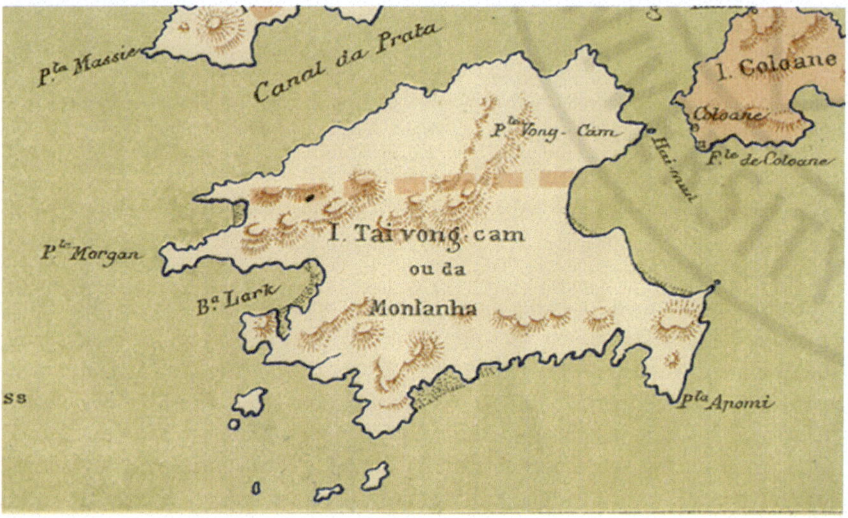

Fig. 12 Partial map of Macau Province, 1909 (*Source* From Macau University of Science and Technology Library)

hence the name Hai-Moon Bay (海門灣) (Fig. 12). There were numerous passing ships here, and the words "環 (Huan)" and "灣 (Bay)" are pronounced similarly in Cantonese, so it was also known as the Guolu Huan (過路環) [13]. The place name

"Guolu Huan" has been found to have been relocated to other places, but the reasons for this remain unclear.

Yangchuan Bay is a deep-water harbor located on the northeast side of Lesser Hengqin, closest to Macau. Since ancient times, it has been a place where passing ships docked to replenish supplies. It was later named Yangchuan Bay because it became the main docking point for foreign ships, and eventually Yanghuan Village (洋環村) was formed.

In the vicinity of Greater and Lesser Hengqin, the key surrounding islands include Greater and Lesser Mangzhou (大小芒洲), Nei Island (內島) (present-day Xiangzhou Zai (香洲仔)), Water Island (水島) (present-day Shilan Zhou (石欄洲)), and Malau Zhou (馬騮洲) (Fig. 13). Except for Malau Zhou, which was incorporated into the Shanwanzi reclamation area opposite it, the other islands were merged with Hengqin during subsequent reclamation projects. Subsequently, except for Malau Zhou which merged with the reclamation area of Duimian Shan (對面山) (present-day Wan Zai (灣仔)), all other islands were integrated into Greater and Lesser Hengqin through later land reclamation projects. The waterway between Greater and Lesser Hengqin was historically known as Xia Jiao (下滘) (Fig. 6) during the Ming and Qing Dynasties, with the Portuguese designation "Canal da Prata" (meaning "Silver Canal") (Fig. 12). Over time, sediment deposition from the river estuary gradually formed a tidal flat in this area, which later became known as Zhongxin Gou (中心溝, meaning Central Channel).

2.2 *The Formation of Villages*

The earliest evidence of human habitation on Hengqin Island dates back over 6,000 years to the Neolithic period, as documented by the Chisha Bay (赤沙灣) archaeological site on the Greater Island's southwestern coast [14]. During the maritime trade period of the Tang and Song Dynasties, Greater and Lesser Hengqin gradually evolved into vital supply points for neighboring islanders and passing merchant vessels. The Complete Maritime Atlas: Guangdong Coastal Map records the existence of Hengqin Village, confirming village settlements that existed on Hengqin Island during the Wanli era (1573–1620) of the Ming Dynasty. However, due to its peripheral location beyond effective central government control, the area historically functioned as an ungoverned zone attracting various marginal groups, including fugitives, pirates, and consequently maintained generally inhospitable conditions for stable habitation [15]. During the Yuan Dynasty, Greater Hengqin was occupied by the pirate Wang Yi (王一) until Ming forces suppressed his rebellion. Soon afterward, the imperial court "declared the area vacant" (詔虛其地), prohibiting civilian settlement and enforcing coastal population relocations inland [16]. The Great Evacuation Edict of 1661 (18th year of the Shunzhi reign (順治年間)) precipitated severe population loss across southeastern coastal regions, including Greater and Lesser Hengqin, reducing cultivated lands to wilderness. It was not until the full lifting of coastal restrictions in 1683 (22nd year of the Kangxi reign (康熙年間)), following

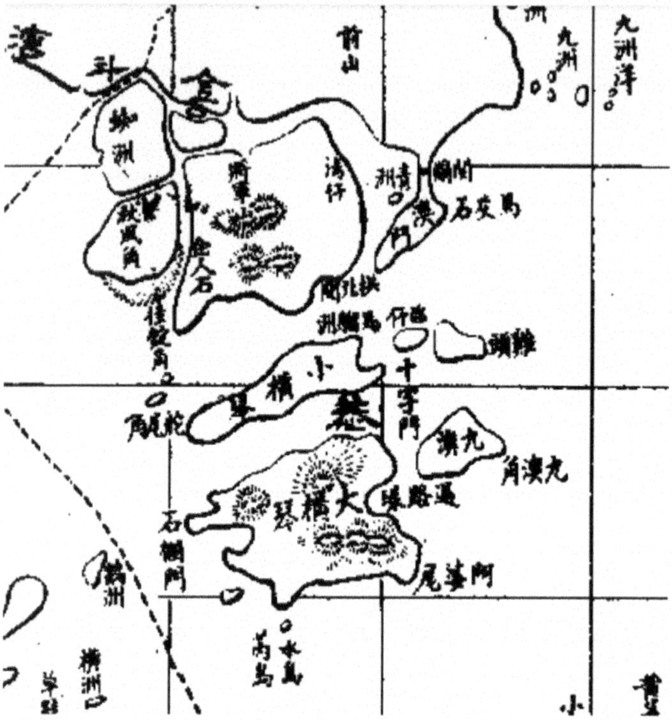

Fig. 13 Partial map of the entire territory of Xiangshan County (*Source* Li Shijin, 1929, Ref. [7])

the Qing conquest of Taiwan, that gradual resettlement occurred and new villages were established.

After the lifting of coastal restrictions, Cantonese-speaking islanders from peripheral regions gradually resettled in Hengqin, exemplified by the establishment of Jiu Village (舊村, means old village) during the Kangxi reign—the earliest recorded settlement on the island. The "Tenant Recruitment for Land Reclamation (招佃承墾)" policy implemented in 1732 (10th year of Yongzheng reign (雍正年間)) initiated large-scale Hakka migration. During the Qianlong reign (乾隆年間), the construction of temples in Xiangshan required a large amount of stone, while the Shizimen area was rich in stone mines, attracting a large number of Hakka stonemasons to this area and thus forming a peak period of Hakka migration to Shizimen [17]. On Hengqin Island, the Hakka villages of Hongqi Village (紅旗村) and Shishan Village (石山村) were formed during the reign of Yongzheng. It is estimated that Cushahuan Village (粗沙環村), which was established by Hakka immigrants from Wuhua (五華), was also formed during this period. The Hakka village of Yanghuan Village (洋環村), which relied on stone quarrying for livelihood, was formed during the Qianlong reign. Subsequently, more villagers continued to migrate and establish villages, eventually evolving into the 12 original natural villages of Hengqin, with

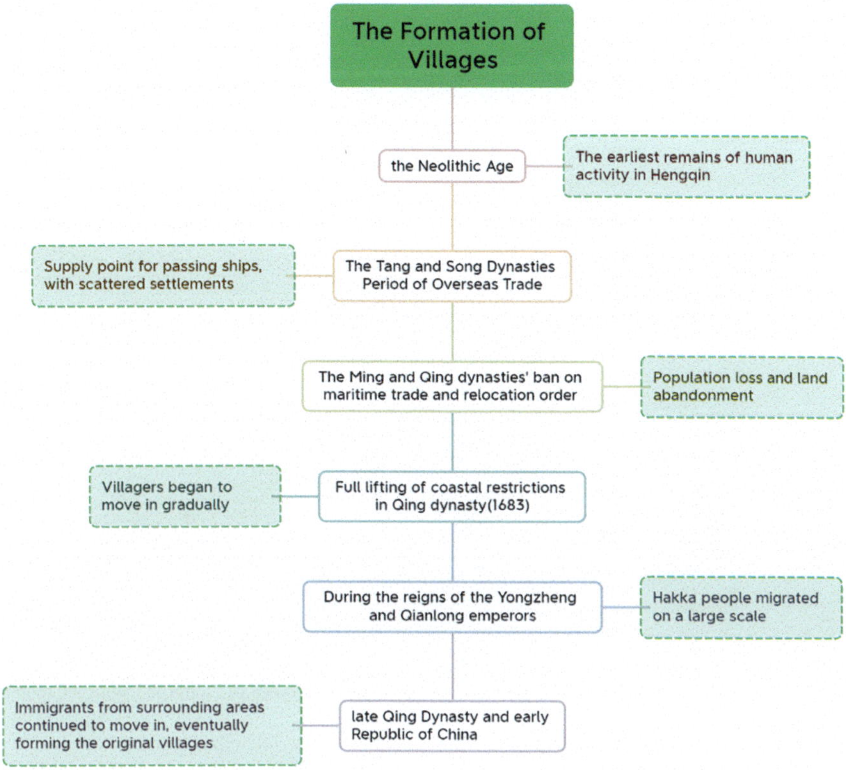

Fig. 14 Analysis of the formation of villages (*Source* Created by the author)

four on Lesser Hengqin and eight on Greater Hengqin (Appendix Table 4, Figs. 14 and 15).

The original villages in Hengqin were relatively small, with each village having a population of only about 100–200 people in the 1990s. These villages were systematically sited following the principle of "backing against mountains while facing water," with structures terraced along slopes in a stepped elevation from front to rear. Hakka villages were predominantly distributed along shallow inlets or semi-mountainous ridges on the eastern coast, while Cantonese settlements primarily occupied inner bays across the island. Together, they fostered an agro-maritime cultural blend, developing fisheries while reclaiming tidal flats for paddy fields and banana plantations. This created a stratified bay morphology of mountains-villages-sea (Fig. 16). Villages situated in areas with relatively extensive flat terrain predominantly exhibit compact block-shaped layouts, characteristic of Cantonese comb-pattern settlements, while those adapted to mountainous topography typically assume linear configurations.

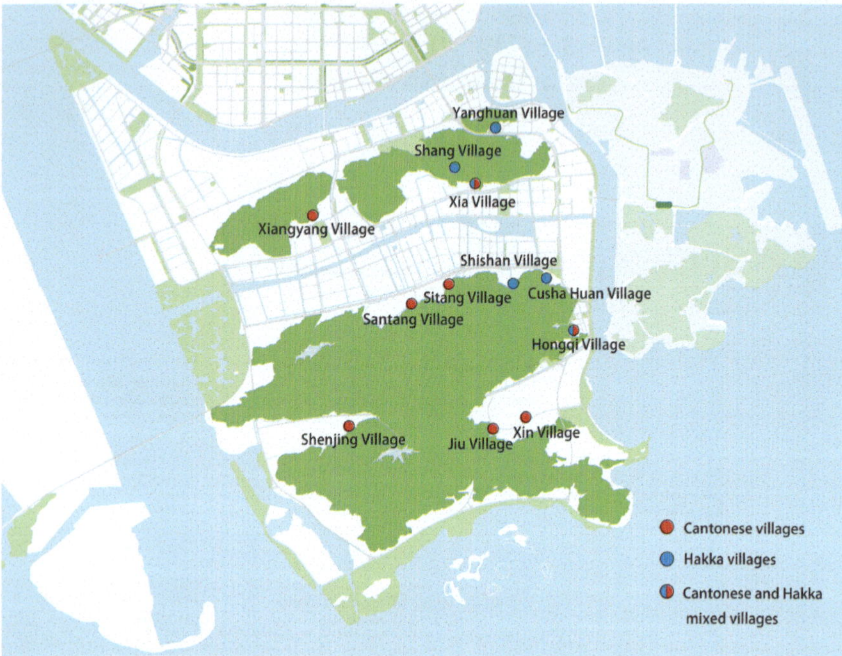

Fig. 15 Distribution map of original natural villages and ethnic types of Hengqin (*Source* Created by the author)

2.3 Settlements and Places Associated with Macau

The coastal areas of Xiangshan (香山), including Sanzao (三灶), Gaolan (高欄), and the Greater and Lesser Hengqin Islands within the Shizimen archipelago, underwent concurrent development during the Kangxi-Yongzheng-Qianlong reigns. Subsequent population migrations and intermarriage in the region fostered close kinship ties among ethnic groups. This is exemplified by Hakka villagers from Shi Shan (石山) constructing and renovating Guanyin Temples (觀音廟) on Coloane Island, as well as the inscribed horizontal plaque donated by Huanyang (環洋) villagers at the Tan Gong Temple (譚公廟) in Coloane [17]. Given its proximity to Macau, Hengqin's ethnic activities maintained close ties with Macau. During the incipient urbanization of Hengqin, key developmental sites were notably influenced by Macanese social factors, with Portuguese activities on the Hengqin islands forming an early "Macau-adjacent interaction zone" (Fig. 17).

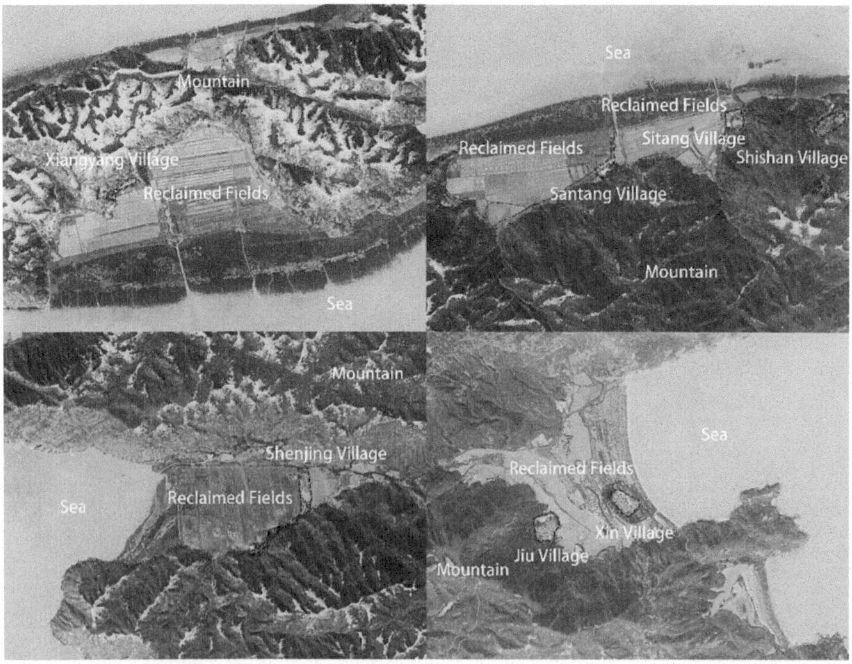

Fig. 16 Stratified bays morphology of mountains-villages-sea (*Source* Analysis based on the 1969 Keyhole Satellite images)

2.3.1 Yangchuan Bay

Prior to the Portuguese occupation of Macau, designated anchorages for foreign merchant ships existed not only at Xiangshan's Langbai (浪白) (present-day Nanshui Town (南水鎮)), Haojing (蠔鏡) (Macau), and Shizimen, but also at other Pearl River Delta locations such as Xinning (新寧), Xinhui (新會), and Dongguan (東莞) [11]. However, after the relocation of Guangzhou's Maritime Trade Office anchorage to Macau in 1535 and its formal establishment as a port [18], the superior natural conditions of Macau's inner and outer harbors—combined with its proximity to Guangzhou—led to the abandonment of other anchorages like Shizimen. As Qu Dajun (1678) recorded, "all other havens fell into disuse, leaving Haojing (Macau) as the sole port for foreign vessels" [19]. Following the Portuguese occupation of Macau in 1557, the Macau-Portuguese authorities prohibited non-Portuguese merchant ships from anchoring in the inner harbor of Macau. Consequently, foreign ships were compelled to anchor temporarily at Lesser Hengqin—the closest viable location near the northern entrance of Outer Shizimen Channel—where freshwater springs and resident communities existed [10]. This led to the gradual emergence of the eastern bay of Lesser Hengqin as a designated anchorage for non-Portuguese ships (Fig. 18), ultimately giving rise to the toponym "Yangchuan Bay" (means Foreign Ships Bay) (Fig. 19).

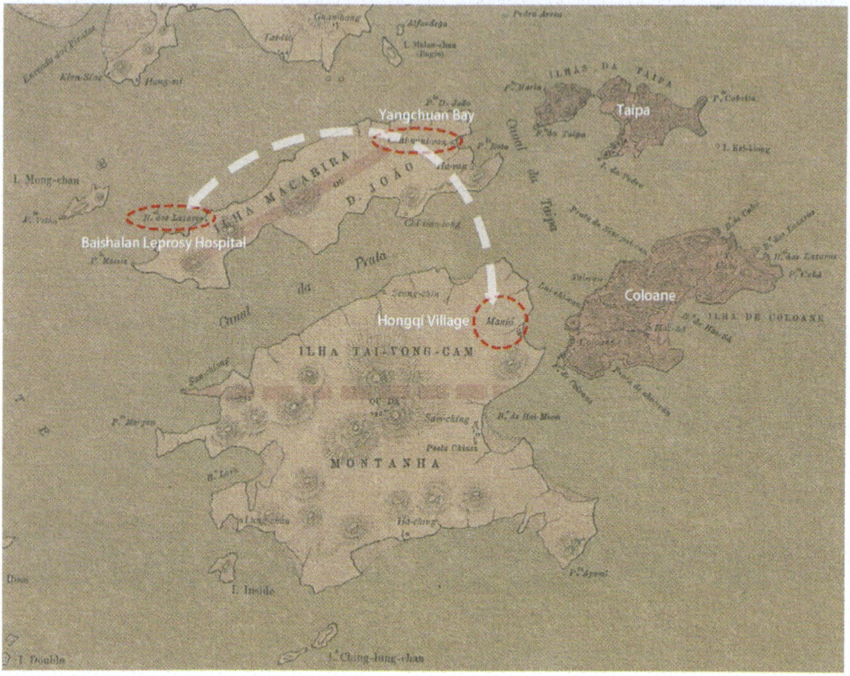

Fig. 17 An early "Macau-adjacent interaction zone" (*Source* Analysis based on the map of Macau Province by the Portuguese Colonial Cartographic Commission, 1912)

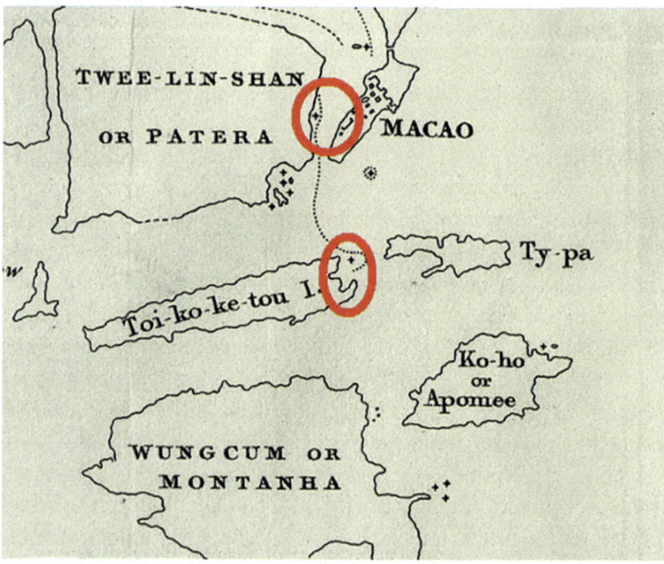

Fig. 18 Map of the relationship between Yangchuan Bay and the Macau Inner Harbor (*Source* Jin Guoping, 2022, Ref. [10])

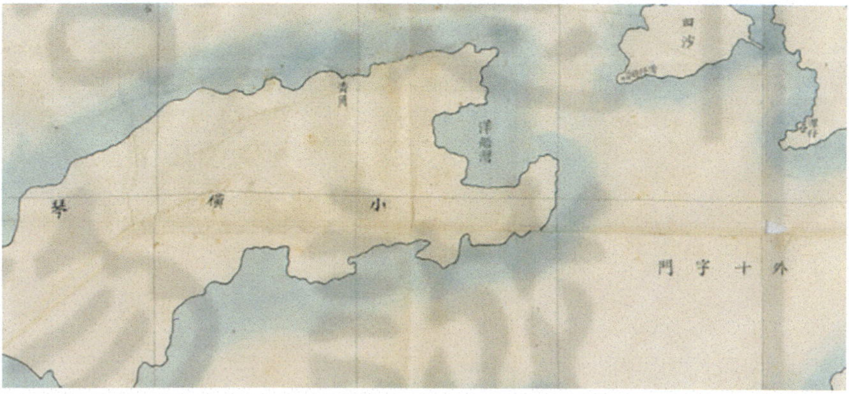

Fig. 19 Map of the Portuguese Concession in Macau, about 1900 (*Source* Collected from the Library of the University of Science and Technology of Macau, Provenance: Academia Sinica)

The bay is bisected by a prominent reef into two sub-bays: the upper deeper section named "Chai-ngui-van" (側魚灣) and the lower "Ha-van" (夏灣)—the latter's Cantonese pronunciation "Ha" (夏) being homophonic with "lower" (下) suggesting its toponymic origin. The basin also features designated Watering Places, Hortas (gardens), and scattered structures, likely representing Huanyang Village's proto-settlement phase (Figs. 20, 21 and 22).

Why did not foreign merchant vessels concentrate in the bay between Lesser Hengqin's southern coast and Greater Hengqin? Jin Guoping's analysis of the *Topography of Zhongshan (Preliminary Draft)* reveals that the southern tip of Lesser Hengqin, named "Chi-sion-long" (長沙欄, means Long Sandy Barrier), derived its toponym from an elongated and narrow sandy bay. This area featured shallow waters with heavy waves, while the adjacent seabed exhibited complex topography unsuitable for anchorage, thus precluding its development as a viable harbor.

In summary, influenced by Macau's port opening and the Macau-Portuguese government's exclusionary policies, Lesser Hengqin's strategic geographic position elevated its status to function as Macau's outer harbor and the originating node of the "Trans-Pacific Silver Silk Route" [10].

2.3.2 Baishalan Leprosy Hospital

With population growth straining Macau's urban space, the Macau-Portuguese administration relocated patients from the Saint Lazarus Leprosy Hospital to Lesser Hengqin Island in 1878, establishing the Baishalan Leprosy Hospital (Hospital dos Lazaros) (Fig. 23). Initially, the Qing government did not interfere due to its tolerance of charitable works. However, the Macau-Portuguese government later used the Hospital dos Lazaros as a tool in its struggle for colonial sovereignty. By the late

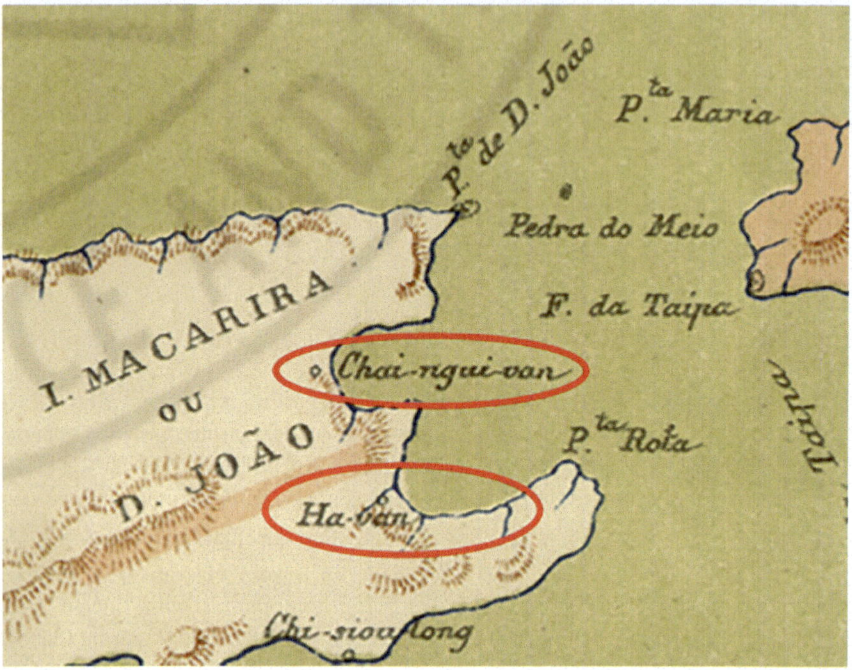

Fig. 20 Partial map of Macau Province, 1909, the part of Yangchuan Bay (*Source* From Macau University of Science and Technology Library)

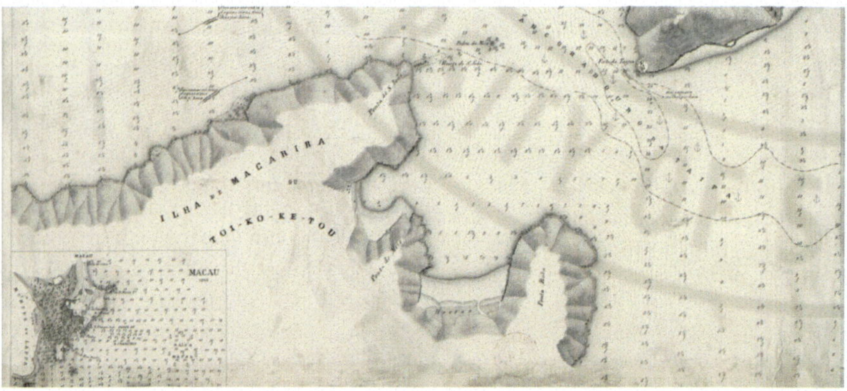

Fig. 21 Macau and its neighboring islands, 1866, the part of Yangchuan Bay (*Source* Collected from Macau University of Science and Technology Library, Provenance: National Library of France)

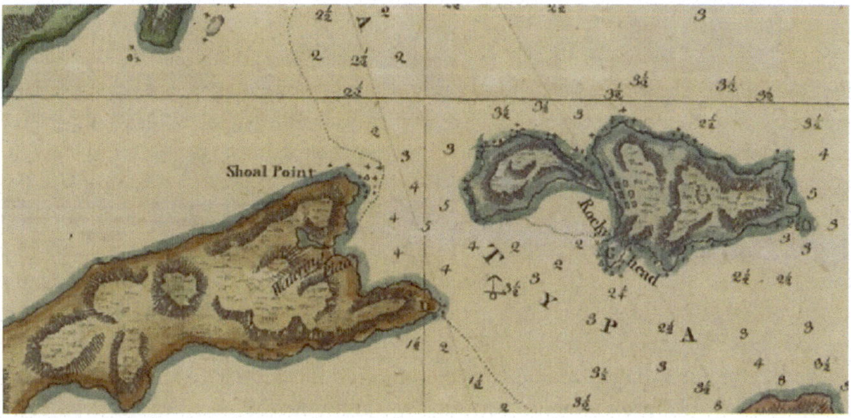

Fig. 22 Partial sketch of the Typa and Macau, 1784 (*Source* Collected from Map Department of Harvard University Library)

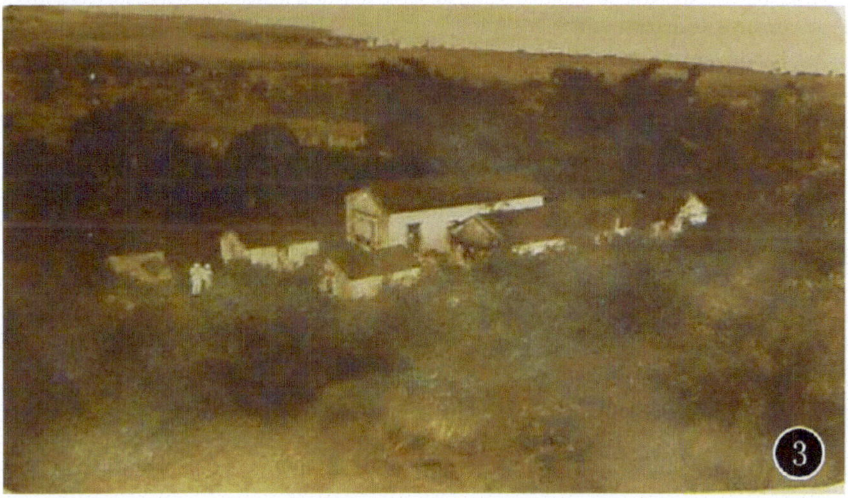

Fig. 23 Baishalan Leprosy Hospital (*Source* Wang Jingquan, 2018, Ref. [21])

nineteenth century, both Qing and Macau-Portuguese forces had constructed military camps near the hospital, triggering repeated sovereignty conflicts [20, 21]. The institution's 75-year operation ceased in 1953 when the patients were transferred to Macau Jiuao Leprosy Hospital after the liberation of New China.

2.3.3 Hongqi Village

Hongqi Village and Macau's Coloane Island are separated by a narrow Shizimen Waterway. Residents on both sides have kept in closed contact, with regular ferry services operating between Coloane and Hengqin except during the Cultural Revolution of China, commonly known as the "Hengshui Ferry (横水渡)." Within Hongqi Village lies a flowing waterfall known as Manió River (馬尿河, present Xiangsi Waterfall), named for its resemblance to a horse urinating during heavy rains and Hongqi Village was formerly known as Manió Village (馬尿村). Due to Macau's scarcity of freshwater resources and its proximity to Hengqin across the water, this area also became a water source for Macau residents. Because of the pier and waterfall, Hongqi Village became an important hub for frequent exchanges between Hengqin and Macau as well as surrounding areas in the early days, eventually evolving into the early urban center of Hengqin.

In addition to water resources, Hengqin Island also has abundant stone and timber resources. Therefore, after the Opium War, the Portuguese began to dispute sovereignty over Hengqin Island and the surrounding waters. The 1891 boundary map indicates that, in addition to the three islands of Macau, the entire island of Lesser Hengqin and the northern part of Greater Hengqin Island north of Hoi Men (near Hongqi Village) should also be allocated to the Portuguese side (Fig. 24), indicating that, from the perspective of the Macau-Portuguese government at the time, the northern parts of Lesser Hengqin and Greater Hengqin were closely integrated with the urban structure of Macau. Today, Portuguese inscriptions left by the Portuguese still remain beside the waterfall and are currently the first historical relics discovered on Hengqin Island that are directly related to the Portuguese Macau.

According to the inscription on the stone tablet, it was carved in June 1937, and the landscape around the waterfall was destroyed when the Japanese occupied Hengqin in 1938. In 1945, the Japanese withdrew from Hengqin Island, and the Portuguese repaired the pool and the stone tablet, continuing to make it available for use by residents of Zhuhai and Macau (Fig. 25).

3 Zhongxin Channel Reclamation Period (1949–1989)

3.1 Zhongxin Channel Reclamation Project

After the establishment of the People's Republic of China, in order to overcome the country's food shortage, expand the cultivated area, and increase grain output, a wave of land reclamation swept across the country in the 1960s. Influenced by ocean and air currents, as well as the reclamation projects at Modaomen (磨刀門) and Baiteng Lake (白藤湖), which have altered the surrounding environment, the waters between the Greater and Lesser Hengqin Islands have accumulated thick layers of tidal flats over time. Some of the silt had even reached depths of 30 m, making it a valuable

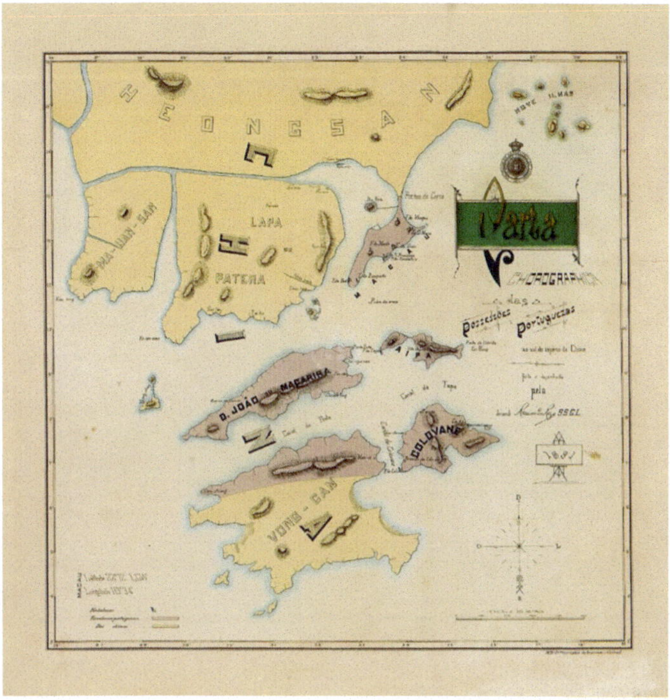

Fig. 24 The map of the boundaries of Macao drawn by the Portuguese government, 1891. (*Source* Collected from Macau University of Science and Technology Library)

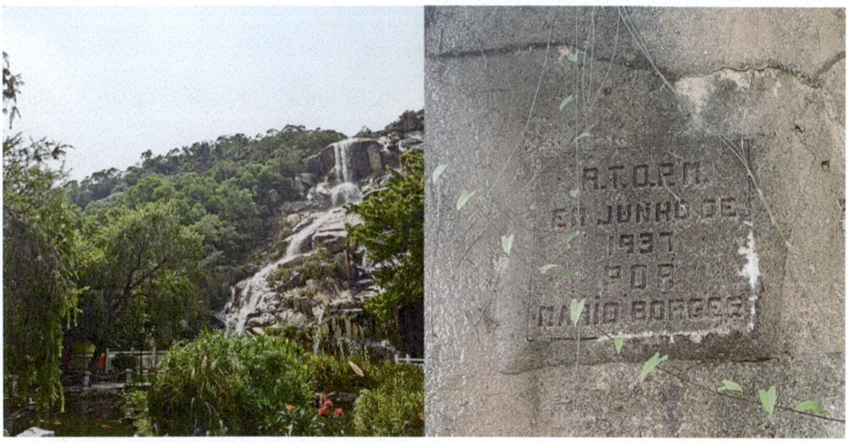

Fig. 25 The Xiangsi Waterfall and the Portuguese stone tablet (*Source* Photographed by the author)

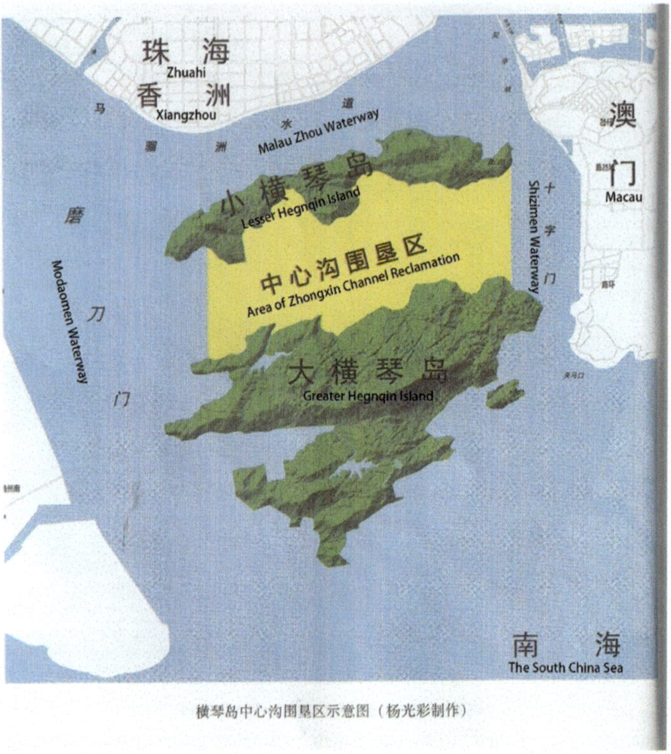

Fig. 26 Area of Zhongxin channel reclamation (*Source* Tan Yuanheng, Zhu Wenbin, and Lu Menghe, 2018, Ref. [1])

land resource awaiting development. This led to the Hengqin Zhongxin Channel Reclamation Plan [1] (Figs. 26, 27 and 28).

Since Zhuhai didn't have enough manpower, material resources, or money to handle this tough project at the time, they asked the Foshan Regional Committee for help. At that time, Shunde (順德) was facing a food shortage, and there was full of rivers and canals. The locals were familiar with water and there was plenty of labor. Additionally, they had rich experience in land reclamation from the historical Sangyuanwei (桑園圍) project. Consequently, in late December 1970, Shunde sent over 3,200 young farmers to Hengqin to participate in the land reclamation project, which officially began on New Year's Day 1971. In the early 1960s, just prior to the reclamation project, the permanent resident population of Hengqin was only 956 [1, 22]. The reclamation of Zhongxin Gou began a period of increasing immigration.

The Zhongxin Channel Reclamation Project adopted a two-sided approach, which involved first constructing dikes on both the eastern and western sides to enclose a section of sea, followed by land reclamation within the enclosed area. The difficulty of this project is more severe and challenging than the usual single-sided dike reclamation method of gradually building embankments from the shore outward (Fig. 29).

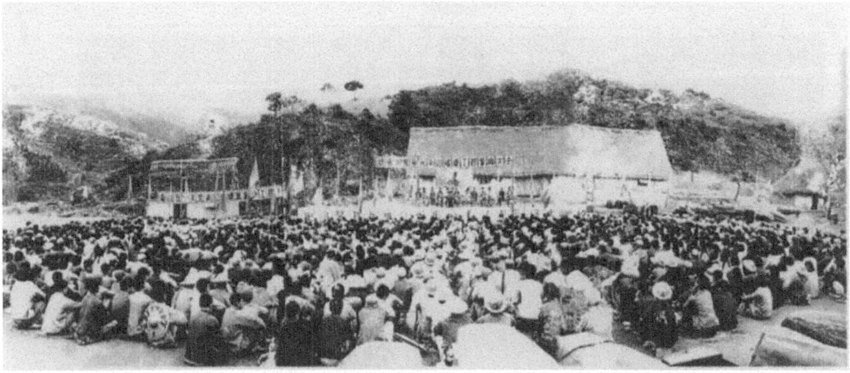

Fig. 27 1971 New Year's Day reclamation construction mobilization meeting (*Source* Collected by Shunde Archives)

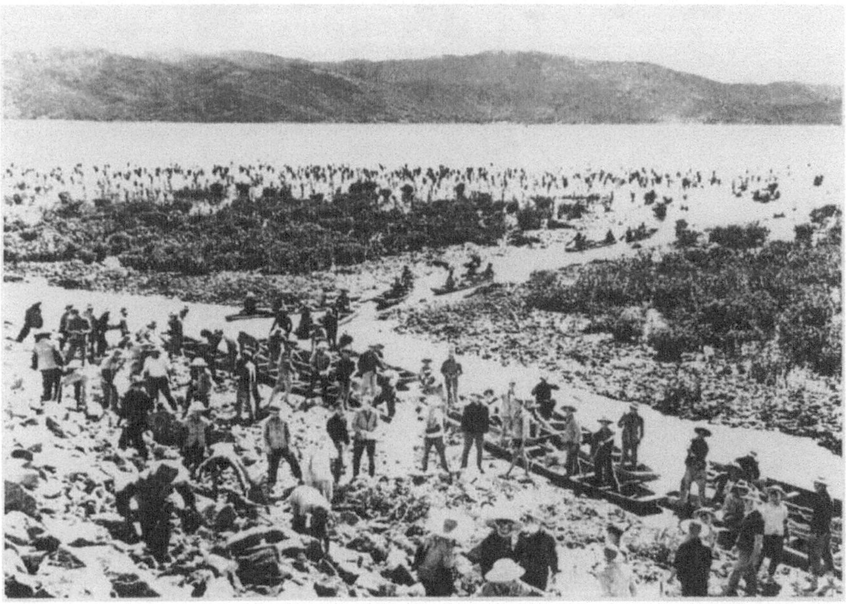

Fig. 28 Pictures of the construction of the dike during the Zhongxin Channel Reclamation (*Source* Collected by Shunde Archives)

After completing the east and west dikes projects, a system for bringing in freshwater and draining salt water was planned to improve the soil quality of the reclaimed land, marking the beginning of the first unified construction plan and development in the history of Hengqin.

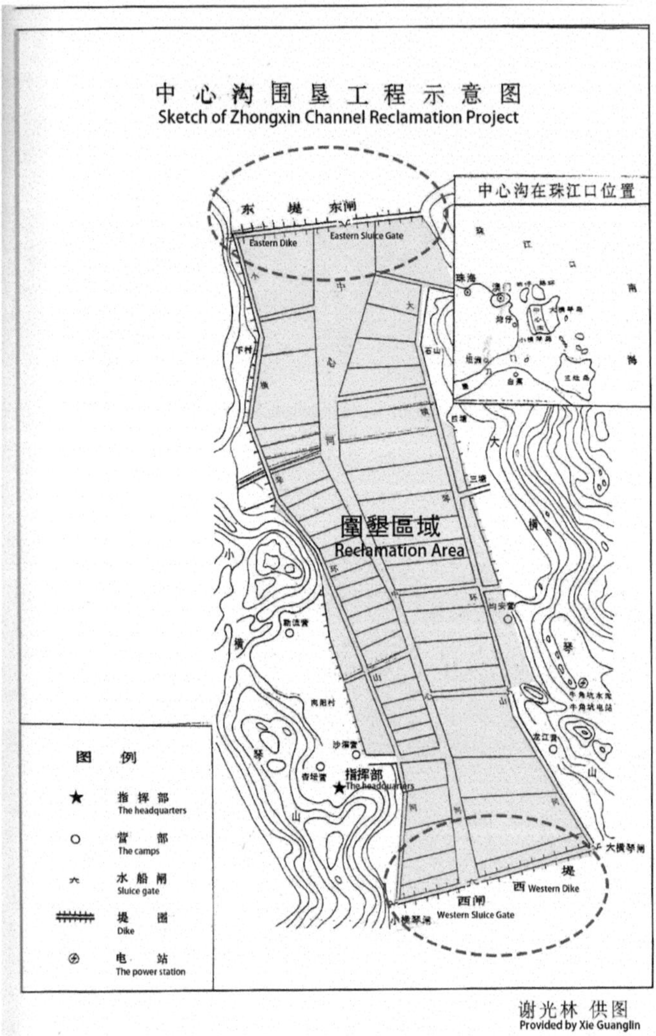

Fig. 29 Sketch of Zhongxin Channel Reclamation Project (*Source* Zhang Jiemin and Ye Chunsheng, 2005, Ref. [22])

3.2 *Planar Morphological Framework Influenced by Agricultural Water Conservancy Systems*

Due to the terrain being higher in the west and lower in the east, the western side had shallow water and more exposed land, while the eastern side had deeper water and less exposed land. The natural shape of the tidal flats was followed in the reclamation (Fig. 30), resulting in the central river being narrow in the west and wide in the east.

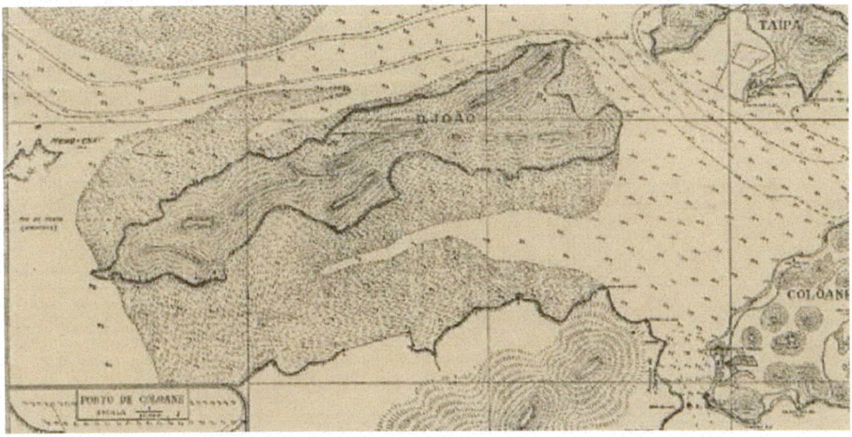

Fig. 30 Partial sea chart of Macao Colony/Mcao Prot Project, 1929 (*Source* Collected from Macau University of Science and Technology Library, Provenance: National Library of Australia)

The Zhongxin Channel agricultural water conservancy construction plan mainly involves the construction of sluice gates and the excavation of four rivers: Huanshan Rivers (the river encircling the mountains) of the Greater and Lesser Hengqin, the Zhongxin River (the river in the center), the flood discharge river, and the irrigation and drainage branch rivers (Fig. 31). Because the terrain in Zhongxin Channel is higher in the west and lower in the east, with 1 m height difference, the water flow in from the west and out from the west. The entire irrigation and drainage system import freshwater from the Xijiang River (西江) into the Huanshan River through the Greater and Lesser Hengqin Sluices, then from the irrigation branch rivers to the farmland, and, subsequently, the saltwater could be led to the Zhongxin River by the drainage branch rivers and discharged eastward to the east dike (Fig. 32). Under the influence of the construction of the agricultural water conservancy system, the morphological framework of the Zhongxin Channel planar pattern was initially formed during this period.

3.2.1 Road System

During this period, waterway transportation is the main mode in Zhongxin Channel. In addition to its irrigation and drainage functions, the Zhongxin River also serves as a transportation route. Overall, it forms a "three horizontal, multiple vertical" framework structure centered on the Zhongxin River. The current road network was established based on this structure, and several flood discharge rivers serve as the prototype for the main roads connecting the Greater and Lesser Hengqin Islands today (Fig. 33). With the subsequent development of transportation from riverways to land routes, the new road network became based on two main roads alongside the Huanshan River, weakening the status of the roads near the Zhongxin River.

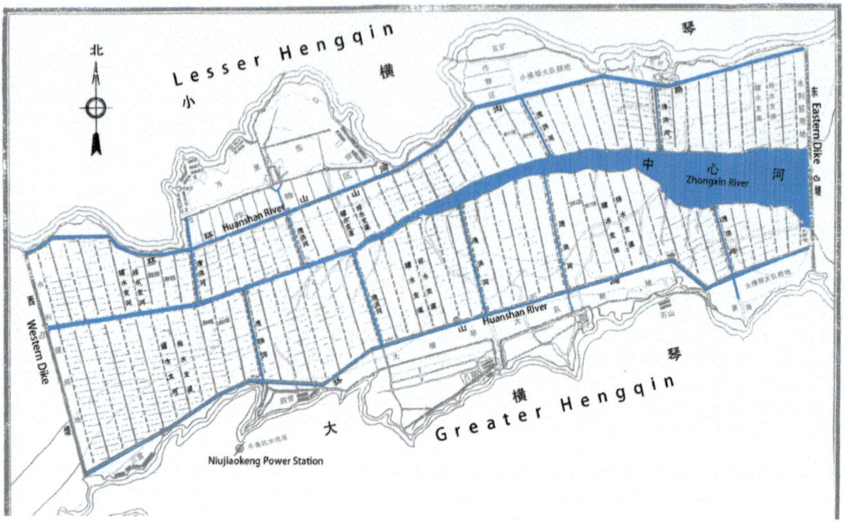

Fig. 31 Analysis diagram of Zhongxin channel farmland water network structure (*Source* Redrawn based on Zhang Jiemin and Ye Chunsheng, 2005, Ref. [22])

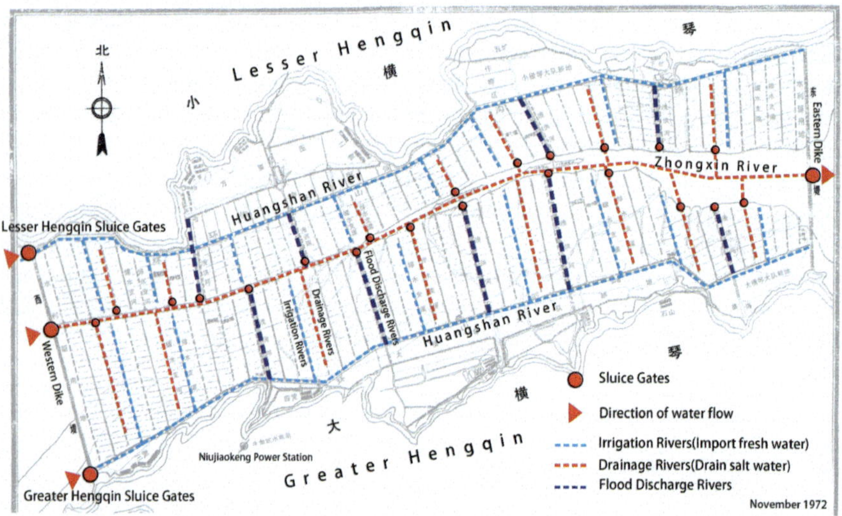

Fig. 32 Analysis diagram of Zhongxin channel drainage and irrigation system (*Source* Redrawn based on Zhang Jiemin and Ye Chunsheng, 2005, Ref. [22])

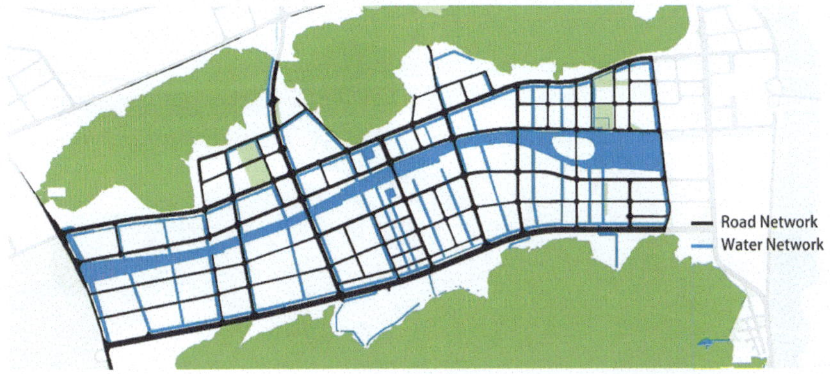

Fig. 33 Present-day roads and water network (*Source* Redrawn based on Google Earth in 2022)

3.2.2 Boundaries of Plots

The water network of the Zhongxin Channel agricultural water conservancy project was the initial boundary of the plots, which cut out the early land plot layout pattern in the Zhongxin Channel (Figs. 34 and 35). After the "Night-Seven Meeting" on September 7, 1972, Zhuhai and Shunde reached an agreement on the distribution of reclaimed land and demarcated the boundaries of different land ownership plots. Originally, it was agreed that, apart from the portion allocated to the Greater and Lesser Hengqin Commune (the original reclaimed land and the appropriate compensation area), Shunde would receive two-thirds of the land in Zhongxin Channel Reclamation, and Zhuhai would get one-third. This one-third would be returned to Zhuhai after Shunde reclaimed Maozhou Island. However, due to various reasons, the reclamation project of Maozhou Island was not implemented. Therefore, most of the land ownership in Zhongxin Channel belonged to Shunde after the Zhongxin Channel Reclamation completed [22]. The issue of land rights on Zhongxin Channel was also an important reason for the marginal development of Hengqin in the 1990s.

3.3 Land Utilization of Zhongxin Channel

In accordance with the policy "Take grain as the key link, develop all-round agriculture," in the early stage of reclamation, rice was mainly grown in the Zhongxin Channel, and crop area was set up on the hillsides to grow aquatic plants, water chestnuts, corn, sweet potatoes, and other crops according to the depth of the water (Figs. 36 and 37). Therefore, the risk of growing rice on the island is relatively high. By the late 1970s, large-scale rice production had gradually shifted to a model focused on commercial crops, with appropriate development of livestock farming. Under the influence of reform and opening-up, rural production and management models were reformed. In the 1980s, the main reclaiming force withdrew back to

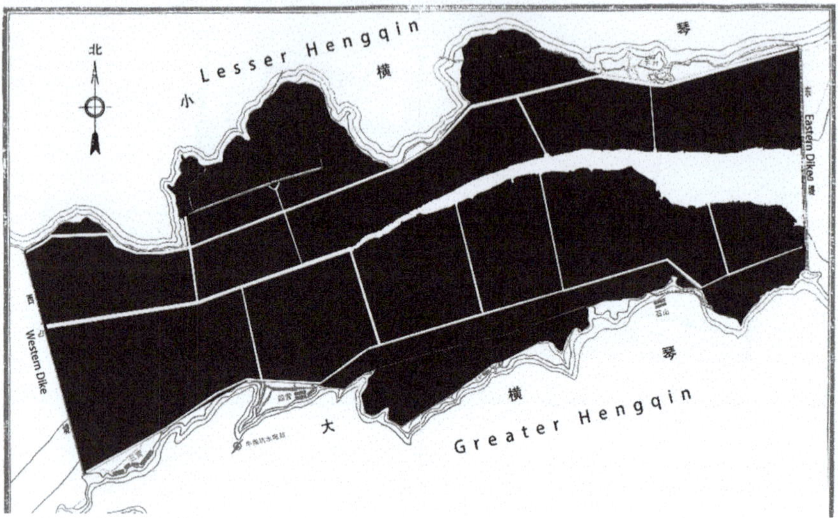

Fig. 34 Analysis map of farmland plot boundaries (*Source* Redrawn based on Zhang Jiemin and Ye Chunsheng, 2005, Ref. [22])

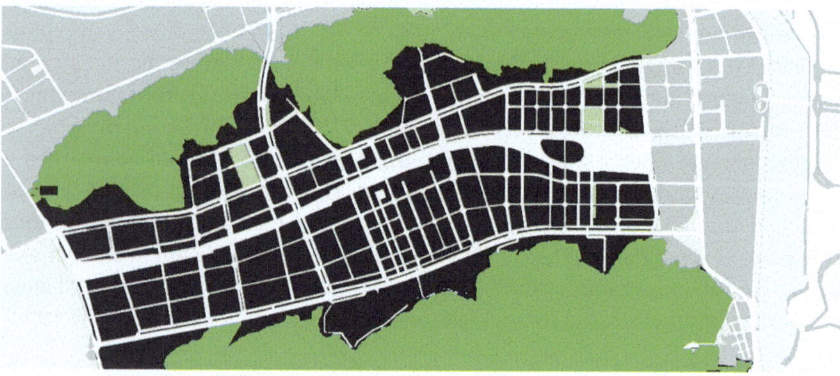

Fig. 35 Analysis map of present-day city plot boundaries (*Source* Redrawn based on Google Earth in 2022)

Shunde, leaving only a few migrant workers on the island. The production mode on the island gradually turned to a contracted fishpond aquaculture production mode [22] (Fig. 38). By the 1990s, the land utilization of Zhongxin Channel was mostly freshwater farm and dike-ponds in the east, large areas of sugarcane land in the western mountainous areas, a small number of paddy fields next to villages in the valley, and fruit orchards, tea plantation as well as forests planted by the reclamation team (Fig. 39).

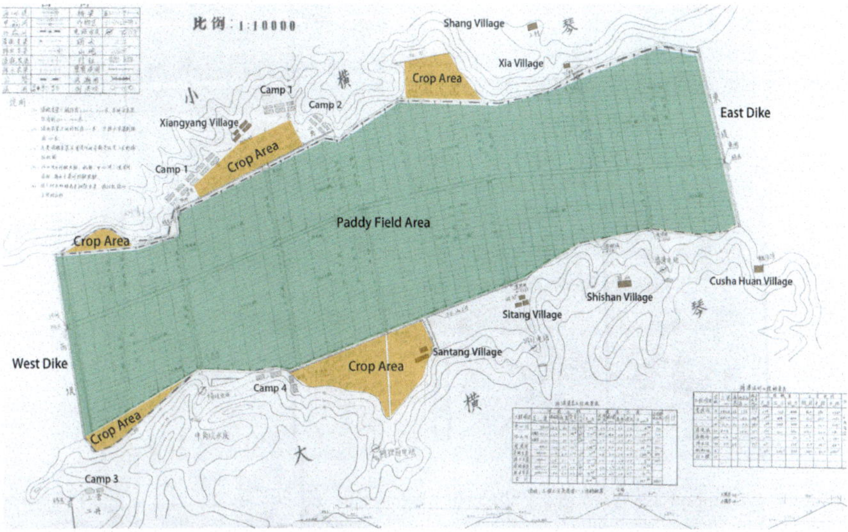

Fig. 36 Analysis map of land utilization of Zhongxin Channel (*Source* Redrawn from Water Conservancy Plan of Zhongxin Channel, 1972.6, collected by Shunde Archives)

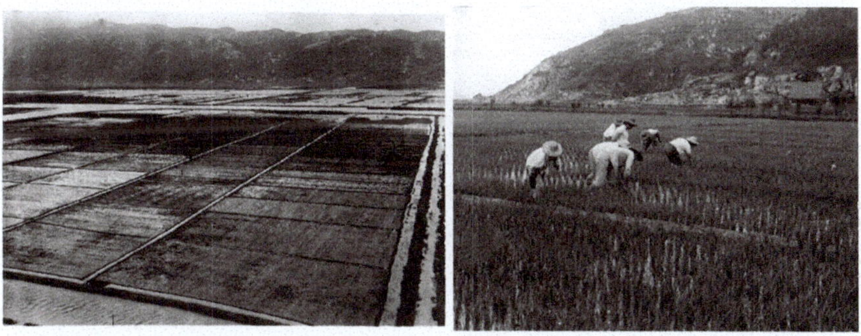

Fig. 37 The standard paddy fields in the reclamation area developed from 1974 to 1980 (*Source* Collected from Shunde Archives)

3.4 *The Evolution of Architectural Fabric*

3.4.1 The Formation of Reclamation Camp Settlements

(1) ***Settlement Distribution Patterns***

Most of the original villages in Zhongxin Channel were located in the east closer to Macau, while the Shunde reclamation troops moved down the Xijiang River and were initially responsible for the construction of the western dike, with Zhuhai responsible for the eastern dike. Consequently, the Shunde reclamation camp was concentrated in

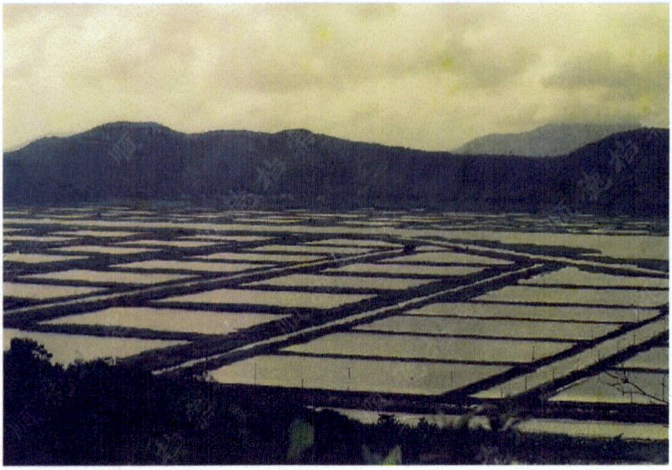

Fig. 38 The fishpond landscape of Zhongxin Channel at the end of the last century (*Source* Collected from Shunde Archives)

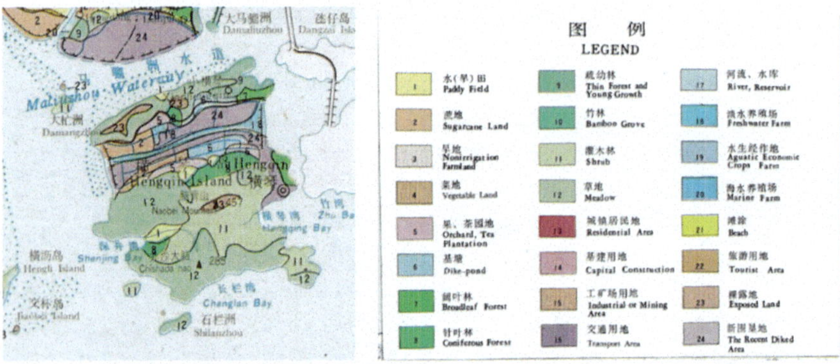

Fig. 39 Partia land utilization map of Zhuhai (*Source* Guangzhou Institute of Geography and Zhuhai Urban Planning Committee, 1991, Ref. [23])

the western mountainous coastal area. After the land division meeting, the locations of Xingtan Camp (杏壇營), Shajiao Camp (沙滘營), and Longjiang Camp (龍江營) were appropriately adjusted, while the overall distribution characteristics remained unchanged (Fig. 40).

(2) ***The Building Form of Settlement***

In the 1960s, the buildings on the island were all made of thatch [24]. Given the limited conditions, the reclamation troops could only stay in simple thatched huts at first, constantly suffering from the annoyance of mosquitoes, snakes, and ants, as well as facing the threat of heavy rains, flooding, and fires. After completing the construction of the east and west dikes in 1973, the reclamation troops began

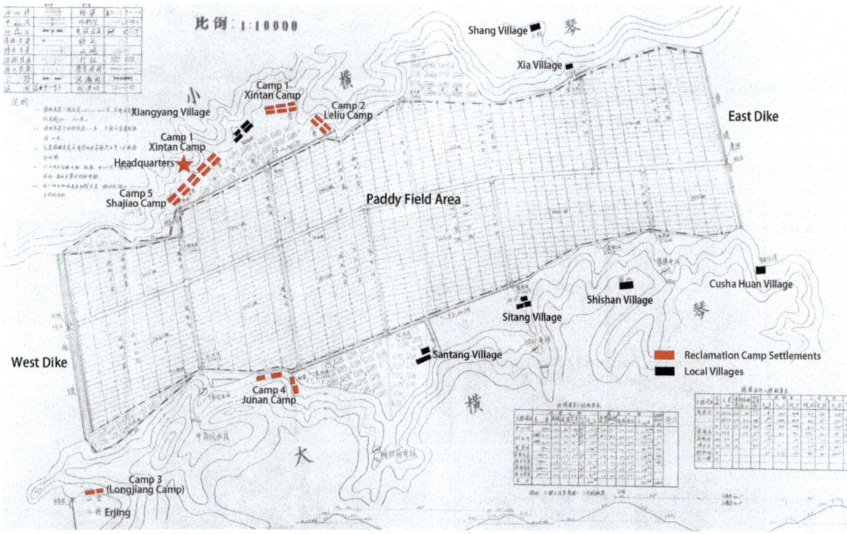

Fig. 40 Distribution map of reclamation camp settlements and local villages (*Source* Redrawn from water conservancy plan of Zhongxin Channel, 1972.6, collected by Shunde Archives)

to improve their camp living conditions on their own. There were abundant stone resources on Hengqin Island, so they carved out mountains and created stones, used the local materials to build more than 100 stone houses in the camp between 1973 and 1974, including headquarters, dormitories, hospitals, docks, banks, and other facilities (Fig. 41). Meanwhile, stone houses constructed by local materials from the same period can also be found on nearby Guishan Island (桂山島), Coloane Island, and Qi'ao Island (淇澳島). This may have been due to the limited availability of building materials on the islands at the time, leading to the development of stone house architecture with a distinctive Pearl River Estuary Island character. The facades and walls of the barracks feature the perforated brick windows and perforated balustrade commonly found in Lingnan folk dwellings, which facilitate ventilation and heat dissipation in the humid and hot environment, reflecting the regional architectural characteristics of the Lingnan coastal area (Fig. 42).

3.4.2 The Relocation of Shang Village Under the Development Plan

Shang Village was originally located on the hillside of Shuliang Gang (薯莨崗) on Lesser Hengqin Island (Fig. 43). With the development of the island, the original site of Shang Village was designated for the construction of a reservoir in 1969. Additionally, with its inconvenient location to external transportation on a hillside, the village was gradually relocated to the new site at Wutang (五塘), retaining the name Shang Village [2]. The current scale of the village is nearly three times larger than it was in 1969 (Fig. 44).

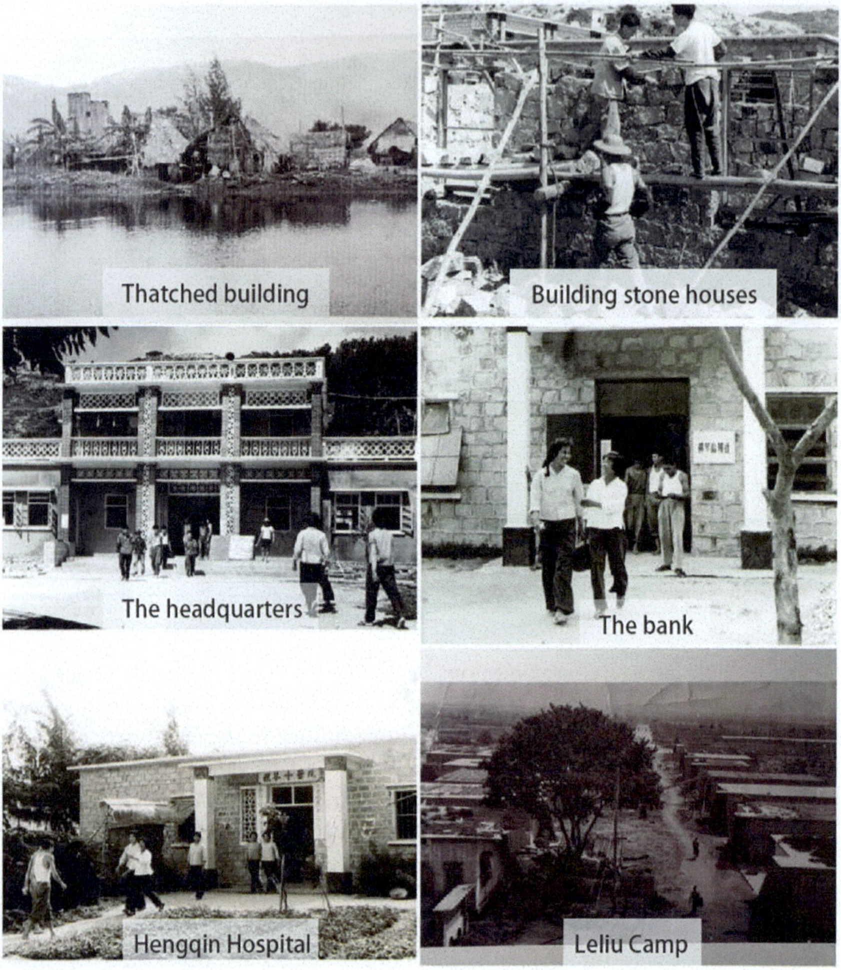

Fig. 41 The buildings of reclamation (*Source* Collected from Shunde Archives)

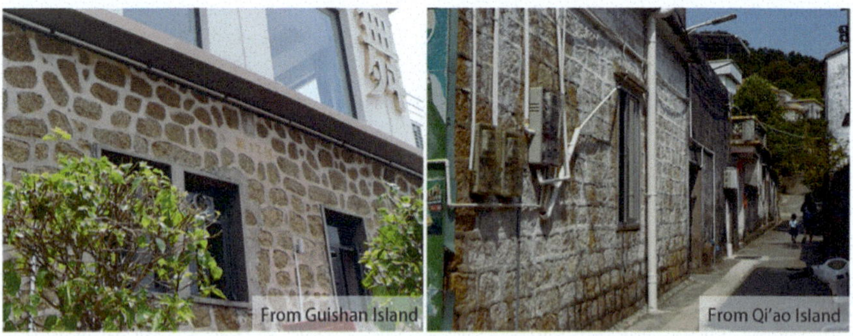

Fig. 42 Stone houses on Guishan Island and Qi'ao Island (*Source* Photographed by author)

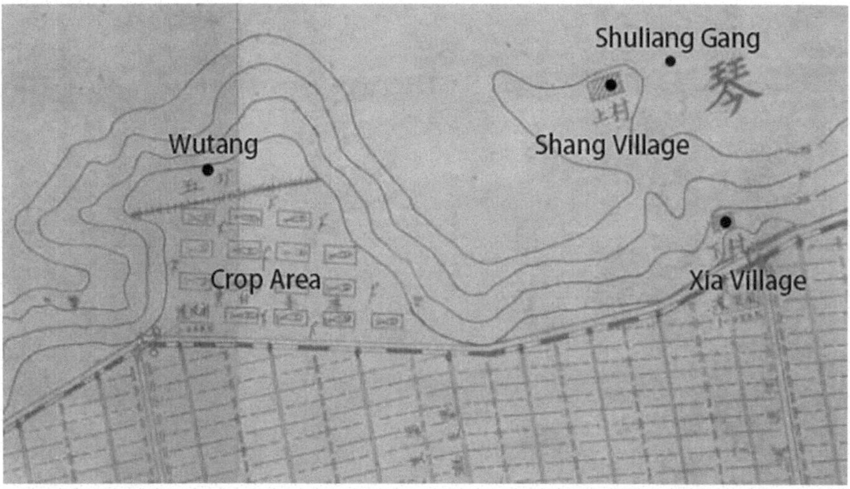

Fig. 43 The location of Shang Village before relocation (*Source* Partial map of water conservancy plan of Zhongxin Channel, 1972.6, collected by Shunde Archives)

4 Period of Development Along the Margins (1990–2009)

4.1 Island Expansion Under the Second Land Reclamation

Under the favorable conditions of reform and opening-up, Hengqin entered a new phase of modernization and development. In 1992, Hengqin was designated by Guangdong Province as one of four key development zones for expanding foreign openness. The Hengqin Economic Development Zone was established the same year, marking the beginning of the second major planning and development phase for Hengqin Island. In 1993, the permanent population was approximately 3,000, with an additional 6,000 newcomers arriving on the island to participate in development projects or engage in business activities. The continued development of Hengqin Island led to another population growth surge [25].

As early as 1991, Zhuhai City's reclamation project included plans to reclaim the areas west and north of Lesser Hengqin and around Greater and Lesser Mangzhou (Fig. 45). Compared with the previous Zhongxin Channel Reclamation Project, this project was led by Zhuhai City and involved modern machinery and construction forces from all over the country. Subsequently, reclamation and development were carried out in the area around the Shizimen Waterway in the east and the region from the Southwest Shenjing to Shilan Island, leading to the shape of Hengqin Island expanded to all directions. The area of the second reclamation project was more than twice that of the first, with the Zhongxin Channel now enclosed at the center [26] (Fig. 46).

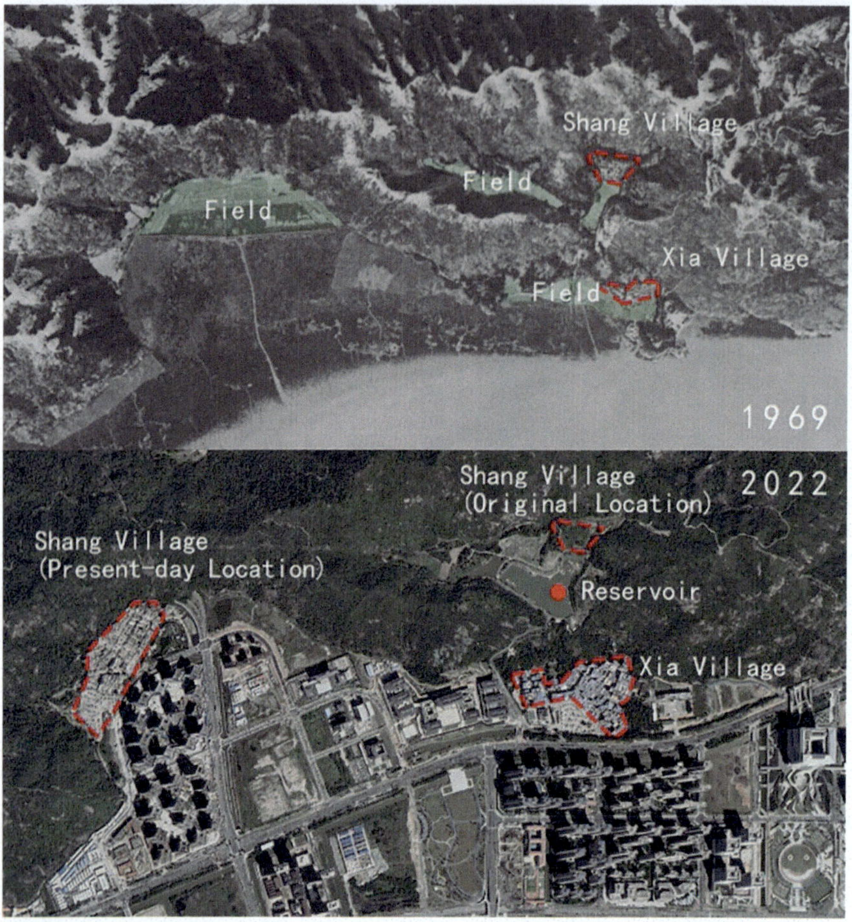

Fig. 44 The relocation of Shang Village (*Source* Redrawn based on satellite images)

4.2 Morphological Characteristics of Development Along the Margins

4.2.1 Urban Spatial Planning Layout

The *Third Round of Master Planning for Zhuhai City in 1995* and the *Second Round of the West Area Urban Development Strategy Plan* led to Hengqin being included in Zhuhai City's master plan for the first time, marking the beginning of Hengqin's urban modernization process. However, Zhongxin Channel was not developed at the same time due to ownership issues and policy and economic obstacles encountered in the 1990s [1]. Therefore, these master plans did not include plans for Zhongxin Channel,

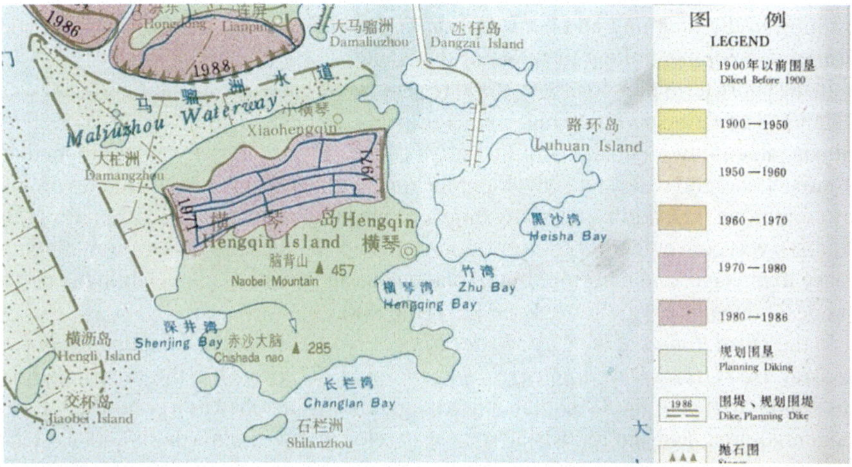

Fig. 45 Partial map of the reclamation process and planning of Zhuhai City in 1991 (*Source* Guangzhou Institute of Geography and Zhuhai Urban Planning Committee, 1991, Ref. [23])

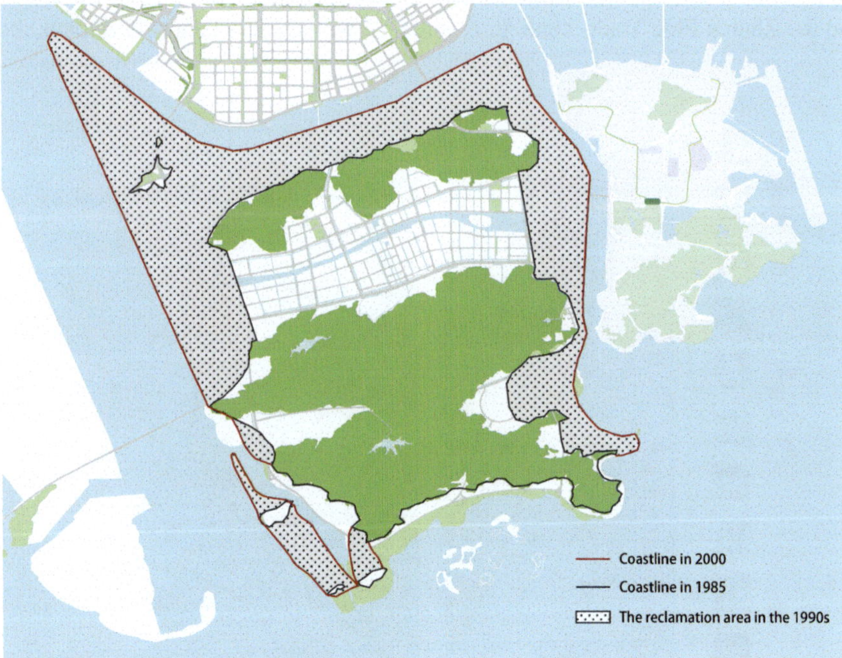

Fig. 46 The reclamation area in the 1990s (*Source* Created by the author)

resulting in the planning and construction at this stage taking on the characteristic form of development along the edge of the island.

In the short-term plan for 1992–2000 (Fig. 47), the eastern side facing Macau was designated as a transportation hub connecting the city center to other areas, as well as a financial, commercial, and administrative center. The southeast was designated as a tourist resort and real estate development zone with its scenic beauty and favorable bay conditions. The northwest Mangzhou area, near the planned Malau Zhou Port (馬騮洲港), was convenient for transportation and had an independent location, making it easy to manage, and therefore it was planned as an industrial zone, while the north was designated as a warehouse zone [25] (Fig. 48).

In the *Third Round of the City Master Plan (1995–2020)* and the *Second Round of the West Area Urban Development Strategy Plan (1995)* (Fig. 49), the overall zoning was similar to that of the 1992 plan, but Mangzhou was rezoned as a commercial and residential area, and the industrial zone was relocated to the northern and western coastal areas, forming a layout where the eastern area is characterized by tourism, commerce, finance, and other consumer-oriented sectors, while the western area is characterized by industrial and warehousing sectors. This layout also sets the tone for the functional layout of the entire Hengqin area in the future. Influenced by Macau and the Zhuhai Free Trade Zone, in accordance with the 1992 short-term plan, the

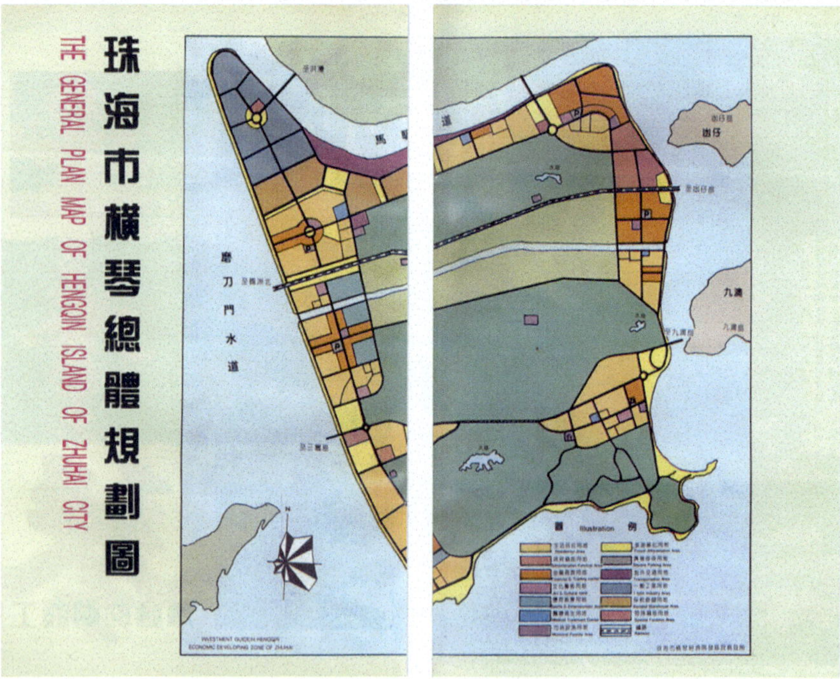

Fig. 47 The General plan map of Hengqin Island of Zhuhai City (*Source* Wu Nansheng, 1993, Ref. [27])

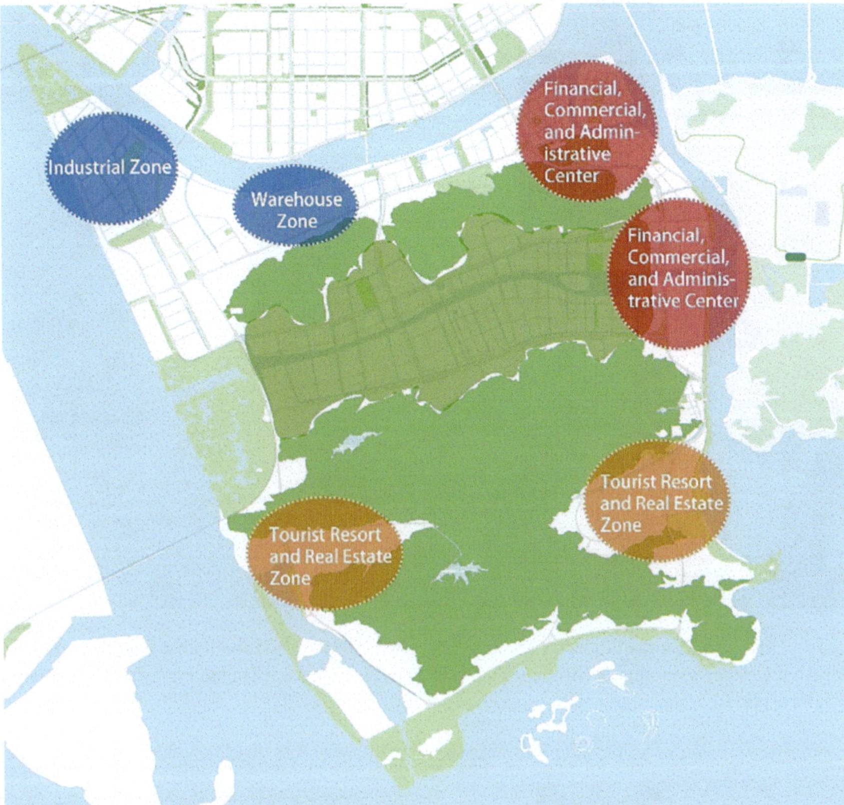

Fig. 48 The short-term planning functional zoning map, 1992–2000 (*Source* Created by the author)

eastern and northern periphery of Hengqin was developed into urban land in 1998 (Fig. 50).

At the beginning of the twenty-first century, with the opening of the Lotus Bridge and the new customs clearance port, the town center of Hengqin showed a trend of moving northward from Hongqi Village and Hengqin Ferry Terminal to the Lotus Bridge. In the long-term planning, the three historic bays (Shenjing Bay, Hengqin Bay, and Yangchuan Bay) and Mangzhou are designated for tourism, while urban land near the new customs clearance port will further extend into the interior of the island (Fig. 51).

4.2.2 The Development of Transportation

In the end of the 90s, with the completion of the Hengqin Bridge and Lotus Bridge, Hengqin's outward transportation connections gradually changed from waterways to land transportation by bridge, strengthening Hengqin's connections with surrounding

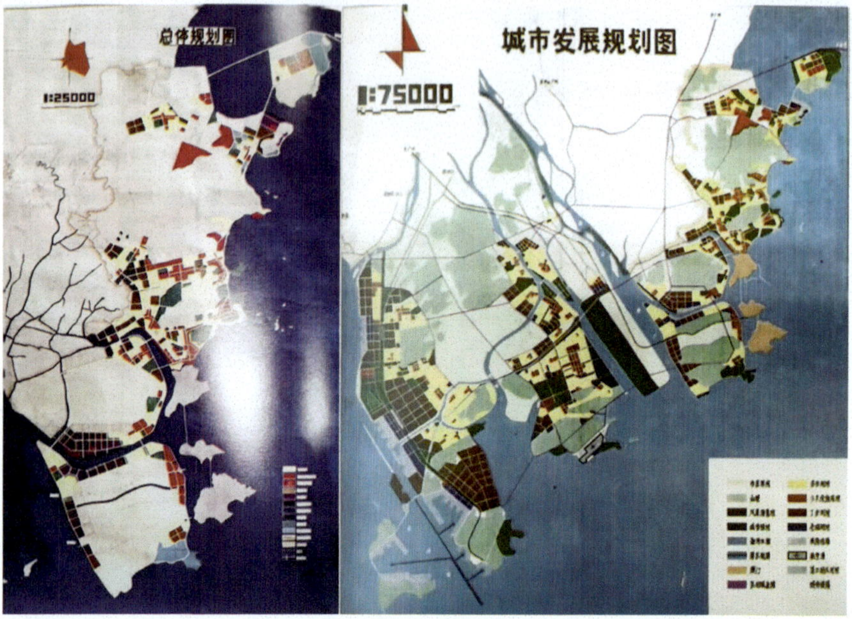

Fig. 49 The Third Round of the City Master Plan (1995–2020) (Left) and the Second Round of the West Area Urban Development Strategy Plan (1995) (Right) (*Source* Zhuhai Planning Department, 2005, Ref. [28])

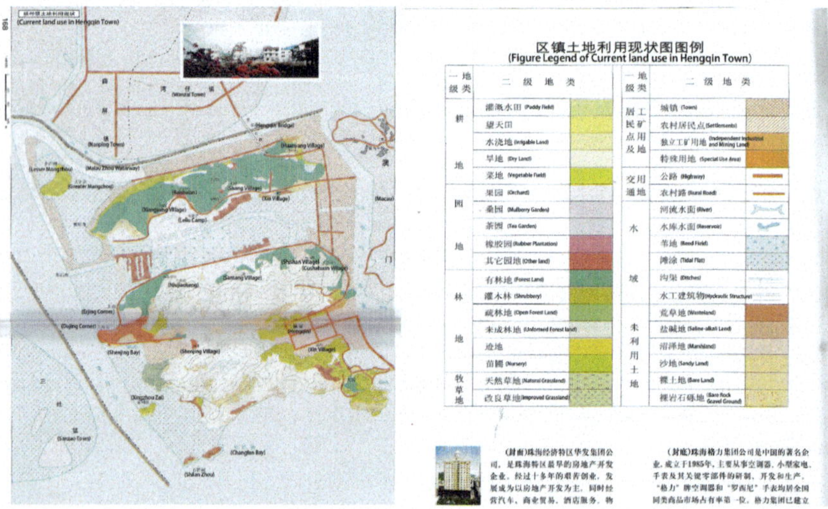

Fig. 50 Current land is used in Hengqin Town in 1998 (*Source* The Editorial Committee of *The Atlas of Zhuhai City*, 1998, Ref. [3])

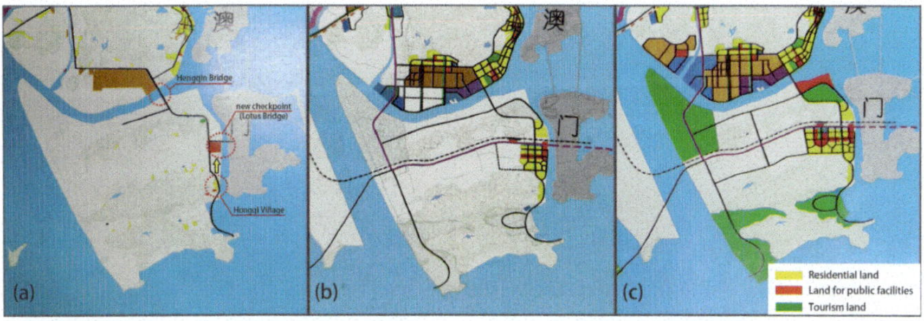

Fig. 51 Analysis map of town center migration. **a** Urban land use in 2000, **b** Short-term master plan map in 2000, **c** Long-term planning in 2000 (*Source* Analysis based on Zhuhai Planning Department, 2005, Ref. [28])

areas. Affected by margin development, the modern transportation network emerged first in the reclaimed area on the eastern part of the island, which spread from east to west. In 1998, two roads were built along the mountainside of Zhongxin Channel, as well as Binjiang Avenue (濱江大道) in the northern reclamation area, which is now the early form of Hong Kong-Macau Avenue (港澳大道), Hengqin Avenue (横琴大道), and Aiqin Road (愛琴路) (Fig. 52).

4.3 The Evolution of the Urban Form of Hengqin Town Center Under the Influence of Land Reclamation

4.3.1 Expansion of the Block Plan

Since the 1980s, a tidal flat has formed around Hongqi Village (Fig. 53). Land reclamation activities since the 1990s have transformed the tidal flat into development and construction land. The Hongqi Village block plan has continued to expand toward the sea, and the coastline has become increasingly straight (Fig. 54). Hongqi Village, adjacent to Macau's Coloane, has always been the area with the most frequent external activities in Hengqin. In the early 1990s, the administrative, cultural, and commercial centers of the island were temporarily set up in Hongqi Village [29]. A large number of multi-story commercial and residential buildings and public buildings have been constructed in the reclaimed area (Fig. 55). Meanwhile, a large number of modern new houses made of reinforced concrete with 2 to 3 floors have been also built in the village (Fig. 56). The changes in building types, roads, and plot scales in the reclamation area have created two morphological regions from the original village sites (Fig. 57).

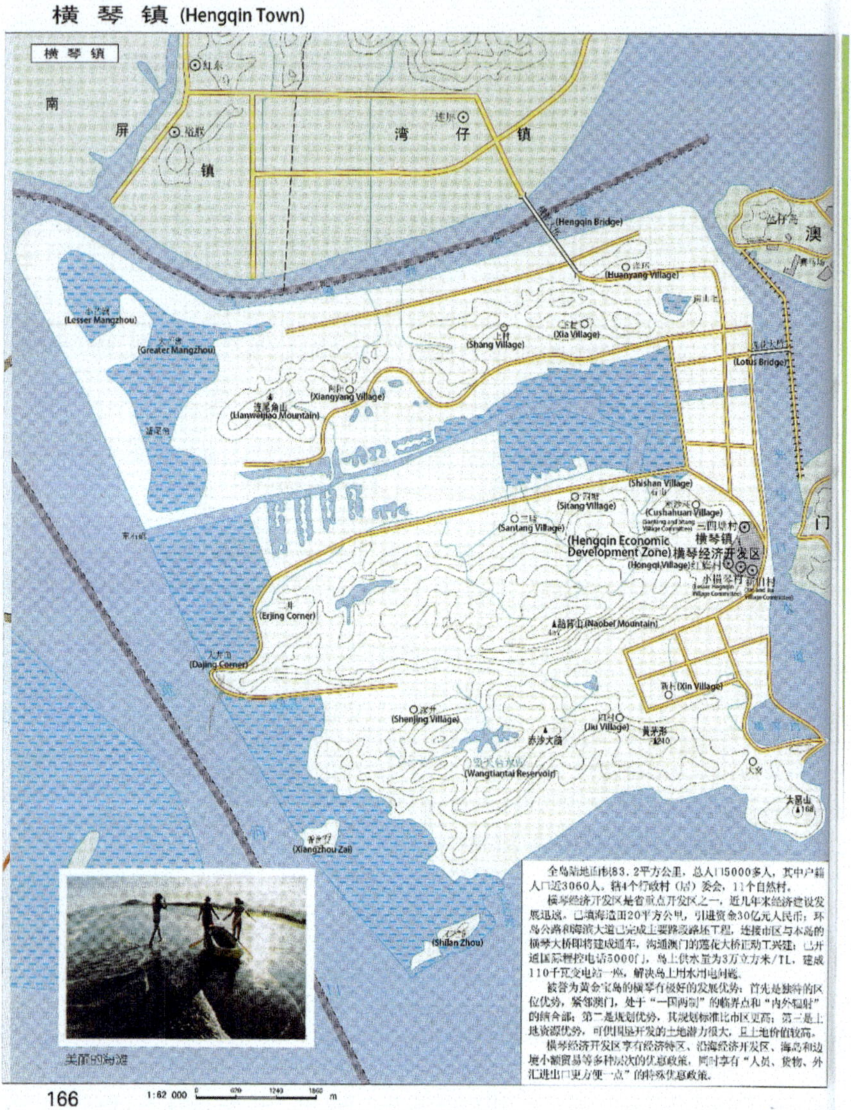

Fig. 52 Map of Hengqin Town in 1998 (*Source* The Editorial Committee of The Atlas of Zhuhai City, 1998, Ref. [3])

4.3.2 The Evolution of Hengqin Ferry Terminal

The outward movement of the coastline led to the relocation of the Hengqin Ferry Terminal in the late 1990s. The old Hengqin Ferry Terminal was located within Hongqi Village, at the site of today's "Asen Food City (阿森美食城)." It was the

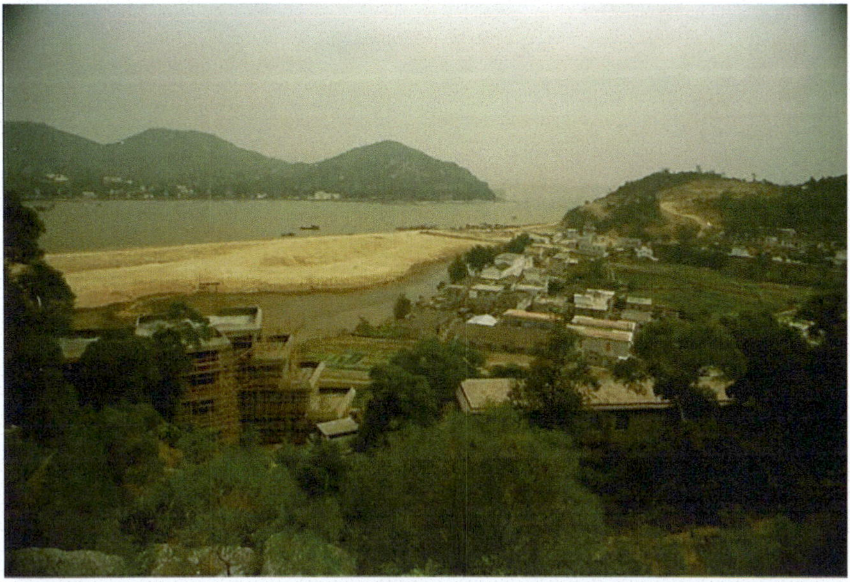

Fig. 53 Hongqi Village before land reclamation (*Source* From Zhuhai Jingzhi Photography Design Co., Ltd.)

predecessor of the Hengqin Passenger Terminal, and the ferry ride to Macau took only 15 min. With the reclamation of the eastern coastal area of Hongqi Village in the 1990s, the "Hengshui Ferry" was relocated to the east side of present-day Huandao East Road (環島東路) (Fig. 57).

The original ferry terminal's waiting hall was just a simple iron shed. In the twentieth century, the city appearance of Hengqin changed rapidly along with the country's development and construction. In 2013, with the opening of the first Zhuhai International Circus Festival and the Zhuhai Chimelong Ocean Kingdom, the ferry terminal on Hengqin Island underwent gradual beautification and renovation, with the demolition of the original iron buildings and the construction of a new terminal. In 2019, the Hengqin ferry Terminal underwent further renovation, with the addition of several new convenient facilities. The exterior facade was also changed to its current European style (Fig. 58).

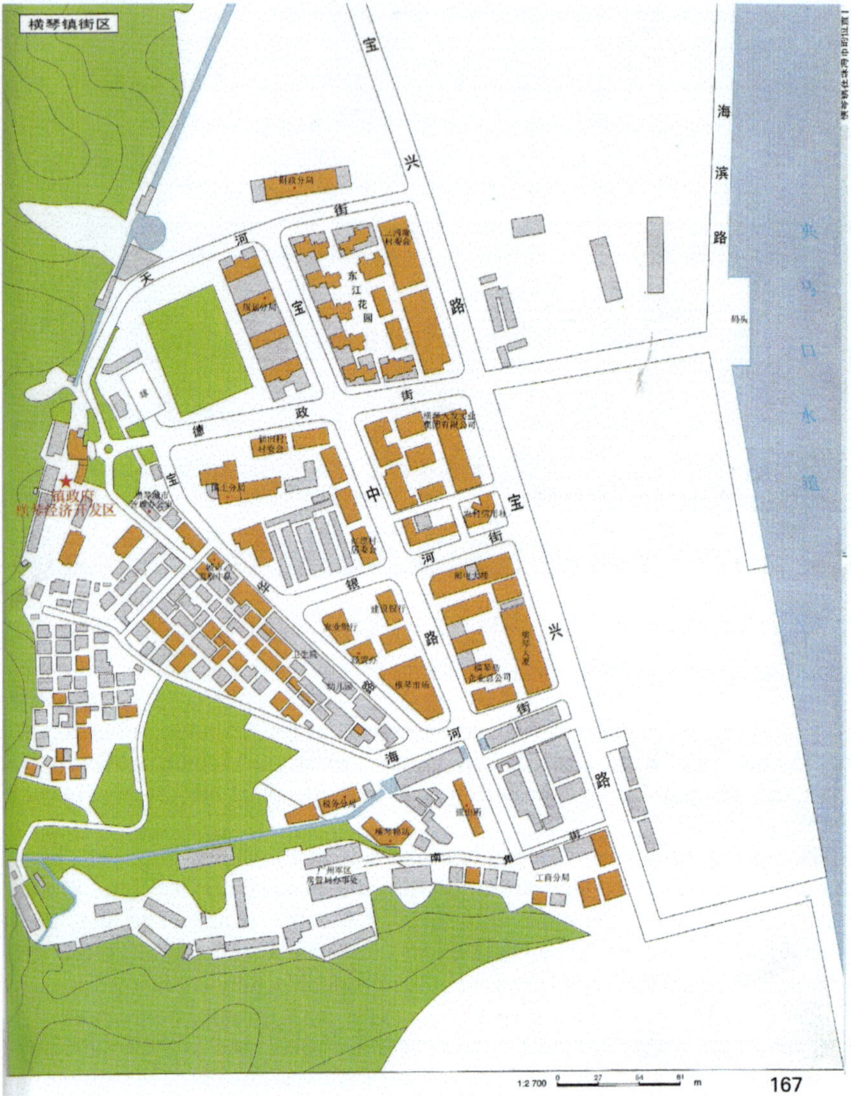

Fig. 54 Plan of blocks of Hengqin Town Center (Hongqi Village) in 1998 (*Source* The Editorial Committee of The Atlas of Zhuhai City, 1998, Ref. [3])

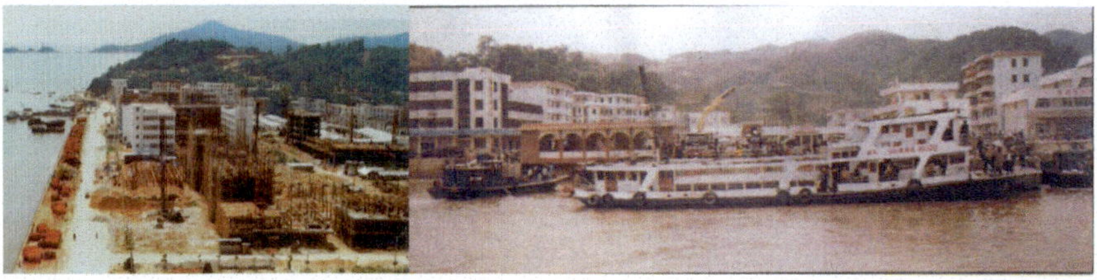

Fig. 55 The eastern urban area (Hongqi Village) and the Hengqin Ferry Terminal under construction in Hengqin in 1992 (*Source* The left one is from Zhuhai Jingzhi Photography Design Co., Ltd., the right one is from Wu Nansheng, 1993, Ref. [27])

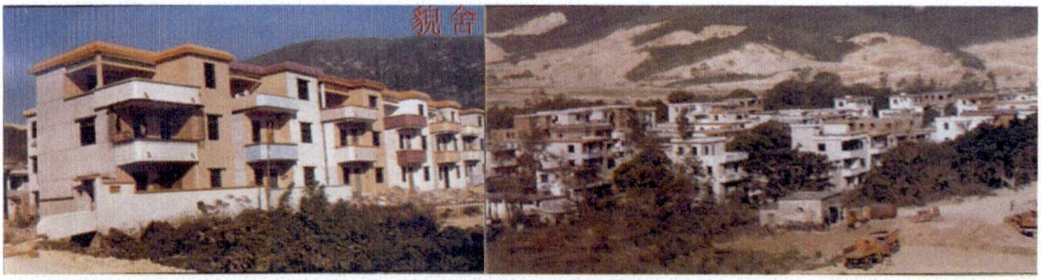

Fig. 56 Modern new houses made of reinforced concrete in the village (*Source* Wu Nansheng, 1993, Ref. [27])

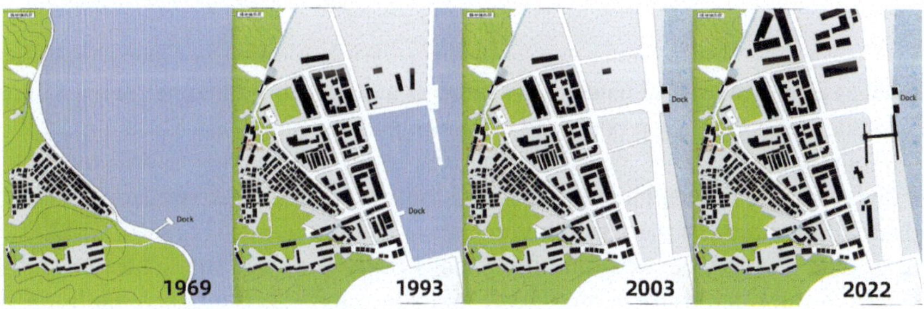

Fig. 57 The evolution of town plans of Hongqi Village (*Source* Created by the author, analysis based on the 1998 map of Hengqin Town Center, and satellite image from 1969 to 2022)

5 Period of Integrated Development (2009–Present)

5.1 The Policies and Plans for Promoting the Integrated Development of Hengqin

In January 2009, the central government announced the official development of Hengqin. On August 14 of the same year, the State Council approved the "The Overall Development Planning for Hengqin New Area" and in December, the Hengqin New

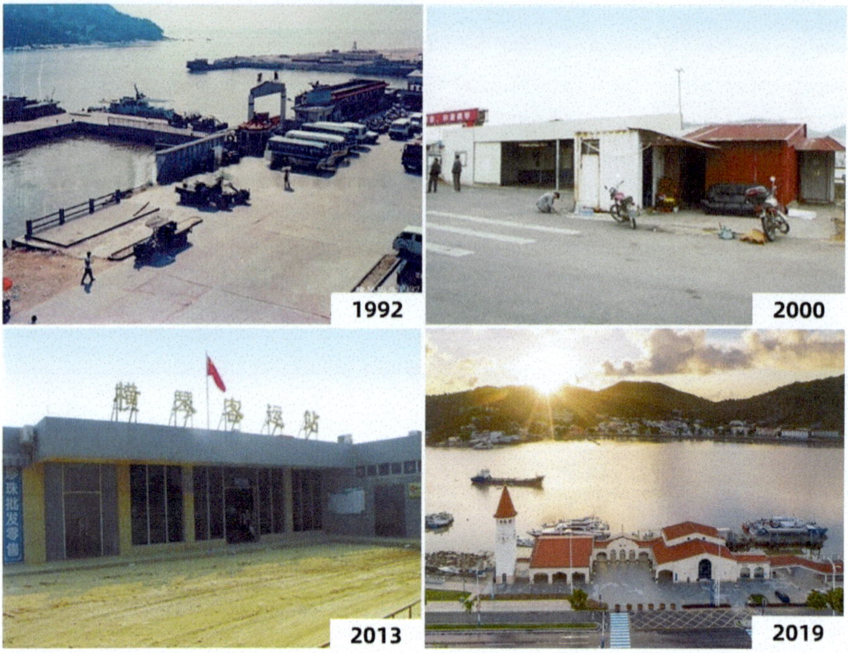

Fig. 58 Hengqin Port construction old photos (1992–2019) (*Source* From Zhuhai Jingzhi Photography Design Co., Ltd., Hengqin Guangdong-Macau Deep Cooperation Zone Official Website)

Area was established. The new area is taken as a demonstration zone for exploring new models of cooperation between Guangdong, Hong Kong, and Macau under the "one country, two systems" principle, a pioneer zone for deepening reform, opening-up, and scientific and technological innovation, and a new platform for promoting industrial upgrading in the western Pearl River Estuary region (Fig. 59). The 2009 master plan divided the area into three zones comprising ten sub-zones: business services, leisure and tourism, and R&D and education. This enriched and refined the diversity of land utilization in Hengqin (Fig. 60). The 2010 Zhongxin Channel property rights transfer unified land ownership in Hengqin. In 2015, the Hengqin Free Trade Zone was established, and in 2018, the "Hengqin New Area and Bonded Zone, Hongwan (洪灣), and Wanchai (灣仔) Area Integrated Development Plan" was officially announced. On September 5, 2021, under the "General Plan for Building a Guangdong-Macau In-depth Cooperation Zone in Hengqin," Hengqin's strategic development status was upgraded again with the establishment of the Guangdong-Macau In-Depth Cooperation Zone in Hengqin. It is defined as a new platform to promote the diversified development of Macau's economy and to gradually establish, enhance, and improve the pattern and mechanism of integrated development between Hengqin and Macau. A series of policies have prompted Hengqin to move from localized development concentrated in Hongqi Village and the vicinity of the new

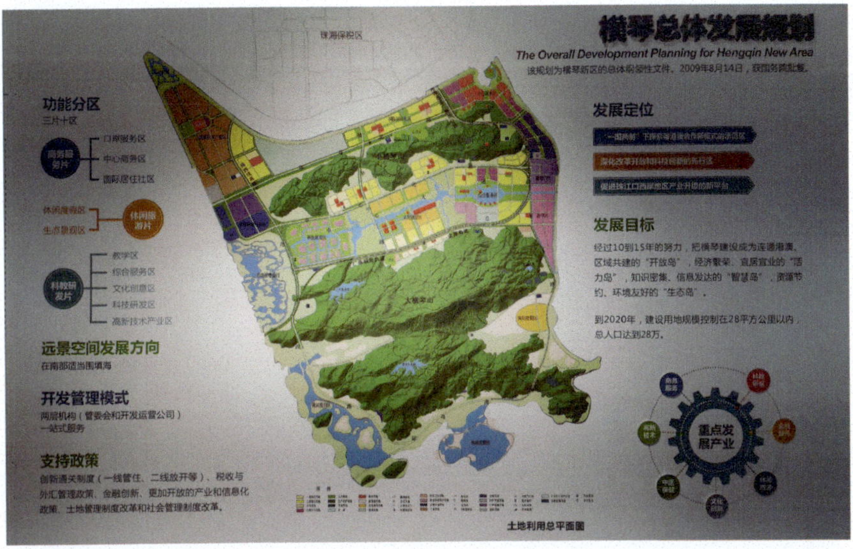

Fig. 59 The overall development planning for Hengqin New Area in 2009 (*Source* Photographed by author)

customs clearance checkpoint to a stage of integrated development across the entire island and its neighboring areas.

5.2 The Evolution of the Island's Plan

5.2.1 The Development Process of Construction Land

Before 2010, construction and development land were scattered around the villages. In the late 1990s, with the opening of Lotus Bridge, construction sites began to increase near the new checkpoint, but development remained scattered. Following the promulgation of the *Hengqin Master Plan in 2009*, under the influence of strengthened cooperation between Hengqin and Macau, a new "Macau-adjacent development zone" along the Hengqin Bridge, Shizimen Financial Island, Lotus Port, and Hengqin Bay has once again taken shape. Projects constructed during this period include the University of Macau, completed between 2009 and 2013, and Zhuhai Chimelong International Ocean Resort, completed between 2010 and 2014. At the same time, on the western side of the island, China National Offshore Oil Corporation moved into Hengqin in early 2000, and in April 2011, the Guangdong-Macau Cooperation Traditional Chinese Medicine Science and Technology Industrial Park was officially established, once again driving the development of the western part of Hengqin. With the transfer of land ownership rights in Zhongxin Channel in 2010, new construction

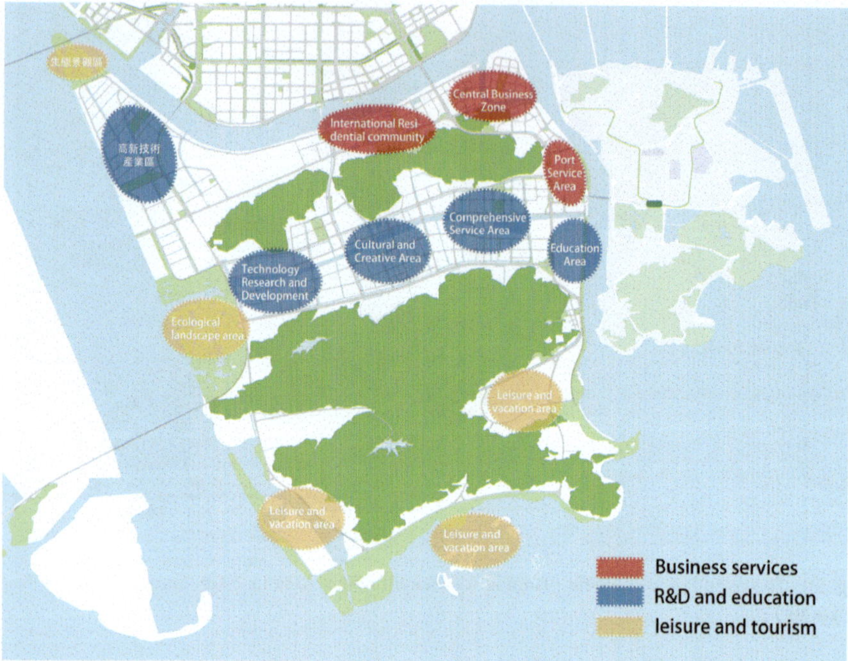

Fig. 60 The functional zoning map of the 2009 Hengqin Master Plan (*Source* Made by author based on The Overall Development Planning for Hengqin New Area in 2009)

land in Hengqin spread westward from the new port on the east side along Zhongxin Channel, forming a T-shaped structure (Fig. 61).

5.2.2 The Evolution of the Local Area's Plan

(1) ***The First Development and the Relocation of Villages in the Area Adjacent to Macau in the East***

Satellite images from 2009 to 2022 have been used to create two maps of construction plots developed after 2009. These maps reveal that during the early stages of Hengqin's overall development, completed projects were concentrated in the eastern area adjacent to Macau and near the Hengqin Port, primarily focused on educational, tourism, residential, and commercial functions. After 2018, new office buildings began to take shape, and the "Financial City" landscape facing the Shizimen Waterfront gradually emerged (Figs. 62, 63 and 64). In the eastern coastal area, as Hengqin Bay was planned to be developed into the Chimelong International Marine Resort, the land where the Xin Village (新村) and Jiu Village (舊村) were located in began to be expropriated and demolished in 2009. Therefore, there are currently only 10 natural villages left in Hengqin (Fig. 65).

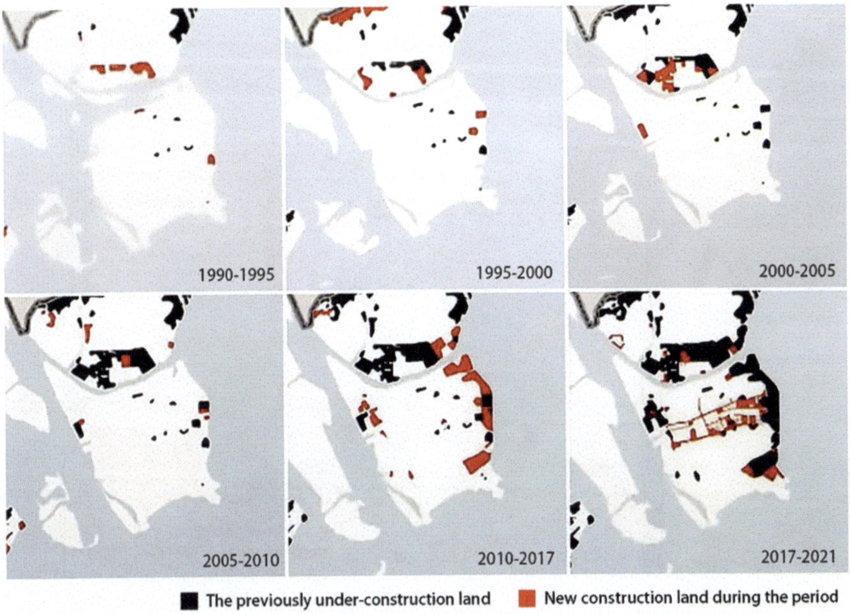

Fig. 61 Development process of construction land from 1990 to 2021 (*Source* Made by author based on Zhuhai Planning Exhibition Hall)

(2) ***The Development and Construction of the Zhongxin Channel and the Demolition of the Reclaimed Buildings***

After 2009, the Zhongxin Channel underwent several years of site formation, and large-scale development and construction began around 2016. The original agricultural landscape of mulberry fish ponds was gradually filled and leveled up into bare construction land. Currently, only a small area of fields remains between Santang Village (三塘村) and Sitang Village (四塘村) (Fig. 66). Some of the original reclamation camps and buildings were also gradually demolished from 2013 onward, while the east and west dikes had also been demolished by the end of 2016 (Figs. 67 and 68). The distribution map of the demolished buildings is shown in Fig. 63. Only the headquarters camp and a few stone houses in the surrounding area have been preserved (Fig. 68), which are currently the only remaining traces of the Zhongxin Channel reclamation project on Hengqin Island. As the area of the Zhongxin Channel waterway decreased significantly due to landfilling, the Zhongxin River grew narrower with a new name as Tianmu River (天沐河). It became not only the most important flood discharge channel between Greater Hengqin and Lesser Hengqin, but also an important landscape corridor for the construction of the Hengqin New Area. The Tianmu River Park (Tianmu River Landscape and Flood Control Project) began in 2017 and completed in May 2019, creating a city waterfront landscape belt along both sides of the river. According to the winning bid for the Hengqin Guangdong-Macau Deep Cooperation Zone Urban Design, six urban corridors will

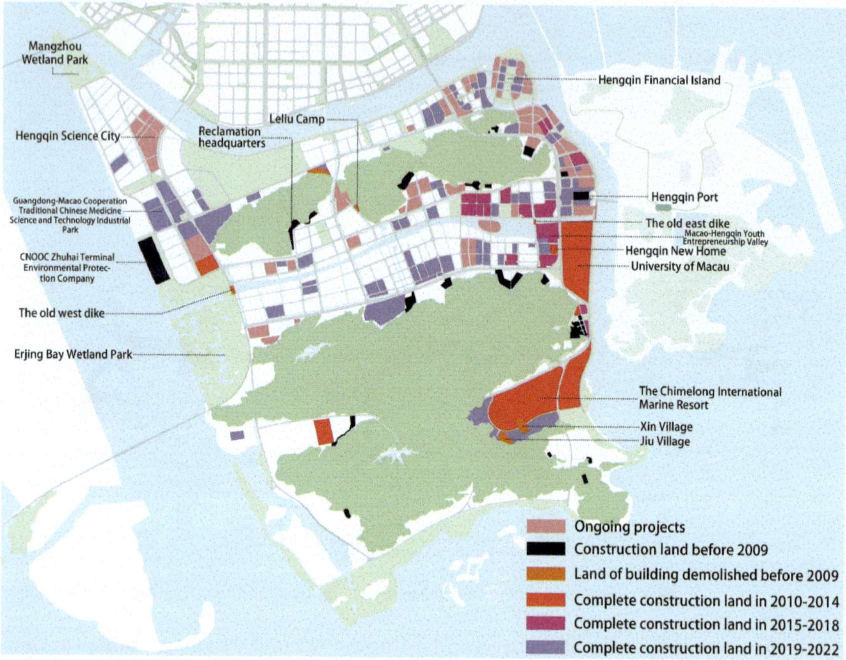

Fig. 62 The evolution map of the construction and demolition of land around 2009 (*Source* Created by the author)

be created in the central area to connect the northern and southern parts of Hengqin Mountain. The central area's dynamic framework will be created along the Tianmu River between the two mountains (Fig. 69).

(3) *The Formation of Natural Landscape Areas in the Northwest and the Erosion of Mangzhou Terrain*

The Hengqin Master Plan clearly proposes that Hengqin should be developed into an "ecological island" where humans and nature coexist in harmony. To this aim, Hengqin has drawn up a comprehensive ecological island development plan comprising nine specialized plans, with the goal of establishing a beautiful bay landscape based on a "one ring, two cores, multiple bays" layout. The "ring" refers to the ecological barrier ring surrounding the waters around Hengqin Island, the "two cores" refer to Mangzhou Wetland Park and Erjing Bay Wetland Park (二井灣濕地公園), and the "multiple bays" include Erjing Bay (二井灣), Shenjing Bay, Changlan Bay (長欄灣), Dadong Bay (大東灣), Hengqin Bay, etc., which seamlessly connect with Macau's "Zhuwan-Heisha Bay (竹灣-黑沙海灣)," providing a sustainable and resilient spatial foundation for the integrated development of Hengqin and Macau. The Hengqin Mangzhou Wetland Park began construction in 2011 and officially opened on New Year's Day 2017. It is positioned as an educational exhibition and a high-quality leisure and social public park with an urban wetland ecological theme.

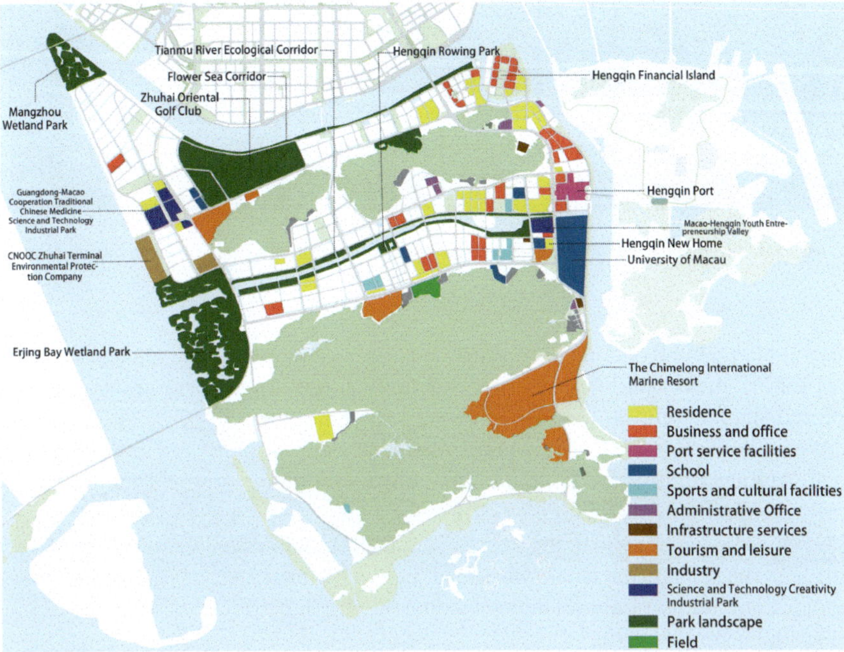

Fig. 63 Function of completed construction land after 2009 (*Source* Created by the author)

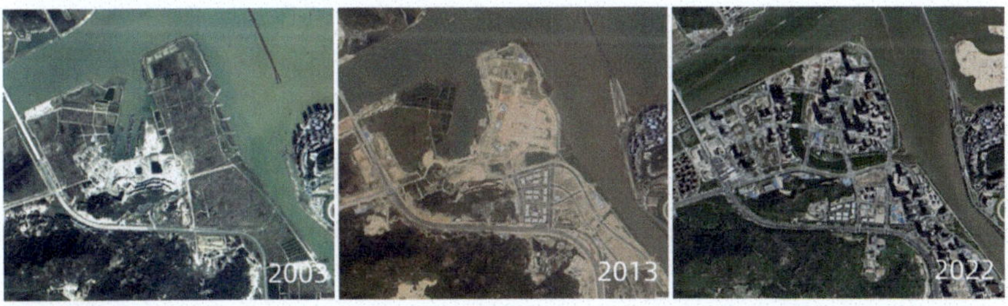

Fig. 64 Satellite image of the evolution of Financial Island construction (*Source* From Google Earth)

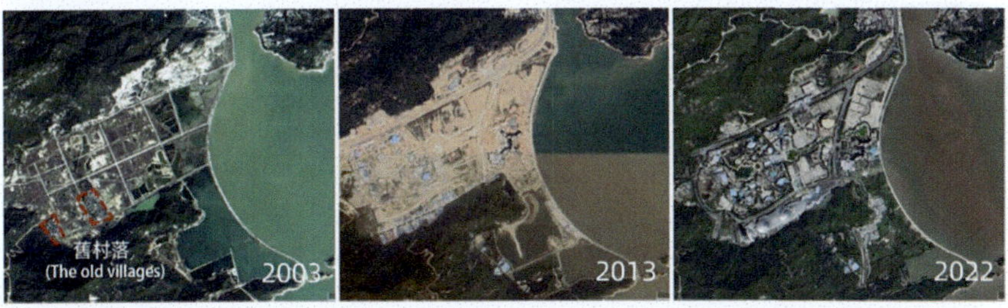

Fig. 65 Satellite image of Hengqin Bay construction evolution (*Source* From Google Earth)

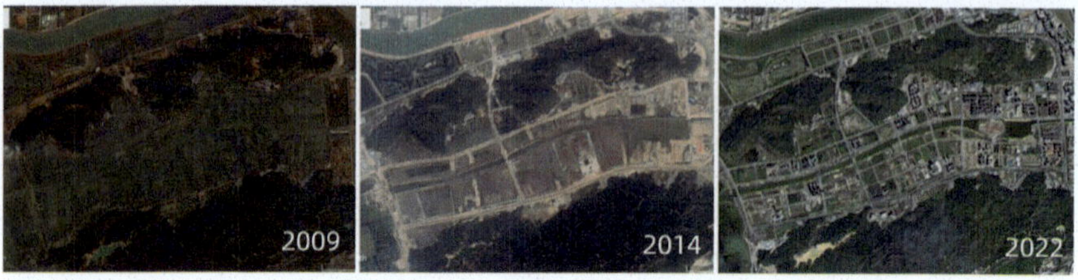

Fig. 66 Satellite image of Zhongxin Channel construction evolution (*Source* From Google Earth)

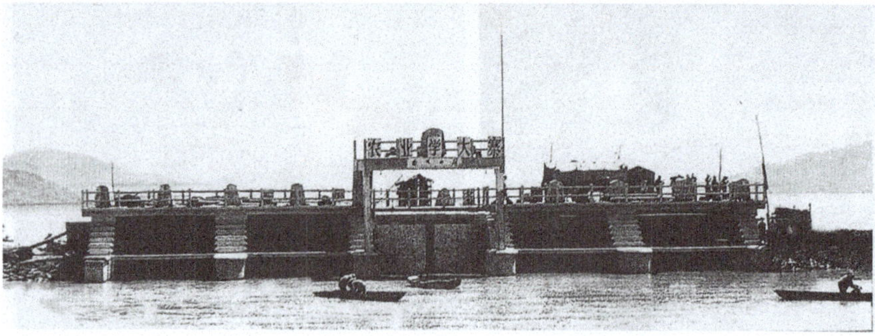

Fig. 67 The west dike is on Zhongxin Channel, 1947 (*Source* Collected by Shunde Archives)

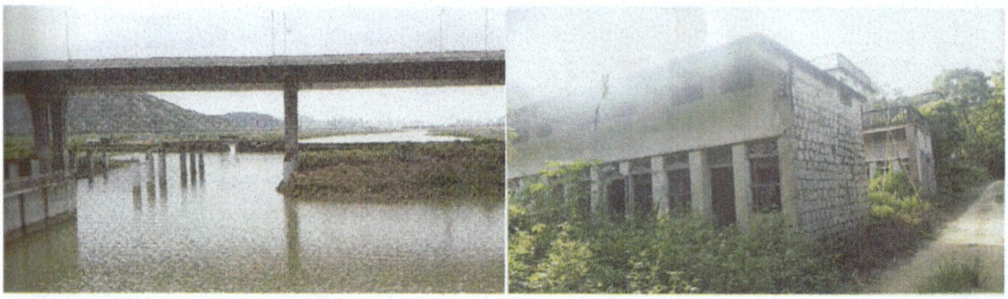

Fig. 68 The old west dike and the new expressway, and the old stone houses, photographed in July 2016 (*Source* Tan Yuanheng, Zhu Wenbin, Lu Menghe, 2018, Ref. [1])

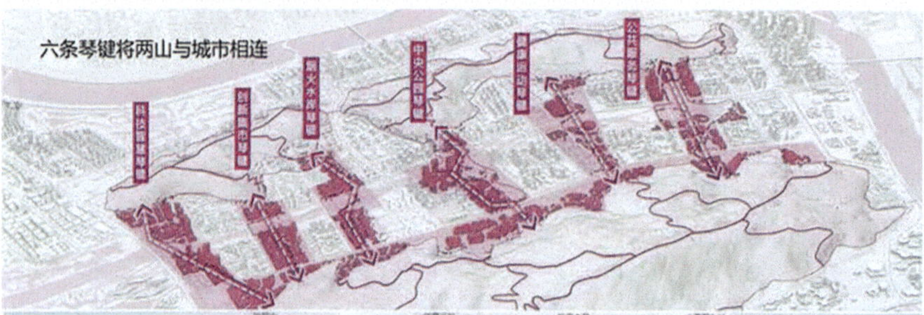

Fig. 69 Six corridors connecting mountains and waterways (*Source* The winning urban design proposal for the Hengqin-Macao In-depth Cooperation Zone)

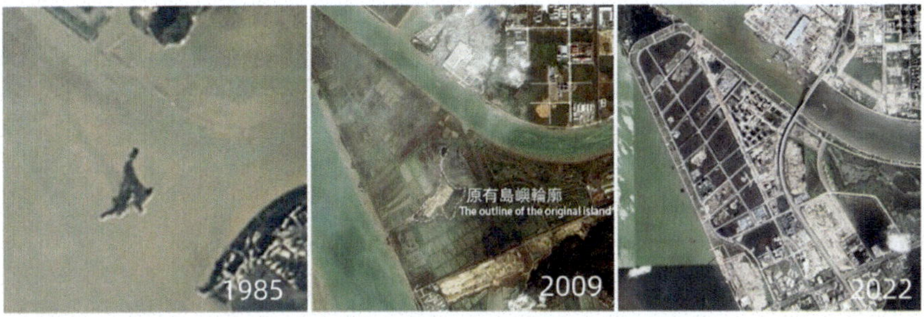

Fig. 70 Satellite image of Mangzhou construction evolution (*Source* From Google Earth)

The construction of Erjing Bay Wetland Park is divided into two phases. The first phase was completed at the end of 2019, focusing on the restoration of coastal mangrove ecosystems, bird conservation, and the construction of a bird watching park. In addition, a coastal flower garden promenade has been developed along the waterfront area of the Malau Zhou Channel in northern Hengqin since November 2010, creating a leisurely, natural, and ecological landscape area (Fig. 63).

The development of the Mangzhou area has led to the erosion of the original terrain of the Mangzhou Islands. The first and second phases of the Hengqin Science City are now located within the original boundaries of the Mangzhou Islands. During the development of Mangzhou, the prominent slopes of the original islands were flattened, and the original outline of the islands disappeared amid the construction of a modern urban road network (Fig. 70).

5.2.3 Roads and Transportation Construction on the Island

During this period, the road network on the island continued to extend westward from the Macau-adjacent zone, with tunnels constructed in the south. At the same time, vertical transportation connections were strengthened, gradually forming a closed ring road network around the island. By the end of 2016, the framework of the main transportation network had basically taken shape (Fig. 71). After 2016, a more detailed street network gradually formed, with further subdivision of street blocks. Under the impetus of the development of the Guangdong-Macao Greater Bay Area and the integration planning of surrounding regions, the Second Hengqin Bridge, Hengqin Tunnel, Phase I of the Zhuhai-Zhuhai Airport Intercity Railway, and the New Hengqin Port have been completed. The Jinhai Bridge (金海大橋) and Phase II of the Zhuhai-Zhuhai Airport Intercity Railway are currently under construction, further strengthening connectivity with surrounding areas (Fig. 72).

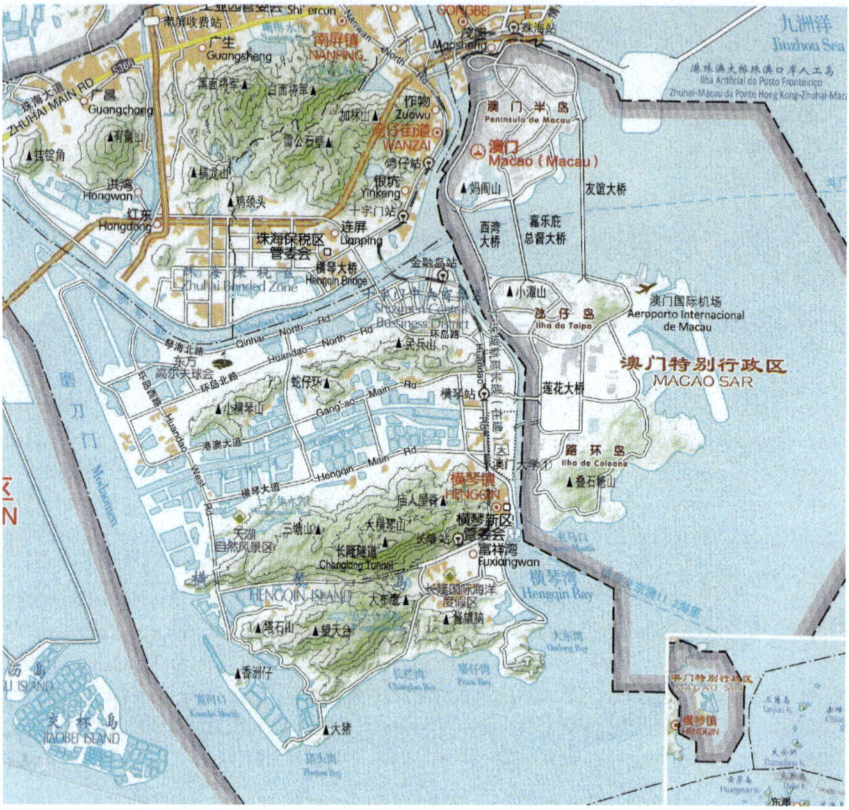

Fig. 71 Map of Hengqin in 2016 (*Source* Department of Land and Resources of Guangdong Province, Municipal Bureau of Land and Resources of Zhuhai, Macao Cartography and Cadastre Bureau, 2016, Ref. [30])

5.3 Building Form

5.3.1 Current Situation of Reclamation Buildings

During the Zhongxin Channel development, reclamation camps such as Leliu Camp (勒流營) and Xingtan Camp (杏壇營) were completely demolished after 2009, with only the reclamation headquarters and some buildings in the nearby Lecong Camp (樂從營) remaining. The preserved area is planned to be developed into a historical and cultural area of Hengqin called "Shunde Zhongxin Channel Town" (Fig. 73 and Table 1). Most of the old stone houses have been converted into restaurants, one is a small shop, and one is still inhabited, while the rest are vacant.

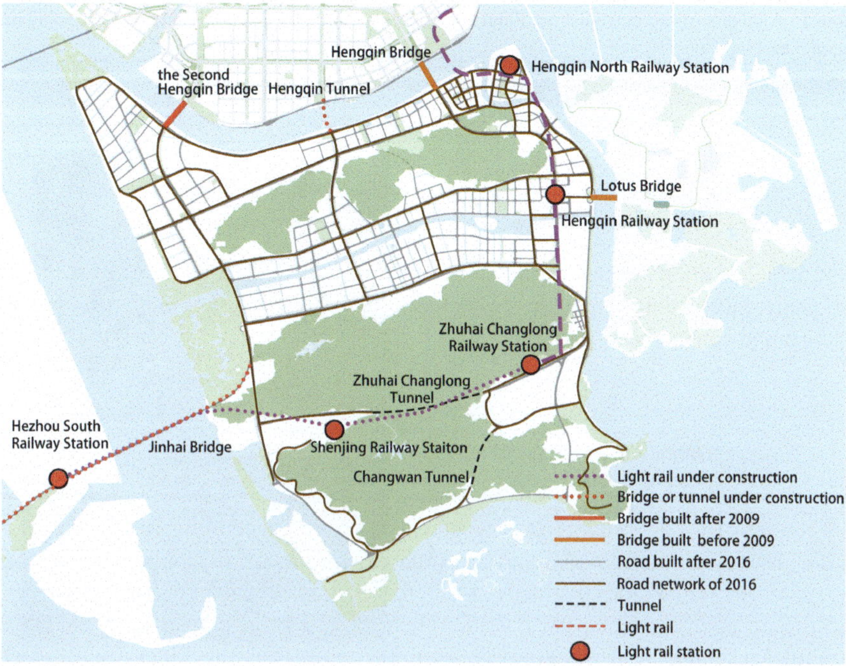

Fig. 72 Analysis map of roads and transportation construction of Hengqin (*Source* Created by the author)

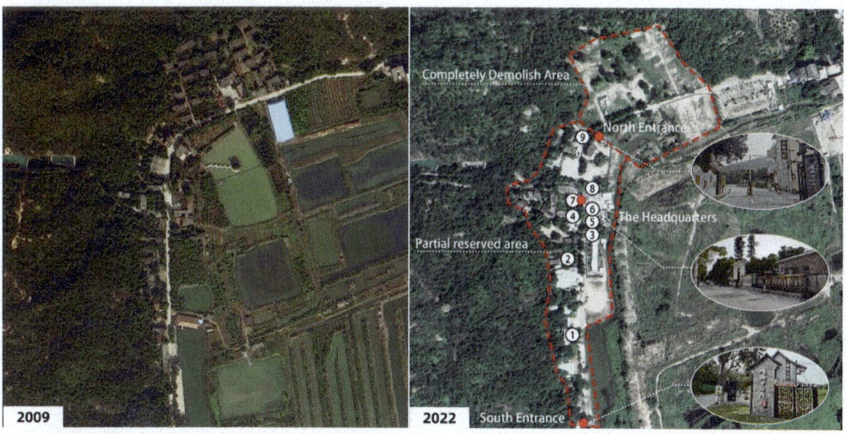

Fig. 73 Satellite image of the evolution of Zhongxin Channel Headquarters area from 2009 to 2022 (*Source* From Google Earth)

Table 1 Analysis of the current functional status of the buildings at the old site of the Zhongxin Channel reclamation

Serial number	Original function	Current function	Current photos
1.	Grain shop	Restaurant (Hengqin Commune)	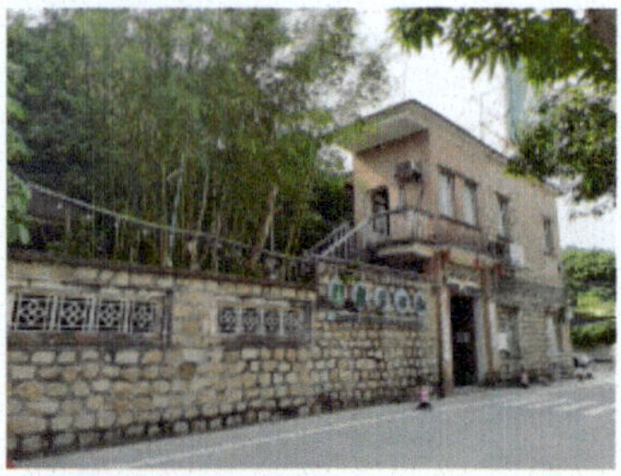
2.	Dormitory (Lecong Camp)	Vacant	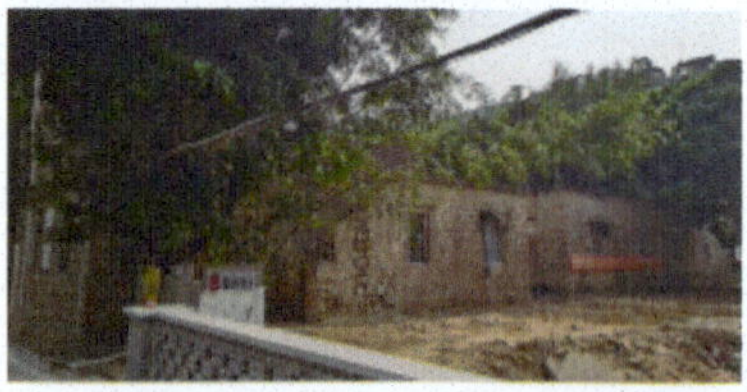
3.	Bank	Vacant	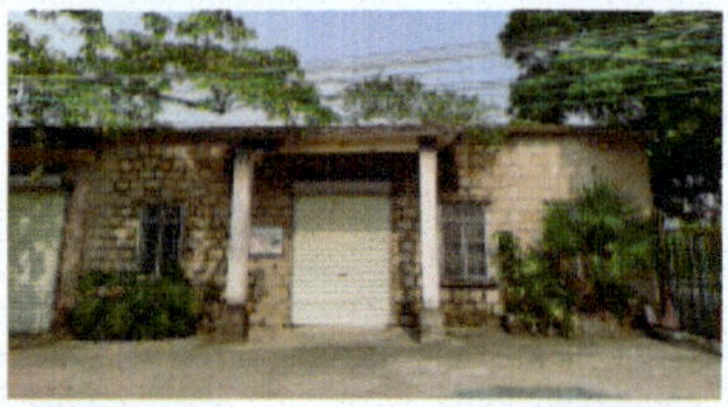
4.	/	Store	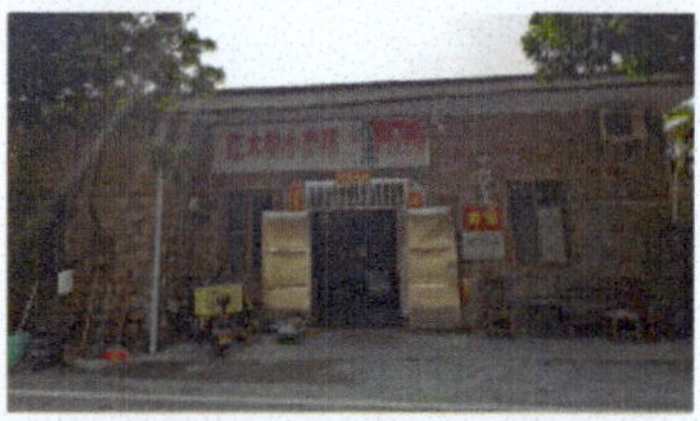
5.	Warehouse	Vacant	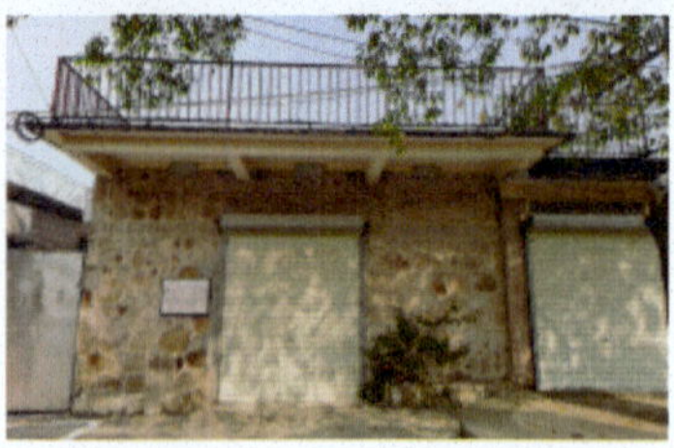

(continued)

Table 1 (continued)

Serial number	Original function	Current function	Current photos
6.	Hengqin Hospital	Restaurant (Lakeside Cuisine)	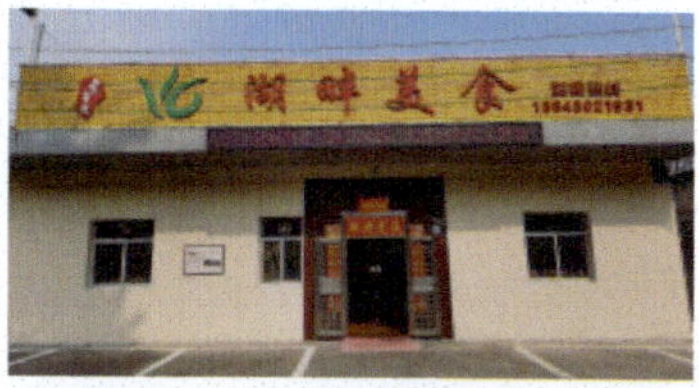
7.	Headquarters	Administrative office work and exhibitions	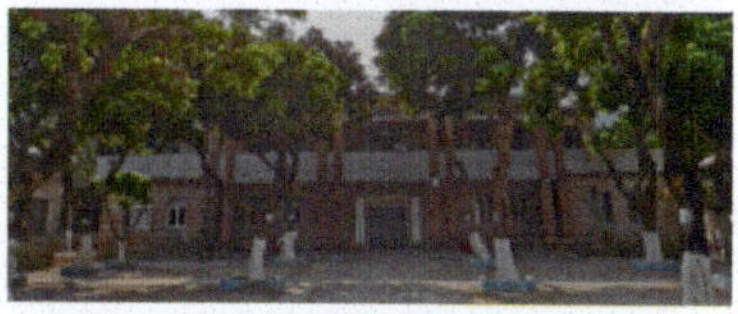
8.	/	Residence	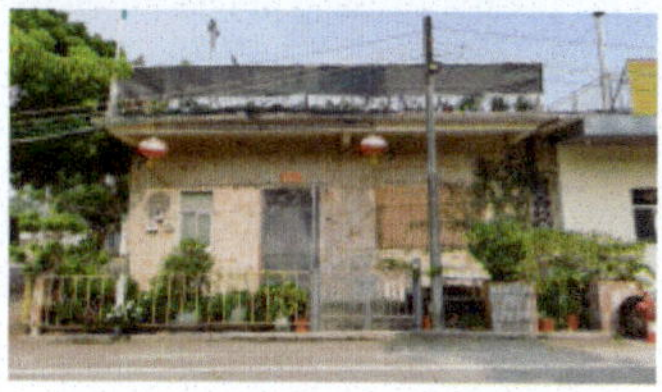
9.	Dormitory	Restaurant (Shunjing Farm)	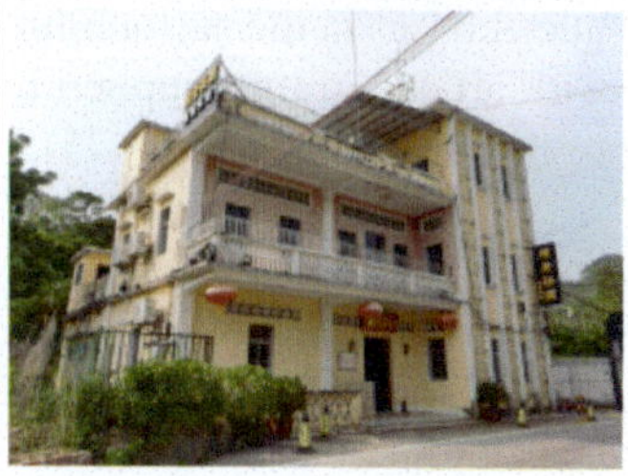

Photographed by author

5.3.2 A Large Number of Self-Built Rental Houses Emerged with the Sharp Increase of Migrant Workers

Before the comprehensive development of Hengqin, the village buildings in Hengqin were mostly low-rise Lingnan-style dwellings (Fig. 74). Since the comprehensive development, there has been new wave of migrant workers moving in, leading to an increase in demand for housing rentals since the comprehensive development. However, with no overall planning for villages, the villagers replaced their original old dwellings with a large number of multi-story rental units driven by economic interests. Although the villagers' incomes have increased, the large-scale demolition

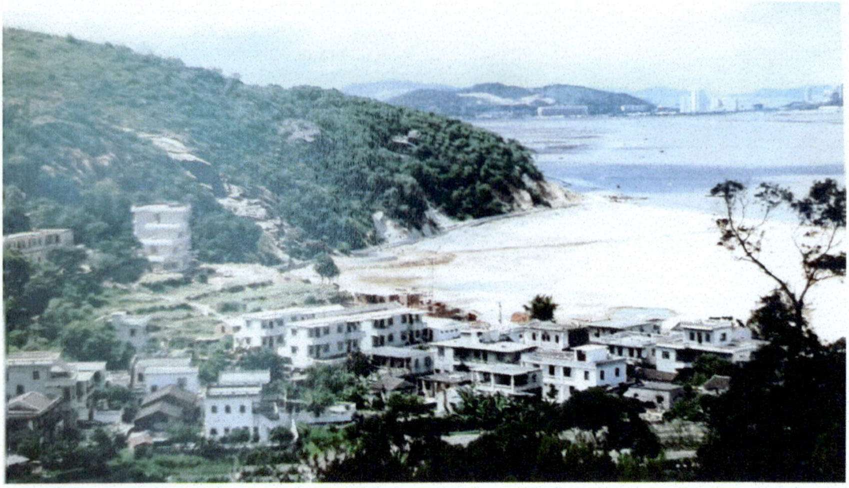

Fig. 74 Old photos of the villages in Hengqin Town at the end of the last century (*Source* Wu Nansheng, 1993, Ref. [27])

of old houses has erased the original characteristics of the southern coastal fishing village, transforming them into some densely populated urban villages, which is not beneficial to the overall urban image and cultural connotations of Hengqin. Following the completion of the Chimelong Resort, houses were successively converted into themed guesthouses in the surrounding villages of Shengjing, Hongqi, Shishan, and Sitang, with artistic graffiti applied to the exterior facades to enhance the fun and interest of the environment (Fig. 75).

5.3.3 Modern Architectural Projects with the Characteristics of the Integration of Hengqin and Macau Under the Integrated Development

The University of Macau Hengqin Campus is the earliest large-scale project to be developed on Hengqin Island since its formal development. Its overall design is based on the traditional Lingnan architectural style, which is characterized by its grand and inclusive nature, the elegance of Confucianism, simplicity and naturalness, and a low-key and reserved aesthetic, while also incorporates elements of Southern European architectural design. It has created a pleasant environment for socializing, reading, and thinking [31], featuring a modern college-style campus that blends Chinese and Western characteristics.

The skyscraper office buildings in the Shizimen Financial Center and near new Hengqin Port echo the Zhuhai International Convention and Exhibition Center to the north and the Macau Tower to the east. The Lijun Pangdu (勵駿龐都) Commercial Plaza across from the Hengqin Port adopts the Manueline architectural style that was

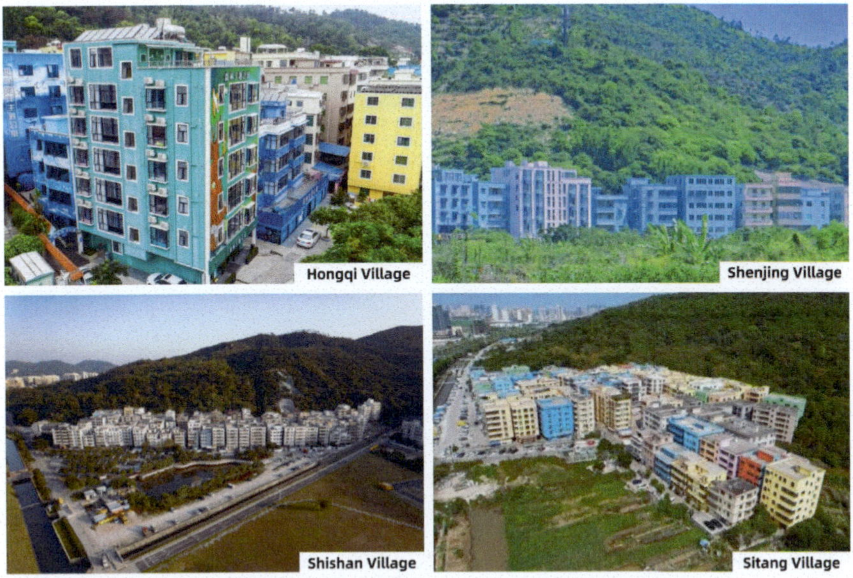

Fig. 75 Current photos of the villages (*Source* From Baidu image)

popular in Portugal during the fifteenth and sixteenth centuries, with some Chinese cultural elements added, continuing Macau's architectural culture that blends Chinese and Western styles. It also reflects the financial and commercial integration and cultural exchange between Hengqin and Zhuhai.

To develop new industries that promote the diversified development of Macao's economy, Hengqin has established projects such as the Macau Youth Entrepreneurship Park, the Guangdong-Macao Cooperation Traditional Chinese Medicine Science and Technology Industrial Park, and the Hengqin Science City, providing a diverse platform for Macau talent to develop in Hengqin.

To provide convenience for Macau residents living and working here, several new residential community projects have been built, such as the "Macau New Neighborhood (澳門新街坊)" project, which integrates with Macau's public services, including education, community services, and healthcare. The design style of some public areas will also incorporate Southern European elements (such as Portuguese-style paving), creating a living environment similar to that of Macau (Fig. 76).

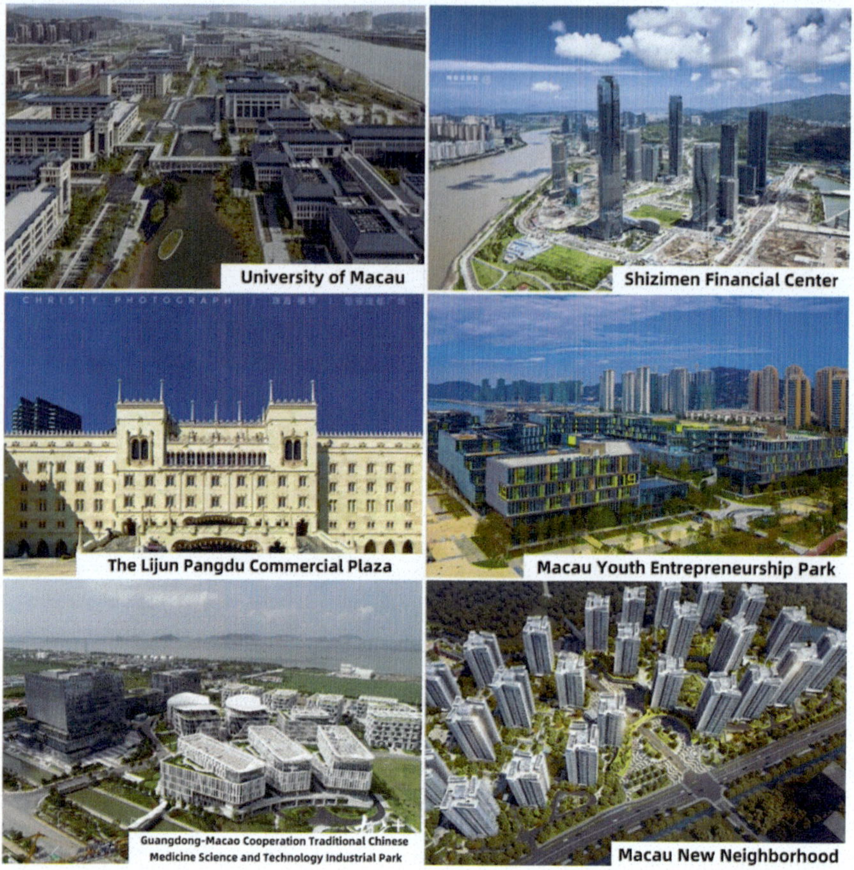

Fig. 76 Modern architectural projects with the characteristics of the integration of Hengqin and Macau (*Source* From Baidu image)

6 Conclusion

6.1 *The Urban Morphology of Hengqin at Different Periods and Influencing Factors*

6.1.1 Inner and Outer Shizimen Period (Before 1949)

During this period, Greater and Lesser Hengqin were separated and formed a regional archipelago with the surrounding islands, with the Shizimen Waterfront serving as the hub. With the dawn of the Age of Discovery in the sixteenth century, Hengqin became a key point for maritime trade between China and Western countries, as well as a national border defense strategic stronghold as an important gateway to the

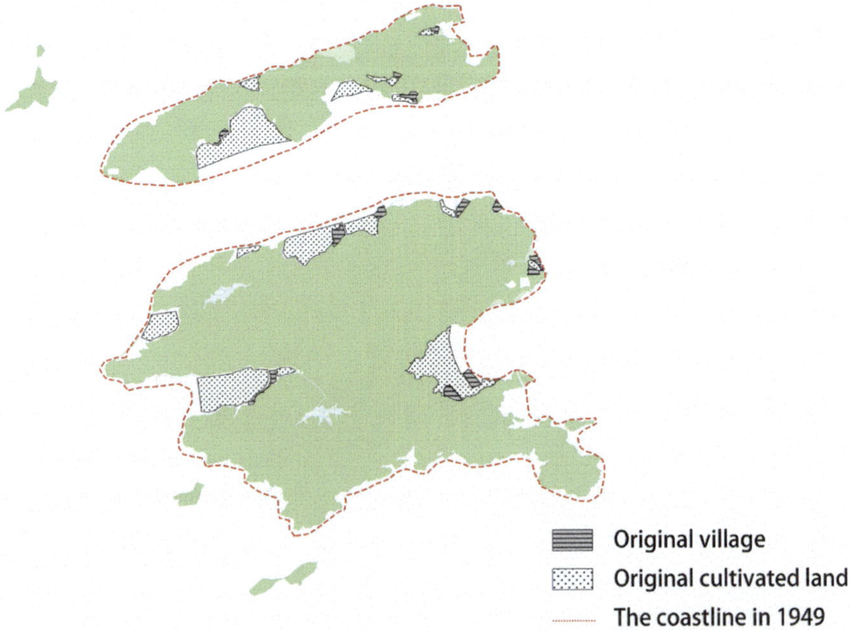

Fig. 77 Plan units of Inner and Outer Shizimen Period (before 1949) (*Source* Created by the author)

Pearl River Delta from the South China Sea. After the opening of Macau trading port, Lesser Hengqin became Macau's outer port, gradually raising its social and historical status and having a certain impact on the development of the original Hengqin settlement. After the Qing Dynasty's policy of coastal evacuation and restoring, it was not until the mid-to-late Qing Dynasty that immigrants began to settle in Hengqin and establish villages, forming early settlements concentrated in various bays. Most of these villages were located near the Shizimen Waterway. Based on maps from 1949 and satellite images from 1969, the plan units map of Hengqin during the Inner and Outer Shizimen period was drawn (Fig. 77). After the foundation of the People's Republic of China, land reclamation projects were both carried out on Hengqin and Macau. With the construction of the Taipa and Coloane Continuity Road in Macau in 1963 and the reclamation of the Zhongxin Channel in Hengqin in the 1970s, the east–west passage of the Outer Shizimen was blocked, ending the Inner and Outer Shizimen Period.

6.1.2 Zhongxin Channel Reclamation Period (1949–1989)

In order to solve the problem of food shortages in the early years of the new China, the Zhongxin Channel Reclamation Project was carried out. This was the first large-scale unified planning and development project in Hengqin. Through this project,

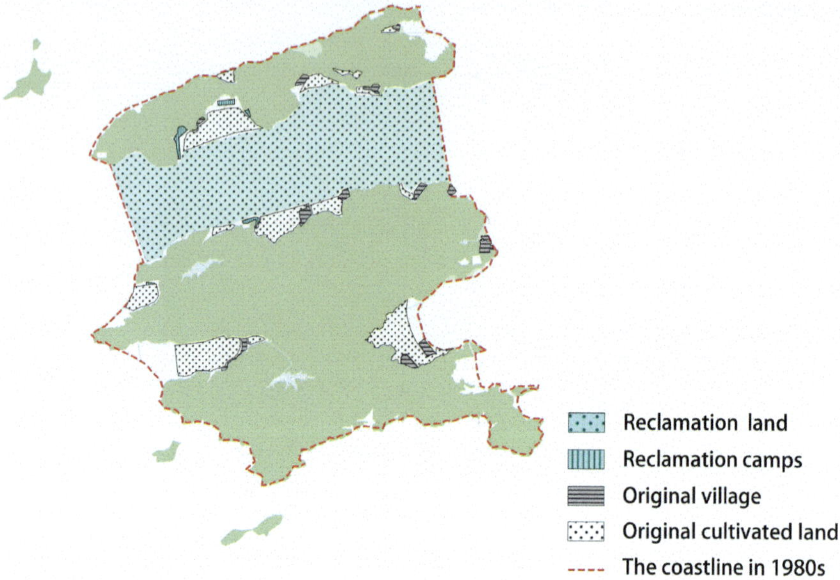

Fig. 78 Plan units of Zhongxin Channel Reclamation Period (1949–1989) (*Source* Created by the author)

Greater Hengqin and Lesser Hengqin were connected, and Zhongxin Channel was transformed from a tidal flat into a picturesque landscape featuring mulberry fish ponds. The construction of the agricultural water conservancy system established the basic framework for the plan layout of Zhongxin Channel, with two mountains sandwiching a river, and the modern urban road system. During this period, the location of Shang Village was also relocated. Based on the reclamation settlements, villages, and reclaimed lands developed during different periods, a map showing the plan units of the Zhongxin Channel reclamation period was created (Fig. 78).

6.1.3 Period of Development Along the Margins (1990–2009)

Driven by reform and opening-up and the construction of economic development zones in the 1990s, Zhuhai unilaterally carried out a second land reclamation project on Hengqin, further expanding the island's boundaries. However, due to complex factors such as policy, economy, and land ownership, Zhongxin Channel was not developed together with the marginal reclamation areas in the 1990s, resulting in the characteristic margin development pattern along the island in the 1990s. After Macau's return to China in 1999, the opening of the Lotus Port drove the development of surrounding land, causing the urban center of Hengqin to gradually move northward from Hongqi Village. The opening of the Hengqin Bridge facilitated transportation between the island and the outside world, leading to a gradual shift in the

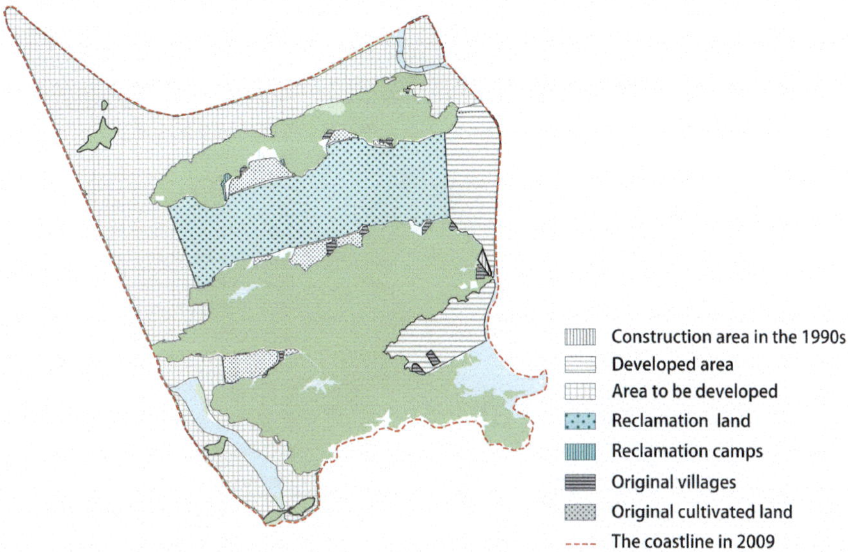

Fig. 79 Plan units of Period of Development Along the Margins (1990–2009) (*Source* Created by the author)

island's primary mode of transportation from waterways to road networks. Based on different levels of development, plot types, and building types, a plan unit diagram for the margin development period was drawn (Fig. 79).

6.1.4 Period of Integrated Development (2009–Present)

Under the influence of national policies, the island's landscape remained relatively stable during this period, and it entered a new phase of urban construction. The development process of the construction plots followed the pattern of point-to-line-to-area evolution. Initially concentrated near the Hengqin Port, the plots later formed a development belt along the border with Macau. After 2017, they gradually spread westward, ultimately forming a planar development. With the construction of new bridges, new ports, and light rail, the connection between Hengqin and the surrounding areas has been strengthened. In terms of architectural style, in addition to the historical reclamation buildings that have been preserved and the local village houses, many new buildings have been constructed with the cultural diversity of Hengqin and Macau. Based on the overlay of information such as different levels of development, planning zones, land use, road networks, and building types during this period, a city plan unit map for the integrated development period of Hengqin was drawn up (Fig. 80).

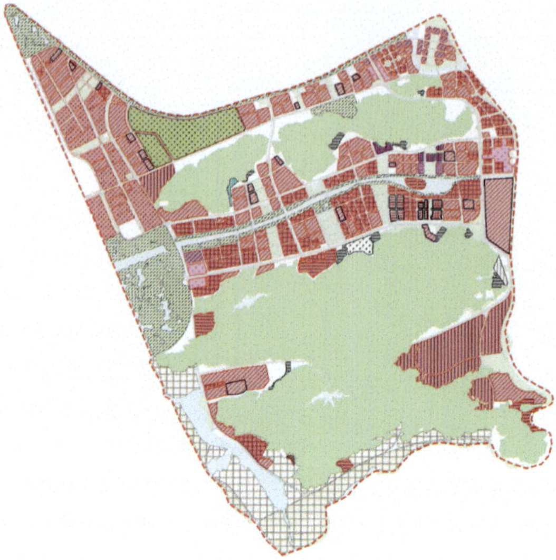

Fig. 80 Plan units of Period of Integrated Development (2009–present) (*Source* Created by the author)

6.2 The Evolution of the Relationship Between Hengqin Urban Morphology and Macau

The evolution of Hengqin's urban form shows a certain degree of diversity and connection with surrounding areas. From a broad view to regional historical spatial evolution, the development of Hengqin's urban form is closely related to Macau.

Hengqin and Macau share similar prehistoric archaeological remains. Neolithic artifacts similar to those from the Black Sand Site (黑沙遺址) in Coloen, Macao, have been unearthed at the Shenjing Mountian Site (深井山崗遺址) in Hengqin and the Chisha Bay Dune Site (赤沙灣沙丘遺址). Therefore, it is speculated that Hengqin and Macao have had similar growth genes since ancient times. Historically, Hengqin and Macau were both located within the peripheral Shizimen Islands, which were under the same border control measures imposed by the central government. Furthermore, the original villages of Hengqin and Macau were composed of Cantonese fishermen and Hakka immigrants from the surrounding areas, and there were certain kinship ties between the villages on both sides of the border. During the Age of Discovery, Hengqin entered its embryonic stage of urban development. Influenced by the opening of Macau trading port and the activities of the Portuguese, Hengqin formed settlements and places associated with it, such as Yangchuan Bay and the Bashilan Leprosy Hospital. For a long time after the founding of new China, communication between two sides decreased due to the ideological differences between them, as Macau was occupied by the Portuguese, and the cities grew independently. Nevertheless, from the 1970s to the early 2000s, the urban form development processes of Hengqin and Macau were quite similar. Both connected the two islands in the north and south into a unified whole by Zhongxin Channel Reclamation in Hengqin and Coloane-Taipa Reclamation Project in Macau, gradually integrating them while the margins continued to expand. With the return of Macau, the Hengqin Bridge and Lotus Bridge were constructed, reconnecting Hengqin's urban development with Macau. In the early twenty-first century, the urban center of Hengqin gradually moved northward from Hongqi Village to the vicinity of the Lotus Bridge (now the Hengqin Port). Following the official development by the central government, Hengqin has once again formed a development belt adjacent to Macau (similar to the area where the Portuguese previously had activities in Hengqin), starting from the area near the Hengqin Port. The process of integrated development was in a radiating pattern from point to line to area, spreading across the entire island. The development correlation between Hengqin and Macau has generally shown a tendency from moderately strong to weak, then back to moderately strong, and finally to strong (Figs. 81 and 82).

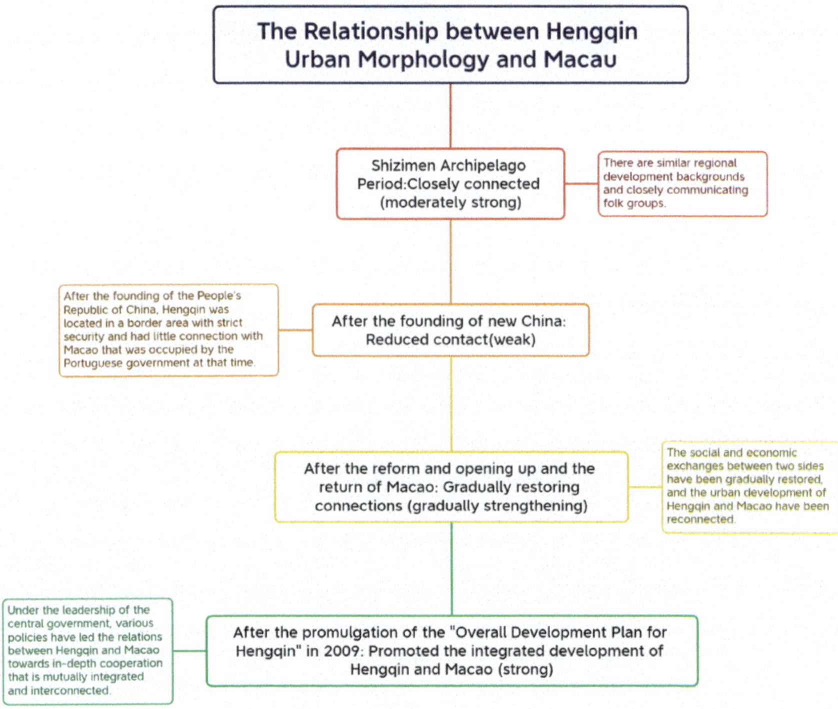

Fig. 81 Analysis diagram of the relationship between Hengqin Urban Morphology and Macau (*Source* Created by the author)

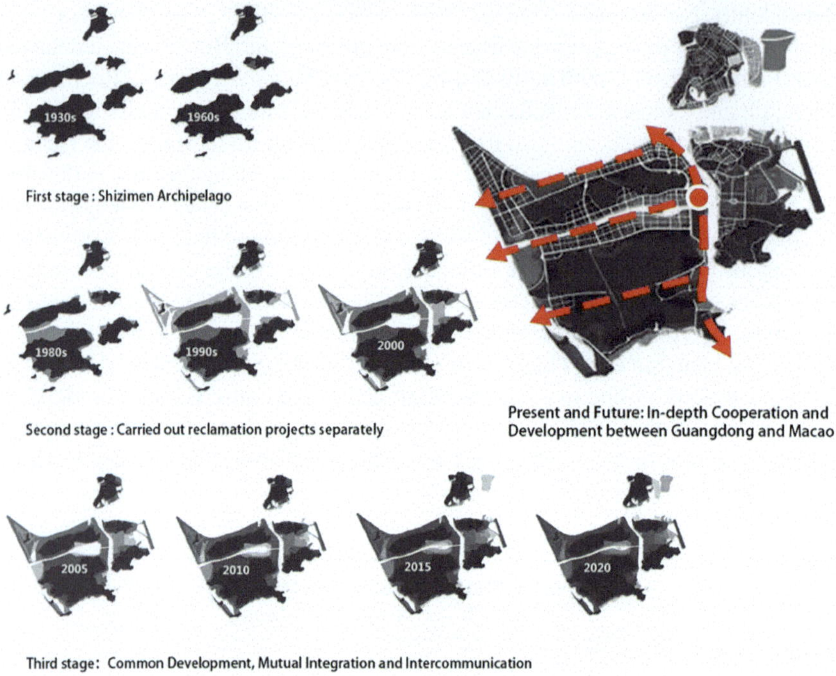

Fig. 82 The morphological development map of Hengqin and Macau at different stages (*Source* Redrawn from the winning urban design proposal for the Hengqin-Macao In-depth Cooperation Zone)

6.3 Concluding Remark

The urban morphology of Hengqin has undergone four morphological periods, with each period of change being driven primarily by local policies toward Hengqin or its surrounding areas during different periods. For example, the Portuguese Macau governments' policy of expelling merchant ships led to the formation of Yangchuan Bay. The timing of the formation of Hengqin Village was closely related to the central government's maritime border policy during the Yuan, Ming, and Qing Dynasties. The Zhongxin Channel Reclamation Project, reform and opening-up, and the construction of economic development zones, which led to two major developments and dramatic morphological changes in Hengqin, were also the result of different periods of central government policy orientation. The overall evolutionary process showed strong continuity in the first three periods, with the final period exhibiting more significant changes. In conclusion, the urban development of Hengqin has experienced different degrees of connection with Macau at each stage, from the early stages of settlement and village formation to the subsequent emergence and migration of town centers. Overall, the degree of connection has gradually weakened and then strengthened again.

Funding This research received no external funding.

Institutional Review Board Statement Not applicable for studies not involving humans or animals.

Informed Consent Statement Not applicable for studies not involving humans.

Conflicts of Interest The authors declare that the research was conducted in the absence of any commercial or financial relationships that could be construed as a potential conflict of interest.

Appendix 1: Common Chinese Historical Place Names in Greater and Lesser Hengqin

See Table 2.

Appendix 2: Common Foreign Historical Place Names in Greater and Lesser Hengqin

See Table 3.

Appendix 3: Village Establishment Information in Hengqin

See Table 4.

Table 2 Table of common Chinese historical place names in greater and lesser Hengqin

Name of document or map	Greater Hengqin place name	Lesser Hengqin place name	Original text/image
Jiajing Period *Xiangshan County (1546)*, Volume 1 (嘉靖 1546《香山縣誌》卷一)	深井山 (Shenjing Mountian); 深井 (Shenjing); 仙女澳 (Xiannu Ao); 井澳 (Jing Ao)	/	深井山即仙女澳也, 亦名井澳, 在横琴下, 宋端宗禦舟曾至此. (Shenjing Mountain is also known as Xiannv Ao, or Jing Ao, located in Hengqin. Emperor Duanzong of the Song Dynasty once visited here by boat)
Maps in Jiajing Period—*Xiangshan County (1546)* (嘉靖 1546《香山县志》圖)	大横琴山 (Greater Hengqin Mountain)/小横琴山 (Lesser Hengqin Mountain)	/	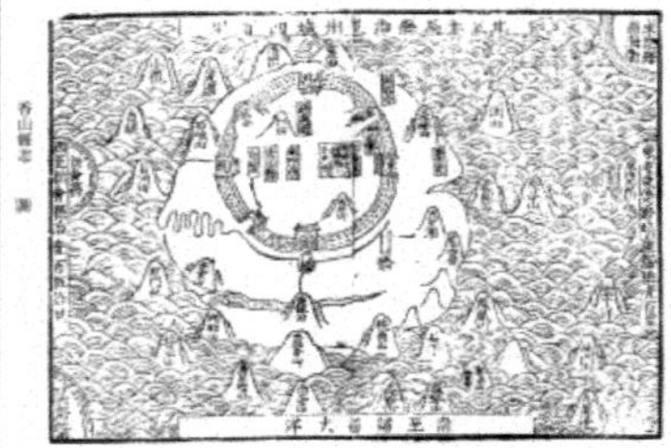
Wanli Period (1573–1620)—*Guangdong General Records* (萬曆《廣東通志》)	井澳 (Jing Ao)	/	横琴山 (有大小横琴)其下有井澳. (There was Jing Ao under the Hengqin Mountain, which included Greater and Lesser Hengqin)

(continued)

Table 2 (continued)

Name of document or map	Greater Hengqin place name	Lesser Hengqin place name	Original text/image
Ming Dynasty *Compilation of Sea Maps—Map of Mountains and Sands along the Guangdong Coast* (明《籌海圖編》之《廣東沿海山沙圖》)	/	小黄琴山 (Lesser Huangqin Mountain)	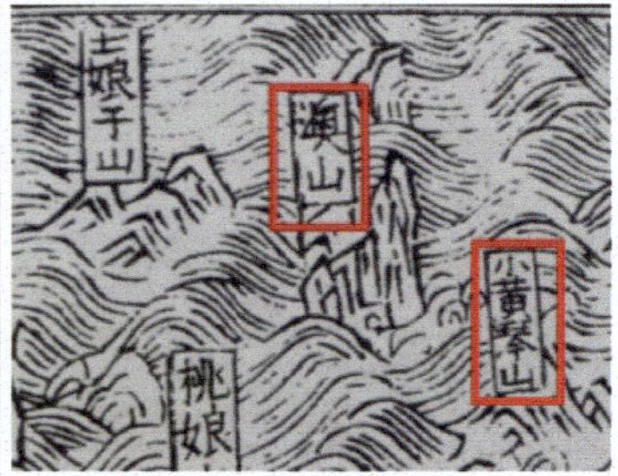 (Image from the article of Jing Guopin, 2022 [10])
Qing Tongzhi Period Map of *Xiangshan County* (清同治《香山縣圖》)	大小横琴山 (Greater and Lesser Hengqin Mountain); 深井 (Shenjing); 二井 (Erjing); 亞婆尾 (Yapo Wei); 過路環 (Guolu Huan)	舵尾 (Duowei)	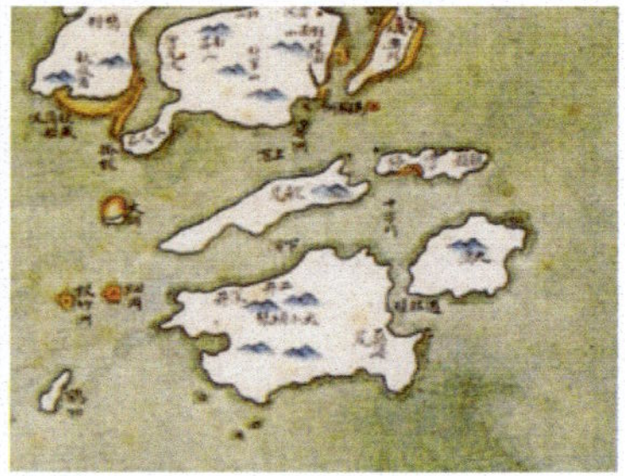 [32]

(continued)

Table 2 (continued)

Name of document or map	Greater Hengqin place name	Lesser Hengqin place name	Original text/image
Qing Dynasty Qianlong Period *Brief Records of Macao—General Map of Maritime Defense* [6] (清乾隆《澳門紀略》之《海防屬總圖》)	深井 (Shenjing); 浅井 (Qianjing)	大角豆 (Toi-ko-ke-tou)	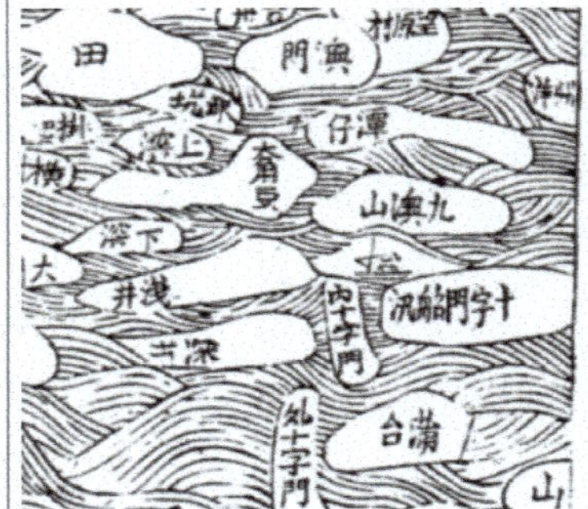 (From Brief Records of Macao,1751[6])
Qing Dynasty Tongzhi Period Map of Guangdong Navy Camp Troop Deployment (Macao Section) (清同治《廣東水師營官兵駐防圖(澳門部分)》)	黃琴洲 (Huangqin Zhou)	淺井 (Qianjing)	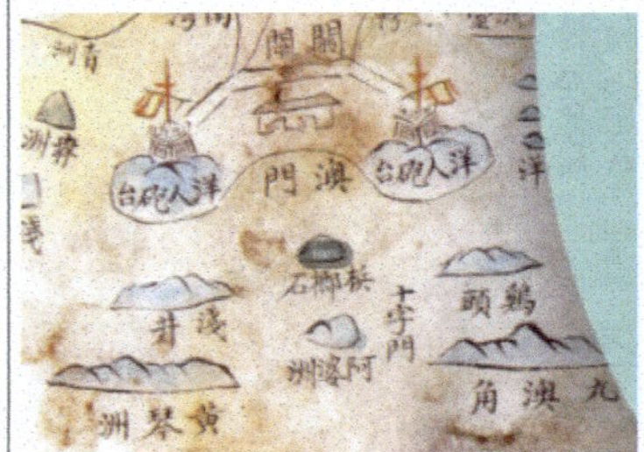 (*Source* Unknown)

(continued)

Table 2 (continued)

Name of document or map	Greater Hengqin place name	Lesser Hengqin place name	Original text/image
Partial map of Guangdong Province in the third year of the Tongzhi period (1864) (清同治三年廣東各府輿圖局部)	十字門洲 (Shizimen Islets)		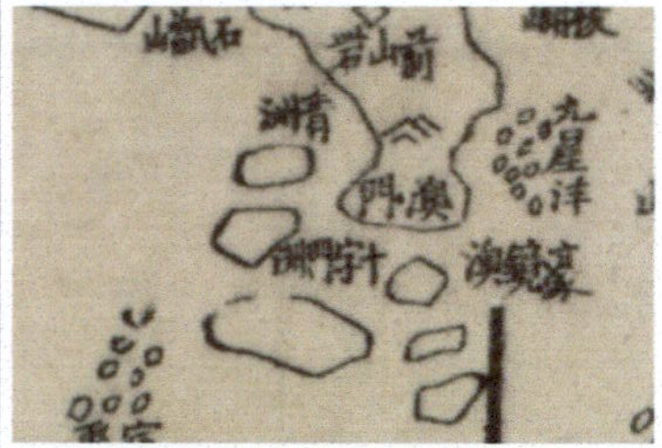 (From Library of Congress website)
Part of the Guangzhou Port section of the Qing Dynasty Guangxu period Map of Key Ports in Seven Coastal Provinces (清光緒《沿海七省口岸險要圖》之 < 廣州口圖 > 局部)	大横琴 (Greater Hengqin)	馬格裡勒 (Macarira)	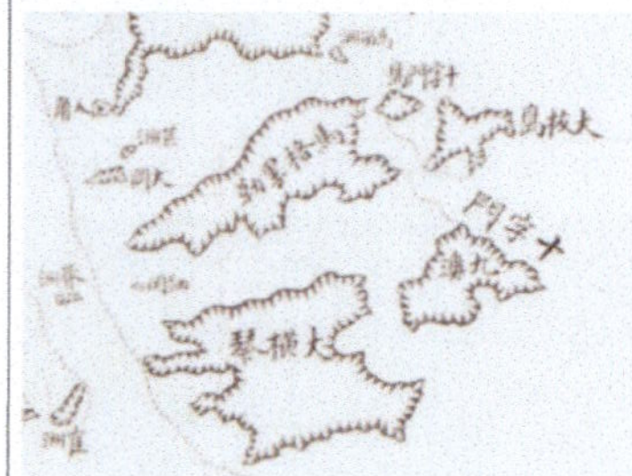 (From the *Selected Historical Maps of Macau* [32])

(continued)

Table 2 (continued)

Name of document or map	Greater Hengqin place name	Lesser Hengqin place name	Original text/image
Partial view of a map of the Guangdong coast during the mid-Qing Dynasty (清中期廣東沿海圖局部)	大黃琴 (Greater Huangqin); 小黃琴 (Lesser Hengqin)	十字門山 (Shizimen Mountain)	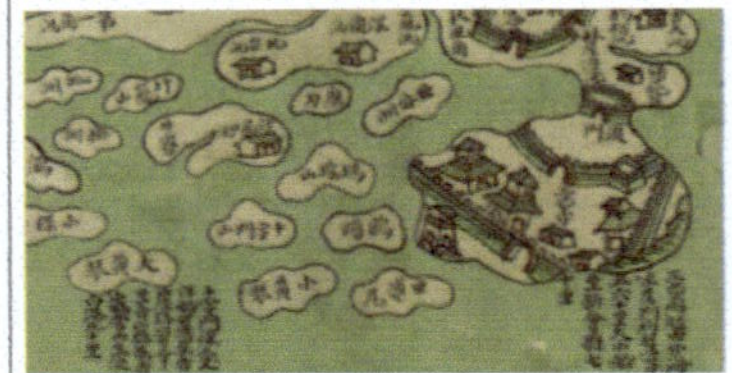 (Collected by the Hong Kong Maritime Museum)
Coastal Map of Seven Provinces of the Prosperous Dynasty (1798盛朝七省沿海圖)	橫琴山 (Hengqin Mountain)	十字門山 (Shizimen Mountain)	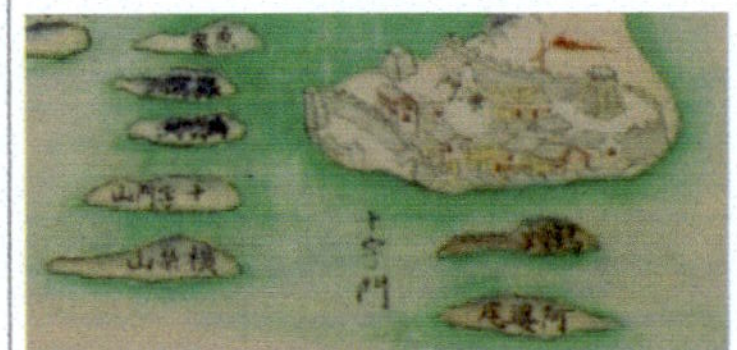 (Collected from the Library of the University of Science and Technology of Macau, Provenance: National Library of China)

(continued)

Table 2 (continued)

Name of document or map	Greater Hengqin place name	Lesser Hengqin place name	Original text/image
1927 Latest Complete Map of Zhongshan County (Partial) (1927 最新中山縣全圖局部)	大橫琴 (Greater Hengqin); 亞當山 (Adam Mountain); 過路環 (Guolu Huan); 摩根角 (Morgan Corner)	小橫琴 (Lesser Hengqin); 長沙欄 (Chi-sion-long); 馬尾角 (Mawei Corner)	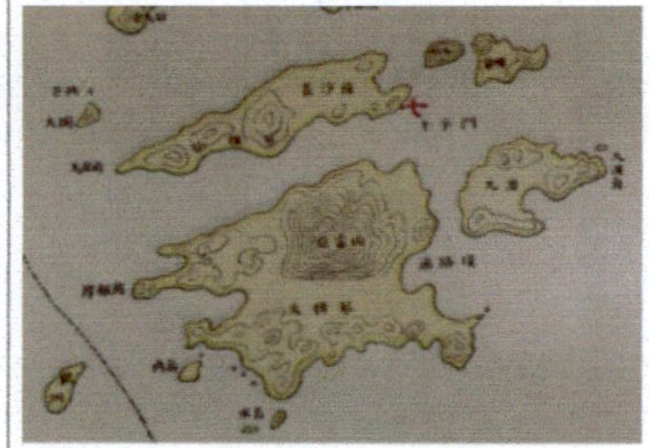 (Collected from the Library of the University of Science and Technology of Macau, Provenance: National Library of China)
The Macau Business Yearbook, published in 1963 (1963年出版的《澳門工商年鑒》)	/	噼囉埠 (Moluobu)	概澳門舊稱濠鏡, 其後改稱澳門. 這一命名的說法有二: 第一, 因澳門南面有氹仔、噼囉埠、路環、橫琴, 四山峙立, 劃海成十字形, 這一"十字門" 合稱為澳門. (Macau was formerly known as Hao Jing, and was later renamed Macau. There are two theories behind this name: first, because Macau is surrounded by four mountains to the south—Taipa, Moluuobu, Coloane, and Hengqin—which form a cross shape across the sea, this "cross gate" is collectively referred to as Macau)

Source Author's statistics

Table 3 Table of common foreign historical place names in greater and lesser Hengqin

Name of document or map	Greater Hengqin place name	Lesser Hengqin place name	Original text/image
1646 Map of the Pearl River Delta and Macau Peninsula	Montanha	Maquareira	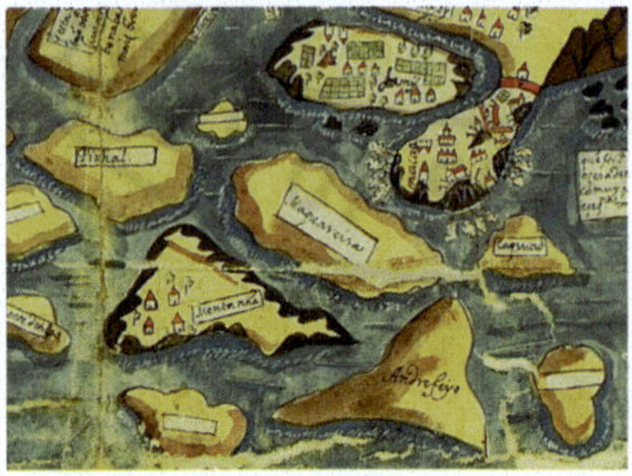 (Collected from the Library of the University of Science and Technology of Macau, Provenance: Library of Ajuda)
Map of islands at the mouth of the Guangzhou River in 1698	Montagna	Macarira	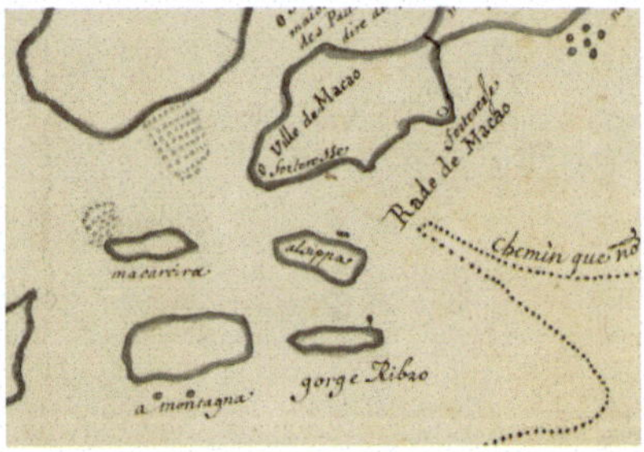 (From Archives of Macau)
1809 The Typa and harbor of Macao	Montanha	Mackkarera	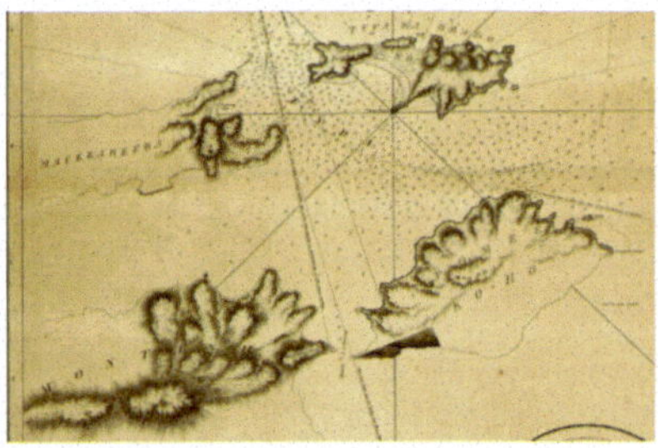 (Collected from Macau University of Science and Technology Library, Provenance: Geographical Society of Lisbon)

(continued)

Table 3 (continued)

Name of document or map	Greater Hengqin place name	Lesser Hengqin place name	Original text/image
1834 map of the Pearl River Estuary	Montanha	Mackkarira	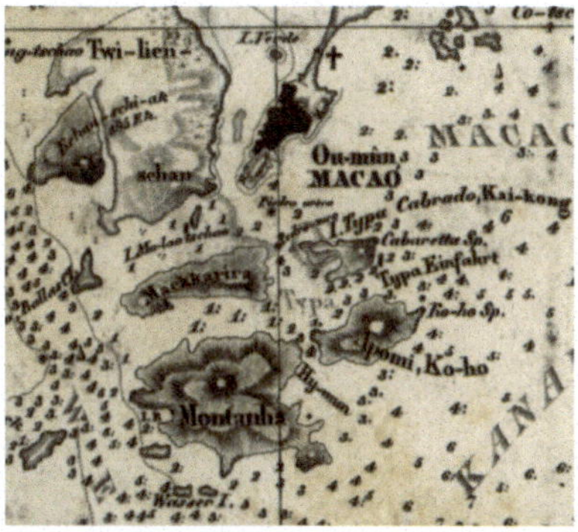 (From Library of Congress, Geography and Map Division)
1866 China, East Coast Macau with the islands and adjacent coasts (partial)	/	Macarira/ Toi-ko-ke-tou	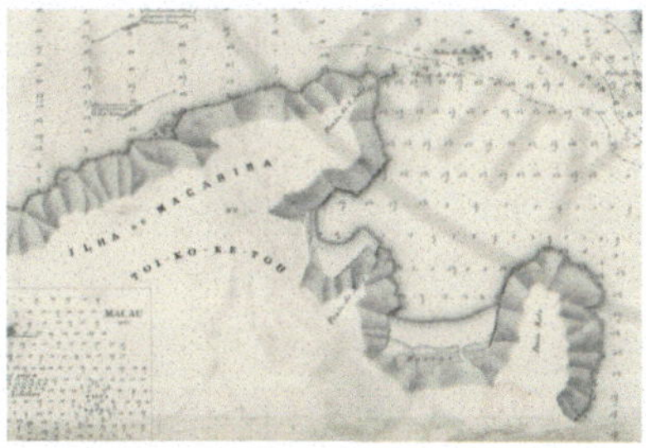 (Collected from Macau University of Science and Technology Library, Provenance: National Library of France)

(continued)

Table 3 (continued)

Name of document or map	Greater Hengqin place name	Lesser Hengqin place name	Original text/image
1891 The map of the boundaries of Macao drawn by the Portuguese government	Vong-can	D. João/ Macarira	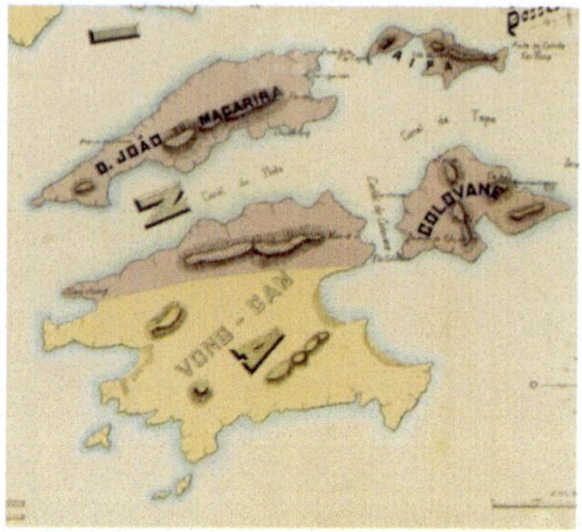 (Collected from Macau University of Science and Technology Library, Provenance: Geographical Society of Lisbon)
Partial annotations of map of Portuguese Colonies, 1901	Tai-vong-cam/ Montanha	D. João	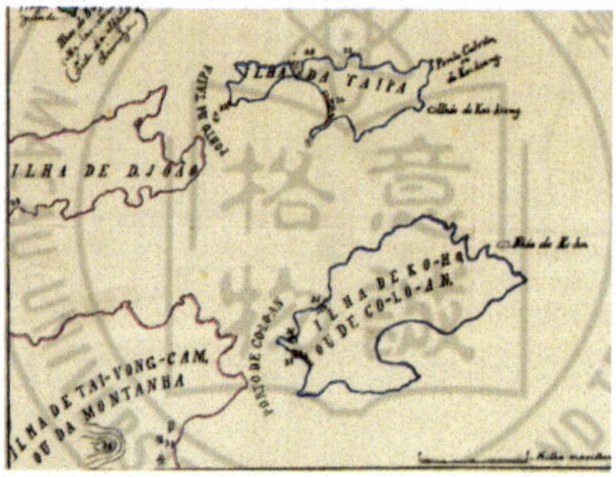 (Collected from Macau University of Science and Technology Library, Provenance: Geographical Society of Lisbon)
Map of Macau Province by the Portuguese Colonial Cartographic Commission, 1909	Tai-vong-cam/ Montanha	Macarira/D. João	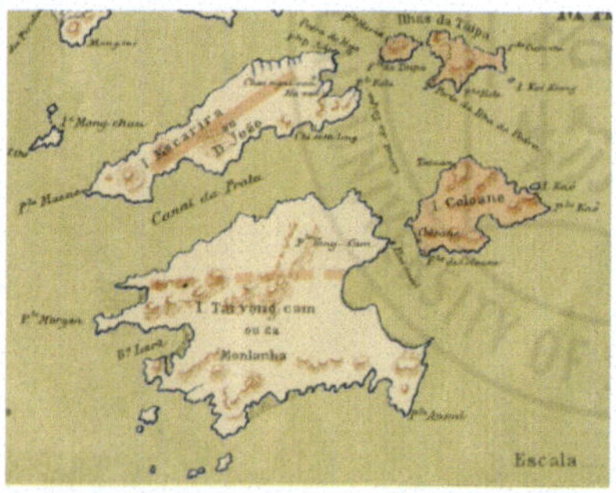 (Collected from Macau University of Science and Technology Library, Provenance: Royal Geographical Society)

Source Author's statistics

Table 4 Table of village establishment information for villages in Hengqin

Location	Village name	Origin of the name	Time of village establishment	The origin of the villagers	Ethnic group	Villagers' livelihood
Lesser Hengqin	Shang Village	Before 1949, it was collectively known as Shuliang Gang Village. In 1962, it was divided into two villages and got their names because the location on the upper and lower slopes of Shuliang Gang. In 1969, the original site of Shang Village was planned to build a pond. Between 1972 and 1976, it was gradually relocated to the new site of Wutang, still known as Shang Village	The Guangxu reign of the Qing Dynasty (1875–1908)	Three brothers surnamed Fan from Xingning came here to build a village	Hakka	Mainly farmed oysters, fish, and also grew rice
	Xia Village			First, farmers surnamed Huang moved in from Guishan, and then farmers from Bao An, Zhongshan, Xinhui, and Yinkeng in Wanzi moved in	Cantonese, Hakka	Mainly farmed oysters, fish, and also grew rice
	Yanghuan Village	Because of the foreign merchant ships gathered here to get the name, also known as Longchuan Huan, Jiefang Village	The Qianlong reign of the Qing Dynasty (1736–1796)	The earliest settlers here were Hakka farmers surnamed Li from Wuhua. Later, farmers surnamed Zhuang from Haifeng moved in	Hakka	At the beginning, he made a living by stone quarrying and fishing. Later, he mainly engaged in oyster farming, and also grew rice, vegetable, and raised fish

(continued)

Table 4 (continued)

Location	Village name	Origin of the name	Time of village establishment	The origin of the villagers	Ethnic group	Villagers' livelihood
	Xiangyang Village	Formerly known as Wanliwei, it was named after the Wanliwei field. Later, it was renamed Xiangyang Village due to its geographical features	The late Qing Dynasty	Farmers from Doumen, Zhongshan, and Xinhui who fled from famine	Cantonese	Mainly planted banana, fish, and rice
Greater Hengqin	Jiu Village	The Jiu village is the earliest natural village on Hengqin Island. The Xin village was formed by the later relocation of the old village. Later, due to the construction of the Chimelong Ocean Resort in Zhuhai, demolition of them began in 2009 and they moved to the New Home (No. 2, Caihong Road, Hengqin New Area) at the end of 2011	The Kangxi reign of the Qing Dynasty (1661–1722)	Those surnamed Su and Huang moved from Zhongshan, and those surnamed Zhong moved from Xinhui to Sanzao and then to Hengqin	Cantonese	Mainly farmed oysters and grew rice
	Xin Village		The Guangxu reign of the Qing Dynasty (1875–1908)	From Jiu Village	Cantonese	

(continued)

Table 4 (continued)

Location	Village name	Origin of the name	Time of village establishment	The origin of the villagers	Ethnic group	Villagers' livelihood
	Hongqi Village	Originally named Manió Village, because there was a waterfall in the north of the village, resembling a horse urinating. The original name was vulgar and it was changed to current name in 1958	The Yongzheng region of the Qing Dynasty (1722–1735)	Mainly from Xingning, Wuhua, Zhongshan	Cantonese, Hakka	Mainly grew rice, and also farmed oysters and fish
	Cusha Huan Village	It is named after a coarse-sand beach several hundred meters long in front of the village	The late Qing Dynasty	From Wuhua	Hakka	Mainly farmed oysters and fish
	Shisha Village	Originally named Wagangzhai, it was later renamed Shishan Village when people moved to a rocky mountain halfway up the mountain	The Yongzheng region of the Qing Dynasty (1722–1735)	From Wuhua	Hakka	Mainly grew rice, and also farmed oysters and fish
	Santang Village	It is named after the sequence of oyster ponds in front of the village	1917	From Zhongshan and Xinhui	Cantonese	Mainly grew rice, and also farmed oysters and fish
	Sitang Village				Cantonese	
	Shenjing Village	This bay is the deepest one on the entire island	1938	From Mutouchong, Tianxin Village, Sanzao, Zhuhai	Cantonese	Mainly planted rice, banana, and oysters

Source Author's statistics by Compilation Committee of *Gazetteer of Geographical Names of Zhuhai City, Guangdong Province*, 1989 [2]

References

1. Yuanheng, T., Wenbin, Z, & Menghe, L. (2018). *History of Reclamation in Hengqin Zhongxin Channel.* Sun Yat-sen University Press.
2. Compilation Committee of *Gazetteer of Geographical Names of Zhuhai City, Guangdong Province.* (1989). *Gazetteer of Geographical Names of Zhuhai City, Guangdong Province.* Guangdong Science and Technology Press.
3. The Editorial Committee of *The Atlas of Zhuhai City.* (1998). *The Atlas of Zhuhai City* Guangdong Map Publishing House.
4. Zhenxing, L. (1999). A Brief History of Hengqin. The Cultural and Historical Materials Committee of the Zhuhai Municipal Committee of the Chinese People's Political Consultative Conference (Chief Editor). *Selected Historical and Cultural Materials of Zhuhai* (60–76). Guangdong People's Publishing House.
5. Shimin, Y. (2021). The ancient cannons and the Yichuan City Museum confront each other, recreating the dignity of the key coastal defense route of Zhuhai's Shizimen. Zhuhai Special Economic Zone Daily. http://zhuhaidaily.hizh.cn/html/2021-04/22/content_1212_4354990.htm
6. Guangren, Y., & Rulin, Z. (1751). *Brief Records of Macao.* Chengwen Publishing House.
7. Shijin, L. (Chief Editor). (1929). *Continuation of the Annals of Xiangshan County.* Chengwen Publishing House.
8. Compilation Committee of *The History of Zhuhai Customs* (1991–2010) (Chief Editor). (2016). *The History of Zhuhai Customs.* China Customs Press.
9. Shengbo, G., Jiaqi, L., & Yanzhao, L. (2009). The evolution of the names of the islands in the southern part of the Pearl River Estuary in Xiangshan County during the Ming and Qing Dynasties. Cultural Magazine (Macao). Issue 72, 2009, 160–161.
10. Guoping, J. (2022). Lesser Hengqin: Interpretation of Chinese and foreign place names and their historical status. Macau Studies (1).
11. Kaijian, T., & Xiaolei, Z. (2020). Research on the "Annotation of the complete sea map: map of the coast of Guangdong" by Song Yingchang of the Ming Dynasties. *Journal of Chinese Historical Geography, 03*, 5–20.
12. Weiche, X. (2022). Summary [Lecture 261 of the Cultural Research Institute] Jin Guoping: The Convergence of East and West in Macao - The Portuguese Coming to the East and the Maritime Silk Road. The official account of the Institute of Humanities and Social Sciences, Peking University. https://mp.weixin.qq.com/s?__biz=MzIzNDQ0MDUwNg==&mid=2247529391&idx=2&sn=85c4c8af9b6f852346bfd92b904387fa&chksm=e8f461fadf83e8ec3faf7a0d09a1c54bc65569142ef8b7bbbfcf88a5136164f2ed4df77334e6&scene=27
13. Longji, D., & Xunling, Y. (2020). *Macao in the global map - Xiangshan volume.* Macau University of Science and Technology.
14. Zhuhai Cultural Relics Management Committee. (1994). *Annals of cultural relics of Zhuhai City.* Guangdong People's Publishing House.
15. Shibao, T. (1996). The legend of "villages" on the Macao Peninsula before the opening of the port. *Cultural Magazine (Macao)* (26), 117–130.
16. Mingyao, T. (Chief Editor). (1873). Tongzhi Xiangshan County Annals. Unknown. Unknown.
17. Wenhui, H. (2022). A study on the construction of villages in Coloane island of Macau and the migration of Hakka people. *Macao Studies Issue, 1*, 2022.
18. Zhiliang, W., & Yunzhong, Y. (Chief Editor). (2005). *Encyclopedia of Macao.* Macao Foundation.
19. Dajun, Q. (1678). *New Writings on Guangdong.* Zhonghua Book Company.
20. Wu., & Yuxian. (2015). The tools of Portuguese colonization in Portuguese Macau during the 19th and 20th centuries: A Study at the Baishalan Leprosy Hospital on Lesser Hengqin Island. *Journal of Shenzhen University (Humanities and Social Sciences Edition), 04*, 149–155.
21. Jingquan, W. (2018). The establishment and evolution of the Macao Leprosy Hospital. *Chinese Journal of Leprosy and Dermatology, 04*, 253–256.

22. Jiemin, Z., & Chunsheng, Y. (2005). *The Land Held Up by Youth*. People's Publishing House.
23. Guangzhou Institute of Geography and Zhuhai Urban Planning Committee. (1991). *Atlas of Zhuhai's Natural Resources and Economic Development Research*. Science Press.
24. Ruichang, G. (2003). *Reclamation of Hengqin Island*. Foshan Writers Association.
25. Senlin, Z. (1993). Seize the opportunity to accelerate the pace of township modernization. Liu Mengyu, Zhou Yingyao (Chief Editors). *The Book of the Hundred Towns of South China (Hengqin Volume)* (1–2). Jinan University Press.
26. Pingbiao, Z. (2022). Fifty Years of Hengqin development - large-scale reclamation and the Hengqin campus of the university of Macau. *Tong Zhou Gong Jin, 02*, 7–10.
27. Nansheng, W. (1993). *The book of the hundred towns of South China (Hengqin Volume)*. Jinan University Press.
28. Zhuhai Planning Department. (2005). *Basic Data Set of "Zhuhai Urban Spatial Development Strategic Plan" (Zhuhai 2030)*. Zhuhai Planning Department.
29. Weilin, H. (1993). Towards the glorious Golden Island of Hengqin. Zhou Yingyao (Chief Editors). *The Book of the Hundred Towns of South China (Hengqin Volume)* (1–2). Jinan University Press.
30. Department of Land and Resources of Guangdong Province, Municipal Bureau of Land and Resources of Zhuhai, Macao Cartography and Cadastre Bureau. (2016). *Atlas of Zhuhai and Macao*. Guangdong Publishing House.
31. Yan, L., & Jun, H. (2011). An ecological and humanistic college integrating Chinese and western elements: A Brief analysis of the college design in the new Hengqin campus of the university of Macau. *New Architecture, 04*, 115–118.
32. Ailian, Z., & Qichang, H. (2000). *Selected historical maps of Macau*. Huawen Publishing House.

Traditional Architecture Types in the Xiangshan Area

The second part focuses on the unique architectural space types in the Xiangshan area, covering ancestral hall architecture, garden landscape, wharf architecture, and the "Pátio" and "Beco" architectural forms.

Ancestral Hall Architecture: The Clan Culture in Tangjiawan Ancient Town

Yinqi Wang

Abstract The ancestral halls of Tangjiawan Ancient Town exemplify Lingnan's clan-based architecture, showcasing regional characteristics in spatial layout, structural systems, materials, and decorative techniques. This study examines well-preserved examples through spatial, structural, and material dimensions. The halls typically follow traditional axial layouts, with layered spaces such as front, central, and rear halls, courtyards, and side corridors forming a coherent ritual order. Structurally, they adopt wood-stone hybrid systems suited to the humid, typhoon-prone climate. Locally sourced materials—Keruing wood, granite, gray bricks, oyster shells—combine with plaster, wood, stone, and brick carvings to enrich aesthetics and symbolism. Artistic elements like murals and ridge sculptures reinforce their multifunctional role in ritual, assembly, and education. These buildings reflect a fusion of local resources, craft traditions, and cultural values, representing the architectural wisdom of Lingnan.

Keywords Spatial layout · Ancestral hall · Structural systems · Materials

1 Introduction

The term "Ci (祠)" originally appeared in ancient Chinese texts as a designation for sacrificial rites. In the Classic of Poetry (Shijing, Xiaoya, Tianbao), it is written: "Yue Ci Zheng Chang, Yu Gong Xian Wang (禴祠烝尝，于公先王)," where "Yue" refers to summer sacrifices, "Ci" to spring sacrifices, "Zheng" to winter sacrifices, and "Chang" to autumn sacrifices [1].

Over time, the meaning of the term gradually evolved into a more general reference to ancestral worship and sacrificial ceremonies. In the Rites of Zhou (Zhouli, Chungong, Xiaozongbo), it is recorded: "In times of great disaster, officials shall offer prayers and sacrifices (祠) to the gods above and below." Here, the character "Shi (示)" is a phonetic variant of "Zhi (祇)," referring to deities of the earth [2].

Y. Wang (✉)
Graduate School of Design, Kyushu University, Fukuoka, Japan
e-mail: ww961120@gmail.com

Y. Chen et al. (eds.), *City-Sea Symbiosis*, Springer Geography,
https://doi.org/10.1007/978-3-032-13725-8_6

The ancestral hall, or "cítáng" (祠堂), is a type of public architecture dedicated to the worship of ancestors who share a common bloodline. It serves as a physical manifestation of ancestor veneration, built and maintained collectively by a particular lineage, clan, or extended family. Variously referred to as "Great Clan Hall (大宗祠)," "Clan Ancestral Hall (宗祠)," "Family Shrine (祖祠)," "Public Ancestral Hall (公祠)," or "Family Temple (家庙)," these spaces were fundamental to the socio-religious structure of traditional Chinese communities [3].

As a quintessential example of vernacular architecture, ancestral halls embody a rich tapestry of historical, cultural, and social significance, offering valuable insights for the study of traditional villages and rural settlements. More than just architectural forms, ancestral halls functioned as symbols of clan identity and unity, serving as the political, cultural, and ritual centers of local lineages.

The ancestral hall was both a sacred space for solemn ancestral rites and a forum for intra-clan governance and decision-making. It was also an important venue for life-cycle ceremonies, most notably weddings, where it served as a ceremonial hall for new couples. Architecturally, the ancestral hall was often the most elaborately constructed and materially refined building within a village, distinguished by its scale, craftsmanship, and choice of materials. The size and sophistication of an ancestral hall stood as a visible testament to a clan's prosperity, influence, and socio-economic standing.

2 Overview of Tangjiawan Ancestral Halls

The residents of the Tangjiawan area, following several waves of migration, eventually settled and took root on this land, where their descendants continue to reside today. Over successive dynasties, government policies that promoted clan-based social structures further encouraged the construction of ancestral halls. As a result, the practice of building clan halls flourished in Tangjiawan. By the Qing Dynasty, ancestral halls had proliferated throughout the towns and villages of the region, emerging like bamboo shoots after the rain.

At present, a significant number of ancestral halls remain preserved in the Tangjiawan area. According to the *List of Immovable Cultural Relics in the High-Tech Zone of Zhuhai City*, 94 of the 119 officially registered immovable architectural heritage sites in the region are historic buildings, among which 37 are ancestral halls.

Today, approximately forty ancestral halls are still extant in Tangjiawan, primarily distributed across nine village clusters, as previously mentioned (Fig. 6.1). Based on field surveys conducted by the author, it was found that most of these ancestral halls date back to the Qing Dynasty. However, many of them have suffered from years of neglect, structural deterioration, or repeated renovations, leading to a loss of original architectural features.

To this study, 24 relatively well-preserved ancestral halls have been selected as representative case studies for detailed analysis and introduction. This chapter

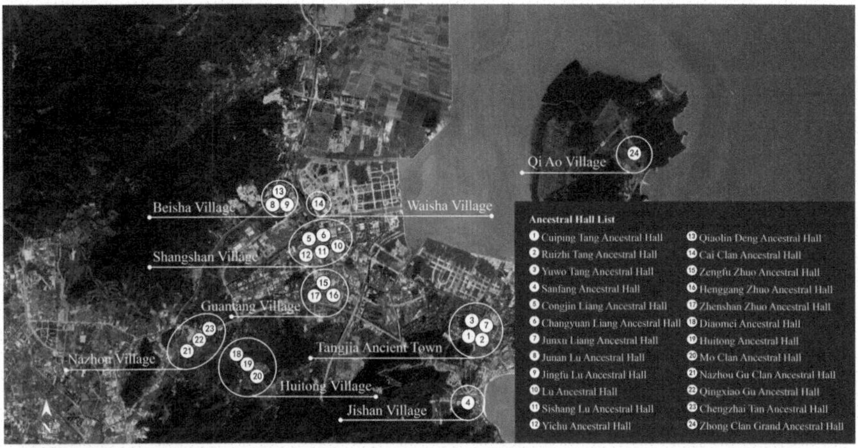

Fig. 6.1 Distribution of ancestral halls in Tangjiawan (*Image source* drawn by the author)

will present the Tangjiawan ancestral halls from three key perspectives: spatial organization, architectural components, and building materials.

3 Spatial Elements and Layout

The ancestral halls of Tangjiawan are primarily composed of several key spatial elements, including the main entrance gate (toumen), front hall, central hall, rear hall, courtyards (tianjing), subsidiary halls (chenci), corridors (lang), and outdoor spaces.

3.1 Main Entrance Gate (Toumen)

The toumen serves as both the ceremonial threshold and spatial starting point of the entire building complex. In Tangjiawan ancestral halls, the toumen typically falls into two principal categories: the Open-Bay Type (Changying Style) and the Recessed-Bay Type (Aodou Style) (Figs. 6.2 and 6.3).

Among these, the Open-Bay Type is more commonly observed. This entrance style belongs to a traditional timber frame construction technique known as "Lumingzao (露明造)," characterized by partially exposed structural elements and a visually open and permeable spatial quality. Such designs offer excellent ventilation and natural lighting, creating a spacious and welcoming ambiance that symbolically invites good fortune and auspicious energy into the ancestral hall. For this reason, the Open-Bay Type has become the predominant entrance form in the Tangjiawan area.

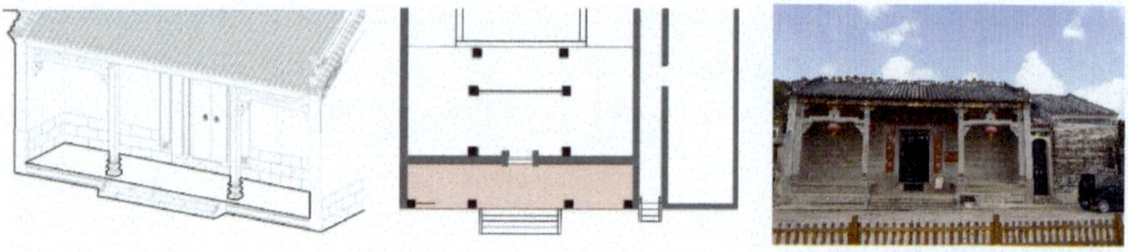

Fig. 6.2 Open-Bay Type Case: Mo Clan Ancestral Hall (*Image source* drawn by the author)

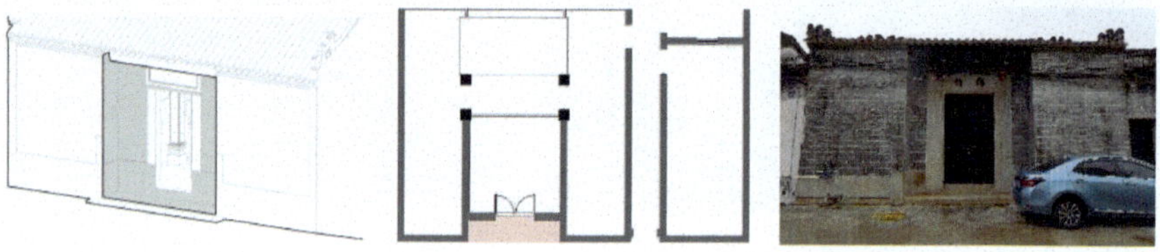

Fig. 6.3 Recessed-Bay Type Case: Xiangyue ancestral hall (*Image source* drawn by the author)

By contrast, the Recessed-Bay Type features a central space, known as the "heart bay (心间)," that is set back and recessed inward, forming a more introverted and enclosed spatial composition. The flanking "secondary bays" (次间) on both sides are typically enclosed by solid walls, creating a more intimate and confined entry space. Architecturally, this configuration conveys stronger symbolic meanings but remains relatively rare in Tangjiawan.

As the first spatial node in the overall architectural sequence, the toumen not only fulfills its functional role as a transitional and welcoming space but also embodies the local community's cultural understanding of feng shui principles, ritual propriety, and spatial ceremonialism.

3.2 *Front Hall (Qiantang)*

The term "Tang (堂)," meaning "hall," has a long historical lineage in Chinese architectural terminology. In classical texts, "Tang" is often equated with "Dian (殿)," signifying a grand or elevated space. Etymologically, it is believed that the original character for "Tang" derived from pictographic representations of sacrificial altars and offerings. By the Han Dynasty, the term was frequently used to describe buildings elevated on platforms above ground level [4].

The three-part spatial sequence of "front gate–central hall–rear chamber (前门–中堂–后寝)" became the standard layout for Guangfu (Cantonese region) ancestral halls from the Ming and Qing dynasties through to the Republican period. This typology remains the most common organizational form among extant ancestral halls today [5]. Its spatial logic is deeply rooted in Chinese cosmology and ritual philosophy, reflecting the belief in the continuity of the soul and the principle of

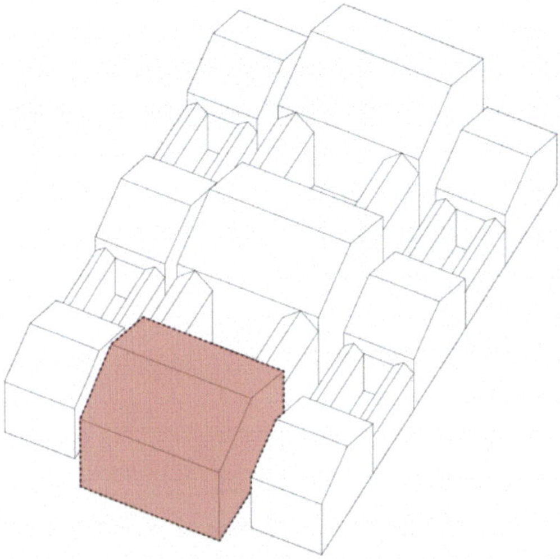

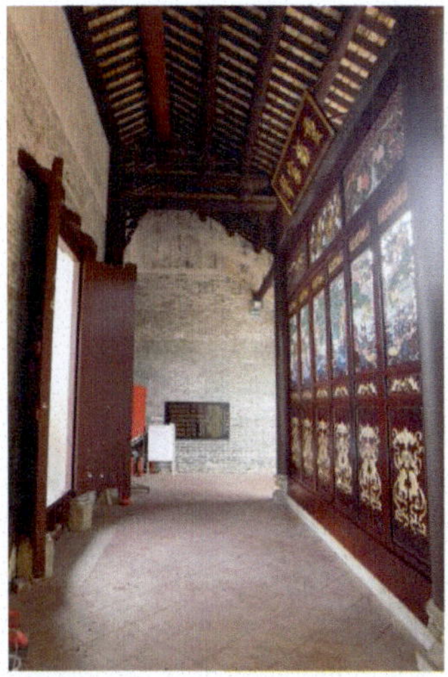

Fig. 6.4 Layout of the front hall (Liang Clan Ancestral Hall) (*Image source* drawn by the author)

serving the dead as though they were alive. The architectural arrangement of ancestral halls thus mirrors traditional residential layouts, with the front gate (前门) marking the entrance, the central hall (中堂) serving as the site for ancestral worship and ritual kneeling, and the rear chamber (后寝) acting as the spiritual sanctuary for enshrining ancestral tablets [6].

The front hall (前堂) is the first interior space encountered after entering through the toumen, acting as a transitional zone between the entrance gate and the courtyard (tianjing) beyond. Functionally, it serves multiple roles, including welcoming guests, providing spatial transition, and displaying the clan's identity and social standing.

In keeping with ritual practice and feng shui principles, a decorative screen door (pingmen 屏门) is often installed between the timber columns. This design element serves to shield the interior from direct external views, maintaining visual privacy while enhancing the sense of spatial progression and sacredness (Fig. 6.4). After passing the screen, visitors enter the open courtyard, where the visual axis extends toward the central hall and rear hall, allowing the entire spatial sequence to unfold layer-by-layer, reinforcing both processional order and the ritual axis of the site.

3.3 Central Hall (Zhongtang)

The central hall (中堂, Zhongtang) represents the most sacred and ritualistically significant space within Tangjiawan ancestral halls. Primarily used for ancestral worship, which is the most important communal activity of the clan, the central hall stands as a spatial and symbolic embodiment of ancestral faith and clan identity.

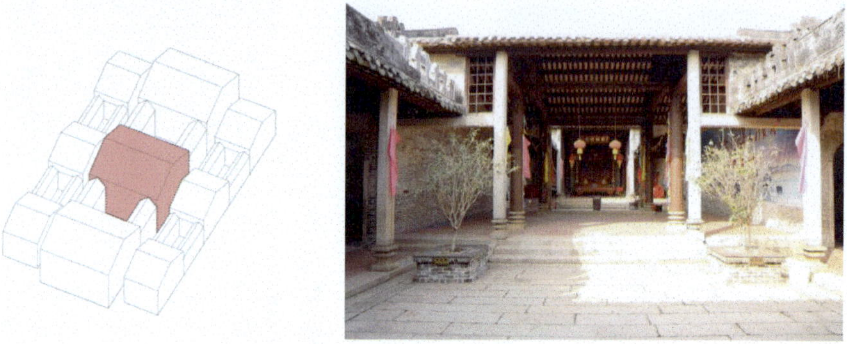

Fig. 6.5 Layout of the central hall (Liang Clan Ancestral Hall) (*Image source* drawn by the author)

In addition to serving as a place for ritual offerings, it also functions as a venue for major clan gatherings and decision-making assemblies (Fig. 6.5).

Architecturally, the central hall is positioned along the main longitudinal axis of the ancestral hall complex, making it the second major spatial node following the courtyard and front hall, and preceding the rear hall. Lateral side corridors (lang) on both sides provide spatial transitions and circulation, linking the central hall to the surrounding ancillary spaces. Together, these elements form a highly ordered and hierarchical spatial axis, characterized by clear sequential layering and ceremonial spatial progression.

In terms of structural hierarchy and spatial relationship, the central hall is considered the "master space" of the entire complex. It plays the dominant role within the architectural composition, while elements such as side corridors and subsidiary halls (chenci) serve as "attendants" or "extensions," fulfilling supportive roles. This master-subordinate spatial order reflects not only traditional clan rituals and social hierarchy but also reinforces the central hall's authority and solemnity as the symbolic heart of the clan.

Structurally, the central hall often adopts an open-ended columnar layout, with no solid walls at the front or rear, creating a visually and physically permeable space. This openness enhances the ceremonial atmosphere and allows for the accommodation of large gatherings during rituals. The open-column space also facilitates the placement of ritual implements and ensures unobstructed circulation flows during ceremonies.

However, in alignment with feng shui principles and ritual decorum, many ancestral halls install a removable screen door (pingmen 屏门) between the columns that separate the central hall from the rear hall. This screen serves to visually and symbolically shield the sacred space of the rear hall, where the ancestral tablets are enshrined, representing the final resting place of the ancestors' spirits.

In everyday use, this screen door typically remains closed, with family activities restricted to the front and central halls, while the rear hall remains a sanctified and inviolable space, reserved for the spiritual presence of the ancestors. This screen thus becomes a symbolic threshold between the realms of the living and the dead, manifesting the clan's reverence for their ancestors and their strict spatial delineation of sacred versus secular zones.

3.4 Rear Hall/Ancestral Shrine (Houtang)

The rear hall (后堂, Houtang) represents the culmination of the spatial sequence within the ancestral hall, serving as the final and most sacred space dedicated to the enshrinement of ancestral tablets and the spiritual housing of the clan's ancestors. As the ritual core of the ancestral worship system, the rear hall embodies the clan's deepest spiritual beliefs and cultural heritage. Within the formalized ritual system of the Ancestral Hall, the arrangement and hierarchical placement of the ancestral tablets are governed by strict conventions, and the rear hall serves as the physical manifestation of this ritual order. Thus, the rear hall functions not only as the architectural terminus but also as the ultimate spiritual destination of the clan's collective faith.

In contrast to the open-plan layout of the central hall, the rear hall is a fully enclosed space, defined by three solid perimeter walls and supported by eight columns, forming a stable and ordered structural system. Typically, the floor level of the rear hall is elevated above that of the central and front halls, reflecting the traditional Chinese spatial philosophy of "height denotes reverence." This design emphasizes the supreme and sacred status of the ancestors, clearly expressing spatial hierarchy and ritual respect.

Inside the rear hall, there is usually an incense burner (xianglu 香炉) and a centralized arrangement of ancestral tablets, positioned at the geometric and symbolic center of the rear wall, establishing a strong visual and ceremonial focal point. Given that worshippers must face the ancestors while offering incense, the rear hall typically remains open toward the central hall or front hall, establishing a formal and directional ritual axis. This spatial configuration creates a solemn spatial dialogue between the worshippers and the ancestral spirits (Fig. 6.6).

Some clans focus on creating opulently decorated rear halls to express their profound veneration for their ancestors. A representative example is the Nanzhou Gu Clan Great Ancestral Hall, where a richly carved canopy (feizhao 飞罩) with floral and avian motifs spans between the central gold-painted columns of the heart bay. The canopy, crafted from rosewood and embellished with gold leaf, features intricate, layered carvings of entwined floral vines, while the gold detailing highlights auspicious creatures such as birds and squirrels, creating an ornate yet solemn artistic atmosphere.

Between the secondary bay columns and the gable walls, hanging panels (hualuo 挂落) featuring interwoven key-fret (huiwen) patterns and bird-and-flower motifs are suspended. These are often accented with gilding, lending a lively and delicate character to the space. Above the panels, decorative horizontal boards (hengpi 横披) adorned with "ruyi" patterns further enhance the ceremonial atmosphere and convey auspicious blessings. Particular attention is also given to the shrine (shenkhan 神龛)

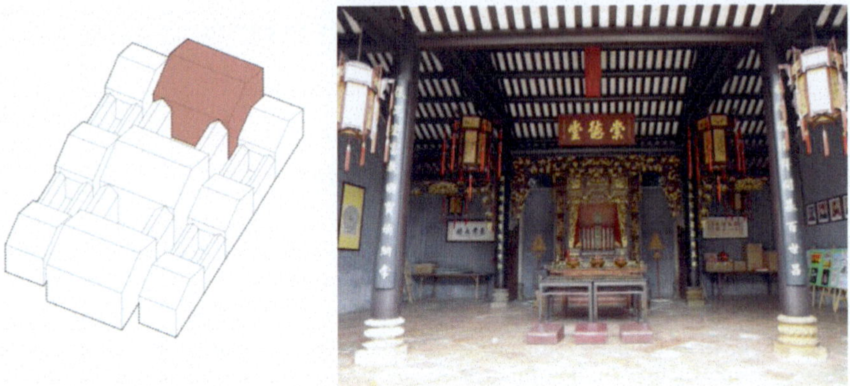

Fig. 6.6 Rear hall (Nazhou Gu Clan Ancestral Hall) (*Image source* drawn by the author)

itself, which is often lavishly ornamented with fully gilded canopies and shrine tops, intricately carved with phoenixes and other auspicious birds, symbolizing dignity, sanctity, and spiritual reverence.

3.5 Courtyard (Tianjing)

The courtyard (天井, Tianjing) is an indispensable and defining spatial element within the ancestral halls of Tangjiawan. Situated at the heart of the building complex, it is enclosed on four sides by the front hall, rear hall, and the flanking side corridors, forming an open-air rectangular void. When viewed from above, its appearance resembles a square well, which gives rise to the term "Tianjing," literally meaning "sky well."

Regardless of the size or scale of the ancestral hall, even on narrow or constrained plots, the inclusion of a courtyard is considered a standard design feature, reflecting the Tangjiawan community's deep cultural emphasis on nature and environmental integration.

In terms of spatial sequencing, ancestral halls featuring a central hall typically incorporate two courtyards: one between the front hall and central hall, and another between the central hall and rear hall, thus creating a distinction between the "front courtyard" (前庭) and "rear courtyard" (后庭). This dual-courtyard layout not only enhances the spatial hierarchy and depth but also significantly improves daylighting, ventilation, and energy flow (qi movement), achieving a harmonious balance between enclosure and openness, or as the Chinese phrase suggests, "internal concentration and external circulation" (内聚而外通) (Fig. 6.7).

From a climatic and functional perspective, Tangjiawan located on the southern coast near the South China Sea, experiences hot, humid weather with frequent rainfall. Under these environmental conditions, the courtyard plays a critical role in promoting

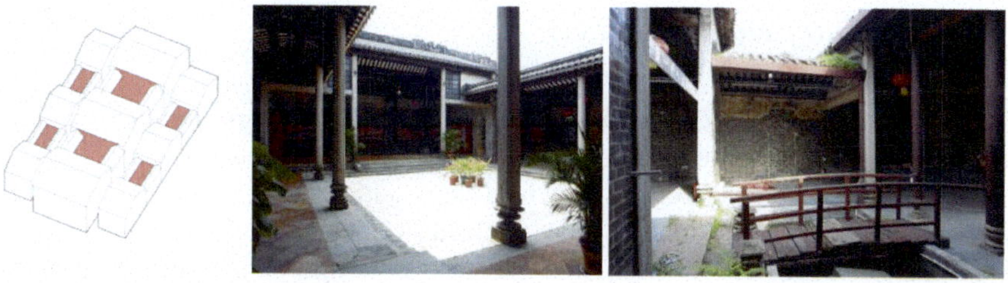

Fig. 6.7 Courtyard (Center: Ruizhi Tang Ancestral Hall; Right: Yuwotang Tang Ancestral Hall) (*Image source* drawn by the author)

air circulation and facilitating rainwater drainage. To prevent water backflow and dampness intrusion, the courtyard floor level is usually set lower than the adjoining front and rear halls and side corridors, allowing rainwater to drain away naturally.

In addition, the courtyard floor is often paved with light-colored stone slabs, which reflect natural light, enhancing indoor brightness in adjacent spaces. This treatment balances both practical needs and aesthetic qualities, creating a soft and diffused daylight effect within the surrounding rooms.

Beyond its physical and environmental functions, the courtyard also carries profound spiritual and symbolic significance. It is widely regarded as a cosmic intermediary space, bridging the realms of "Heaven, Earth, and Humanity (天地人)." Its open-to-sky design invokes a sense of awe toward nature, reverence for ancestors, and spiritual aspiration toward the divine. Standing within the courtyard, one experiences an emotional alignment with the cosmos, gazing upward to the heavens and bowing low in ancestral worship—an embodiment of the traditional Chinese worldview of "unity between heaven and humanity (天人合一)," fostering a solemn and sacred ritual atmosphere.

Some ancestral halls have employed innovative courtyard designs to further enhance their symbolic and ritual functions. For example, in the Yuwo Tang Gong Ancestral Hall in Tangjia Ancient Town, the courtyard in front of the central hall is transformed into a water pond, spanned by a wooden bridge connecting the front and rear halls, creating a distinctive and immersive spatial experience. According to feng shui principles, water symbolizes both "energy (qi)" and "wealth," while the bridge signifies "connection" and "continuity." This design echoes the auspicious imagery of "Four Waters Converging Toward the Hall (四水归堂)," expressing the clan's aspiration for prosperity and fortune accumulation, while deepening the courtyard's ritual significance as a space of reverence for heaven, deities, and ancestors.

As a core spatial element within Tangjiawan ancestral hall architecture, the courtyard seamlessly integrates environmental adaptation, symbolic meaning, and spiritual expression. It not only regulates the microclimate but also embodies the clan's cultural imagination and emotional connection to the cosmos, their ancestors, and their own destiny, showcasing the ingenious spatial design philosophy of Lingnan architecture.

3.6 Subsidiary Halls (Chenci)

Some scholars have suggested that the practice of adding side rooms or subsidiary halls (chenci 衬祠) to either side of the main hall in Guangfu ancestral architecture during the Ming and Qing periods may be a continuation of the ancient "Jia Shi (夹室)" system, wherein subsidiary spaces evolved as spatial extensions or derivatives of the main ancestral hall form [7]. In the spatial framework of Tangjiawan ancestral halls, chenci serve as primary auxiliary spaces, fulfilling multiple functional and spatial support roles. They often act as storage areas for ritual paraphernalia, or in some cases, as kitchens for food preparation during clan gatherings.

Compared with the relatively rare appearance of "rear annexes (houzuo 后座)," subsidiary halls are far more prevalent, making them an integral and indispensable component of Tangjiawan ancestral architecture. Although not part of the core ceremonial space, chenci play a critical supporting role in spatial organization, functional extension, and even visual composition (Fig. 6.8).

Typically, chenci are arranged parallel to the central axis, flanking the main halls on either side, with a narrow alleyway (xiangdao 巷道) preserved between the side corridors and the main hall. This configuration maintains the dignity and ritual purity of the main worship space, while also enabling flexible spatial division and circulation management. The functions of chenci were never rigidly standardized; they were adaptively used as guard rooms, storage rooms, kitchens, or other secondary spaces. In some ancestral halls with frequent clan activities, these subsidiary spaces were also utilized for pre-ritual preparations, funeral or wedding ceremonies, or general clan affairs.

The wealthier clans often enhanced their chenci, not only to expand the usable space of the ancestral hall but also to visually magnify the building's scale and grandeur, indirectly signaling the family's prosperity and social prestige. Despite their functional importance, chenci have always remained subordinate spaces, and their architectural treatment is generally more restrained and modest. In terms of plan layout, they are usually single-bay structures, contrasting with the grand three-bay composition of the main hall, thereby highlighting the architectural prominence of

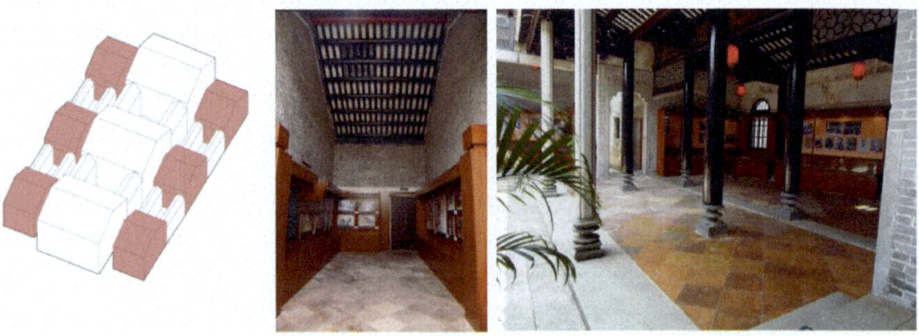

Fig. 6.8 Subsidiary halls (Case: Ruizhi Tang Ancestral Hal) (drawn by the author)

the primary ritual space. In elevation, the ridge height of roofs is often deliberately kept lower than that of the main three-hall sequence, further emphasizing spatial hierarchy and visual dominance.

Some ancestral halls went further by breaking down the conventional enclosed boundaries between the main hall and subsidiary spaces. Instead of enclosing walls, open corridors (lang 廊) were introduced to create a more fluid and integrated "hall-corridor" spatial configuration. This design approach enhanced spatial flexibility and ceremonial atmosphere, allowing for greater spatial openness and connectivity.

A representative example is the Ruizhi Tang Gong Ancestral Hall, where the subsidiary halls on both sides feature a three-bay depth, connected to the front and rear halls through side corridors and a "Qingyun Alley (青云巷)." Notably, both the side corridors and Qingyun Alley are set at a slightly lower floor level than the rear hall, a deliberate spatial strategy to reinforce the hierarchical relationship between primary and secondary spaces, reflecting traditional notions of spatial hierarchy and ritual propriety. Furthermore, the courtyard space along the central axis at Ruizhi Tang Gong Ancestral Hall lacks lateral boundary walls, fostering a horizontally open and interconnected spatial relationship between the central route and the side routes. This design significantly strengthens spatial fluidity and interactivity. The side corridors thus become crucial circulation channels, linking the front hall, rear hall, and Qingyun Alley, allowing for unobstructed movement of people, sightlines, and functional flows across multiple axes. This results in a more three-dimensional, dynamic, and efficient spatial system, greatly enhancing usability and supporting the diverse functional needs of clan activities.

3.7 *Corridors and Alleys (Langdao and Xiangdao)*

Within the spatial system of Tangjiawan ancestral halls, corridors (廊道, langdao) and Alleys (巷道, xiangdao) serve not only as crucial components of circulation networks but also as architectural expressions of aesthetic sensibility and cultural symbolism. These spatial elements interconnect the front hall, central hall, rear hall, and subsidiary halls, forming a clear and hierarchical transitional system that ensures a harmonious balance between movement and stillness, as well as fluid spatial organization and ritual order throughout the ancestral complex (Fig. 6.9).

The side corridors (celang 侧廊) are typically located on both sides of the courtyards (tianjing), functioning as key passageways connecting the three principal halls. Although the bay width of the side corridors is generally narrower than that of the secondary bays in the main halls, they play an indispensable structural and functional role. As users primarily experience these spaces in motion, side corridors are considered quintessential "dynamic viewing spaces," with modest decorative treatments at the roof eaves and overhangs to enrich visual rhythm and layering (Fig. 6.9).

The roofs of side corridors in Tangjiawan ancestral halls are often constructed in the barrel-vaulted style (juanpengding 卷棚顶), featuring gently curved undersides that create soft, flowing lines. This design enhances spatial cadence and rhythm,

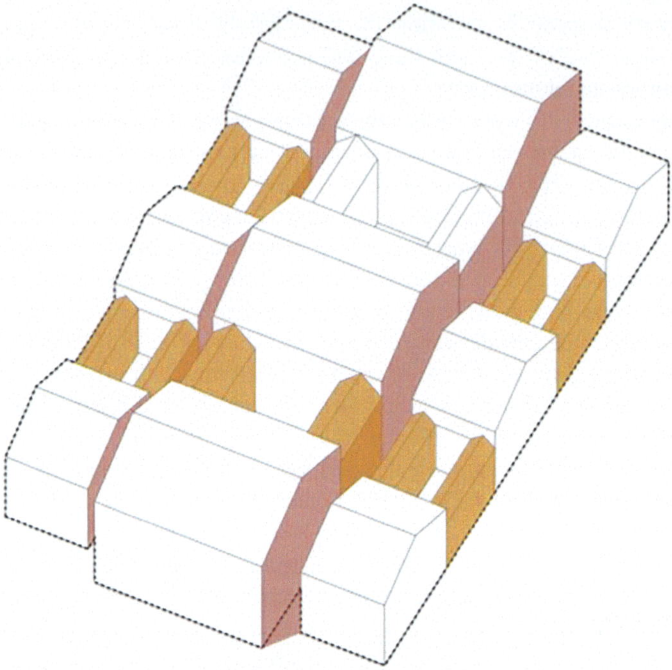

Fig. 6.9 Corridors and alleys (Langdao and Xiangdao) (drawn by the author)

while also amplifying the psychological sense of solemnity and reverence during ritual processions (Fig. 6.10).

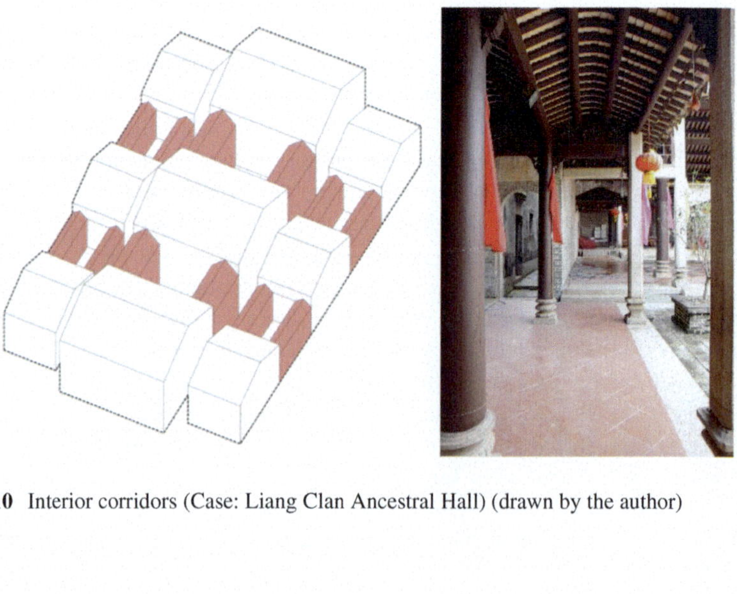

Fig. 6.10 Interior corridors (Case: Liang Clan Ancestral Hall) (drawn by the author)

In some ancestral halls, to extend this spatial rhythm into the ritual zone, the eaves of the rear hall are also designed with barrel-vaulted roofing, creating a space referred to as a "veranda corridor (xuanlang 轩廊)." This not only provides shelter for ritual participants but also serves as a formal transition between sequential spatial nodes. Beyond symbolic and aesthetic functions, the barrel-vaulted roof offers excellent climatic adaptability, facilitating shade, rain protection, and ventilation, reflecting Tangjiawan builders' climatic wisdom in responding to the region's hot and humid subtropical environment.

Complementing the side corridors are the "Qingyun Alleys (青云巷)," which run longitudinally between the three main halls and the subsidiary spaces. These linear alleys, positioned on both sides of the central axis, are designed to follow the site's natural topographical gradient. Given that most Tangjiawan ancestral halls exhibit a "lower front and higher rear" floor elevation pattern, the Qingyun Alleys typically rise gradually from the front hall toward the rear hall, symbolically conveying the aspiration for "ascending to the clouds" (青云直上)—a metaphor for success, social mobility, and the clan's hopes for their descendants' official and scholarly achievements.

With a typical height-to-width ratio of approximately 7:1, the tall bounding walls on both sides create an elongated and narrow spatial character, resulting in limited sunlight penetration and consistently shaded conditions. The increased air pressure and accelerated airflow within the alley produce a distinct microclimate with lower ambient temperatures compared to surrounding spaces. Locally, these Alleys are colloquially known as "Cold Alleys (Leng Xiang 冷巷)," reflecting the unique thermal environment created by the architectural form. This phenomenon illustrates the sensitive environmental awareness and responsive design strategies inherent in this ancestral architecture.

Furthermore, inscriptions engraved above the alley entrances often bear auspicious and didactic phrases, such as "Dragons Soaring and Phoenixes Rising (蛟腾凤起)" or "Enter with Filial Piety, Exit with Fraternal Duty (入孝出悌)," reflecting the clan's moral expectations and cultural hopes for their descendants. These inscriptions imbue the Alleys with additional layers of ethical and symbolic meaning, reinforcing the interplay between space, ritual, and Confucian values (Fig. 6.11).

3.8 Outdoor Spaces: Forecourts and Water Features

In Tangjiawan ancestral architecture, outdoor spaces are far more than mere transitional zones leading to the main buildings. They carry deep cultural symbolism, functional significance, and environmental adaptability, reflecting the architectural wisdom of the local community. The forecourt (Ditang 地堂), located at the front of the Ancestral Hall, and the surrounding water features and Pan Ponds (泮池) together form a distinctive outdoor spatial configuration unique to Tangjiawan's ancestral hall complexes.

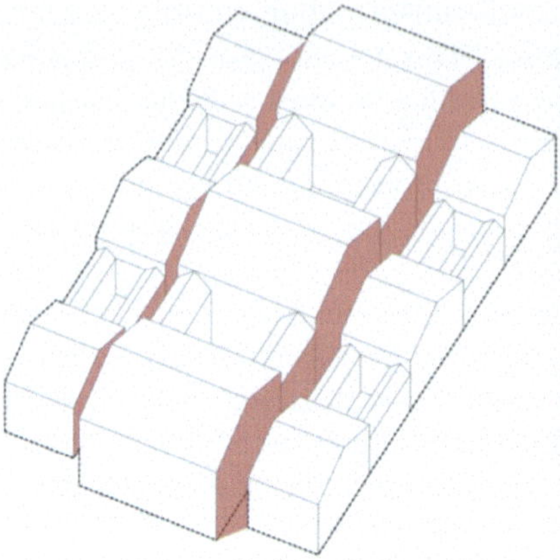

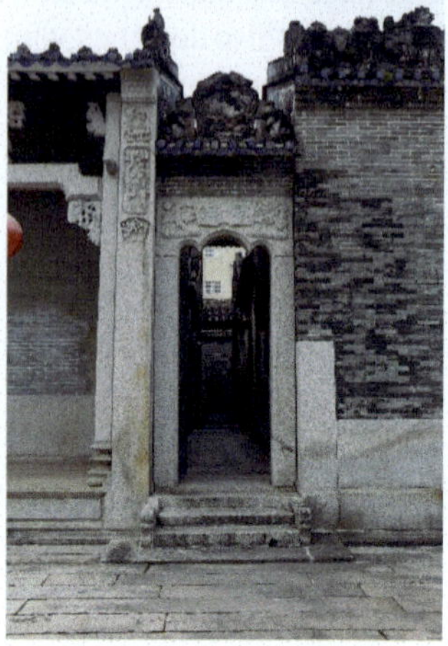

Fig. 6.11 Qingyun Alleys (Case: Jingfu Lu Ancestral Hall) (drawn by the author)

Ditang is typically situated outside the toumen (main entrance gate) and is characterized by a spacious stone-paved plaza, serving not only as the physical pathway leading into the hall but also as an important visual and spatial transition space. Functionally, the Ditang connects the ancestral hall to village streets and alleyways, accommodating both circulation and community gatherings. Spatially, its open expanse and dignified stone paving create an atmosphere of solemnity and reverence, encouraging a sense of ancestral respect and clan belonging even before entering the sacred interior spaces. Beyond its role in enhancing spatial layering and ritual ambiance, the Ditang also accommodates key functions during major clan events, such as ancestral rites, weddings, funerals, and celebrations, serving as a gathering, waiting, and preparation area. More importantly, the Ditang is a vital site for expressing clan honor and the transmission of ancestral values. If a clan member achieved distinction in the imperial examination or held an esteemed official post, memorial steles or flagpole stones (wei-gan shi 帷杆石) were often erected at the front of the Ditang, inscribed with the individual's name and accomplishments, serving as a source of collective pride and a moral exemplar for future generations. Such commemorative installations transform the Ditang from a functional outdoor space into a symbol of family spirit and cultural confidence (Fig. 6.12).

Surrounding the Ditang, water features and Pan Ponds form another essential layer of outdoor spatial expression. Located within a network of streams and waterways, many Tangjiawan ancestral halls were built adjacent to natural water bodies, creating picturesque compositions that harmonize architecture with the surrounding hydrological landscape. This siting not only aligns with feng shui principles but also reflects the traditional Southern Chinese philosophy of human-nature coexistence.

For ancestral halls unable to sit directly by natural watercourses due to topographical or feng shui constraints, artificial Pan Ponds (泮池) were often ingeniously constructed, recreating the symbolic spatial relationship of "hall facing water (面水

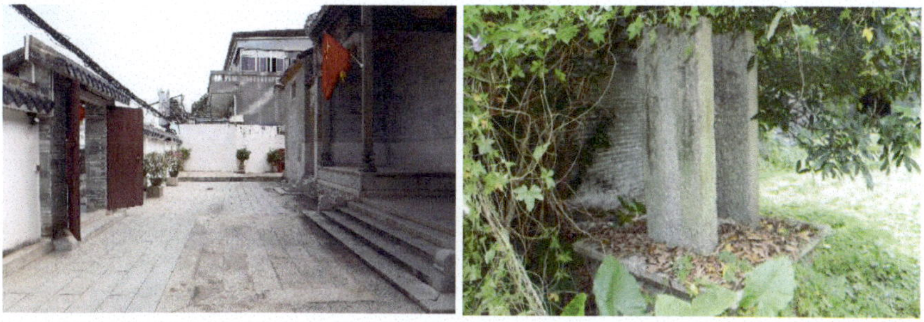

Fig. 6.12 Left: ground hall of Ruizhi Ancestral Hall; Right: flagpole base stone on the ground hall of Zengfu Zhuo Ancestral Hall (*Image source* photographed by the author)

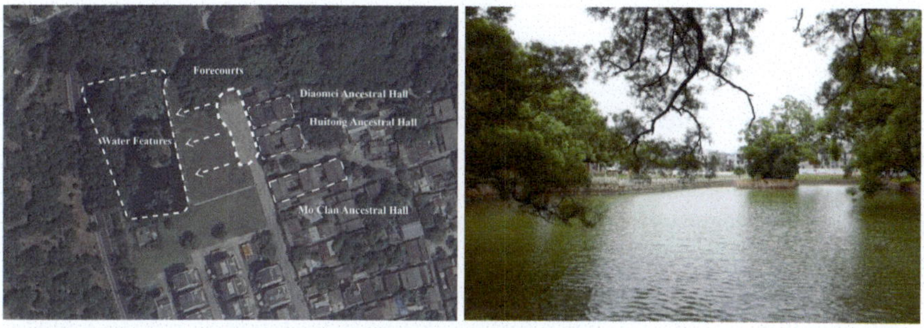

Fig. 6.13 Left: front plaza and the pond of the three Ancestral Halls in Huitong Village; Right: pond (*Image source* photographed by the author)

立祠).” In these contexts, water symbolizes wealth, prosperity, and good fortune, while also conveying deep-rooted feng shui and cultural meanings (Fig. 6.13).

In particular, the Tangjiawan community places great emphasis on the layout concept of “water in front (前有水)”, symbolizing both “gathering water to accumulate wealth and blessings” (聚水成财,纳福兴旺) and creating a visually dynamic and ceremonially charged dialogue between the shimmering water surface and the solemn architectural façade of the ancestral hall. The Pan Pond is more than a decorative landscape feature; it also serves as a practical drainage and disaster prevention system, essential for mitigating flood and fire risks in Lingnan’s rainy subtropical climate.

Ultimately, water and the Pan Pond contribute not only to the spatial atmosphere but also to the emotional and cultural memory of the community, embodying their aspirations for a prosperous life and the enduring transmission of clan heritage across generations.

4 Fundamental Elements of Spatial Typology

4.1 Basic Concepts of Bay (Jin), Route (Lu), and Bay Unit (Jian)

Tangjiawan's spatial organization of ancestral halls has undergone a long process of historical layering and typological evolution, resulting in diverse, adaptable, and con-textually responsive layouts. The plan configurations of these halls vary significantly depending on factors such as clan size, site conditions, and functional needs, demonstrating a spatial wisdom deeply rooted in contextual sensitivity and cultural logic.

4.1.1 Bay (Jin)

The term "Jin (进)" refers to the depth-wise progression of spatial units, measured along the longitudinal axis (meridional direction), with the baseline running parallel to the front facade (toumen) and the path axis following the direction of increasing spatial depth. Influenced by traditional cosmological concepts such as "Harmony between Heaven and Humanity" (天人合一) and "Yin-Yang Unity" (阴阳合德), Tangjiawan ancestral halls typically adopt a "one courtyard, one bay depth (一院一进)" modular system, layered sequentially to extend the archetype of Chinese courtyard-based architecture. As the scale of the ancestral hall expanded and functional requirements increased, additional spatial modules were added along the axis, leading to a progressively enclosed and centripetal spatial composition.

The minimum spatial unit for Tangjiawan ancestral halls is the two-bay layout, consisting of a front hall, courtyard, and rear hall. In two-bay halls, the central and rear hall functions are often merged, resulting in a more compact spatial sequence. The three-bay layout includes the front hall, front courtyard, central hall, rear courtyard, and rear hall, providing clear spatial zoning and fully articulating the ritual order of ancestral worship: the front hall marks the ceremonial starting point, the central hall (serving as the main ritual hall) is the core worship space, and the rear hall houses the ancestral tablets, symbolizing the clan's roots and bloodline continuity.

4.1.2 Route (Lu)

In contrast, the concept of "Lu (路)," refers to the lateral organization of spatial units, measured along the transverse axis (latitudinal direction), with the baseline running parallel to the building's depth (i.e., along the gable wall direction). A series of architectural units and courtyards aligned along a single depth axis is referred to as "one-route (一路)." However, in the local Tangjiawan context, the term "Lu" denotes not only the architectural sequence along the central axis but also includes the Qingyun Alleys (青云巷) situated between the main hall and the subsidiary halls (chenci).

In the "Three-Route Layout (广三路)," the Qingyun Alleys function as both circulation corridors and ritual-symbolic spaces. According to Confucian spatial etiquette, visitors enter the hall complex through the "Li Gate (礼门)" on the left side and exit via the "Yi Route (义路)" on the right, reflecting an integration of spatial movement and ritual propriety.

Based on spatial scale and organizational form, Tangjiawan ancestral halls can be broadly categorized into three typological types (Figs. 6.14 and 6.15):

(1) One-Route Halls (一路祠堂): These contain only the central axial buildings, with no subsidiary halls or Qingyun Alleys. Typically smaller in size, this layout was commonly used for Public Ancestral Halls (gongci 公祠).
(2) Two-Route Halls (二路祠堂): These include the main halls, along with subsidiary halls and Qingyun Alleys on one side. The positioning of the chenci (left or right) varies based on site conditions, demonstrating Tangjiawan's flexible and context-sensitive approach to spatial planning.
(3) Three-Route Halls (三路祠堂): These feature the main hall flanked by two parallel subsidiary halls, creating the typical "Guangfu Three-Route Layout (广三路)" widely recognized across the Lingnan region.

4.1.3 Bay Unit (Jian)

The concept of "Jian (间)," referring to bays or structural modules, is defined by the distance between two upright columns. The main entrance bay of the front hall is called the "central bay (心间)," followed sequentially by the "secondary bay (次间)," together constituting the total frontage width (通面阔).

Most Tangjiawan ancestral halls adopt a three-bay layout, a choice influenced not only by functional needs but also by ritual and regulatory constraints of the Qing dynasty. According to the *Comprehensive Ritual Code of the Great Qing* (大清通礼), the number of bays was a symbolic marker of social rank and official status:

> The ancestral temples of Princes and Dukes shall have seven bays, with five central bays serving as the main hall and two flanking bays as side rooms.
>
> The temples of Beile and Beizi shall have five bays, with three central bays for the main hall.
>
> Temples for officials of the First to Third Rank also have five bays, with three central bays for the hall.
>
> Temples for Fourth to Seventh Rank officials shall have three bays, with one central bay and two side bays.
>
> Temples for Eighth and Ninth Rank officials shall also have three bays, though the central bay shall be larger and the side bays narrower.

This historical regulation underscores how the number of bays not only expresses architectural scale but also encodes social hierarchy, ritual propriety, and cultural order (Fig. 6.16) [8].

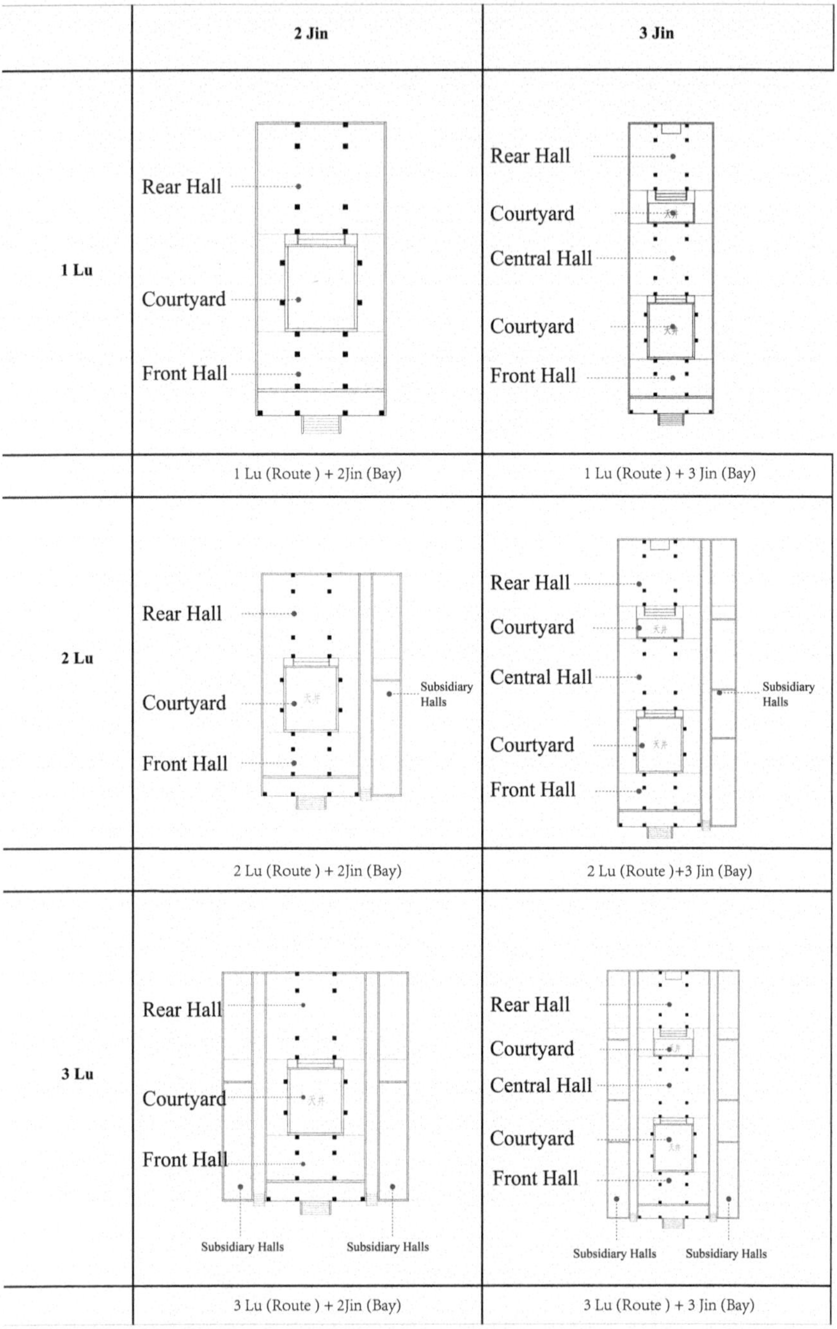

Fig. 6.14 "Lu-Jin" combination layout types (*Image source* drawn by the author)

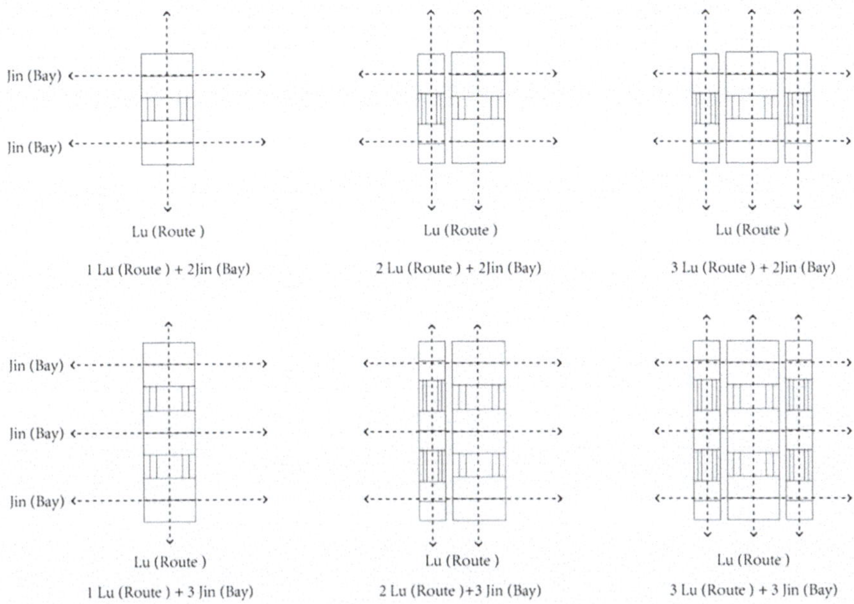

Fig. 6.15 Six combinations of "Lu-Jin" (axis-depth) layouts (drawn by the author)

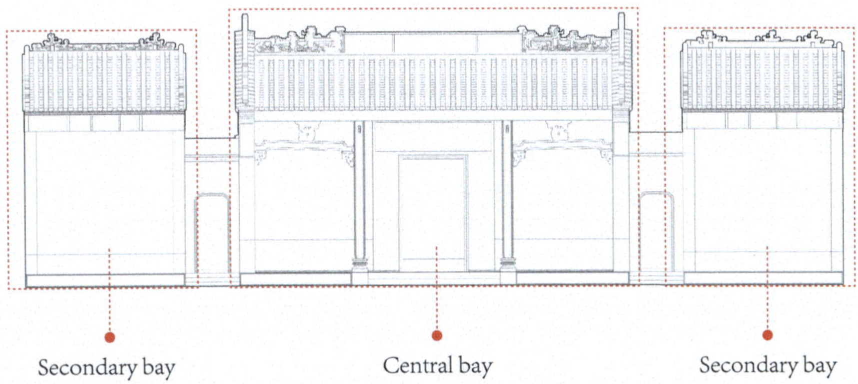

Fig. 6.16 Three-bay facade (*Image source* drawn by the author)

The spatial typology of Tangjiawan ancestral halls reflects both context-driven and functionally rational design strategies, while simultaneously inheriting the ritual spatial order and hierarchical values of traditional Chinese culture. The diversity of "one-route, two-route, and three-route" layouts, combined with the codified three-bay facade composition, has enriched the architectural morphology of Lingnan clan halls, providing a profound spatial and cultural legacy.

4.2 *Case of Tangjiawan Ancestral Halls Layout*

Overall, the architectural typology of Tangjiawan ancestral halls can be broadly categorized into three fundamental layout types:

- Two-Route, Two-Bay, Three-Bay-Wide Layout (二路两进三开间)
- Two-Route, Three-Bay, Three-Bay-Wide Layout (二路三进三开间)
- Three-Route, Three-Bay, Three-Bay-Wide Layout (三路三进三开间)

Each of these spatial configurations represents a nuanced response to the clan's social structure, ritual order, and everyday functional needs. Regardless of whether the layout follows a one-route, two-route, or three-route composition, the sequential layering of halls creates a dynamic interplay of spatial density and ceremonial hierarchy. This ensures that the ancestral halls can simultaneously accommodate daily communal gatherings and provide a dignified setting for solemn ancestral rites, thereby sustaining both social cohesion and intergenerational continuity within the clan.

Of particular note is the presence of unique spatial variants, such as the "Twin Ancestral Halls (孖祠, Ma Ci)," which appear in Tangjiawan as distinct departures from conventional layouts. These special configurations break the traditional monotony of standard typologies, adding spatial diversity and flexibility to the architectural landscape of the village.

Such variations often emerged in response to internal clan divisions, where separate family branches required individual worship spaces, or they resulted from site constraints and specific functional considerations. These adaptations reflect Tangjiawan's acute spatial sensitivity and adaptive design strategies in addressing the complexities of clan structures and the diverse needs of the local community.

Ultimately, these pluralistic and sophisticated spatial compositions are not only a microcosm of Lingnan architectural culture but also a direct spatial manifestation of Tangjiawan's village social fabric, vividly illustrating the interwoven relationship between family lineage and humanistic values as shaped by local geography and collective memory.

4.2.1 Two-Route, Two-Bay, Three-Bay-Wide Layout: A Case Study of Lushi Twin Ancestral Halls

Within the spatial typology of ancestral halls in Tangjiawan, the "two-axis, two-depth, three-bay" (二路两进三开间) layout represents a more complex configuration that balances functional diversity and ritual solemnity. This type is typically found in large-scale, high-status clan halls, often composed of a central route and one side route (left or right), each serving distinct roles. While functionally independent, the two axes are interconnected via transitional spaces such as Qingyun Lane, creating a well-ordered, multilayered spatial structure. The layout reflects not only a sophisticated understanding of function and ceremony, but also the clan's deep commitment to spatial hierarchy and ritual form.

A representative example is the Lu Clan Twin Ancestral Halls located in Beisha Village, Tangjiawan (Fig. 6.17). This complex consists of two symmetrically arranged ancestral halls—Jingfu Lu Gong Hall and Xunan Lu Gong Hall—forming a unique twin-hall layout. Unified in structure and proportion, it exemplifies regional excellence in spatial planning and architectural expression.

The Xunan Lu Gong Hall features a three-bay, two-depth composition. The front courtyard (ditang) is spacious, leading to the first hall (front hall), which adopts a 13-rafter deep timber structure with a traditional insert-beam (charred beam) system.

The rear section of the front hall extends nine rafters in depth, separated by internal partitions. The wall sections preserve remnants of painted decoration, especially on lintels above doorways between the central and side bays. The second hall (main rear hall) sits atop a raised Sumeru-style platform, about five steps above courtyard level, emphasizing its sacred status. The right axis of Xunan Hall extends across three functional zones: the first and third bays house side rooms separated from Qingyun Lane by walls, while the second bay opens directly to the lane, reinforcing spatial connectivity.

In summary, the Lushi Twin Ancestral Halls represent an advanced model of dual-axis spatial organization. Through the interplay of independent yet connected pathways, graded elevations, and carefully layered spaces, this complex reflects the depth of ritual consciousness and architectural mastery characteristic of elite clan halls in Tangjiawan.

4.2.2 Two-Route, Three-Bay, Three-Bay-Wide" Layout: Ritual Hierarchy and Spatial Refinement in Tangjiawan Ancestral Halls—A Case Study of the Liang Twin Ancestral Halls in Sancun Village

Within the architectural system of Tangjiawan ancestral halls, the "Two-Route, Three-Bay, Three-Bay-Wide" layout (两路三进三开间) represents one of the most elaborate and hierarchically sophisticated spatial typologies. Typically adopted by prominent clans with extensive ceremonial needs, this configuration emphasizes ritual-centered axial planning coupled with functional zoning, embodying the Lingnan tradition's pursuit of spatial order, symbolic clarity, and ceremonial dignity.

A representative example is the Liang Twin Ancestral Halls in Sancun Village, Tangjiawan Town, Zhuhai, comprising the Changyuan Liang Gong Ancestral Hall and the Congjin Liang Gong Ancestral Hall. Constructed as a unified "Twin Hall (孖祠)" complex, they maintain functional independence while forming a monumental and cohesive composition—a spatial expression of clan unity and legacy (Fig. 6.18).

Among them, the Changyuan Liang Gong Ancestral Hall, completed in 1891 (Guangxu 17th year), exemplifies this typology. Facing east, the hall spans approximately 20.98 m in width and 40.2 m in depth, covering a total floor area of about

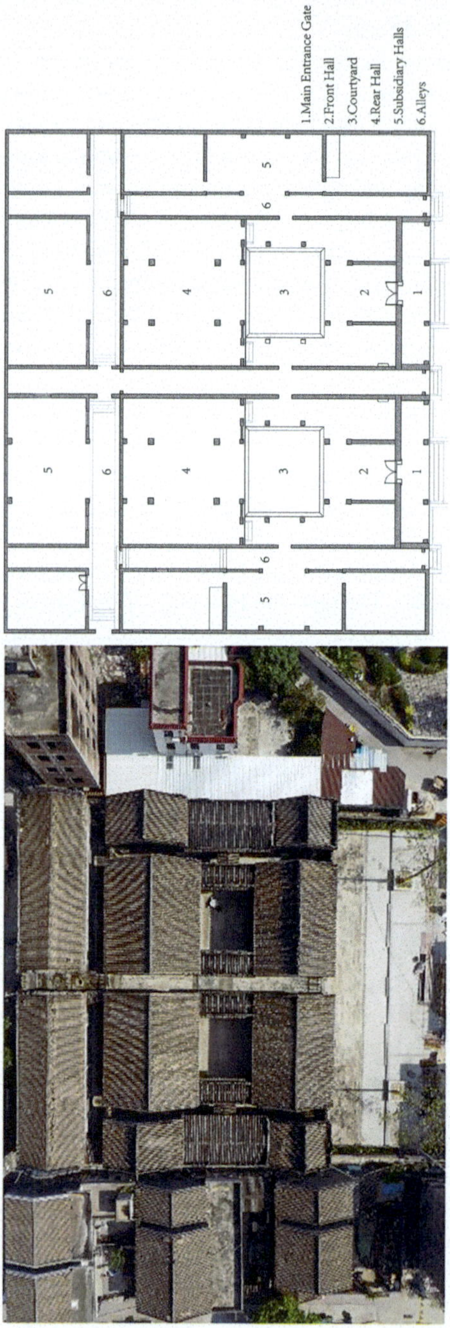

Fig. 6.17 Lu Clan Twin Ancestral Halls (*Image source* drawn by the author)

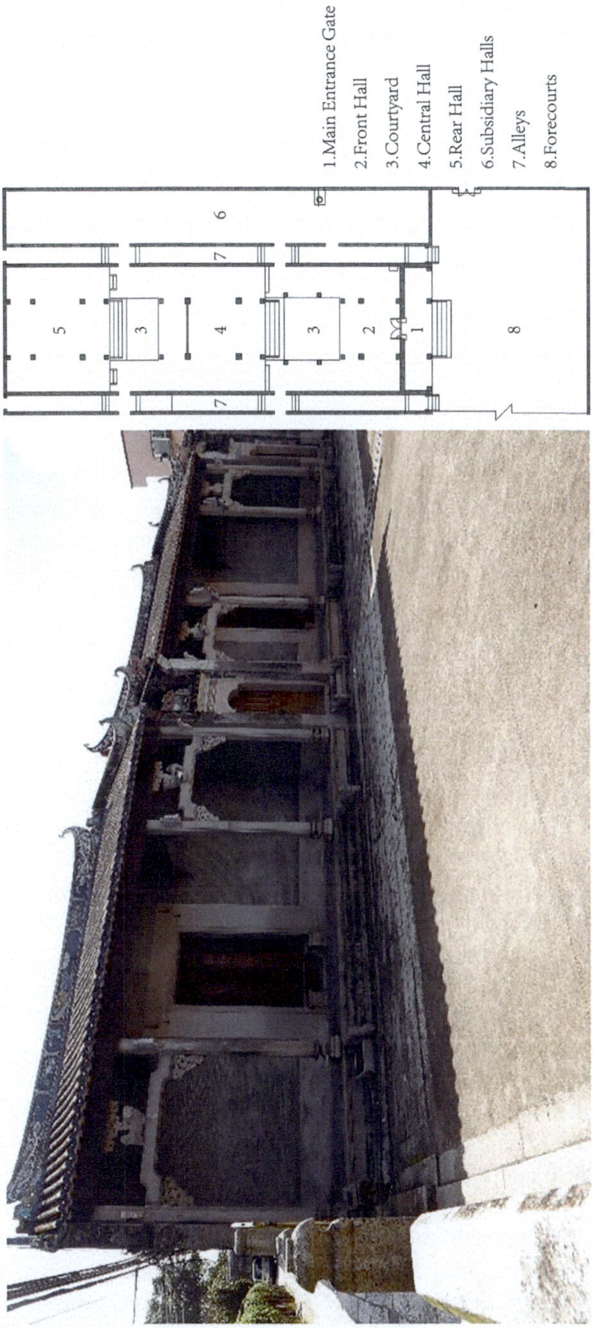

Fig. 6.18 Liang Twin Ancestral Halls (*Image source* drawn by the author)

843.4 square meters. The building adopts a "Three-Bay with Two Courtyards (三进夹两天井)" configuration along a central ceremonial axis (zhonglu), flanked by a left-side auxiliary route (zuolu). The two routes are connected via Qingyun Alley, allowing for flexible circulation and dynamic spatial layering.

The first bay (Toumen or front hall) serves as the formal entrance and adopts an Open-Bay structure with an insert-beam timber frame (插梁式木构架) spanning eleven purlins in depth. A five-step platform with paired drum-shaped stone pedestals reinforces the hall's ceremonial formality and axial prominence.

The second bay (central hall) acts as the ritual core of the complex. With a beam system of thirteen purlins, it features a broad, open interior supported by a Boguliang framework. Its architectural symmetry and proportional clarity underscore its role as the spiritual and ceremonial heart of the ancestral hall.

The third bay (rear hall) marks the spatial and symbolic culmination of the axis. Elevated on a high platform, the space deepens in form and significance. The hall connects to side corridors via stone steps, with flanking elements reinforcing hierarchical transitions and ritual boundaries.

In summary, the Liang Twin Ancestral Halls exemplify how Tangjiawan clan architecture integrates spatial hierarchy, functional separation, and ritual symbolism into a coherent whole. The "Two-Route, Three-Bay" layout, as seen here, reflects a matured architectural logic grounded in ancestral worship, social order, and regional aesthetic values.

4.2.3 Three-Route, Two-Bay, Three-Bay-Wide Layout: Ritual Order Within Symmetrical Composition

Among the typological system of ancestral halls in Tangjiawan, the "three-axis, two-depth, three-bay" layout represents a highly structured and strongly symmetrical spatial configuration. This type is commonly found in medium-scale ancestral halls associated with prominent clans. It emphasizes axial symmetry and a clear hierarchy, with the central axis forming the core sacrificial space flanked by two auxiliary wings, effectively expressing the clan's pursuit of ritual order and spatial discipline.

The Cuiping Tang Ancestral Hall, located on Bianshanzhijie Street in Tangjia Village, is a representative example of this layout type. Constructed in 1901 (Guangxu 27th year of the Qing Dynasty), it faces north with a total area of approximately 418.74 m^2, a frontage of 21.43 m, and a depth of 19.54 m. The structure is compact and proportionally balanced, showcasing the maturity and craftsmanship of Tangjiawan ancestral hall construction.

The central axis features a two-hall configuration with a courtyard (tianjing) flanked by side corridors. The first hall (front hall), elevated on a high platform, has an 11-rafter-depth timber frame and a spacious, open structure, reinforcing its ceremonial function. The second hall (main hall), with a 13-rafter depth and a wider span, serves as the primary sacrificial space. The central courtyard paved with stone not only improves drainage and lighting but also enhances spatial openness and layering.

The left- and right-side wings serve as auxiliary spaces for daily gatherings, hosting duties, and clan affairs. With a shallower depth (approx. 15.59 m) and lower elevation compared to the central axis, these wings function to highlight the central space's hierarchy and significance. The three axes are connected by the Qingyun Lane, approximately 1.1 m wide and equipped with three steps, accommodating site elevation changes while emphasizing spatial gradation and symbolic progression.

Overall, the Cuiping Tang Ancestral Hall demonstrates a well-organized, solemn, and symbolically rich spatial system. Through its clearly defined axial composition and functional hierarchy, it exemplifies the architectural characteristics and advanced spatial organization typical of ancestral halls in Tangjiawan (Fig. 6.19).

5 Architectural Components of the Ancestral Hall

5.1 Ground Components

5.1.1 Stylobate (Taiji) and Music Platform (Shutai)

The ground-level components of Tangjiawan ancestral halls primarily consist of the stylobate (台基, Taiji) and the music platform (塾台, Shutai). These elements serve both structural and aesthetic purposes. The foundation system refers to the architectural components closest to ground level, including the stylobate and associated steps.

The stylobate, as the lowest architectural foundation, exhibits relatively minimal variation in form, typically maintaining a height between 20 and 35 cm. Given Tangjiawan's coastal location, characterized by high humidity and elevated salt content in the air, granite is commonly used as the primary material, providing effective moisture resistance and structural support for the superimposed timber frame, walls, and roof loads.

The steps, as an integral extension of the stylobate, may appear simple in form, yet they are composed of multiple components, including tread stones, side moldings (牵边), side panels (侧板), and step skirts (台阶垂带). In some more elaborately crafted ancestral halls, miniature drum-shaped stone ornaments (抱鼓石) are carved onto the step skirts, enhancing decorative richness and visual impact.

The shutai, positioned above the stylobate, is a distinctive feature of Open-Bay entrance halls (敞楹式头门), typically located beneath the secondary bay (次间) of the front hall. Its primary function is to serve as a raised platform for musicians performing ritual music during ancestral ceremonies, earning it the alternate name "Drum Platform (鼓台)."

The shutai is not a mandatory component in all ancestral halls, typically found only in halls built by wealthier clans. It is usually constructed integrally with the stylobate, carved from single blocks of stone, and often adopts the form of a Sumeru

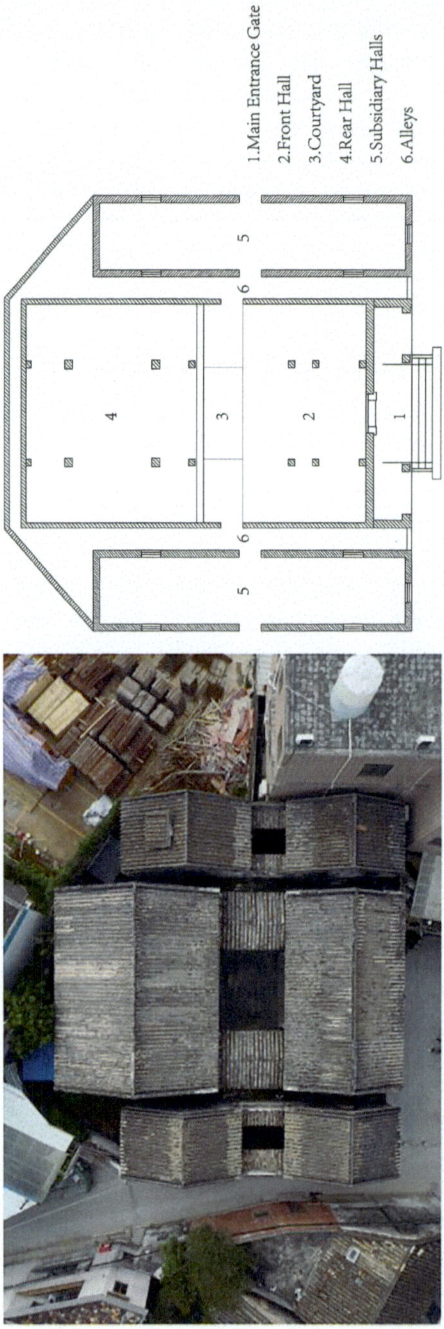

Fig. 6.19 Cuiping Tang Ancestral Hall (*Image source* drawn by the author)

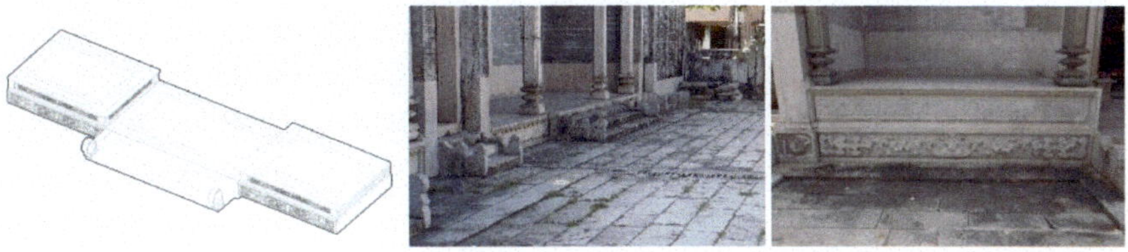

Fig. 6.20 Ground Components Case: (Center: Steps of Congjin Liang Ancestral Hall; Right: Music platform of Ruizhi Ancestral Hall) (*Image source* drawn by the author)

pedestal (须弥座形式). Structurally, the shutai consists of three horizontal sections: the upper frame (上枋), the waist section (束腰), and the lower frame (下枋).

Carving techniques primarily employ relief carving, with the front elevation framed by decorative Xiahun line panels (枭混线画框). Within these frames, motifs such as scrolling grass patterns (卷草纹) or auspicious themes are commonly featured. For instance, the shutai at Ruizhi Ancestral Hall (瑞芝祠) exemplifies exceptional craftsmanship: the stylobate section employs deep relief carving, while the shutai portion features shallow relief, creating a distinct visual contrast that emphasizes the solidity of the base and the lightness of the music platform, achieving a refined sense of layered composition and spatial articulation (Fig. 6.20).

5.1.2 Paving

The paving of Tangjiawan ancestral halls can generally be classified into two main types: strip-stone paving and large clay brick paving. These two paving methods are spatially differentiated based on exposure conditions and functional zones.

Strip-stone paving is typically used in outdoor spaces, such as courtyards (tianjing 天井) and forecourts (ditang 地堂). In contrast, large clay bricks (dajiézhuan 大阶砖) are predominantly applied to indoor areas, including the interior of halls, side corridors, and subsidiary halls (chenci 衬祠).

Given Tangjiawan's subtropical climate, characterized by frequent rainfall, outdoor paving materials are selected for their resistance to moisture and erosion. Accordingly, granite, known for its durability in damp environments, is commonly used for strip-stone paving. Each stone slab is typically elongated in form, and the laying direction is generally perpendicular to the longitudinal "route (lu 路)" axis, a design technique that enhances the perceived horizontal extension of the space, thereby widening the spatial experience.

For interior flooring, the preferred material is large clay bricks, made from locally sourced red clay and kiln-fired without glazing, resulting in a porous texture with excellent water absorption properties, making them highly suitable for Tangjiawan's humid conditions. These large floor bricks are typically square-shaped, with side lengths of approximately 36 to 42 cm centimeters and a thickness ranging from 25 to 50 mm.

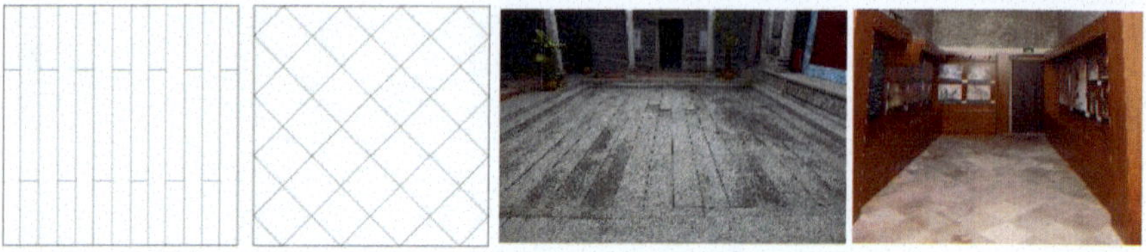

Fig. 6.21 Strip-stone paving and large clay brick paving. Center: Jingfu Lu Ancestral Hall; Right: Ruizhi Ancestral Hall (*Image source* drawn by the author)

A defining characteristic of their installation is the diagonal laying pattern, wherein bricks are set at a 45-degree angle, creating a diamond-shaped visual effect. This not only enhances the visual dynamism of interior spaces but also contributes to the ceremonial atmosphere through patterned floor articulation (Fig. 6.21).

5.2 *Columns*

5.2.1 Type of Column

Within the architectural system of Tangjiawan ancestral halls, the column system serves as the core structural framework for load-bearing as well as the primary element for spatial definition. Each column not only plays a critical role in transferring roof loads to the foundation but also visually and symbolically articulates spatial hierarchy and ritual order through its placement, material selection, and formal expression.

The column structures in Tangjiawan ancestral halls typically consist of two main components: the column shaft (柱身) and the column base (柱础). Based on positional context and functional role, columns can be broadly categorized into eave columns (yan zhu 檐柱) and interior columns (jin zhu 金柱) (Fig. 6.22).

This distinction between column types reflects both structural logic and ritual significance, reinforcing the spatial sequencing and ceremonial zoning that defines the architectural identity of the ancestral hall.

5.2.2 Eave Columns (Yan Zhu)

Eave columns (yan zhu 檐柱) are positioned along the edges of the roof eaves, serving as primary structural supports for the overhanging roof sections and defining the architectural interface between interior and exterior spaces. They represent a critical transitional element in the overall spatial composition, marking the threshold between enclosed ritual spaces and open courtyards.

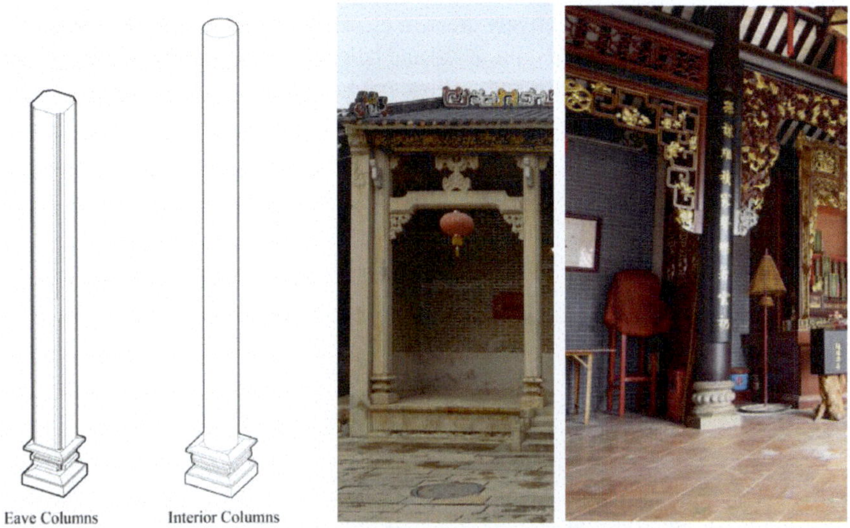

Fig. 6.22 Eave columns and interior columns. Left: Huitong Ancestral Hall; Right: Gu Clan Ancestral Hall (*Image source* drawn by the author)

In Open-Bay front halls (changying-style toumen 敞楹式前堂), the main façade typically features four eave columns, creating a pronounced and rhythmic vertical articulation that gives the façade its most distinctive spatial cadence.

Inside each hall, two additional eave columns are generally positioned at the interior eave line, working in tandem with the roof framing system (梁架) to provide essential structural support for the roof overhang (挑檐部分).

5.2.3 Interior Columns (Jin Zhu)

Distinct from eave columns, the interior columns, commonly referred to as "golden columns" (jinzhu 金柱), are predominantly placed within the central zones of each hall, serving as the primary load-bearing elements and forming the structural backbone of the ancestral hall. Typically, each hall contains four interior columns, arranged in two longitudinal rows along the depth axis of the building.

In the front hall, the front pair of interior columns is often embedded against or partially integrated into the front wall, sharing the structural load with the wall itself. Similarly, in the rear hall, the rear pair of interior columns frequently forms a structural continuum with the rear wall, enhancing overall rigidity and stability.

The interior columns collectively create a clear column grid, which supports the entire roof framing system (liangjia 梁架) and helps establish a solemn, orderly, and rhythmically articulated interior space (Figs. 6.23 and 6.24).

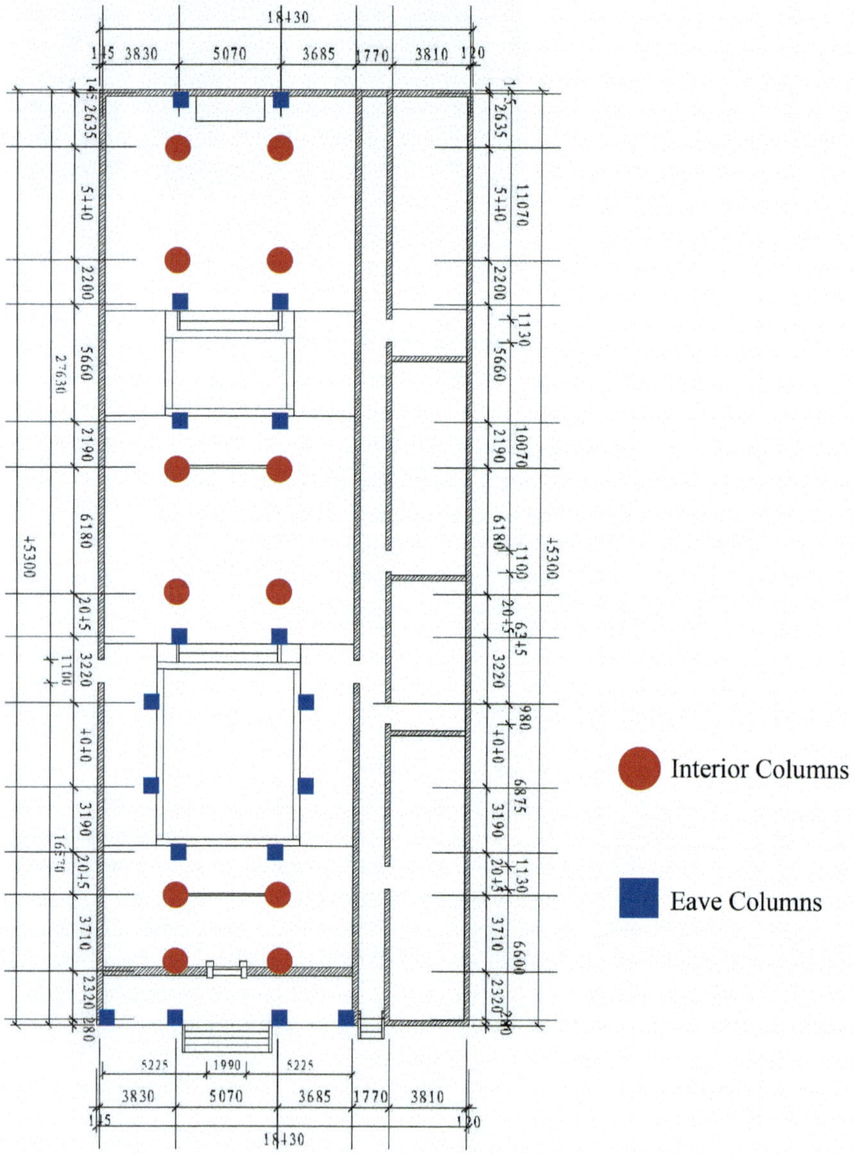

Fig. 6.23 Layout of columns (*Image source* drawn by the author)

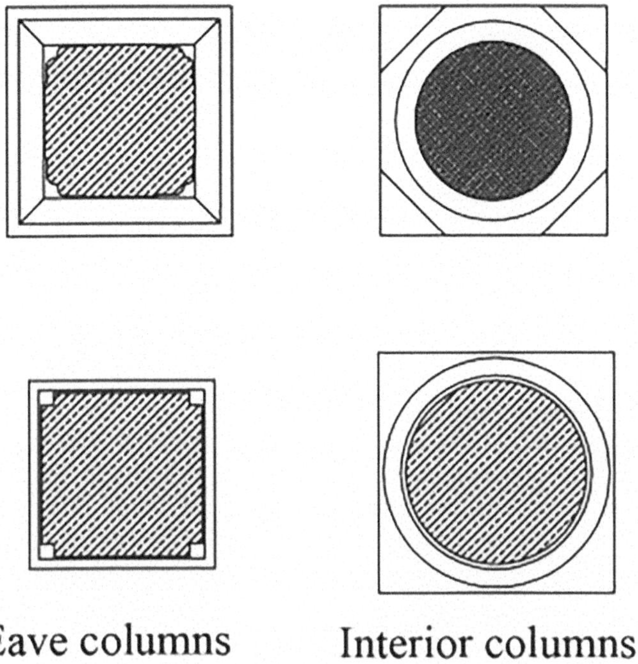

Fig. 6.24 Section of columns (*Image source* drawn by the author)

The column layout reflects not only structural necessity but also ritual spatial hierarchy. Eave columns define the external spatial interface, while interior columns establish the core axial order of the interior, reinforcing clear distinctions between interior and exterior, primary and secondary, and noble and subordinate spaces.

Particularly within the three-hall spatial configuration, the rhythmic arrangement of the interior columns directly correlates with ritual processional routes, the placement of screen doors (pingmen 屏门), and other spatial behaviors linked to ceremonial practice, making the jinzhu one of the most essential architectural elements in the materialization of ancestral hall spatial ethics.

5.2.4 Column Bases (Zhuchu)

The column base (柱础, Zhuchu), serving as the transitional component between the column shaft and the ground, plays a dual role in structural performance and aesthetic expression. In Tangjiawan ancestral halls, the column base is not only a fundamental element within the traditional timber construction system, but also a visual marker of spatial function and ritual hierarchy, as revealed through variations in form, ornamental motifs, and positional correlation with different column types (Fig. 6.25).

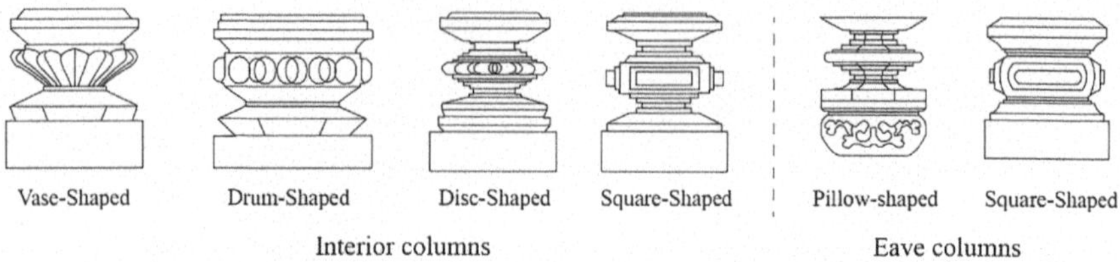

Fig. 6.25 Shaped of column base (*Image source* drawn by the author)

(1) **Functional Roles**

Within the structural system, the column base fulfills two critical practical functions.

Moisture Prevention: Situated along the humid, rain-prone Lingnan coastal region, Tangjiawan's ancestral halls face constant challenges of ground moisture. By elevating the column footings away from direct contact with the ground, the column bases protect timber columns from water absorption, decay, and potential structural failure, serving as an essential measure for enhancing the durability of timber architecture.

Load Distribution: Functioning as an intermediary between the column and the foundation, the column base increases the bearing surface area, allowing the structural loads from the columns to be evenly distributed to the ground, thereby enhancing overall stability and compressive strength.

(2) **Aesthetic Significance and Formal Variation**

Positioned at the juncture between the column shaft and the paved flooring, the column base naturally becomes a visual focal point within the spatial composition. Given its visibility within the viewer's eye level, the column base holds significant decorative value.

In Tangjiawan, column bases are generally composed of two distinct parts: the base body (础身) and the base pedestal (础座), with the base body serving as the primary locus of visual and formal variation.

The base body forms can be categorized into four principal types;

(1) Vase-Shaped (花瓶形): Featuring graceful curves and refined contours, this type is commonly used for interior columns (jinzhu 金柱).
(2) Drum-Shaped (铜鼓式): Full-bodied and symbolically resonant, this form denotes prestige, often reserved for primary ritual columns.
(3) Disc-Shaped (碟形): Flatter in profile, typically applied to eave columns or secondary structural elements.
(4) Square-Sectioned (方形): With a tiered and gradually diminishing profile, this type provides a visually stable and grounded appearance.
(5) The base pedestal is usually simpler in form, commonly adopting square or pillow-shaped (枕形) profiles, occasionally embellished with shallow relief carvings, serving to accentuate the sculptural character of the base body.

(3) **Hierarchical Differentiation by Spatial Position**

The design of column bases in Tangjiawan ancestral halls not only reflects aesthetic sensibilities but also ritual hierarchy and spatial function, creating clear distinctions between functional zones:

(4) **Eave Column Bases**

Eave column bases in Tangjiawan frequently adopt the double-waisted profile (双束腰形式), with the base body typically disc-shaped or square-shaped. Ornamentation often includes bamboo joint patterns or starfruit-like petal divisions, with square bases commonly featuring stepped, diminishing layers. These relatively flat and lightweight designs contribute to a visually airy and elegant appearance, often seated on square or pillow-shaped pedestals, occasionally adorned with shallow relief patterns (Fig. 6.26).

(5) **Interior Column Bases (Jinzhu Bases)**

The bases of interior columns (金柱柱础) are generally larger, heavier, and more elaborately decorated, often adopting drum-shaped, vase-shaped, or pan-shaped profiles. This monumentality and ornamental richness reflect the central role of interior columns in the structural and ritual hierarchy of the ancestral hall (Fig. 6.27).

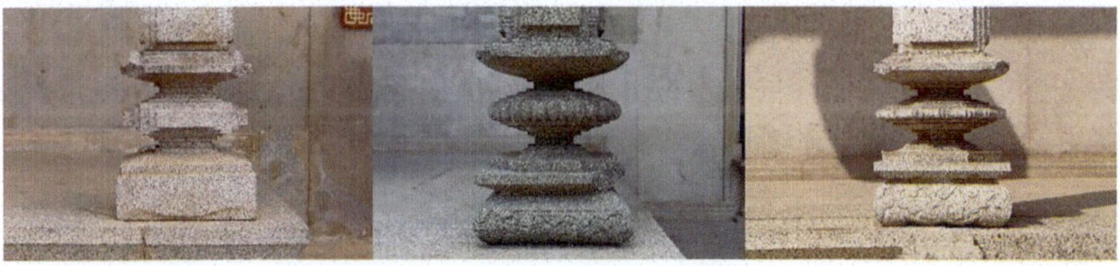

Fig. 6.26 Eave column bases (*Image source* photographed by the author)

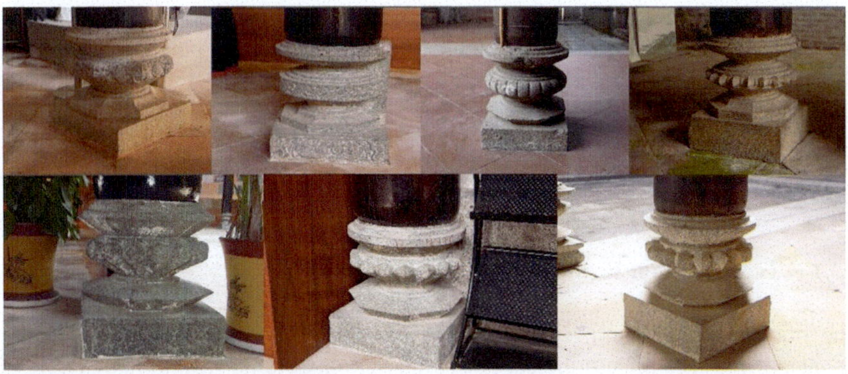

Fig. 6.27 Interior column bases (Jinzhu Bases) (*Image source* photographed by the author)

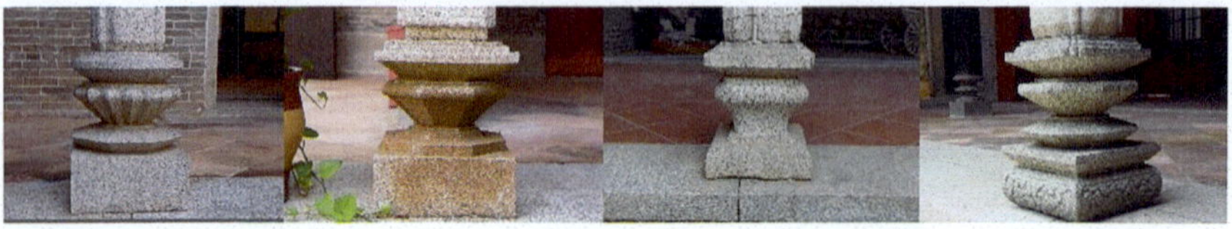

Fig. 6.28 Side Corridor and lateral route column bases (Auxiliary Spaces) (*Image source* photographed by the author)

Notably, the drum form (铜鼓式) holds deep symbolic significance, historically associated with sacred power, authority, wealth, and social status. As referenced in historical texts:

> Bronze drums were not only used for ceremonial music but also symbolized divine power and prestige among their users. Fine bronze drums were often considered price-less. [9, 10]

This symbolism aligns with the ancestral hall's spiritual emphasis on reverence (尊), making the drum-shaped base particularly fitting for interior columns.

The interior column bases often feature petal motifs, auspicious symbolic patterns, or three-dimensional zoomorphic designs (auspicious beasts) carved around their perimeters. The lower sections, where the base body meets the pedestal, frequently transition into hexagonal or octagonal plate-like forms, while the pedestals themselves are typically square-shaped, occasionally embellished with scrollwork (juancao) shallow relief carvings, adding to the visual weight and richness of the interior columns.

(6) **Side Corridor and Lateral Route Column Bases (Auxiliary Spaces)**

The columns within side corridors and lateral routes occupy a subordinate spatial status. To visually emphasize the primacy of the interior columns and entrance halls, these auxiliary columns typically rest on simpler and smaller bases. Their base bodies are often plain vase-shaped forms, while the pedestals are unadorned square blocks, with a minority adopting pillow-shaped profiles.

This hierarchical variation in column base design not only reflects structural necessity but also materializes the spatial ethics and ceremonial hierarchy that underpins Tangjiawan ancestral hall architecture (Fig. 6.28).

5.3 Roof Framing Systems

5.3.1 Types of Roof Framing Structures

The roof framing systems (梁架, liangjia) of Tangjiawan ancestral halls can generally be categorized into three principal types: the Melon-Column Beam Frame (瓜柱梁架), the Antique-Style Beam Frame (博古梁架), and the Shrimp-Bow Beam (虾弓梁).

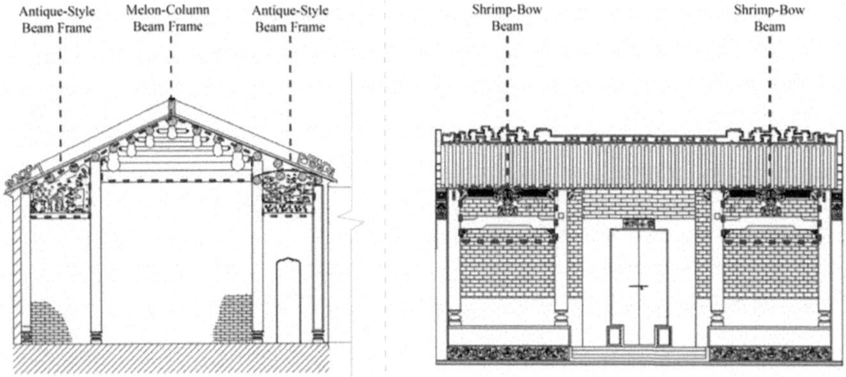

Fig. 6.29 Framing systems (Case: Left Ruizhi Ancestral Hall; Right: Huitong Ancestral Hall) (*Image source* drawn by the author)

Among these, the Melon-Column Beam Frame serves as the primary structural system within the ancestral hall, bearing the main roof loads. This framing type is extensively applied across key interior spaces, including the front hall, central hall, and rear hall. Renowned for its structural robustness and stability, it functions as the core load-bearing framework of the building's central zones.

The Antique-Style Beam Frame functions as a secondary support system. While it also possesses a certain load-bearing capacity, its shorter span makes it suitable for auxiliary spaces, such as side corridors or the front eaves of the main halls. In addition to its structural role, the bogu frame often serves a decorative purpose, enhancing spatial detailing and architectural articulation.

The Shrimp-Bow Beam (xiagong liang 虾弓梁) is a regionally distinctive framing form unique to Tangjiawan, frequently found in the front hall. Primarily decorative in function, its gracefully curved and dynamic profile has become one of the most visually recognizable elements of Tangjiawan ancestral hall architecture, contributing to the hall's spatial identity and aesthetic character (Fig. 6.29).

Beyond these primary beam types, the roof framing system also incorporates several subsidiary structural and decorative components. At the junction between the melon-column beams and the supporting columns, beam-end elements (liangtou 梁头) are commonly installed, serving both as structural transitions and ornamental focal points within the framework.

Furthermore, at the lower intersections where beams meet columns, bracket armatures known as "Queti (雀替)" are typically set in place. These decorative corbel-like elements provide auxiliary support for the beam ends, enhancing structural stability, while simultaneously enriching the spatial and aesthetic experience of the interior.

5.3.2 Primary Structural System—Melon-Column Beam Frame (Gua Zhu Liangjia)

The Melon-Column Beam Frame (瓜柱梁架, gua zhu liangjia) forms the principal structural system for most Tangjiawan ancestral halls, especially within the three main halls (front, central, and rear halls). This framing system is composed of melon columns (瓜柱) and beams (梁).

The Melon-Column Beam Frame commonly employed in Tangjiawan ancestral halls represents a hybrid structural typology, combining the elevated beam frame system (抬梁式) typical of northern Chinese architecture with the post-and-beam frame system (穿斗式) more prevalent in the south, resulting in a regionally adaptive and structurally efficient solution [3].

Structurally, each purlin (檩条) is supported by a short post (melon column), which rests atop a transverse beam below. This transverse beam, in turn, is notched into the vertical support columns at both ends, creating a layered, triangulated roof truss system resembling a human character (人字形屋架). This construction method ensures robust beam-column connectivity and enhances the overall wind resistance of the roof structure [11].

The term "melon column" (瓜柱) derives from the distinctive shape of these short posts. The column head is rounded to receive the upper purlin, while the column body swells in the middle, tapering at both ends, resembling a melon-like or bell-shaped form. Locally in Tangjiawan, they are also colloquially referred to as "Golden Bell Covers (金钟罩)." This form arises from both structural necessity—where the lower end is inserted into the transverse beam and the upper end supports the purlin—and from an aesthetic desire to harmonize with the circular beam structure, giving the columns a sense of visual fullness and elasticity (Fig. 6.30).

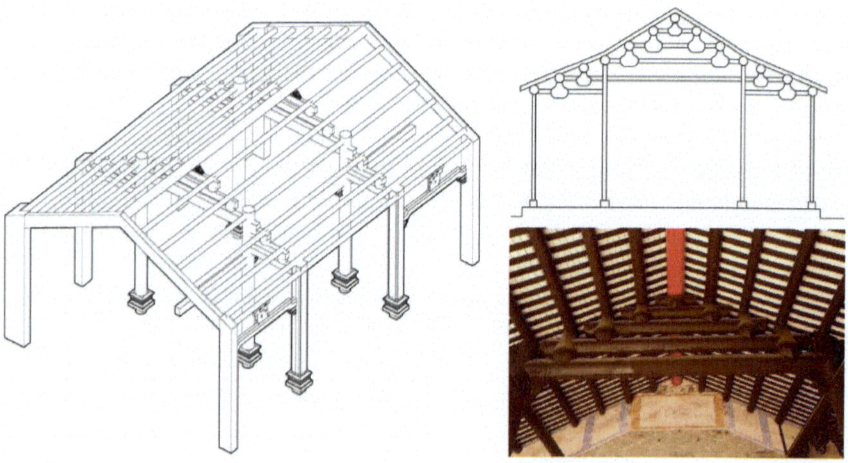

Fig. 6.30 Melon-Column Beam Frame (Case: Junan Lu Ancestral Hall) (*Image source* drawn by the author)

The widespread use of the Melon-Column Beam Frame in Tangjiawan ancestral halls can be attributed to several key advantages:

High Structural Stability: Given Tangjiawan's coastal Lingnan location, where typhoons are frequent during summer months, buildings require exceptional wind resistance. The even distribution of columns and beams, large contact areas at joints, and a low overall center of gravity significantly enhance seismic and wind load performance, making this system ideally suited to coastal climate challenges.

Improved Ventilation and Spatial Expansion: The layered beam-column configuration creates large, column-free spans, facilitating natural air circulation and thermal dissipation, thereby mitigating indoor heat accumulation in Tangjiawan's hot and humid environment and enhancing occupant comfort.

Unified Aesthetic Expression: The full-bodied, fluid lines of the melon columns, often shaped gourd-like or inverted-conical, provide a smooth and visually cohesive appearance. This soft yet dignified form aligns perfectly with the local aesthetic preference for solemn yet unpretentious ancestral hall spaces, making the melon-column frame an architectural component that seamlessly integrates both functionality and beauty.

5.3.3 Auxiliary Beam Frames for Corridors—The Bogu Beam Frame (Bogu Liangjia)

The Bogu Beam Frame (博古梁架) is a commonly observed decorative beam system in Tangjiawan ancestral halls, characterized by its single-plank timber construction and its dual role in providing both structural support and artistic ornamentation. This beam type holds considerable visual prominence within interior spaces and plays an important role in enhancing spatial atmosphere [11].

Unlike the Melon-Column Beam Frame, which prioritizes load-bearing capacity, the Bogu Beam Frame is primarily aesthetic in function, designed to enrich the spatial character and visual articulation of the hall interior.

Due to the span limitations inherent to single-plank construction, the Bogu Beam Frame is unsuitable for large-span spaces. Consequently, it is typically employed in auxiliary architectural zones, such as the front or rear eaves of the three main halls, or within side corridors, where it functions as a supplementary or complementary beam structure.

Bogu beams often adopt a curved arch profile, echoing the linear flow of the corridor space and enhancing spatial rhythm and circulation experience. The Bogu Beam Frame, though secondary in structural role, plays an indispensable part in enriching the ornamental and spatial narrative of Tangjiawan ancestral halls, embodying the Lingnan architectural tradition's emphasis on both structural logic and aesthetic refinement (Fig. 6.31).

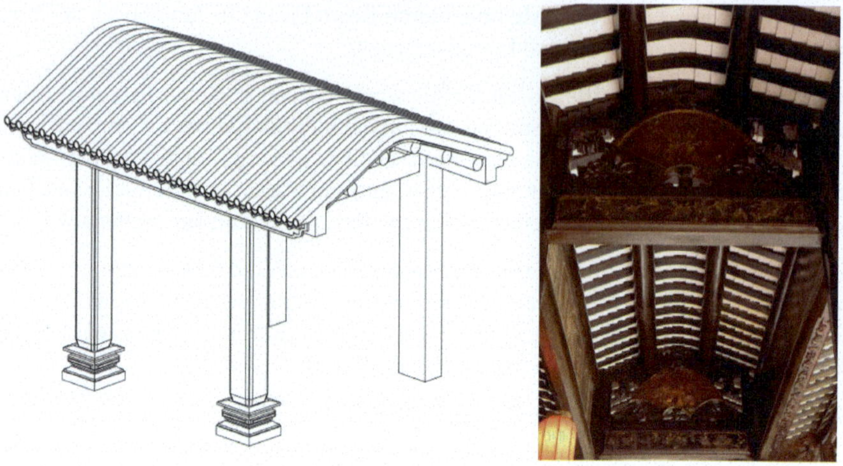

Fig. 6.31 Beam frames for side corridors (Case: Ruizhi Ancestral Hall) (*Image source* drawn by the author)

5.4 *Entrance Hall Beam Frame*

5.4.1 Shrimp-Bow Beam (Xiagong Liang)

In the ancestral hall architecture of Tangjiawan Town, a regionally distinctive beam form known as the Shrimp-Bow Beam (虾弓梁, xiagong liang) is commonly found within Open-Bay front halls (changying-style toumen 敞楹式前堂). Typically installed above the secondary eave columns (次间檐柱) of the front hall, this beam is characterized by downward-curving ends and a horizontally flat center, resulting in a profile resembling the arched back of a shrimp, hence its name.

Beyond its distinctive visual identity, the Shrimp-Bow Beam plays significant structural and ceremonial roles within the architectural hierarchy.

Given the considerable overall height of Tangjiawan ancestral halls, the Shrimp-Bow Beam acts as a visual separator across the secondary bays of the front façade, effectively accentuating the central bay's vertical prominence and spatial importance. This contributes to reinforcing the ritual spatial order, where "centrality equals authority (中为尊)."

From an architectural aesthetics perspective, the contrast between the beam's curving ends and its flat center introduces rhythmic façade layering and depth modulation. Structurally, the Shrimp-Bow Beam, through its lateral spanning form, effectively re-distributes the loads borne by the front eave columns, channeling the roof's overhanging weight down to the foundation, thereby enhancing the stability of the front eave cantilever (Fig. 6.32).

At the center of the Shrimp-Bow Beam, an ornamental component is typically installed. Common motifs include floral carvings or stone lions, collectively referred to as "Stone Shrimp-Bow Beam with Golden Floral Lions (石虾弓梁石金花狮子)."

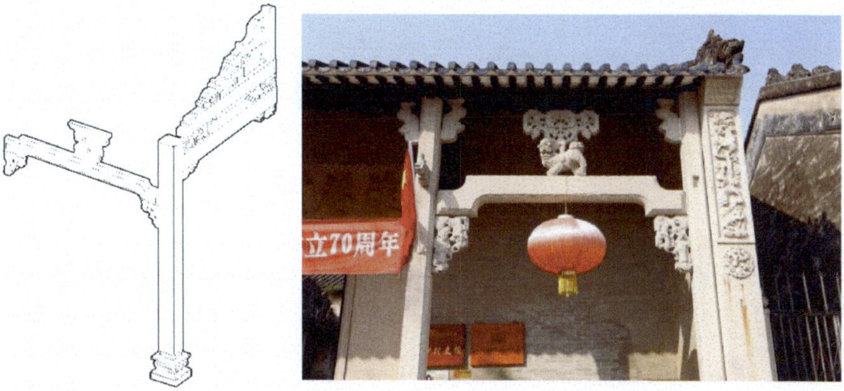

Fig. 6.32 Shrimp-Bow Beam (Case: Mo Clan Ancestral Hall) (drawn by the author)

This decorative element originates from the camel-hump dougong (斗栱) brackets found in traditional timber roof framing. However, given the ceremonial prominence of the front hall and its elevated decorative requirements, the Tangjiawan region often replaced the original wooden bracket forms with stone-carved equivalents, matching the materiality of the beam to enhance both visual mass and long-term durability.

In terms of specific stylistic variations, two common configurations are observed: (1) a lower-level stone lion combined with an upper-level stone floral motif, with the lion symbolizing guardianship and noble dignity. (2) Symmetrical floral motifs on both the upper and lower sections, creating a balanced and continuous visual rhythm (Fig. 6.33). As a hybrid of decorative and structural function, the Shrimp-Bow Beam exemplifies the Tangjiawan ancestral hall's harmonious integration of structural logic, spatial hierarchy, and ornamental aesthetics, reinforcing the unity between form and ritual meaning within Lingnan architectural tradition.

5.4.2 Auxiliary Components of the Roof Framing System—Beam Heads (Liangtou) and Bracket Arms (Queti)

Within the roof framing systems of Tangjiawan ancestral halls, auxiliary components serve not only critical structural reinforcement functions but also represent concentrated expressions of Lingnan carpentry craftsmanship and decorative artistry. Among these, the beam ends (梁头, liangtou) and bracket arms (雀替, queti) stand out as key architectural elements, embodying both structural logic and ornamental tradition, and reflecting the region's unique building wisdom and aesthetic sensibilities.

The Beam Heads (liangtou) refers to the exposed portion of the beam that extends beyond the column face. In Tangjiawan ancestral halls, beam ends serve not only a structural role but also reflect local climatic adaptations. Given Tangjiawan's southern coastal location and its humid, rain-prone environment, traditional timber beam ends

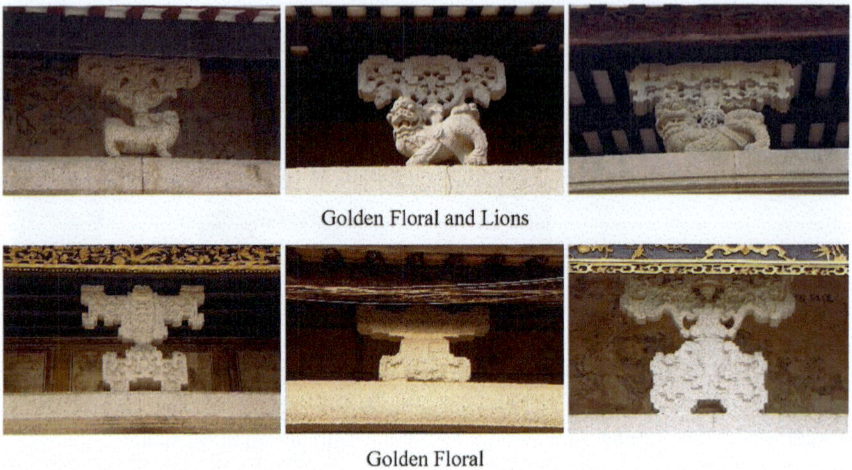

Fig. 6.33 Stone Shrimp-Bow Beam with Golden Floral Lions (*Image source* photographed by the author)

exposed to the elements would have been highly susceptible to moisture-related deterioration, eventually leading to rot and structural compromise.

To address this issue, local craftsmen developed the practice of replacing the exposed wooden beam ends with stone extensions, resulting in the distinctive "Stone Beam Heads (石梁头)." This innovative technique effectively blocks moisture transmission, preserving the longevity of the internal timber structure.

Beyond functionality, stone beam ends also serve as important decorative focal points. They are often carved in low relief, featuring themes ranging from immortals and folk tales to auspicious motifs such as "Prosperous Business (生意兴隆)" and "Abundant Wealth (财源广进)." These decorative programs vividly reflect Tangjiawan's historical identity as a significant mercantile hub in southern China, embodying the community's cultural narratives and artistic expression.

The bracket arm (queti 雀替), typically positioned at the triangular junction between column and beam (or transverse strut), performs both structural and aesthetic functions. Structurally, the bracket arm plays a key role in mitigating shear stress transfer between vertical and horizontal members, preventing uneven load distribution, tilting, or deformation at the beam-column interface, thereby enhancing the overall stability of the roof framing system.

In Tangjiawan ancestral halls, nearly every roof beam frame (liangjia) is fitted with queti at its lower joints, predominantly made from timber. However, at architecturally prominent locations, such as the junction between the shrimp-Bow Beam and the eave columns, stone bracket arms are often used to maintain material continuity and visual coherence with adjacent stone structural members.

In terms of proportions, the height-to-width ratio of bracket arms is typically 2:1, with common dimensions around 500 mm in height and 300 mm in width. Through the integration of structure and ornament, beam ends and bracket arms in Tangjiawan

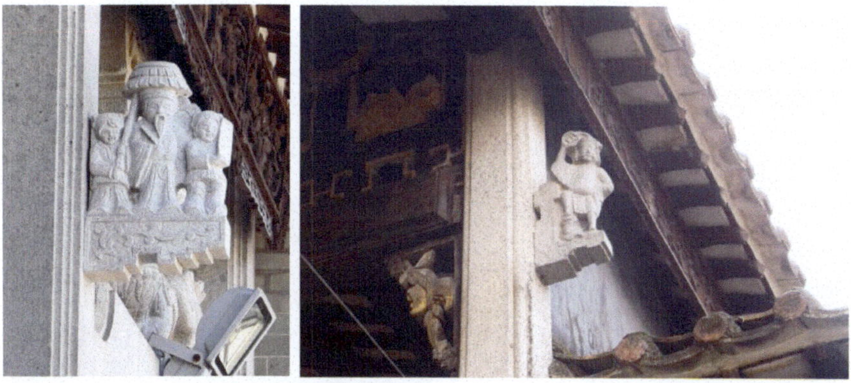

Fig. 6.34 Beam Heads (Left Congjin Liang Ancestral Hall; Right: Cuiping Tang Ancestral Hall) (*Image source* photographed by the author)

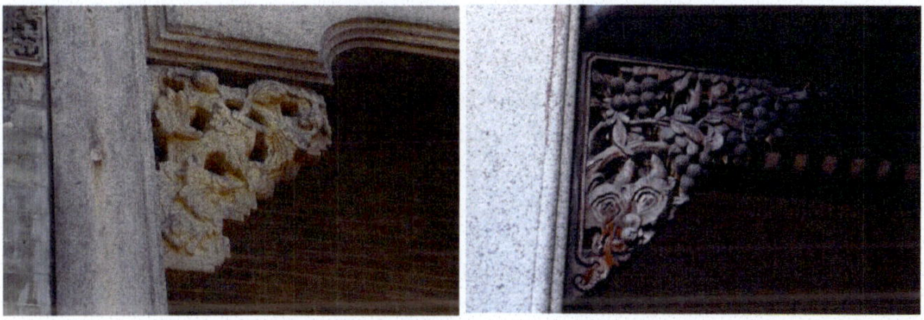

Fig. 6.35 Bracket arm (queti). Left: Bracket carving under Bogu beam structure (Yichu Ancestral Hall); Right: Bracket carving under Xiagong beam (Changyuan Liang Ancestral Hall) (*Image source* photographed by the author)

ancestral halls stand as tangible manifestations of Lingnan regional building intelligence, merging functional necessity with cultural symbolism and artistic flourish (Figs. 6.34 and 6.35).

5.5 *Roof Structures*

The roof structure of Tangjiawan ancestral halls consists of several key components: the roof surface (屋面), the main ridge (正脊), the gutter ridges (垂脊), the parapet ridges (看脊), and the eave closure boards (封檐板).

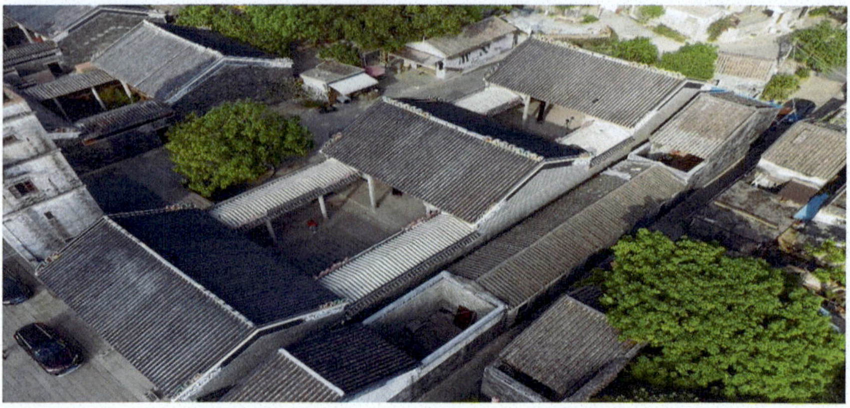

Fig. 6.36 Gable roof form (Case: Mo Clan Ancestral Hall) (photographed by the author)

5.5.1 Roof Surface

All ancestral halls in Tangjiawan adopt the gable roof form (硬山顶, ying shan ding), a direct response to the region's frequent rainstorms, typhoons, and the high-density comb-like village layout (梳式布局). This roof form effectively protects the building from rain and typhoon impacts, while also limiting the spread of fire along the roofline, thereby enhancing the hall's fire resistance.

Given the austere and straightforward form of the gable roof, most architectural variation and decorative attention are focused on the ridge system (脊饰) and the eave closure boards, rather than the roof surface itself (Fig. 6.36).

5.5.2 Main Ridge (Zhengji)

The main ridge (正脊, zhengji) runs along the intersection line between the two sloping planes of the roof, traditionally concealing and waterproofing the joint. In Tangjiawan ancestral halls, the main ridge is also a visual symbol of family prestige and economic power, prompting some clans to lavishly decorate the ridge as an expression of wealth and status.

The primary decorative technique used on the main ridge is gray plaster sculpture (灰塑, hui su). Through this distinctive craft, the otherwise simple roof form is animated and lightened, making the main ridge the most visually striking and ceremonial element of the hall's façade composition.

Based on formal characteristics, Tangjiawan main ridges are broadly classified into two types: the Dragon Boat Ridge (龙舟脊) and the Bogu Ridge (博古脊).

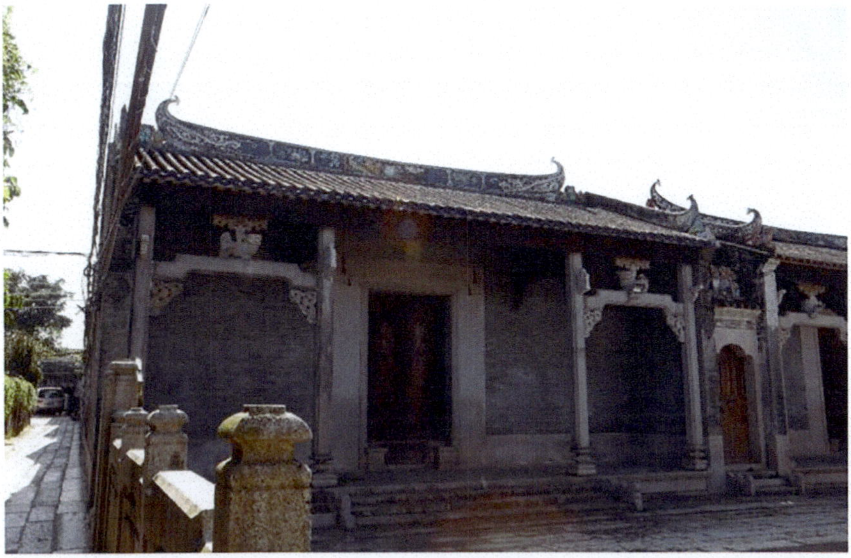

Fig. 6.37 Changyuan Liang Ancestral Hall (*Image source* photographed by the author)

5.5.3 Dragon Boat Ridge (Longzhou Ji)

As the name suggests, the Dragon Boat Ridge features a profile reminiscent of a dragon boat. Historically, Tangjiawan was crisscrossed by rivers and waterways, and boats were integral to the daily life and livelihood of local inhabitants. This cultural reverence for boats led to the abstraction and incorporation of boat forms into ridge design. The Dragon Boat Ridge is characterized by gently up-turned ends resembling boat prows, with some examples incorporating scrollwork motifs (卷草纹) into the upward-curving sections. Compared to other ridge types, the Tangjiawan Dragon Boat Ridge is more modest in scale and ornamentation, with relatively low ridge height and simple scroll decorations painted along the ridge body.

A notable feature is the "Boat Support (船托)," typically installed beneath the up-turned ends of the ridge, often shaped in gray plaster representations of fruits and vegetables, adding a layer of auspicious symbolism.

Overall, the Dragon Boat Ridge, with its combination of boat imagery and scroll motifs, serves as an architectural manifestation of Tangjiawan's maritime cultural identity (Figs. 6.37 and 6.38).

5.5.4 Bogu Ridge (Bogu Ji)

Emerging later in historical development than the Dragon Boat Ridge, the Bogu Ridge typically reaches greater heights, averaging between 50 to 100 cm. The Bogu

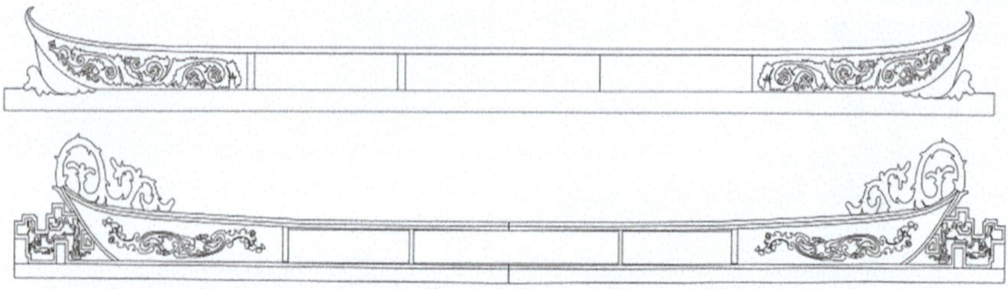

Fig. 6.38 Dragon Boat Ridge (*Image source* drawn by the author)

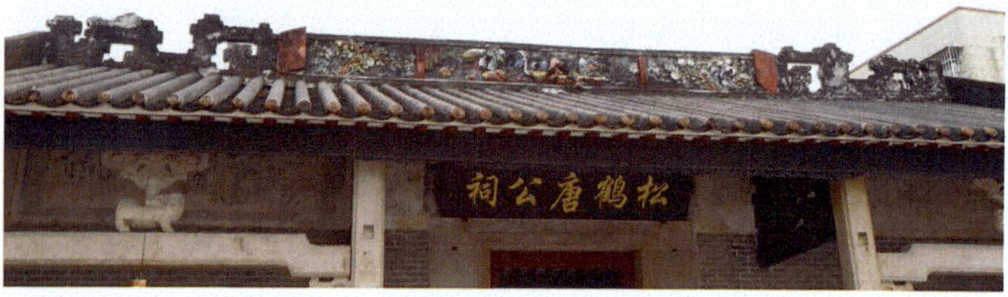

Fig. 6.39 Songhe Ancestral Hall (*Image source* photographed by the author)

Ridge derives its name from the ornamental "Bogu motifs" located at both ridge ends, which are visibly about one-third higher than the central ridge section (Fig. 6.39).

Structurally and compositionally, the Bogu Ridge is more elaborate and decorative, usually comprising several key elements: the Bogu Heads (博古头) at both ends, secondary decorative panels (小品), lattice windows (花窗), and a central pictorial panel (主画). The main pictorial sections and secondary panels are executed in gray plaster relief, typically rendered in single-panel, triptych (three-panel), or openwork formats (镂空). The pictorial content often features auspicious flora and fauna, such as cranes, roosters, pine trees, carp, peonies, and chrysanthemums, arranged into traditional auspicious themes, including "Longevity with Pine and Cranes (松鹤延年)," "Prosperity with Blooming Flowers (花开富贵)," and "Success and Wealth (功名富贵)."

The flanking secondary panels (小品) typically feature symbolic motifs, such as flower baskets or vases, while the openwork variants introduce window-like perforations that enhance spatial lightness and reduce wind pressure on the ridge, demonstrating Tangjiawan's context-sensitive environmental adaptation.

The Bogu Heads, anchoring both ends of the ridge, are especially ornate, often filled with vibrant decorative reliefs. In many ancestral halls, fruits such as peaches, oranges, and occasionally Lingnan-specific tropical fruits like mangoes, are half-embedded in gray plaster within the Bogu motifs, enhancing both symbolic auspiciousness and regional distinctiveness (Figs. 6.40 and 6.41).

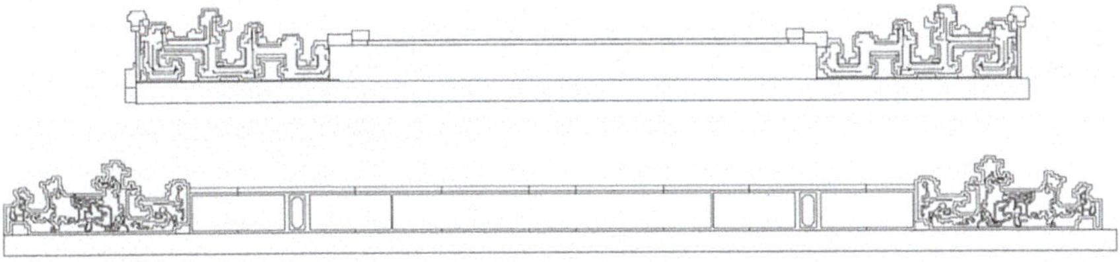

Fig. 6.40 Bogu Ridge (*Image source* drawn by the author)

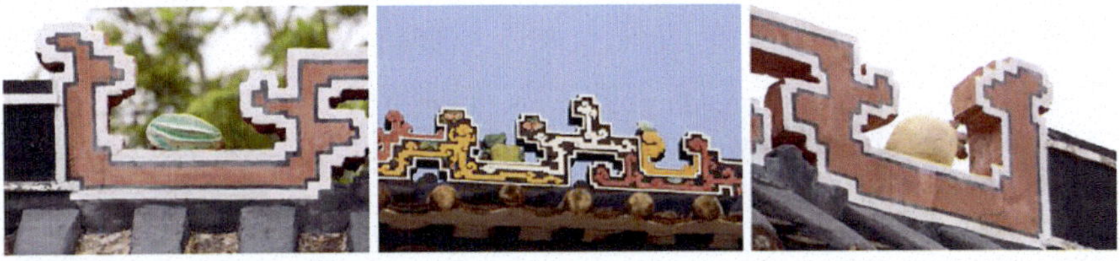

Fig. 6.41 Bogu Heads (*Image source* photographed by the author)

5.5.5 Gable Ridge (Chui Ji)

The gable ridge (垂脊, chui ji) refers to the decorative ridge element positioned along the intersection line of the gable walls and the roof slope, typically serving as a secondary and supportive visual feature relative to the main ridge (正脊).

In most cases, the design language of the gable ridge is coordinated with that of the main ridge. For instance, when the main ridge adopts the dragon boat form (船脊形式), the gable ridge will likewise follow a similar boat-shaped profile; the same applies for bogu-style main ridges, where the gable ridge will feature bogu-themed motifs.

For boat-shaped gable ridges, the decorative techniques mirror those used on the main ridge, with the most distinctive feature being the "boat support (船托)" installed at the tail end, symbolizing Tangjiawan's maritime cultural identity.

In the case of bogu-style gable ridges, the ridge tail typically features relief-carved bogu motifs, often accompanied by polychrome gray-plaster decorations within the interior of the motif, creating a visual dialogue with the associated "Wind-Blocking Walls (搏风)" located on the gable wall below (Fig. 6.42).

5.5.6 Decorative Viewing Ridge (Kan Ji)

The decorative viewing ridge (看脊, kan ji) represents a rarer form of ridge ornamentation within Tangjiawan ancestral halls, typically employed only in halls constructed by wealthier clans.

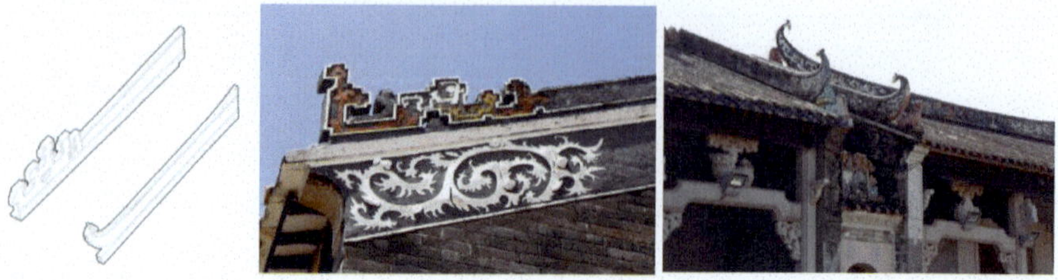

Fig. 6.42 Gable ridge. Boat-shaped (Case: Changyuan Liang Ancestral Hall). "Antique Display Ridge" (Bogu Ridge) vertical ridge (Case: Huitong Ancestral Hall) (by the author)

Located above the Qingyun Alley (青云巷) entrance lintel or above the eave lines of side corridors, the primary function of the viewing ridge, as its name implies, is visual enjoyment and close-up inspection.

Unlike the main ridge, which is primarily viewed from afar, the viewing ridge is positioned much closer to the viewer's line of sight, often directly overhead, necessitating even finer craftsmanship and more intricate detailing. In terms of form, the viewing ridge usually presents a single continuous pictorial composition, with floral, avian, or landscape themes being most common.

A representative example is found at the Lushi Twin Ancestral Halls (卢氏孖祠) in Beisha Village, where both the Qingyun Alley lintel and the side corridor eaves feature viewing ridges, making this hall a pinnacle of ridge decoration within Tangjiawan. The Qingyun Alley viewing ridge presents a narrative scene of a large lion with a playful cub, symbolizing the Confucian concept of "Grand Tutor and Junior Tutor (太师少师)," conveying both authority and nurturing lineage. Inside the hall, along the side corridors, the viewing ridge themes were once equally elaborate, primarily featuring landscape motifs. Unfortunately, due to extensive damage over time, much of this intricate work has been lost and is no longer decipherable (Fig. 6.43).

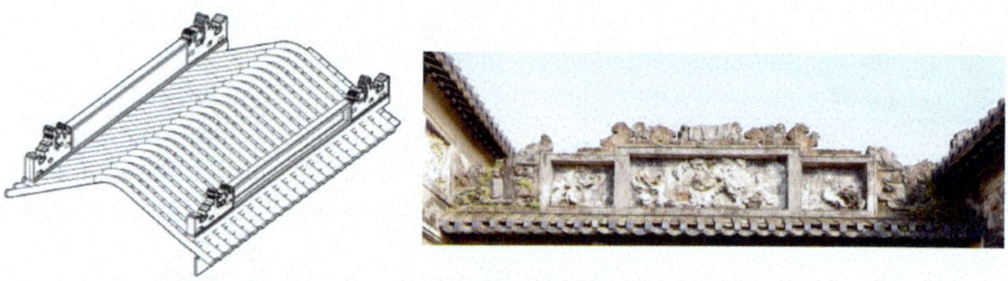

Fig. 6.43 Viewing ridge Case: ridge decorations on Qingyun Alley and side corridors of the Lu Clan Twin Ancestral Halls (*Image source* drawn by the author)

5.5.7 Eave Closure Boards (Fengyan Ban)

The eave closure boards (封檐板, fengyan ban) refer to the timber boards used to seal off the underside of the eaves, located beneath the eave line and visible along both the three main halls and the side corridors.

The primary functional role of the eave closure boards is to shield the purlins and rafters from direct exposure to sun and rain, thereby preventing timber decay and pro-longing structural longevity.

In Tangjiawan ancestral halls, common thicknesses for eave boards range from 35 to 40 mm, with typical heights between 250 and 400 mm. Given their highly visible location at the leading edge of the eaves, eave boards are frequently adorned with elaborate woodcarving, earning them the alternative local name "flower boards (花板)." Owing to their elongated horizontal format, carving techniques typically adopt a panoramic narrative style, with common motifs including flowers and birds, fruits, auspicious birds and beasts, and folk stories. Both ends of the eave board are generally terminated with floral scroll patterns, while the bottom edge is often framed with a continuous band of scrollwork (卷草纹).

A particularly outstanding example is the eave board of the Mo Clan Great Ancestral Hall (莫氏大宗祠), which is regarded as one of the finest surviving works in Tangjiawan. Its surface relief carvings are further accentuated with layers of gold leaf, lending the entire composition an opulent and ceremonial aesthetic (Fig. 6.44).

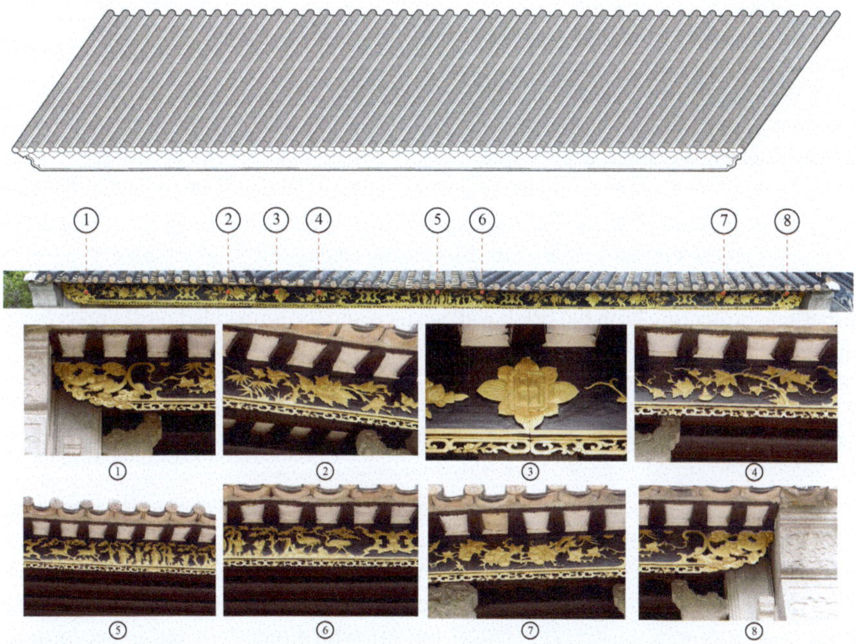

Fig. 6.44 Eave closure boards Case: Mo Clan Ancestral Halls (drawn by the author)

5.6 Walls

The wall system in Tangjiawan ancestral halls primarily comprises three distinct types: the central wall of the entrance hall (toumen zhongqiang 头门中墙), the partition walls along Qingyun Alleys (青云巷隔墙), and the gable walls (山墙). The primary construction material for these walls is gray brick (青砖), a staple in Lingnan architectural tradition, known for its subdued gray-blue tone, which imparts a sense of elegance and solemnity to the ancestral hall façades.

Decorative and structural elements commonly found on hall walls are primarily concentrated in three areas.

(1) Wind-Blocking Bands (搏风) on both the interior and exterior sides of the gable walls.
(2) Wall Friezes (墙楣) along the upper edges of the wall surfaces.
(3) Wall-End Caps (墀头) at the terminal ends of the wall structures.

These three elements serve not only protective and connective functions within the building's structure but also play a crucial visual role in defining the ornamental layering and façade articulation of the ancestral hall.

5.6.1 Wind-Blocking Bands (Bofeng)

Given that most Tangjiawan ancestral halls adopt a double-pitched gable roof (双坡硬山顶), the gable walls naturally present a triangular or gable-end profile (人字形山墙). Along the upper edge of these gable walls, a black decorative band, known as the Wind-Blocking Band (搏风, bofeng), is typically installed.

The lower edge of the bofeng is often outlined with a fine white lime line, while the base section is enhanced with simple patterns executed in gray plaster (灰塑). At the junction between the bofeng and the wall-end cap (墀头), scroll motifs (卷草纹) are commonly painted, featuring flowing and expansive curves, with the decorative range typically occupying about one-third of the total gable wall length.

Constructed from lime-based materials, the bofeng offers excellent waterproofing properties, protecting both the brick wall surface and the embedded purlin ends from wind and rain erosion (Fig. 6.45).

5.6.2 Wall Friezes (Qiang Mei)

The wall frieze (墙楣, qiang mei) refers to the decorative band positioned along the upper section of the wall, typically located approximately 0.5 m below the eave line, making it a major visual focal point within the façade composition.

In Tangjiawan ancestral halls, wall friezes are typically executed in either painted mural form or as gray-plaster relief (灰塑, hui su), with dimensions varying according to wall length. Generally, frieze lengths range from 2 to 4 m, while heights are approximately 80–120 cm.

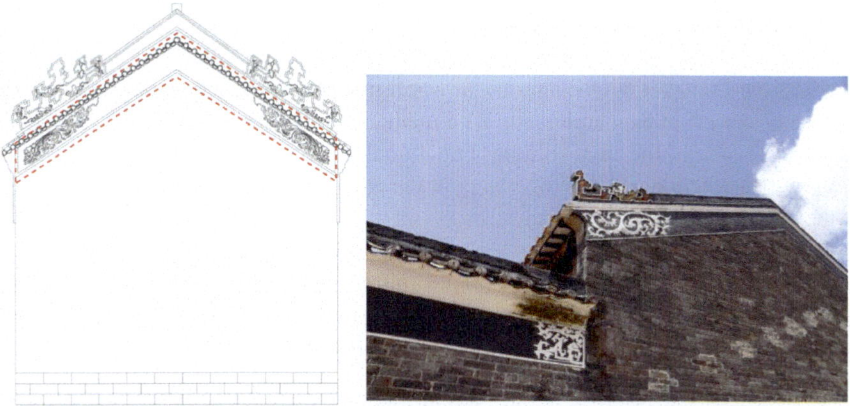

Fig. 6.45 Wind-Blocking Bands Case: Diao Mei ancestral hall (drawn by the author)

The most common form of wall frieze decoration is flat-surface painting, widely distributed across both interior and exterior surfaces of the entrance hall's central wall (toumen zhongqiang), as well as along the interior walls of side corridors. These painted friezes often depict narrative figure scenes, landscapes, flowers, and birds, rendered in vibrant colors and rich in symbolic content.

In contrast, gray-plaster wall friezes are more frequently found on the rear wall of the rear hall or along the interior walls of the ancillary shrines (衬祠). These are typically themed around floral and avian motifs, executed using semi-relief carving techniques, resulting in pronounced three-dimensional layering and strong textural effects. The production process for gray-plaster friezes is technically demanding, requiring skilled craftsmanship, extensive labor, and higher material costs.

As a result, the presence of gray-plaster wall friezes was often a marker of a clan's economic strength, serving as a symbol of social status and family prestige within the Tangjiawan ancestral hall tradition (Fig. 6.46).

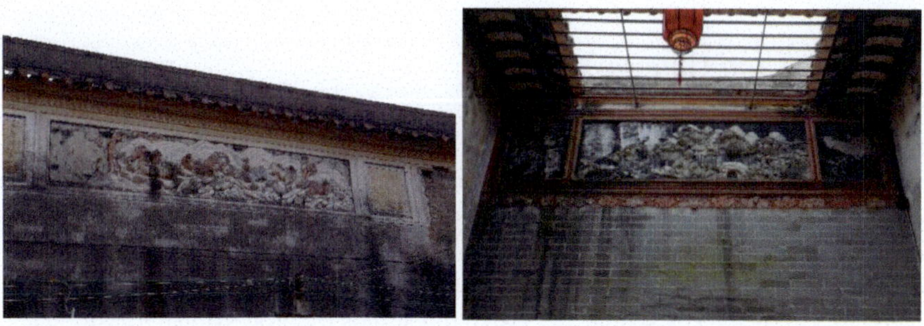

Fig. 6.46 Rear wall plaster relief frieze of the rear hall (Case: Lu Clan Twin Ancestral Halls). Interior plaster relief frieze of the side ancestral hall (Case: Ruizhi Ancestral Hall) (*Image source* photographed by the author)

5.6.3 Wall-End Brackets (Chitou)

The wall-end bracket (墀头, chitou), also referred to as the "corner support stone (附角石)," is a projecting architectural element extending from the gable wall toward the exterior side of the eave column. Functionally, the chitou plays a structural role in supporting the longitudinal overhang of the eaves, both at the front and rear edges of the roofline.

Given its highly visible placement at the interface between the gable wall and the eave structure, the chitou has evolved into an indispensable decorative feature within Tangjiawan ancestral hall architecture. It is widely deployed along the front eave façades of various ancestral halls, contributing significantly to façade layering, articulation, and overall spatial rhythm.

The typical construction of the chitou consists of three primary components, arranged vertically from top to bottom:

Chitou Cap (墀头顶): Slightly projecting outward with an approximate downward inclination of 30°, serving the load-transferring function of supporting the eaves.

Chitou Body (墀头身): This is the principal decorative zone, accounting for approximately 60% of the total chitou surface area. The chitou body is typically framed with ornamental border lines, creating a composition reminiscent of a picture frame, thereby enhancing visual focus and compositional clarity.

Chitou Base (墀头座): Occupying the lower 20%, serving as the structural termination and visual grounding point.

The decorative motifs are diverse and rich in symbolism, with common themes including floral compositions, fruiting plants, auspicious birds, and mythical creatures. These motifs convey blessings of prosperity and well-being, while also embodying the clan's cultural identity and spiritual aspirations [12].

As a visual focal point anchoring the transition between gable wall and eave column, the chitou embodies a dual function of structural performance and decorative expression, making it one of the most representative architectural details within the façade composition of Tangjiawan ancestral halls (Fig. 6.47).

5.7 *Doors and Gateways*

Given the overall spatial character of Tangjiawan ancestral halls, which emphasize external enclosure and internal openness, the number of external door and window openings remains relatively limited, ensuring both security and spatial privacy. The door system primarily consists of the main entrance gate (头门, toumen), the door sill stones (门枕石, men zhen shi), and the Internal Screen Doors (屏门, pingmen)—together forming the key spatial and visual thresholds for processional transition and ceremonial rhythm control within the hall.

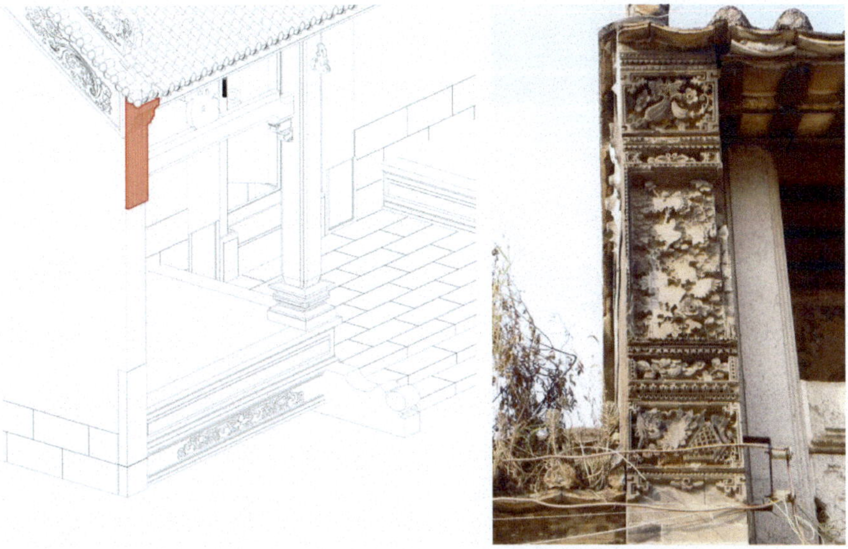

Fig. 6.47 Chitou Case: Yuwo Tang ancestral hall (*Image source* drawn by the author)

5.7.1 Main Entrance Gate (Toumen)

The main entrance gate (头门) serves as the primary architectural threshold, typically located at the fourth purlin position from the eave line, directly in front of the front hall (前堂). Functionally and symbolically, the toumen marks the critical boundary between external and internal space, mediating both physical access and ceremonial sequence. The main components of the toumen include door sill stones (门枕石); door frame (门框); threshold (门槛); double door panels (门板); name plaque or lintel inscription (门额).

In Tangjiawan, entrance gates typically feature a modest and unembellished appearance, with door frames built from long granite blocks, and wooden double-leaf doors designed for practical durability. The door lintel (门额) often displays the hall name (堂号), serving as a symbolic marker of clan identity. Decorative techniques vary from incised lettering (阴刻) and relief carving (阳刻) to suspended wooden name plaques (木质联匾). The wall bases flanking the door frame are usually constructed with the same granite material as the frame, achieving façade consistency and structural continuity. While the exterior side of the main entrance typically emphasizes decorative display, reflecting the clan's public image and status, the interior side focuses more on structural stability and spatial continuity, often treated with minimal ornamentation to enhance the solemnity and sequential progression of the interior spaces (Fig. 6.48).

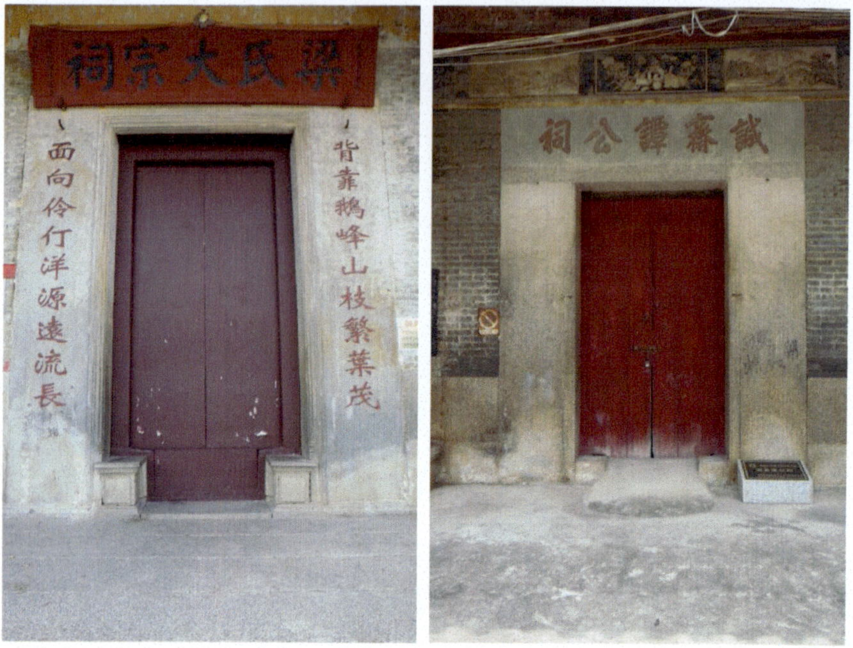

Fig. 6.48 Incised hall name (Case: Chengji Tan Ancestral Hall). Wooden plaque with hall name (Case: Liang Clan Ancestral Hall) (*Image source* photographed by the author)

5.7.2 Bearing Stones (Men Zhen Shi)

The bearing stones (门枕石) are essential subsidiary elements of the main entrance gate, located at both lower ends of the gate opening, serving to support the door pivots and facilitate the door's opening and closing mechanisms.

Structurally, each door sill stone consists of two parts: an interior semi-cylindrical stone block, and an exterior square stone base. Given the heavier load-bearing function on the exterior side, the outer stone block is typically twice the size of the inner component, resulting in a distinctive "inner round, outer square" composition (内圆外方).

As one of the visual focal points on the main façade, many door sill stones feature elaborate carvings, commonly rendered in Sumeru pedestal (须弥座) form, projecting a sense of ritual gravitas and stability.

Decorative emphasis typically centers on the middle and corner sections of the sill stones, with traditional motifs such as bamboo joint patterns (竹节纹) and scroll designs (卷草纹) being frequent choices.

A notable example is found in the Ruizhi Ancestral Hall (瑞芝祠), where the waist section of the door sill stone is carved with a depiction of the Jiaotu (椒图)—one of the Nine Sons of the Dragon, known for its association with door guardianship. This motif embodies the folk belief in protection and closure, symbolizing "Always

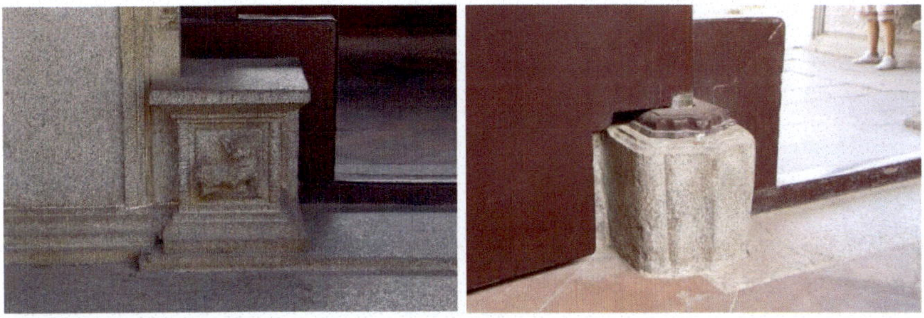

Fig. 6.49 Left: Outer Side of door threshold stone; Right: Inner Side of door threshold stone (Case: Ruizhi Ancestral Hall) (*Image source* photographed by the author)

Close the Door (随手闭门)," blending symbolic significance with cultural humor (Fig. 6.49).

5.7.3 Internal Screen Doors (Pingmen)

The Internal Screen Doors (屏门, pingmen) are generally installed between the golden columns (金柱) of the front hall or central hall, as well as between the side corridors and ancillary shrines (衬祠). Their primary functions include blocking direct sightlines, regulating spatial permeability, and modulating ceremonial progression and environmental conditions.

During ordinary days, the screen doors remain closed, partitioning interior spaces for privacy and environmental control. During ancestral rituals, they are fully opened, allowing the front and rear halls, corridors, and other axial spaces to visually and physically connect, creating a unified and solemn ceremonial arena.

Even when closed, the screen doors often allow for natural ventilation and daylight penetration, contributing to indoor comfort.

Typically, Tangjiawan ancestral halls adopt six-panel or eight-panel screen doors, with the number of panels proportionate to the building's span and spatial scale.

Beyond functional and environmental roles, screen doors also hold important Feng Shui significance, acting as a "filter" to repel malevolent forces (煞气) and invite auspicious energies (吉气). Therefore, door leaf surfaces are frequently decorated with auspicious motifs, such as:

Upside-Down Bats (倒蝠)" symbolizing "arrival of good fortune (福到),

Hidden Eight Immortals (暗八仙)" signifying protection and warding off evil.

For example, the Liang Clan Great Ancestral Hall (梁氏大宗祠) features screen doors filled with bat motifs, symbolizing prosperity flooding the household. The Nanzhou Gu Clan Great Ancestral Hall (那洲古氏大宗祠) presents Taoist-themed "Hidden Eight Immortals" patterns, conveying wishes for peace and spiritual protection (Fig. 6.50).

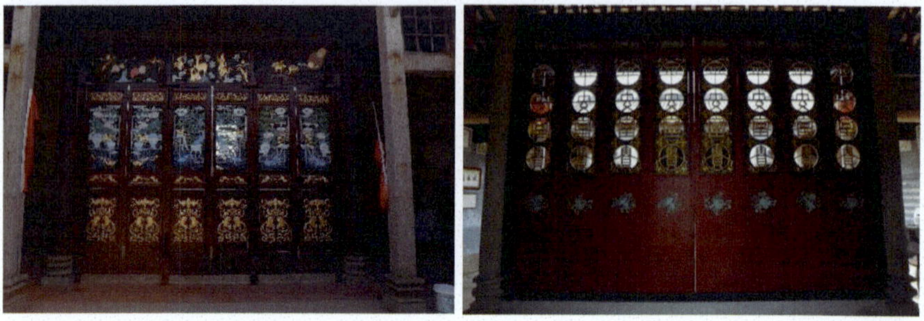

Fig. 6.50 Internal Screen Doors of Ancestral Halls: Case: Liang Clan Ancestral Hall Screen Door Gu Clan Ancestral Hall in Nazhou Screen Door (*Image source* photographed by the author)

5.8 Windows

As a building type primarily oriented toward ritual and ancestral worship, Tangjiawan ancestral halls exhibit a spatial design philosophy that emphasizes internal enclosure and minimizes external openness. Consequently, these buildings rarely feature large-scale exterior windows.

This architectural restraint not only helps to foster a solemn and reverential ceremonial atmosphere but also reflects concerns for privacy, security, and climatic protection.

In certain cases where ancillary shrines (衬祠) have been repurposed as kitchens or utility spaces, small functional windows are installed along the exterior walls, serving purely pragmatic needs for ventilation.

Within the interior spaces, however, windows are primarily decorative architectural elements, designed to introduce limited natural light and controlled airflow, without compromising the overall sense of enclosure and ritual solemnity.

Two principal types of decorative windows are found in Tangjiawan ancestral halls (Fig. 6.51): Horizontal Transom Windows (横披窗, hengpi chuang); and Stained Glass Windows (彩玻窗, caibo chuang).

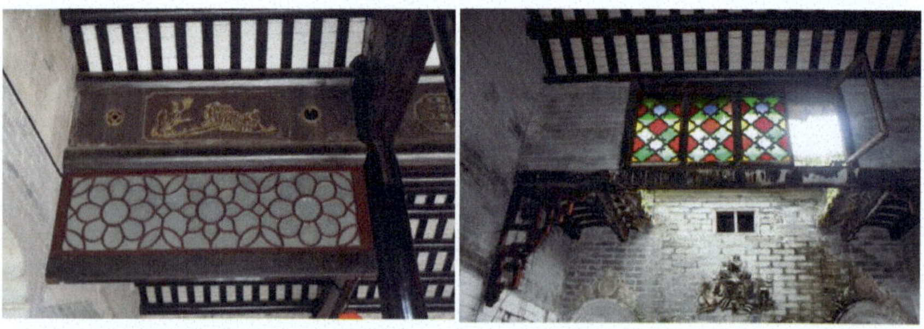

Fig. 6.51 Left: Transom Window of Ruizhi Ancestral Hall; Right: Stained Glass Window of Yichu Ancestral Hall (*Image source* photographed by the author)

5.8.1 Horizontal Transom Windows (Hengpi Chuang)

The Horizontal Transom Window is the most common decorative window type in Tangjiawan ancestral halls. It is typically installed horizontally between two vertical columns, positioned above the beam-and-strut system (梁枋).

Common installation locations include above the Internal Screen Doors (屏门); between beams alongside corridors (侧廊); frames are constructed from timber, with interior lattices (木棂格) often arranged in traditional patterns such as "Begonia Blossom Motif (海棠纹)" or "Cracked Ice Pattern (冰裂纹)."

These windows are non-operable and fixed in place, serving primarily as auxiliary elements for ventilation and daylighting. To further control light diffusion and insect intrusion, the lattice interiors were traditionally covered with thin fabric (薄纱), creating a soft and serene interior ambiance [13].

Due to the limited durability of fabric materials, from the late Qing to early Republican periods, some ancestral halls began replacing fabric with glass panes for improved longevity. A representative example is the Transom Windows in the side corridors of Ruizhi Ancestral Hall (瑞芝祠), which reflect this historical material transition.

5.8.2 Stained Glass Windows (Caibo Chuang)

The Stained Glass Window represents a rarer and more distinctive window form within Tangjiawan ancestral halls, notable for its integration of Western materials and techniques into an otherwise conservative architectural typology.

This phenomenon reflects the "Chinese adaptation of Western elements (洋为中用)" cultural attitude prevalent in Tangjiawan's late modern period, under the influence of broader societal modernization trends.

Field surveys reveal that the Yichu Ancestral Hall (易初祠) in Shangzha Village (上栅村) adopted stained glass windows, skillfully combining imported colored glass with Lingnan's traditional wooden window craftsmanship [14].

The stained glass compositions integrate multiple vibrant colors such as red, yellow, blue, and green, inlaid within begonia-patterned wooden latticework.

The resulting effect is a harmonious blend of traditional structural forms and exotic stylistic influence, producing a visual expression that is both refined and richly layered, embodying a dialogue between classical tradition and modern aesthetic sensibilities.

6 Materials and Craftsmanship

6.1 Timber

Located in Lingnan's humid, rain-prone coastal environment, Tangjiawan faces unique climatic challenges that make timber selection critical to the durability of ancestral halls. With persistent termite threats, high moisture levels, and the extensive use of timber components in traditional architecture, careful material selection is required to ensure resistance to decay, moisture control, and structural strength.

Furthermore, due to the historical prominence of comprador culture (买办文化) and the active overseas trade connections maintained by many Tangjiawan families, high-quality imported timber from Southeast Asia became a common feature in local Ancestral Hall construction. These imported hardwoods, often with slightly acidic properties, exhibit excellent physical performance and strong resistance to insect infestation, making them widely applicable in columns, beams, roof framing systems, doors, windows, and other critical structural and decorative elements.

Among the various timber species used, *Eusideroxylon zwageri* Teijsm. & Binn. (坤甸木, also known as Tiemu or Belian) was particularly prevalent. Native to the city of Pontianak (坤甸市) in western Kalimantan, Indonesia, a region with a significant Chinese diaspora, Kalimantan Ironwood features a yellowish-brown to dark brown heartwood and a golden-hued sapwood. Its exceptional hardness and natural resistance to pests and rot made it the preferred material for golden columns (金柱), gua-column roof framing (瓜柱梁架), and Bogu beam structures (博古梁架) (Fig. 6.52).

Another commonly used timber was Keruing (格木, also known locally as Tokyo Wood), imported from Vietnam, characterized by its tall, straight growth (up to 25 m) and reddish heartwood, noted for outstanding durability against decay. Keruing was often used for main entrance door panels and Bogu beams, especially in areas subject to heavy wear or environmental exposure.

In terms of craftsmanship, Tangjiawan wood carving techniques were deeply influenced by the "Guangzuo (广作)" tradition of the broader Guangfu region, renowned for its rich and intricate craftsmanship hierarchy [15].

Tangjiawan wood carving can broadly be classified into two primary techniques.

Openwork carving (透雕, tou diao): Also known as pierced carving, this technique is commonly found in thinner components such as bracket arms (雀替) and suspended panels (挂落), utilizing multilayered openwork methods to produce exquisite and complex three-dimensional effects.

Relief carving (浮雕, fu diao): Typically applied to larger, compositionally significant components such as Bogu beams, this technique employs varying depths of carving to create a dynamic sense of spatial layering and visual impact, highlighting the extraordinary craftsmanship and cultural richness embedded within Tangjiawan's timber architectural ornamentation.

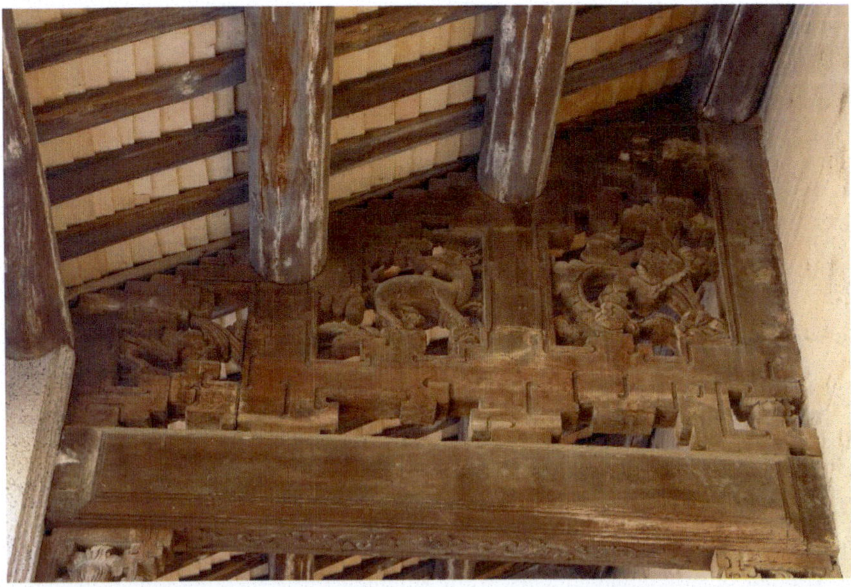

Fig. 6.52 *Eusideroxylon zwageri* Teijsm. & Binn. Beam structure of Wanfu Lu Ancestral Hall (*Image source* photographed by the author)

6.2 Stonework

In Tangjiawan ancestral hall architecture, the selection and application of stone materials reflects a careful consideration of both environmental conditions and structural durability. Among all stone types, granite, locally known as "mashi (麻石)," is by far the most prevalent building material. Composed primarily of silicon dioxide, granite exhibits exceptional resistance to weathering and erosion, making it particularly suited to the rainy and coastal climate of the Lingnan region.

Characterized by its dense and robust texture, granite has been widely employed across a variety of structural, paving, and edge-finishing applications within the Ancestral Halls. Common areas of granite utilization include foundations (台基); ritual platforms (塾台); stone steps (石阶); bracket arms (石雀替); shrimp-bow beams (虾弓梁); eave columns (檐柱); column bases (柱础); door frame stones (门边包石); tie columns (夹杆石); even courtyard paving stones (天井铺地石材).

These stone elements serve not only load-bearing and structural functions but also contribute to the hierarchical ordering of space and enhance the architectural layering of ceremonial zones. In terms of carving techniques, Tangjiawan stone craftsmanship can be broadly categorized into two principal methods:

Round Sculpture (圆雕, yuan diao): This technique is typically used for three-dimensional, stand-alone elements, such as beam ends, bracket arms, and the "Golden Flower Lions (金花狮子)" frequently found adorning shrimp-Bow Beams and brackets.

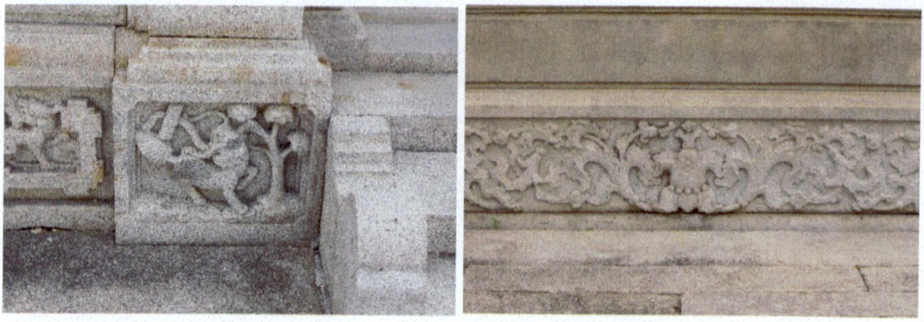

Fig. 6.53 Stone Carving Case: Ruizhi Ancestral Hall (photographed by the author)

These sculptures feature full-bodied, dynamic forms, conveying a sense of vitality and solemnity.

Relief Carving (浮雕, fu diao): Predominantly found on foundations, ritual platforms, door sill stones, and column bases, relief carving in Tangjiawan ancestral halls typically employs shallow relief techniques. This approach emphasizes the profile and contour of structural elements, while simultaneously incorporating ornamental motifs and symbolic patterns, adding rich cultural layers and aesthetic refinement to the architectural composition.

Through the skillful integration of material durability and artistic expression, stone elements within Tangjiawan ancestral halls stand as testaments to both structural ingenuity and the enduring legacy of local craftsmanship (Fig. 6.53).

6.3 *Brickwork and Brick Carving*

In Tangjiawan's traditional architecture, gray bricks (青砖) and brick carving (砖雕) constitute an essential part of wall construction and floor paving, embodying both practical functionality and a distinct regional aesthetic identity.

Locally, dark-toned gray bricks are predominantly used, valued for their restrained and dignified color palette as well as their thermal properties, such as heat absorption and cooling performance, a critical consideration given the humid and rainy Lingnan climate. The standard brick dimensions are typically 25 cm × 10 cm × 5 cm, and the prevailing masonry technique is known as "full-stretch bond (全顺砌)." In certain cases, individual bricks are surface-polished to achieve cleaner edges and a visually uniform façade finish.

Structurally, some ancestral halls employ brick arch systems (砖拱结构) above main entrance openings, with the "arch-and-inverted-arch (券 + 伏)" configuration being especially common. This form enhances load-bearing capacity and improves compressive resistance, exemplifying traditional Chinese structural ingenuity in masonry design.

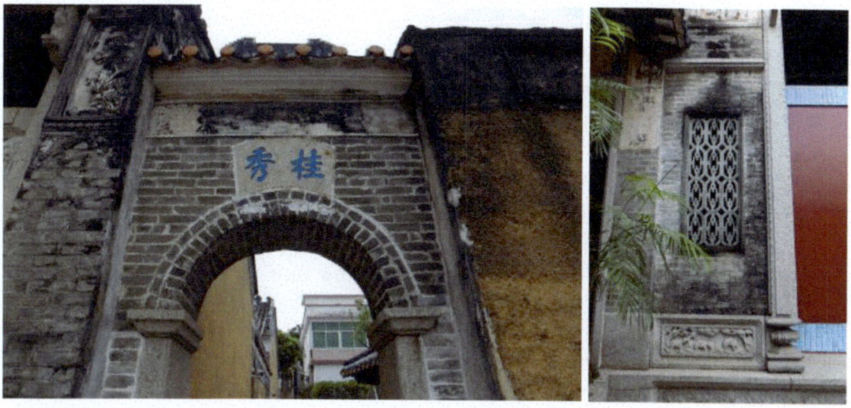

Fig. 6.54 Brickwork and brick carving (*Image source* photographed by the author)

As a decorative element, brick carving (砖雕), though less intricate than wood carving, conveys a unique artistic tension through its simplicity and clarity. Brick carvings are mostly applied at key architectural nodes, such as wall-end brackets (墀头), and are rarely found on large wall surfaces or window frames. The primary carving techniques include relief carving (浮雕) and pierced carving (镂雕), with motifs often centered around auspicious symbols and traditional themes. These designs strike a balance between simplicity and refinement, expressing regional craftsmanship rooted in under-stated elegance (Fig. 6.54).

6.4 Tiles

Located along the Lingnan coastal region, Tangjiawan frequently experiences typhoons and severe rainstorms, making roof stability and wind resistance a central concern in architectural design.

To enhance roof load-bearing capacity and wind resilience, local traditional buildings typically employ heavy barrel-shaped gray clay tiles (辘筒灰瓦) for roof coverings. These dense and moderately weighted tiles, once laid, create a stable and robust roof surface, effectively withstanding high winds and demonstrating the region's climate-adaptive construction wisdom.

6.4.1 Tile Types

Roof tiles in Tangjiawan ancestral halls can be broadly categorized into two main types. Earthenware Tiles (土瓦); Ceramic Tiles (陶瓦). The standard width is approximately 240 mm.

(1) Curved tiles (筒瓦, tongwa): These semi-cylindrical tiles are placed over the joints between two rows of flat tiles, serving both waterproofing and reinforcement functions.
(2) Ceramic tiles (陶瓦) are primarily used for eave decoration and drainage, with two key components:
(3) Tile Ends (瓦当, wadang): Usually circular in shape, functioning as the visual terminus for roof rows.
(4) Drip Edges (滴水, dishui): Designed as protruding pointed elements, these guide rainwater away from the wall face, preventing moisture infiltration and wall damage.

It is noteworthy that not all ancestral halls in Tangjiawan are equipped with drip edges. Some features only tile ends without drip elements, often reflecting the economic capacity of tfeaturestructing family—those affording drip features were typically from wealthier clans.

6.4.2 Tiling Methodology and Construction Process

The installation of roof tiles in Tangjiawan ancestral halls follows a highly systematic sequence, with particular attention to both technical performance and aesthetic effect.

The general tiling procedure adheres to the "bottom-to-top and side-to-center" principle, following four key steps.

(1) Laying the Base Tiles: Applying the "Press Six, Expose Four (压六留四)" overlapping method. The underside of each tile is brushed with limewater to enhance moisture resistance.
(2) Eave Construction: The eave drip edges extend approximately 8 cm beyond the eave line, typically stacked in three tile layers for increased stability and visual appeal.
(3) Covering with Cap Tiles: Following the "Press Seven, Expose Three (压七留三)" technique. Mortar is applied beneath the tiles to prevent water penetration and secure tile placement.
(4) Surface Gray Finishing (上色灰处理): Finally, the entire roof surface is coated with "grass-fibered lime mortar (草筋灰)," a traditional mixture containing plant fibers and lime, to enhance waterproofing, seal tile joints, and ensure a visually smooth finish through natural drying [15].

This comprehensive construction sequence not only reflects Tangjiawan's architectural prioritization of roof performance, but also embodies the Lingnan region's traditional building techniques, where pragmatism and aesthetic refinement are carefully balanced within the roofing craft (Fig. 6.55).

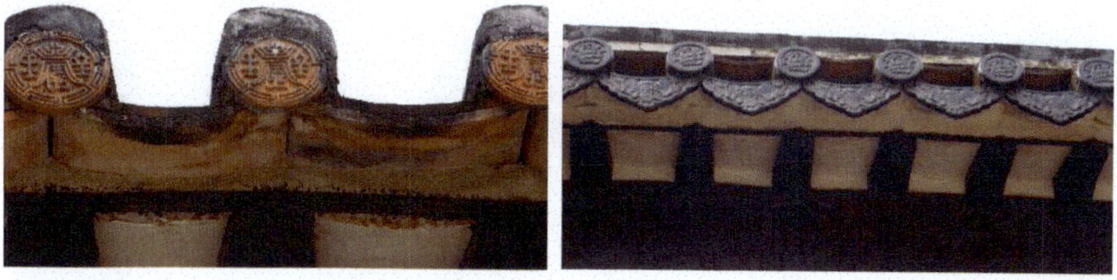

Fig. 6.55 Eaves Tiles and Glazed Decorations (Example: Sanfang Ancestral Hall) (*Image source* photographed by the author)

6.5 Regionally Materials

6.5.1 Oyster Shells

Before the early Qing period, and especially during the mid-to-late Ming dynasty, oyster shells were widely used as a building material for ancestral halls and residential dwellings across the Pearl River Delta, particularly in the Cantonese Area of Lingnan [16, 17]. By the mid-Qing period, however, the use of oyster-shell wall construction gradually diminished.

One of the earliest documented references to oyster-shell walls appears in the Tang dynasty text *Lingbiao Luyi* (岭表异录) by Liu Xun (刘恂), which states: "During the time of Lu Xun's rebellion in Guangzhou, after defeat, his remaining followers fled to the sea islands, lived in the wild, subsisting on oysters and piling up oyster shells to form walls." [18]. The popularity of oyster-shell walls during the late Ming period was primarily due to several factors: exceptional material hardness and durability (locally described as "A thousand-year brick cannot match a ten-thousand-year shell") [19].

In Tangjiawan, a coastal settlement on the western shore of the Pearl River Estuary, the use of oyster shells reflects both a pragmatic adaptation to local resources and a continuation of traditional construction methods. Located near the oyster-rich waters of Haojing'ao (蚝镜澳) adjacent to Macau, Tangjiawan's builders had ready access to oyster shells, making them a staple material for wall construction.

Oyster-shell walls in Tangjiawan are valued not only for their structural resilience and weather resistance but also for their natural decorative texture. The glossy, variegated surfaces create a visually appealing façade, striking a balance between functional durability and aesthetic expressiveness (Fig. 6.56).

6.5.2 Rammed Earth

On the other end of the wall construction spectrum, the Nazhou Gu Clan Ancestral Hall (那洲古氏大祠堂) demonstrates the architectural diversity of Tangjiawan by adopting a rare rammed-earth wall system.

The hall employs a traditional "Sanhe Soil (三合土)" composite, comprising primarily yellow earth, river sand, and lime. To enhance compressive strength

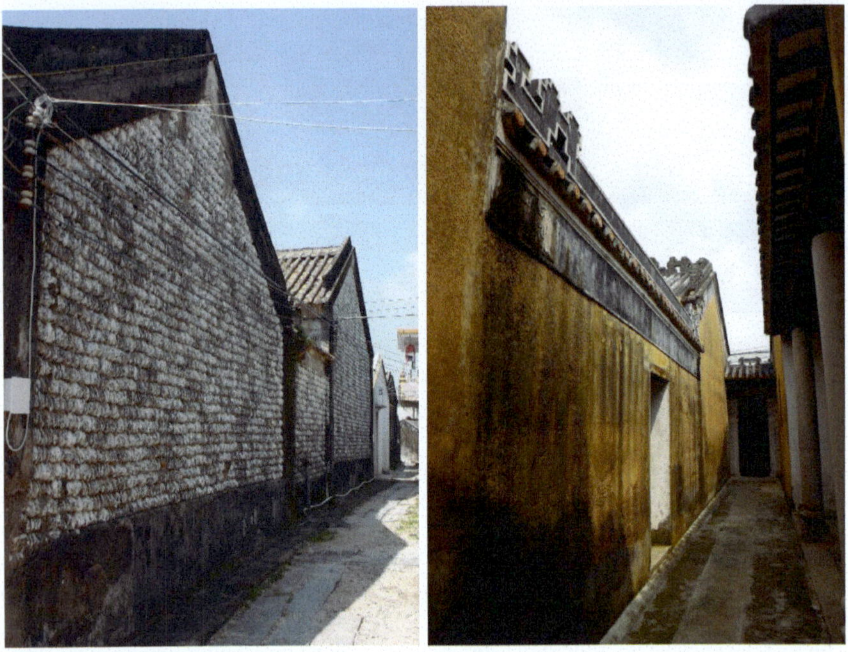

Fig. 6.56 Oyster shell wall (Lu Ancestral Hall). Rammed Earth Wall (Gu Clan Ancestral Hall). (*Image source* photographed by the author)

and long-term durability, natural organic binders such as glutinous rice paste and brown sugar syrup are often added during the mixing process, combining empirical knowledge with traditional material science [20, 21].

The construction technique for rammed-earth walls is highly rigorous and typically involves five main steps:

(1) Pre-wetting and aging of raw materials such as yellow earth to ensure material cohesion.
(2) Foundation compaction, creating a stable base layer.
(3) Incorporation of strengthening agents (e.g., glutinous rice paste), followed by thorough mixing with the Sanhe Soil.
(4) Formwork installation and layer-by-layer compaction executed using the "sequential ramming" method to achieve densification and structural integrity.
(5) Surface finishing and waterproofing treatments, improving weather resistance and aesthetic appearance.

From the contextual use of oyster-shell walls to the experimental exploration of rammed-earth structures, Tangjiawan architecture presents a unique synthesis of traditional wisdom and environmental responsiveness.

These practices exemplify the region's distinctive construction culture and craftsmanship heritage, where material innovation and local adaptation have been key to architectural longevity and cultural identity (Fig. 6.56).

6.6 *Lime Sculpture (Huisu)*

As one of the most representative traditional decorative arts of the Lingnan region, lime sculpture (灰塑, Huisu) plays a prominent role in the architectural expression of Tangjiawan.

Its widespread application not only reflects local craftsmanship and aesthetic sensibilities, but also reveals the deep interplay between architecture, environment, and cultural identity. Situated along the coast of southern China, Tangjiawan experiences a hot, humid, and rainy climate.

Lime-based materials, given their resistance to acids and alkalis and ready availability, have become an ideal choice for both decorative and structural purposes.

Moreover, under the influence of Xiangshan culture (香山文化), residents developed a preference for lively, figurative, and expressive artistic styles, granting Huisu decoration a greater degree of freedom and vitality in visual expression.

Types of Huisu Based on Location and Projection Depth: Depending on the installation location and vertical projection depth on the façade, Huisu craftsmanship in Tangjiawan can be broadly categorized into three main types:

Shallow Relief (平面做, ≤ 5 cm): Primarily used for attached wall decorations, such as gable eaves (博风) and wall friezes (墙楣).

The production process begins with draft sketching, followed by latex application, layered modeling, surface leveling, corner finishing, and finally polychrome painting. The focus here is on flatness, vibrant coloration, and clarity of auspicious patterns and traditional motifs.

Medium Relief (半边做, 5–20 cm): Typically applied to more prominently exposed locations, such as ridge-end decorations (看脊) and wall friezes, this category balances visual depth with structural stability. To enhance adhesion and prevent detachment, artisans commonly embed iron nails or copper wires as internal support, anchoring the Huisu blocks securely to the substrate.

Full Three-Dimensional Sculptural Relief (立体做, > 20 cm): Representing the most expressive and visually impactful form of Huisu, these works project significantly from the wall surface, allowing for multi-angled viewing and detailed appreciation. Often found on roof ridges, this type typically depicts auspicious beasts, fruits, figures, and other symbolic subjects.

The construction process involves creating a steel rod framework, followed by layered modeling using coarse lime-mortar mixed with red clay (土朱). Prior to coloring, the surface is moistened to improve pigment absorption, ensuring color longevity and uniformity.

In summary, Huisu in Tangjiawan ancestral halls serves as both an aesthetic manifestation and a vessel of cultural memory [22]. It embodies the artistic skill and

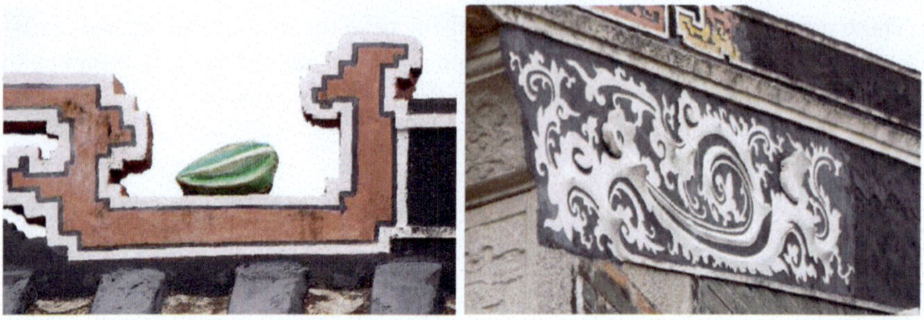

Fig. 6.57 Flat Relief Plasterwork (Example: Diaomei Ancestral Hall). Three-Dimensional Plasterwork (Example: Changyuan Liang Ancestral Hall). (*Image source* photographed by the author)

environmental responsiveness of local craftsmen, translating regional identity and traditional symbolism into visually engaging architectural ornamentation (Fig. 6.57).

6.7 *Wall Murals and Polychrome Painting*

In Tangjiawan ancestral halls, murals serve as a key decorative art form, typically installed at critical visual focal points such as the upper section of the central wall in the front hall, the gable ends, and along the side corridors.

These painted surfaces function not only as cultural conveyors, articulating social values and moral teachings, but also contribute to the atmospheric and aesthetic character of the interior spaces.

The typical mural height ranges between 60 to 120 cm, and in response to the humid subtropical climate of southern China, walls are pre-treated with a lime-based plaster ground layer (灰地仗层) to shield the paintings from moisture and pro-long their lifespan.

6.7.1 Thematic Content

Mural subjects within Tangjiawan ancestral halls generally fall into two primary thematic categories:

Moral and Social Education Themes (社会教化题材): A common example is the "Transmission of the Classics (传经图)," which often depicts historical figures from the Han dynasty, such as Chao Cuo (晁错), Fu Sheng (伏生), and his granddaughter, using historical narratives to convey Confucian ideals of loyalty, filial piety, and reverence for classical learning.

This theme is prominently featured in halls such as the Ruizhi Tang Ancestral Hall (瑞芝祠), Cuiping Tang Ancestral Hall (翠屏唐公祠), and the Mo Clan Grand Ancestral Hall (莫氏大宗祠).

Aspirational and Secular Themes (世俗生活题材): These murals express desires for prosperity, peace, and auspicious fortune, with popular motifs including "Peace and Harmony (安逸祥和)" and "Deer Banquet with Auspicious Greetings (鹿鸣嘉宴)."

Animal symbols like rabbits and deer frequently appear, representing abundance, tranquility, and good omens, reflecting the rural community's longing for an idealized life.

6.7.2 Composition Styles

In terms of composition, Tangjiawan murals integrate traditional Chinese painting and scroll-mounting aesthetics, with careful attention to wall gradients and structural features, resulting in unique spatial-artistic layouts. The main mural composition types include:

Central Wall Compositions (正墙构图): Positioned at the central axis of the hall, featuring prominent themes that serve as the visual focal point of space.

Gable Wall Compositions (山墙构图): Designed to follow the incline of the gable wall, often arranged in diagonal or inverted-triangle formats, enhancing the rhythmic articulation of the façade.

Multi-panel Horizontal Narratives (多段并连长幅构图): Applied to elongated wall surfaces, allowing for continuous visual storytelling across multiple sequential scenes.

Multiple-Frame Single Wall Layouts (单墙多幅构图): Featuring odd-numbered groupings, such as one, three, five, or seven frames per wall, emphasizing visual rhythm, contrast, and dialogue between pictorial elements [23].

Through this diversity in theme and composition, the murals of Tangjiawan ancestral halls offer not only visual enrichment, but also serve as narrative instruments of cultural memory, moral instruction, and aesthetic engagement, making them an integral component of the region's architectural identity (Fig. 6.58).

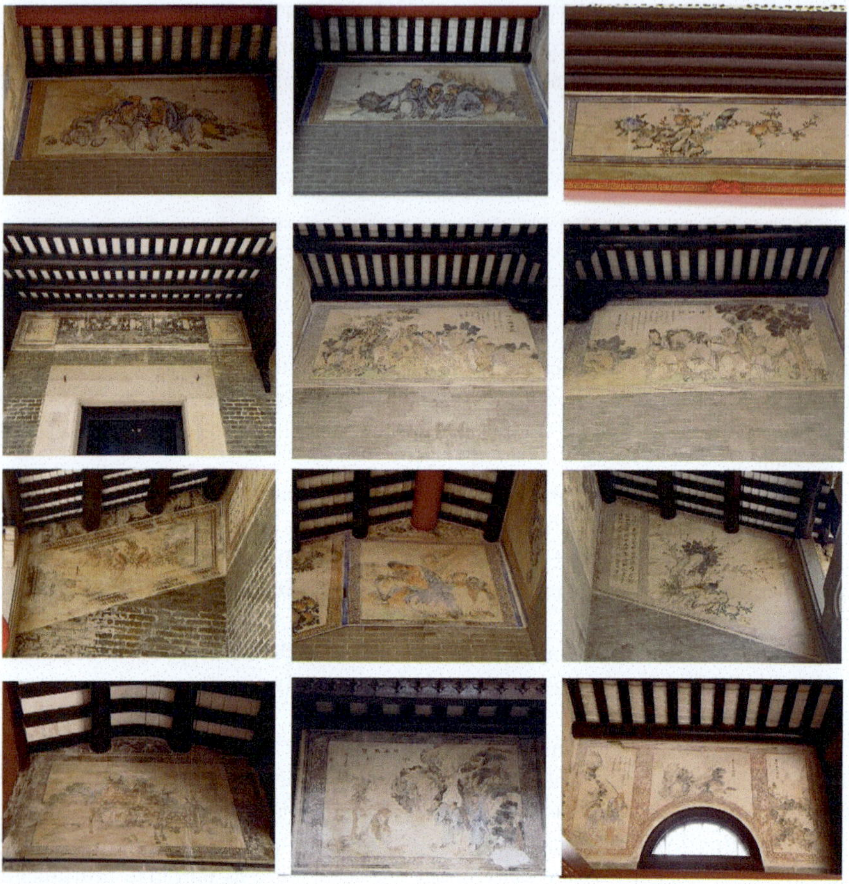

Fig. 6.58 Multiple-frame murals. (*Image source* photographed by the author)

Funding This research received no external funding.

Institutional Review Board Statement Not applicable for studies not involving humans or animals.

Informed Consent Statement Not applicable for studies not involving humans.

Conflicts of Interest The authors declare that the research was conducted in the absence of any commercial or financial relationships that could be construed as a potential conflict of interest.

References

1. Gu Hanyu Cidian Bianxiezu (Ed.). (1998). Gudai hanyu cidian [Ancient Chinese Dictionary] (p. 59). Commercial Press.
2. Cihai [Comprehensive Dictionary]. (1979). Shanghai Lexicographical Publishing House (p. 1586).
3. Lai, Y. (2010). Zhujiang sanjiaozhou Guangfu minxi citang jianzhu yanjiu [Study on ancestral hall architecture of the Guangfu ethnic group in the Pearl River Delta]. Doctoral dissertation, South China University of Technology (p. 145)
4. Zhang, Y. B. (2004). Mingtang zhidu yanjiu—Mingtang zhidu de yuanliu [Research on the Mingtang system: Its origin and development]. Doctoral dissertation, Jilin University.
5. Liu, Z. P. (2000). Zhongguo juzhu jianzhu jianshi: chengshi, zhuzhai, yuanlin: fu Sichuan zhu-zhai jianzhu, Yunnan yi keyin [A brief history of Chinese residential architecture: cities, houses, gardens; including Sichuan and Yunnan] (pp. 55, 194) China Architecture & Building Press.
6. Sun, D. Z. (1993). Zhongguo gu jianzhu daxi: lizhi jianzhu [Encyclopedia of Chinese Ancient Architecture: Ritual Architecture] (p. 130). China Architecture & Building Press.
7. Feng, J. (2010). Ming Qing Guangzhou fu de kaiken, juzu er ju yu zongzu citang de yanbian yanjiu [Reclamation, clan settlement, and the evolution of ancestral halls in Ming-Qing Guangzhou]. Doctoral dissertation, South China University of Technology (pp. 192–193).
8. Feng, J. (2010). Ming Qing Guangzhou fu de kaiken, juzu er ju yu zongzu citang de yanbian yanjiu [Reclamation, clan settlement, and the evolution of ancestral halls in Ming-Qing Guangzhou]. Doctoral dissertation, South China University of Technology (p. 166).
9. Sun, D. Z. (2004). Zhongguo gudai jianzhu shi: Qingdai jianzhu [History of Ancient Chinese Architecture: Qing Dynasty Architecture] (p. 36). China Architecture & Building Press.
10. Luo, Y. L., & Zhang, N. Q. (2008). Yuan lishi zaori zhanxian zhenyan—Ji "Nanyuewang gongyuanli jiachuantai" lunzhenghui [May history reveal its true face soon: A record of the "Nanyue King Palace Garden Pseudo-Platform" Symposium]. *Huazhong Architecture, 12*, 164–168.
11. Wang, P. (2018). *A study on traditional timber structures in the Lingnan Guangfu region.* Doctoral dissertation, South China University of Technology.
12. Liang, M. (2014). *A study on decorative elements of ancestral halls in the Guangfu region.* Master's thesis, South China University of Technology.
13. Huang, X. (2013). *A study on windows in traditional Lingnan architecture.* South China University of Technology.
14. Yan, L., & Chen, Y. (2021). Stained Glass in Lingnan Historic Buildings and Its Application in Manchuria Windows. *Journal of Physics: Conference Series, 1885*(3), 032052. https://doi.org/10.1088/1742-6596/1885/3/032052
15. Li, M. (2013). *A study on brickwork in the Canton region during the Ming and Qing dynasties.* Master's thesis, South China University of Technology.
16. Cao, J., Chen, Y., & Guo, Y. (2021). Characteristics and application of regional building materials in Lingnan area: a case study of Casa da Cheang in Macao. In *E3S Web of Conferences* (vol. 284, p. 05003). EDP Sciences. https://doi.org/10.1051/e3sconf/202128405003
17. Ji, H., Wu, S., Ye, B., Wang, S., Chen, Y., & Deng, J. Y. (2023). Exploring the implementation path of passive heat-protection design heritage in Lingnan Buildings. *Buildings, 13*(12), 2954. https://doi.org/10.3390/buildings13122954
18. Shang, B., & Pan, B. (1988). *Lingbiao yilu jiaobu* [Annotated edition of *Lingbiao Yilu*] (p. 60). Guangxi Nationalities Publishing House.
19. Tang, G. H. (2005). *Lingnan shi re qihou yu chuantong jianzhu* [Lingnan's hot-humid climate and traditional architecture] (p. 86). China Architecture & Building Press.
20. Wang, Y., Song, J., Zhang, J., Huang, Y., & Yang, S. (2025). Analysis of Rammed earth wall Erosion in traditional village dwellings in Zhuhai City. *Coatings, 15*(5), 526. https://doi.org/10.3390/coatings15050526

21. Song, J., Chen, Y., Zheng, L., & Tang, Q. (2025). The evolution of traditional ancient towns in South China: Tangjiawan Ancient Town in Zhuhai city. *Journal of Asian Architecture and Building Engineering, 24*(3), 2114–2132. https://doi.org/10.1080/13467581.2024.2349754
22. Lin, C. (2011). *A study on the craftsmanship and conservation of architectural lime sculpture in the Lingnan Guangfu region*. Doctoral dissertation, South China University of Technology, Guangzhou.
23. Xie, Y. (2018). *A study on murals in traditional architecture of the Lingnan Guangfu region*. Doctoral dissertation, South China University of Technology.

Garden Landscape: A Fusion of Chinese and Western Cultures in Tangjiawan

Yuzhou Ling, Junxin Song, and Jianyi Zheng

Abstract This study investigates the modern garden landscapes of Tangjiawan Town, Zhuhai City. Employing field research, typological analysis, and documentary review, it systematically examines their spatial organization, landscape elements, and cultural significance. The research identifies three categories of modern gardens in Tangjiawan: ancestral hall gardens, residential gardens, and large-scale gardens. Their gardening techniques integrate Lingnan traditions with Western influences, resulting in distinctive characteristics: a dominant axial layout, a blend of Chinese and Western decorative elements incorporating water and stone features, and a repurposing of planting configurations. These landscapes, shaped by adaptation to the natural environment, the inheritance of Lingnan gardening principles, and the infusion of Western gardening culture, exhibit a defining feature of harmonious coexistence between traditional frameworks and Western motifs. This study addresses a gap in the geographical subdivision research of Lingnan Gardens and offers a theoretical foundation for conserving modern garden heritage and preserving local culture.

Keywords Garden space form · Tangjiawan · Modern gardens · Landscape features

Y. Ling
CCTEG Chongqing Engineering (Group) Co., Ltd, Chongqing, China

J. Song (✉)
Institute for Research on Portuguese Speaking Countries, City University of Macau, Macau, China
e-mail: a23092100150@cityu.edu.mo

J. Zheng
Faculty of Humanities and Arts, Macau University of Science and Technology, Macau, China

Y. Chen et al. (eds.), *City-Sea Symbiosis*, Springer Geography,
https://doi.org/10.1007/978-3-032-13725-8_7

1 Introduction

1.1 Research Background

Historical and cultural villages and towns in China constitute invaluable heritage, preserving multifaceted records of political economy, social life, customary practices, and cultural landscapes. Since the Reform and Opening-up era, both national and local governments have prioritized the protection of historical-cultural assets. Beginning in the 1980s, significant ancient structures in rural settlements were designated as National Key Cultural Relics Protection Units, with corresponding conservation measures and regulations progressively refined. The 2003 revision of the *Law of the People's Republic of China on the Protection of Cultural Relics* formally established protections for historically significant towns, streets, and villages. Subsequently, the state has consecutively recognized multiple cohorts of Famous Historical and Cultural Cities, Towns, and Villages. Tangjiawan Town in Zhuhai, distinguished by its rich historical legacy, attained status as a National Famous Historical and Cultural Town in 2007 and was concurrently included in Guangdong Province's Ancient Villages and Towns inventory [1, 2]. The Zhuhai municipal government has actively advanced conservation-oriented development planning for this site. Situated at the periphery of the Cantonese Cultural Sphere, Tangjiawan's distinctive humanistic-geographic context has fostered unique modern garden landscapes [3]. These represent a significant Lingnan garden typology that not only permeates local daily life but also embodies owners' cultural aesthetics while reflecting traditional folk customs and lifestyles. Current scholarship on Tangjiawan's modern gardens remains fragmented and unsystematic, with extant documentation often partial in scope. Field investigations, however, reveal that their sophisticated spatial organization, landscape elements, and artistic richness demonstrate exceptional vernacular craftsmanship—offering critical insights for regional landscape studies. Consequently, systematic analysis of Tangjiawan Modern Garden's spatial configuration and landscape components, along with distillation of its characteristic features, is imperative for the perpetuation and sustainable utilization of this cultural heritage.

1.2 Literature Review

Garden landscape research has gone through a development process from traditional form description to interdisciplinary theory construction. Early research focused on the material form of landscape and the expression of artistic conception. For example, Deng Qisheng defined the garden as "a naturalized garden in a house within an artificially enclosed space" in *A History of Gardens*, emphasizing the integration of artificial beauty and natural beauty [4]. In the early twenty-first century, research gradually combined with urban planning and ecological design. Yan Junsheng et al. explored the environmental creation of residential gardens and integrated traditional gardening techniques into modern residential spaces [5]. Recent decades have witnessed further expansion into landscape restoration and ecological enhancement. Peng Yigang's seminal *Analysis of Classical Gardens* deconstructed traditional spatial logic, establishing critical theoretical foundations for regional landscape design [6]. According to China National Knowledge Infrastructure (CNKI) data, 173 publications addressing "garden landscape" have appeared since 1984. Research themes have progressively shifted from initial emphasis on aesthetic interpretation toward diversified applications in urban spatial planning, ecological functionality, and related domains. Nevertheless, systematic investigations of modern regional gardens remain notably underdeveloped.

Research on Lingnan gardens commenced in the 1960s when Xia Changshi and Mo Bozhi first articulated their architecture-dominant spatial characteristics in *On Lingnan Gardens*, establishing functional adaptation to daily life as fundamental and identifying architecture-nature integration as the core gardening technique [7]. Subsequent scholarship has developed along three primary dimensions: Yang Honglie systematized Guangzhou classical gardens' hydrology principles as "channeling, proximity, and harmony with water" [8], while Wu Yingfeng analyzed the material properties and construction methodologies of Lingnan rockery design [9]; Zhou Linjie documented 137 cases in *History of Modern Gardens in Guangdong*, revealing the transformation from private to public garden typologies [10]. Lu Qi subsequently traced the genealogy of private gardens in *Gardening and Aesthetics in Lingnan*, highlighting western-influenced "multicultural assimilation" as defining modern Lingnan gardens [11]; while Meng Zhaozhen foregrounded their ecological philosophy of "natural symbiosis" [12]. Collectively, these studies constitute a comprehensive framework spanning technical to cultural dimensions. Nevertheless, the geographical specificity of Lingnan garden studies remains underexplored, requiring deeper regional segmentation analysis.

Despite substantial scholarship on Lingnan gardens, thematic research specific to Tangjiawan remains critically underdeveloped. Existing studies predominantly address the macro-Lingnan context, exemplified by Lu Yuanding's comprehensive analysis of regional architectural culture [13], yet lack systematic investigation of Tangjiawan's distinct geographical attributes. Limited local examinations focus narrowly on either individual structures or settlement conservation—such as Huang Guan's preliminary assessment of Huitong Village's architectural significance [14], or Ruan Yisan's preservation plan emphasizing land-use and transportation infrastructure while neglecting garden landscapes' cultural dimensions [15]. Documentation within the *Zhuhai High-Tech Zone Immovable Cultural Relics Inventory* further reflects this gap, offering only fragmentary garden descriptions without substantive analysis of spatial configurations or landscape components. Addressing these limitations, this study establishes for the first time a tripartite typology of Tangjiawan's modern gardens—ancestral temple, residential, and large-scale gardens—and employs morphological analysis to reveal their "symbiotic integration of traditional forms and Western elements." This approach achieves three critical advances: (1) fills the geographical subdivision void in Lingnan garden studies, (2) provides a targeted theoretical foundation for modern garden heritage conservation, (3) decrypts the spatial logic underpinning Sino-Western landscape synthesis.

1.3 Problem Statement and Objectives

This thesis takes the modern garden landscape of Tangjiawan as the research object, through the field research of the existing intact typical garden, in-depth analysis of its spatial layout, landscape elements, and cultural connotations, aiming at the systematic investigation of the modern garden landscape of Tangjiawan, the use of the typological perspective to summarize and classify and analyze the typical cases, summarize the gardening techniques and landscape characteristics, and at the same time, to explore factors affecting the creation of garden landscapes and to distill the landscape characteristics and design connotations. The study has both theoretical and practical significance.

The study has both theoretical and practical significance. Theoretically, as an important type of Lingnan gardens, the modern garden landscape of Tangjiawan lacks systematic research, and this thesis can provide new additions to the study of Lingnan gardens and landscape, expanding the scope of the study. In reality, on the one hand, it can help to protect the modern garden of Tangjiawan, which is facing the impact of urbanization, excavate its landscape and cultural value, and provide a reference for the protection and inheritance; on the other hand, it can refine the characteristics of the modern garden; meet the diversified needs of modern people for the landscape; realize the temporal and spatial connection of the ancient and modern landscapes; and help to inherit the local culture, improve the environmental landscape of Tangjiawan, and avoid the loss of regional characteristics.

2 Research Methods and Analytical Framework

2.1 Research Methods

Based on the complexity and cross-cultural characteristics of modern garden landscapes in Tangjiawan, this study adopts a multi-dimensional research methodology system, systematically deconstructing their spatial patterns, gardening techniques, and cultural influence mechanisms through field research, typological analysis, documentary evidence, and interdisciplinary theoretical integration. The research not only focuses on the inheritance of traditional garden research methods but also combines the special context of modern Chinese and Western cultural fusion to form an analytical framework applicable to regional characteristics, providing a scientific basis for the protection and theoretical expansion of gardens.

(1) Field Survey Method. Through systematic field surveys and data collection, we obtain the material form and existing condition of the garden landscape.
(2) Typological Analysis Method. Based on morphological features and cultural attributes, we construct a classification research framework for modern gardens in Tangjiawan.
(3) Literature Analysis Method and Interdisciplinary Theoretical Integration. We construct the theoretical foundation of research through multi-dimensional literature study and theoretical grafting.

2.2 *Research Objects and Classification System*

Based on the multi-dimensional analysis of functional attributes, spatial forms, and cultural characteristics, the modern gardens in Tangjiawan can be systematically classified into three major types: ancestral temple gardens, residential gardens, and large-scale gardens, which show significant differentiation in spatial layout, gardening techniques, and combination of landscape elements. The three types of gardens can be systematically classified into three major types: ancestral temple gardens, residential gardens, and large-scale gardens, each of which shows significant differentiation in spatial layout; gardening techniques; and combination of Tangjiawan's modern ancestral temples and gardens integrate regional, Lingnan, and religious characteristics, forming a unique spatial structure and landscape elements that satisfy both secular appreciation and religious aesthetics. Its origins can be traced back to the Southern Song Dynasty temples, with ancestral halls flourishing during the Ming and Qing dynasties, divided into clan ancestral halls; lineage ancestral halls; and temples dedicated to natural deities, eminent figures and sages, and Buddhist folk deities. Tangjiawan ancestral temples are classified into early, middle, and late periods according to the sequence of construction time. The earliest appearance in Tangjiawan area was temples, which can be traced back to the Southern Song Dynasty, and ancestral halls were allowed to be built only in the Ming and Qing Dynasties. In the early stage, temple buildings appeared one after another, with a rich variety including temples dedicated to worshiping natural deities, blessing good weather, temples of eminent figures and sages, and temples of Buddhism and folk gods. In the mid-to-late period, temple buildings gradually decreased, while ancestral hall buildings flourished during the same period scape elements, which provides a framework for analyzing the logic of landscape evolution of "traditional—East meets West—Western."

2.2.1 Ancestral Temple Garden

Tangjiawan's modern ancestral halls and temple gardens integrate regional, Lingnan, and religious characteristics, forming a unique spatial structure and landscape elements, which satisfy both secular viewing and religious aesthetics [16–19]. Its origin can be traced back to the temples of the Southern Song Dynasty. Ancestral halls flourished during the Ming and Qing Dynasties, divided into clan ancestral halls, lineage ancestral halls, and temples dedicated to natural deities, celebrated figures and sages, and Buddhist and folk deities. Tangjiawan's ancestral halls and temples can be divided into early, mid-, and late periods according to the construction time sequence. Temples first appeared in the Tangjiawan area, dating back to the Southern Song Dynasty, and ancestral halls were allowed to be built only in the Ming and Qing Dynasties. Early temple buildings emerged one after another, with rich categories including temples dedicated to natural deities to bless good weather, temples of celebrated figures and sages, and temples of Buddhism and folk deities. In the mid- and late periods, temple buildings gradually decreased, while ancestral hall buildings flourished in the same period.

Clan halls did not appear until 1783. According to the *Tangjiawan Town Chronicle*, during the Ming and Qing dynasties, the population increased dramatically, and Central Plains immigrants migrated south. Villagers in Tangjiawan began farming and fishing. With the transformation of production methods, the culture also gradually changed. People settled and multiplied here, successively building clan halls and forming the local clan culture. Therefore, in the mid- to late Qing Dynasty, a large number of clan halls began to be built. Some clan halls expanded horizontally based on the three-bay two-courtyard layout by adding side halls or building "Guang San Lu," making the plane form of the clan hall expand horizontally, and the elevation and decoration also became richer. In the late Qing Dynasty, the "twin clan hall" form became a way to expand the scale of clan halls. The elevation and decoration continued the characteristics of the clan halls in the mid-Qing Dynasty. The decorative themes, colors, and techniques of the elevation became rich and diverse, and the style became complex and redundant.

According to *Immovable Cultural Relics in Zhuhai High-Tech Zone*, Tangjiawan has 37 ancestral halls and 13 temple courtyards. The existing ones are relatively typical. Through field visits and consulting relevant materials, the author found that most of the existing modern ancestral hall and temple gardens are located in Tangjiawan Ancient Town, while the more distant areas such as Qi'ao, Huitong, and Jishan generally have fewer remaining ancestral hall and temple gardens (Fig. 1).

Tangjiawan people highly value ancestral temples and gardens, which are the main places for residents' religious sacrifices and moral education and are the main material embodiment of Tangjiawan residents' life customs and cultural beliefs. This chapter selects the courtyards of the Tangjia Three Temples, the Ruizhi Tang Ancestral Hall, and the Beisha Lu Clan Twin Temples, which have typical characteristics and are well-preserved, in a time series to further analyze and explore the landscape construction characteristics of the ancestral temple gardens in modern Tangjiawan.

(1) Tangjia San Miao (Three Temples of Tangjia)

Tangjia San Miao is located on the northwest side of Datong Road in Tangjiawan Town, composed of the Wenwu Emperor Temple, the Shengtang Temple, and the Jinhua Temple arranged side by side, hence the name "Tangjia San Miao." It is the largest and oldest temple architecture existing in the Zhuhai area, possessing

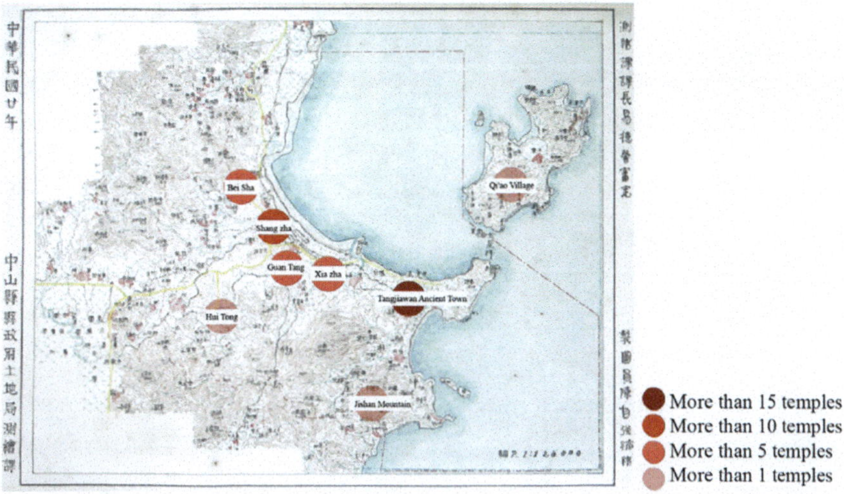

Fig. 1 Tangjiawan temple garden distribution map. (*Image source* Based on the renovation provided by the Cultural Center of Zhuhai High-Tech Zone)

profound historical and cultural connotation and relatively high architectural artistic value. The courtyard landscapes of the three temples each have their own characteristics and distinct themes (Fig. 2). Because the current appearance of the temple was renovated according to the situation in the second year of Tongzhi (1863), this paper considers it to be among the modern ancestral temple gardens of Tangjiawan.

The plan layout of Tangjia Three Temples is strictly symmetrical, with both buildings and courtyards using rectangles as the basic element, forming a typical architectural layout around the courtyard. The overall orientation of the buildings is northwest to southeast, with clear distinction between primary and secondary structures. Among them, the layouts of both Shengtang Temple and Jinhua Temple adopt the "three-bay two-courtyard with two corridors flanking one courtyard" plan, while the Wenwu Emperor Hall adopts the "three-courtyard with two corridors flanking two courtyards" plan. The total width of the Tangjia Three Temples is 32.72 m, the total depth is 29.6 m, and the total area is approximately 1500 square meters. The three temples are separated by Qingyun Alley, and each of the three temples has a front courtyard built in front of its main entrance. The front courtyard is about 1 m higher than the ground, with low walls on three sides, and a length of about 13.5 m. Festival activities are held in the front courtyard, which serves as a public activity space for villagers (Fig. 3).

The Tangjia Temple of Three Shrines is composed of three courtyard spaces. The Shengtang Temple and Jinhua Temple belong to side-wall type courtyards, while the central courtyard of the Wenwu Emperor Hall is a side-wall type courtyard, and the back courtyard is a side-corridor type courtyard. The Shengtang Temple and Jinhua Temple create the theme of the courtyard through the two-entry pavilion, which is the landscape core of the two temples (Fig. 4).

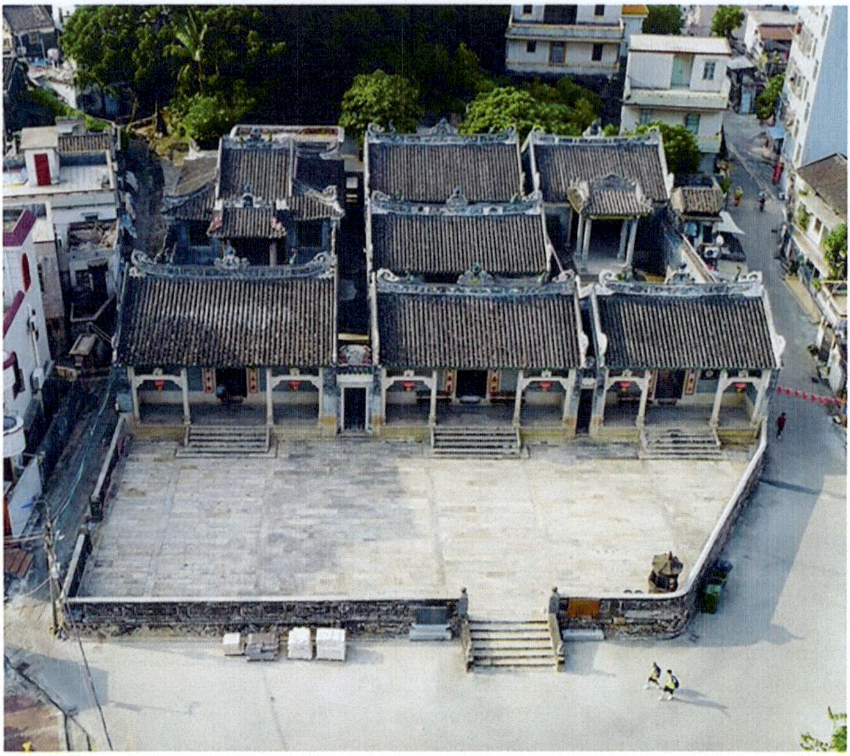

Fig. 2 Aerial photograph of the three temples of Tang family. (*Image source* Courtesy of Zhuhai High-Tech Zone Cultural Center)

The courtyards of the Tang Family's Three Temples have independent entrances and exits and are internally connected by archways. In the horizontal space, the three courtyards are connected in series through the archways, making the space have a sequential and permeable relationship (Fig. 5).

(2) Ruizhi Tang Ancestral Hall

Ruizhi Tang Ancestral Hall is located on Shanfang Road, Tangjia Community, Tangjiawan Town, High-Tech Zone. It was built during the Guangxu period of the Qing Dynasty and is the ancestral hall of Tang Ruizhi's family, a famous comprador and national entrepreneur. It is now a well-preserved ancestral hall in Tangjiawan Town with the architectural style of Lingnan in the Qing Dynasty (Fig. 6). Ruizhi Ancestral Hall has a total width of approximately 25.38 m, a total depth of approximately 28.88 m, and a total building area of approximately 736.4 square meters.

Ruizhi Ancestral Hall faces southwest to northeast, with a symmetrical plane layout along a three-way central axis. The middle axis of the ancestral hall has a width of three bays, with two sections, two corridors, and one patio, flanked by three sections on the left and right paths, with a front courtyard in front of the front door.

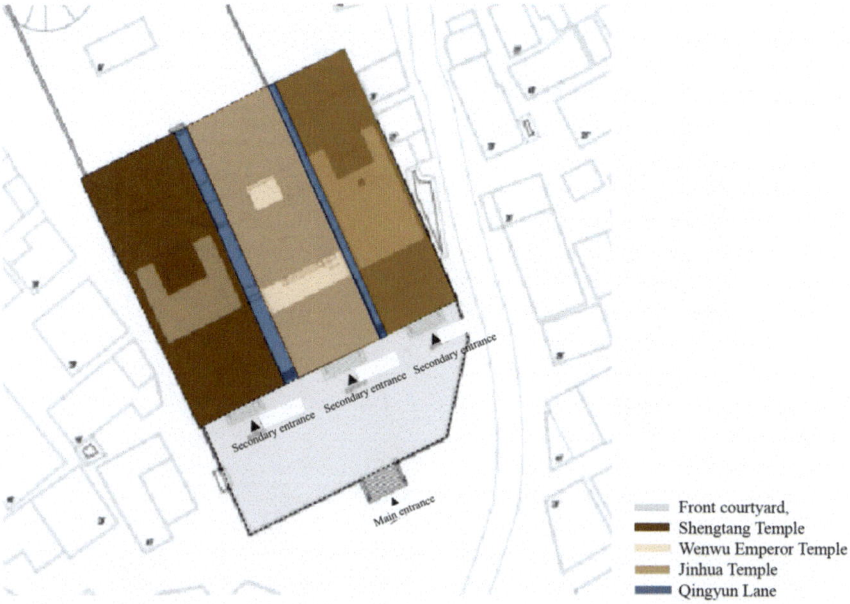

Fig. 3 Top view of Tang family's three temples. (*Image source* Adapted from Gu Xueping's *Research on Cantonese region Temple Architecture*)

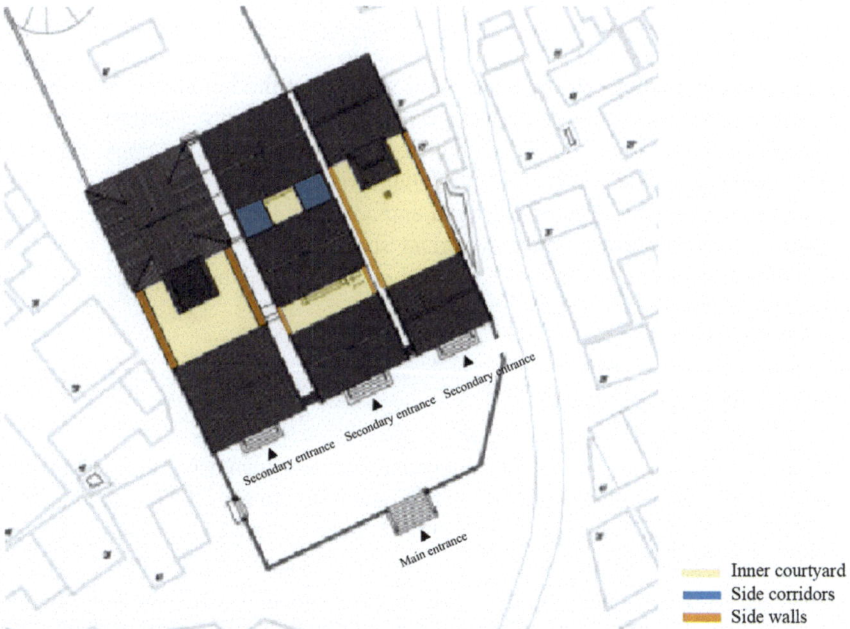

Fig. 4 Side wall and side corridor courtyards. (*Image source* Adapted from Gu Xueping's *Research on Cantonese region Temple Architecture*)

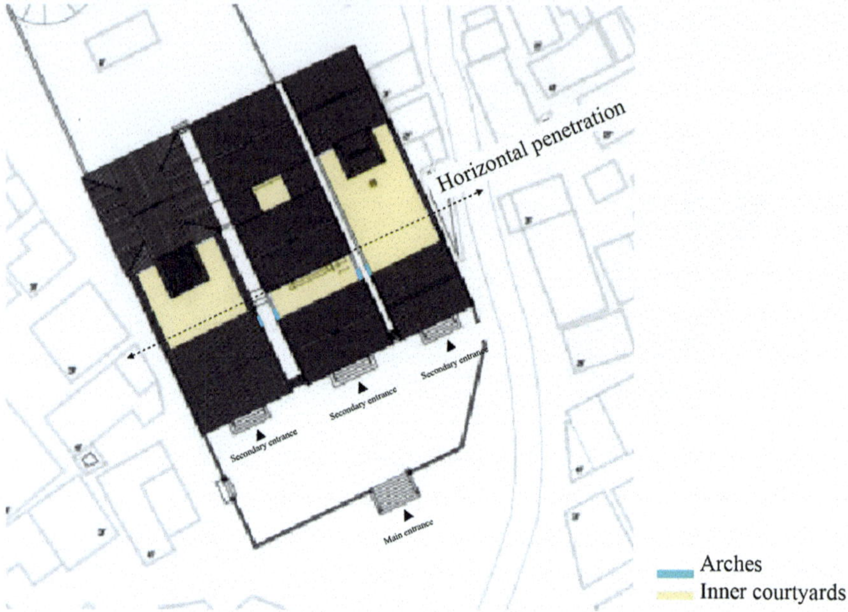

Fig. 5 Horizontal infiltration of interior space. (*Image source* Adapted from Gu Xueping's *Research on Cantonese region Temple Architecture*)

The front door and rear hall of the middle path are separated from the left and right paths by Qingyun Alley. The first and third sections of the left and right paths are both 11-ridge rooms, with a small patio courtyard in front (Figs. 7 and 8).

The entire Ruizi Ancestral Hall comprises six courtyard spaces, consisting of a front courtyard, a middle courtyard plus four atrium courtyards, forming a courtyard cluster space. This type of clustered and arranged courtyard generally adopts a symmetrical approach, placing important spaces on the longitudinal axis, and arranging secondary spaces on the left and right sides of the transverse axis for design (Fig. 9).

2.2.2 Residential Garden

Residential buildings are places built by people for their own habitation, which can meet the needs of people's daily life and production. And residential gardens are mostly attached to the front, rear, or sides of residences, arranged in conjunction with daily life. Modern residential gardens in Tangjiawan can be divided into two types based on landscape elements and architectural forms: the first type is the continuation of traditional styles, and the second type is a combination of Chinese and Western styles.

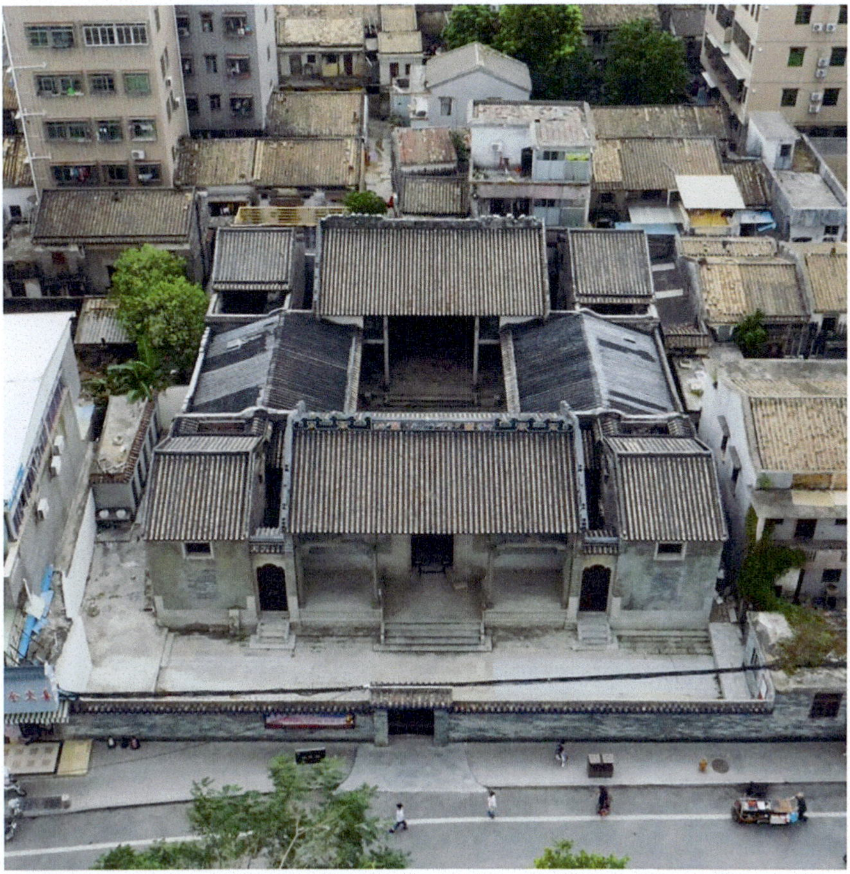

Fig. 6 Aerial photograph of the Ruizhi Tang's Ancestral Hall. (*Image source* Courtesy of Zhuhai High-Tech Zone Cultural Center)

Lingnan residential gardens originated in the Nanyue period, from the earliest Nanyue Royal Palace Imperial Garden site to the Lychee Garden in the Tang Dynasty, and the West Garden in the Song Dynasty. Until the Ming and Qing Dynasties, especially after the Qing Dynasty, there are relatively rich historical materials and existing gardens of residential gardens, mainly concentrated in Guangzhou, central Guangdong, and eastern Guangdong. With the development of cultural life, Lingnan people had further requirements for the design of residential gardens, primarily considering comfort and interest. At this time, various residential gardens suitable for living emerged one after another, such as small gardens mainly for residential leisure or gardens mainly for flowers, trees, and fruit trees. The development of Lingnan residential gardens reached its peak during the Ming and Qing Dynasties. According to records, there were 50 to 60 famous gardens in the Guangzhou area alone. They

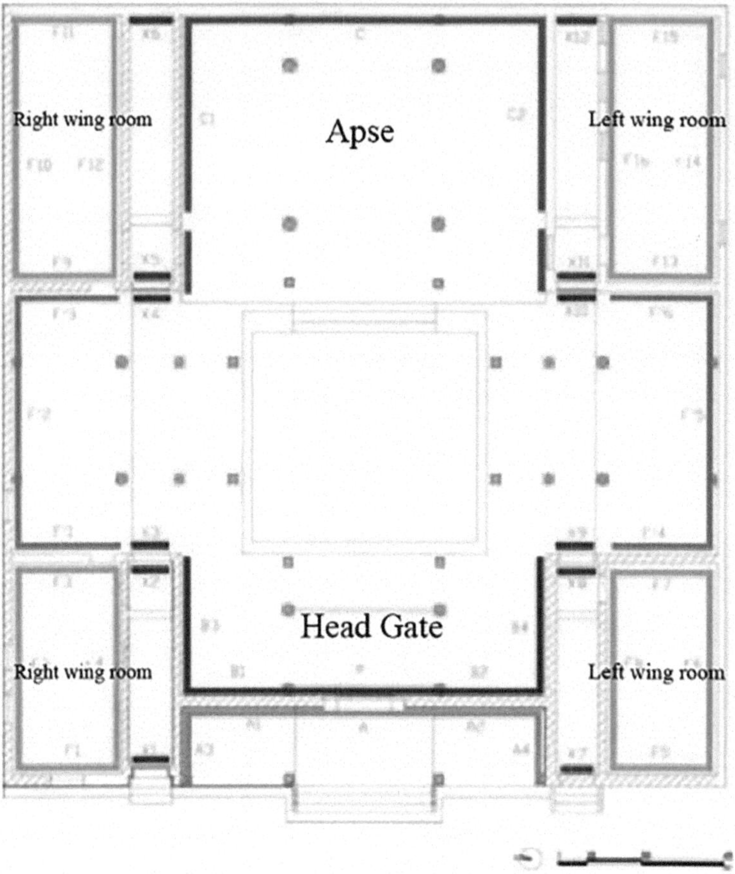

Fig. 7 Ruizhi Temple courtyard plan. (*Image source* author adapted from Zhu Bin's *Research on Traditional Architectural Murals in Cantonese region, Lingnan*)

varied in size, including large mansions of aristocratic families and private gardens distributed in the suburbs.

In modern times, the construction of residential gardens was mainly funded by individual proprietorships, such as overseas Chinese, military and political dignitaries, and wealthy merchants. Simultaneously, residential gardens were influenced by Western culture in terms of layout, gardening techniques, and landscape aesthetics, and developed independent Western-style buildings, row houses, bamboo tube houses, Xiguan mansions, and arcade buildings in residential architecture.

The formation and evolution of modern residential gardens in Tangjiawan are influenced by multiculturalism. Tangjiawan is under the jurisdiction of Guangdong Province. According to the division of the three major ethnic groups in Guangdong, it can be subdivided into Cantonese culture, Hakka culture, and Chaozhou culture. The Cantonese culture mainly exists in the central and western regions of Guangdong,

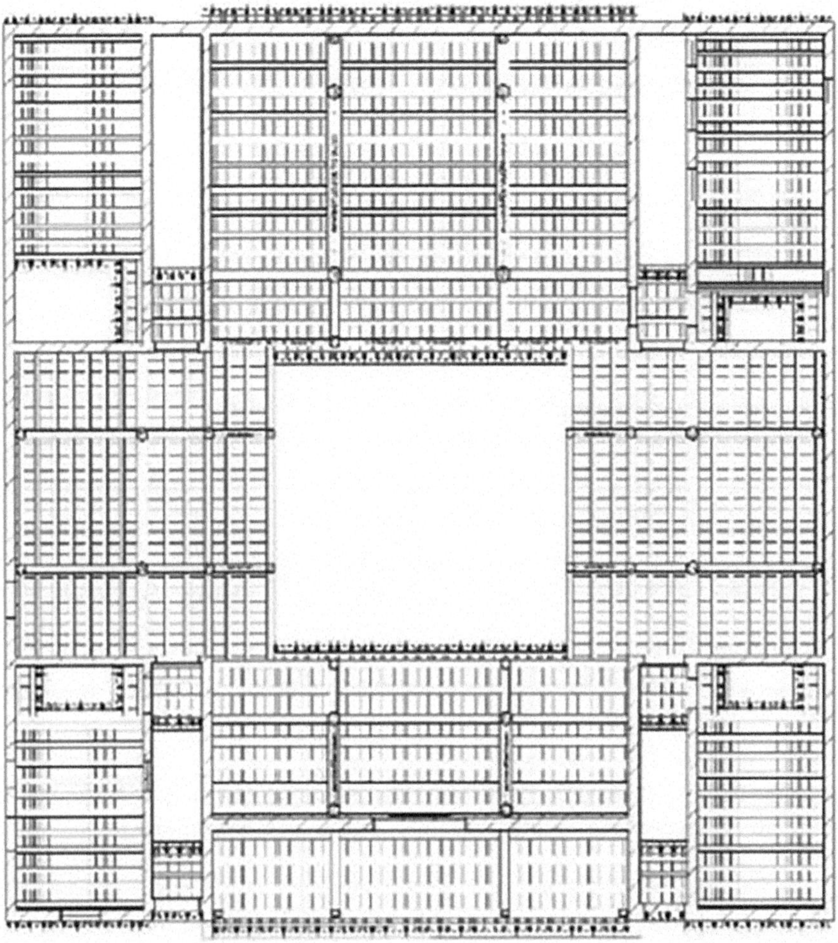

Fig. 8 Roof plan of Ruizhi Temple courtyard. (*Image source* Zhu Bin's *Research on Traditional Architectural Murals in Cantonese region, Lingnan*)

forming distinct characteristics in many cultural fields such as architecture, religion, and art in the region. Therefore, the evolution of modern residential gardens in Tangjiawan is influenced by Cantonese culture, which is reflected in villages with comb layout and three-bay two-corridor courtyard-style residences. At the same time, Tangjiawan is close to Macau and is also influenced by Western culture, and a group of compradors appeared, becoming the historically famous Xiangshan comprador group. The wealthy Xiangshan people returned to their hometowns to build houses. While maintaining traditions, they also absorbed Western Garden construction techniques and cultural characteristics. Western Garden symbols can be seen in many

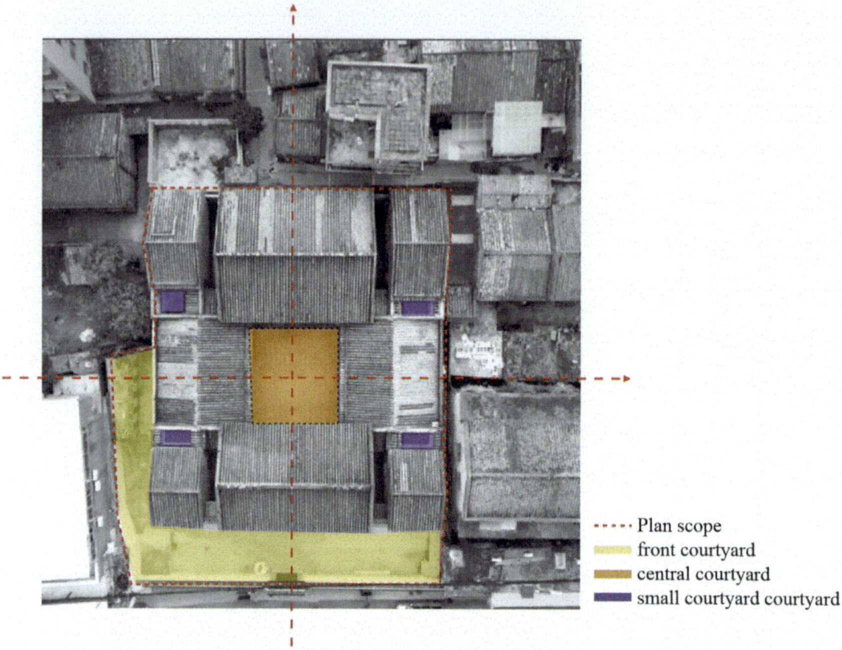

Fig. 9 Space organization of Ruizhi Temple courtyard group. (*Image source* author adapted from Zhuhai High-Tech Zone Cultural Center)

residential buildings and garden landscapes. It is thus evident that modern residential gardens in Tangjiawan are significantly influenced by Western culture.

(1) Traditional Style Residence—Gu Yuan's Former Residence

Traditional style is typically represented by the former residence of Gu Yuan. Gu Yuan's former residence is located at the intersection of Huachang Road, Sancun Village, Nazhou Village, Tangjiawan. It was built in 1912 during the Republic of China period by Gu Wan Jian, Gu Yuan's father. The residence faces northwest from southeast and consists of three parts: the main building, the front yard, and the side garden, covering an area of approximately 367.04 square meters. Gu Yuan's former residence is a typical Lingnan three-bay, two-corridor residential structure, that is, the main building presents a three-bay format, with a courtyard consisting of two corridors and a patio in front. This type of residence is the most common in the Tangjiawan area, and Gu Yuan's former residence is a well-preserved example of this type of residence. It was announced as a cultural relic protection unit of Xiangzhou District in 2006 and announced as the seventh batch of cultural relic protection units in Guangdong Province in May 2012 (Fig. 10).

Gu Yuan's former residence is a Lingnan traditional residential "three-bay two-corridor" style building, which is the most common traditional layout in Tangjiawan. The "three-bay two-corridor" house consists of a main building with three bays,

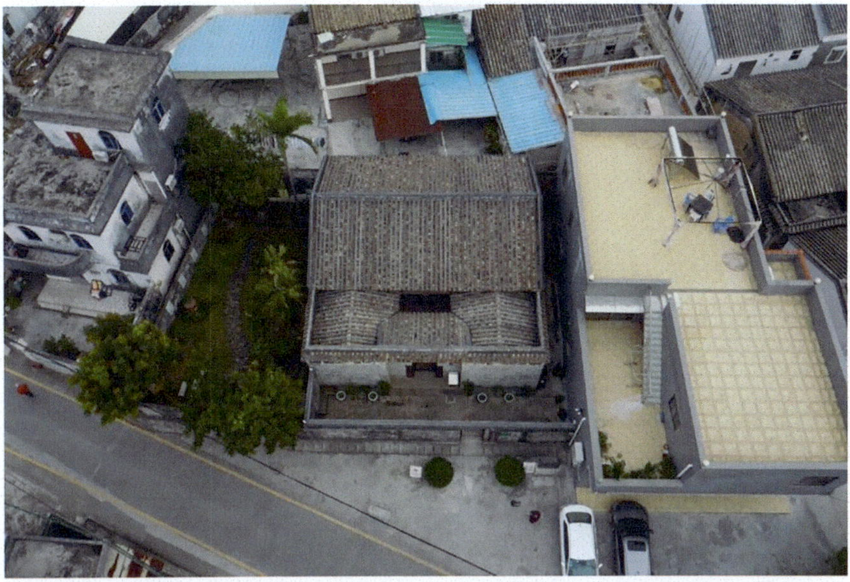

Fig. 10 Aerial photo of Gu Yuan's former residence. (*Image source* Courtesy of Zhuhai High-Tech Zone Cultural Center)

with a courtyard formed by two corridors and a patio in the front. Gu Yuan's former residence has a width of 12.04 m and a depth of 12.75 m, with a building area of approximately 153.51 square meters. The main house is a single-story brick and tile house with a gable roof and earth-wood structure. The main hall is in the center of the main building, with a height of four meters. Inside the hall, there is a shrine and a shrine room. The side rooms are on both sides, and there is a patio in front of the hall. The corridors on both sides of the patio are the kitchen and storage rooms, respectively. In the front courtyard of the former residence, an L-shaped courtyard space at the entrance is enclosed by walls, and the side courtyard is entered on the left side after entering the building's main entrance (Fig. 11).

The courtyard of Gu Yuan's former residence is composed of a front yard, a patio, and a left-side yard. The front yard is located at the main entrance, using walls to enclose the entrance and exit to form an L-shaped courtyard space. The yard gate is not directly opposite the building's main door. On the left side of the front yard, there is an arched small door leading directly to the left-side yard. The left-side yard has rich scenery; the entire yard is covered with lawn, leaving only small paths less than one meter wide. Stone benches are placed in the yard, and all the plants are planted by the garden owner. On one side of the yard, the exterior wall of the residence can be seen, directly facing the first-floor windows. The small patio located indoors is made into a sunken type with drainage holes, providing ventilation, drainage, and lighting functions for the residence (Fig. 12).

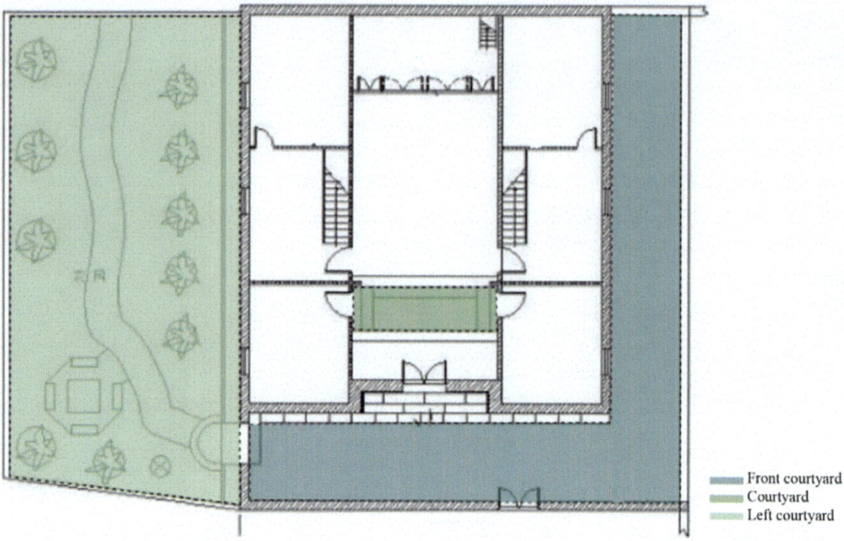

Fig. 11 Plan of Gu Yuan's former residence. (*Image source* redrawn according to *Immovable Cultural Relics of Zhuhai High-Tech Zone*, pp. 236)

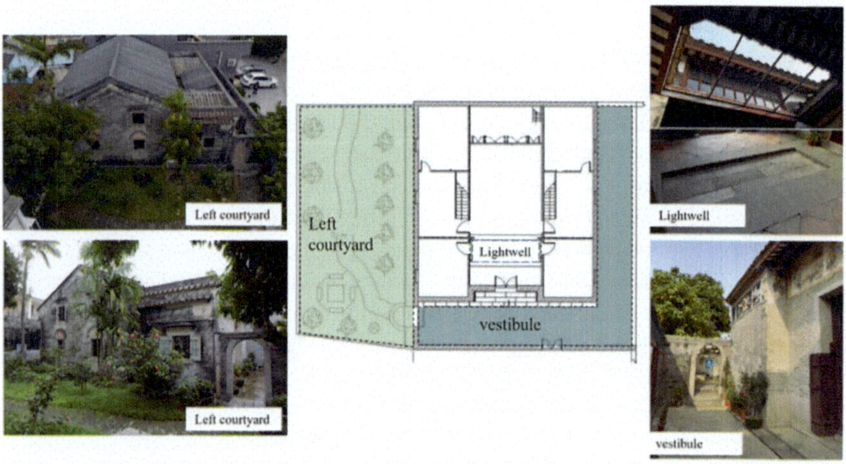

Fig. 12 Courtyard layout of Gu Yuan's former residence. (*Image source* redrawn according to *Immovable Cultural Relics of Zhuhai High-Tech Zone*, pp. 236)

Functional: Although the courtyard space of Gu Yuan's former residence is not large, it emphasizes practicality in landscape layout, meeting the aesthetic, labor, and social needs of the owner within a limited area. To meet the daily functional needs of rest, play, drying, and planting, the courtyard space of Gu Yuan's former residence is arranged with various functional zoning spaces, with clear guidance.

Walls or corridors separate the front yard from the side yard, providing multiple uses for the owner. The greenery in the side yard separates the open lawn space, forming a guiding channel, which is convenient for passage and various functional needs. Stone benches are placed along the walls, which are the main outdoor activity places. The front yard is decorated with potted plants carefully planted by the owner, and the narrow passage on the right side is also used by the owner to place potted plants. Thus, the courtyard landscape space is fully utilized by the owner, realizing the maximization of the courtyard landscape value (Fig. 13).

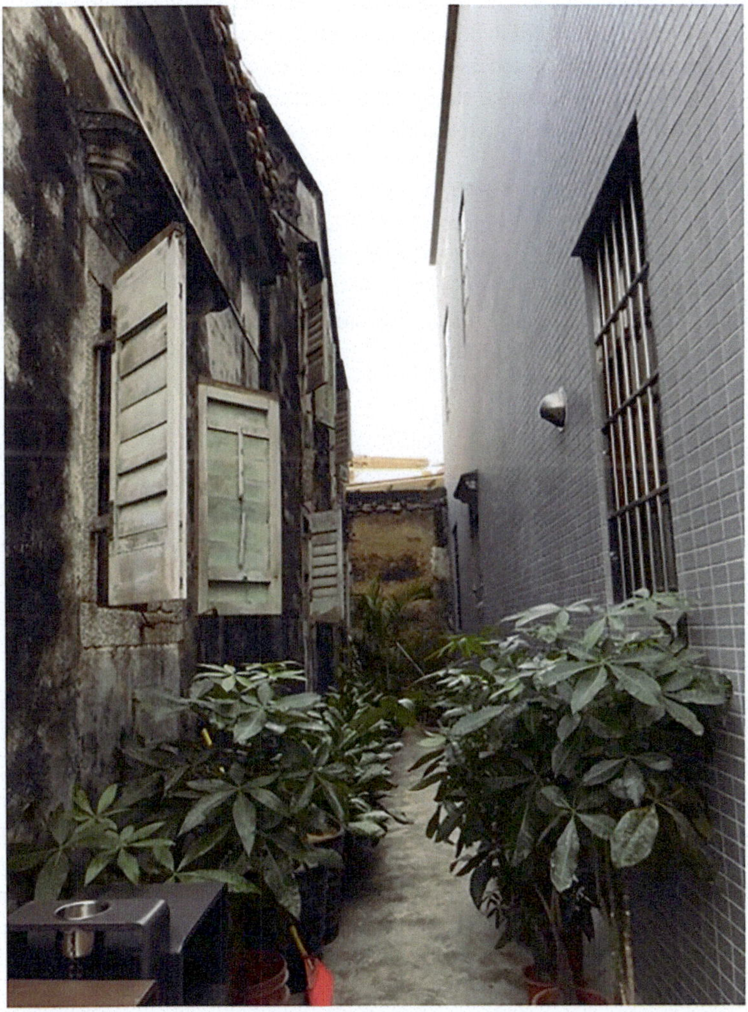

Fig. 13 The vestibule space of Gu Yuan's former residence. (*Image source* Photo by the author)

Adaptability: The courtyard plan of Gu Yuan's former residence is mainly a free-corner style. However, the axially symmetrical structure of the main building is rigorous, arranged symmetrically along the axis, with an internal patio. The free-corner landscape of the courtyard is located in the gaps between the front and side of the house. The landscape form is trapezoidal, using local materials with green plants. Gu Yuan's father, Gu Wanjian, liked to grow melons and flowers. This courtyard combines aesthetics and practicality. In addition, the windows are cleverly used to present the landscape changes, effect of borrowing scenery outside the garden, and conveying emotions through the scenery.

Proportion and Form: The courtyard of Gu Yuan's former residence belongs to a small residential courtyard. Therefore, he loved to adorn potted plants and small rockeries in smaller areas such as corridors, corners, and patios. Taking the human scale as a basis, it embodies the garden-making principle of "exquisite in appropriateness." The plants selected are low shrubs and bonsai, not large in size, but winning with their delicacy.

In terms of morphology, the Gu Yuan former residence's courtyard is an ancillary space to the building, composed of residential architecture and surrounding walls, forming a clear enclosed spatial form. Therefore, the building constitutes the enclosing element of the courtyard, and the architectural layout restricts the morphology of the courtyard. Thus, the Gu Yuan former residence embodies a quadrangular, square form, with the length-to-width ratio changing with the enclosure of the building, generally reflecting site-specific, flexible, and varied landscape morphological characteristics (Fig. 14).

(2) Front House, Back Garden Courtyard Landscape

The plan layout of the former residence of Tang Shaoyi presents a quadrilateral shape, enclosed by walls to form a closed residential courtyard space (Fig. 15), effectively creating a private space that meets security and defense needs, and prevents the intrusion of the external environment, making the environment in the garden harmonious and comfortable. As can be seen from the figure (Fig. 16), the former residence has two entrances, the main entrance is close to the residential area, which is left from the original layout, and the secondary entrance is added during the later expansion. The main entrance to the secondary entrance forms the main flow line, all leading to Building A and serving it, while the remaining secondary routes lead to the kitchen, dining room, gatehouse, and Building B.

The former residence of Tang Shaoyi is comprised of two primary structures, Buildings A and B, in addition to a spacious backyard (Fig. 17). Building B is situated on the left side of the courtyard and primarily served as living and office space for the owner. The backyard on the right was mainly used for recreation and leisure. Constructed during the Qing Dynasty by Tang Shaoyi's grandfather, Building B faces north. It employs a mixed "chuandou tailiang" (穿斗抬梁) structure, a granite foundation, a gable roof, and blue brick walls. Building A was added by Tang Shaoyi in 1929 and features a composite structure. The right compartments of Building A lean on both the upper and lower floors, and a water trough is installed at the intersection of the rooftops.

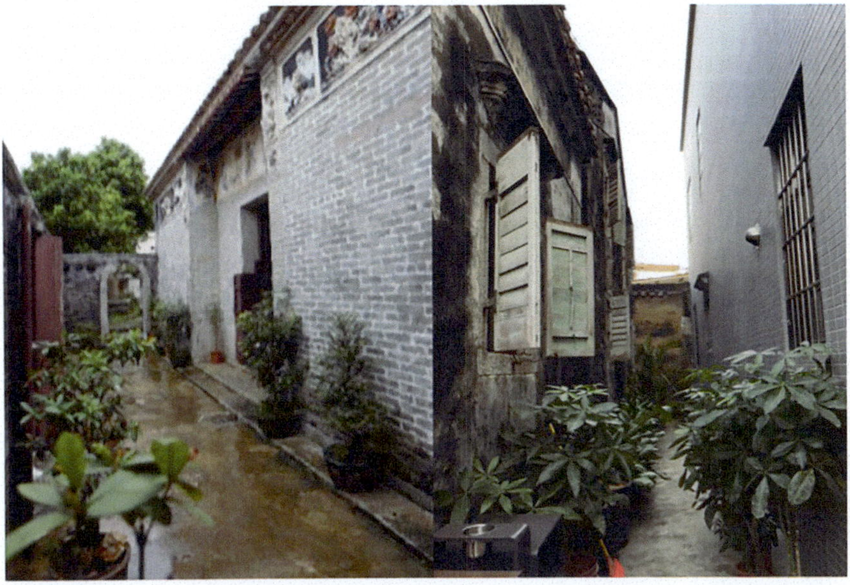

Fig. 14 The courtyard layout of Gu Yuan's former residence is subject to certain restrictions. (*Image source* Photo by the author)

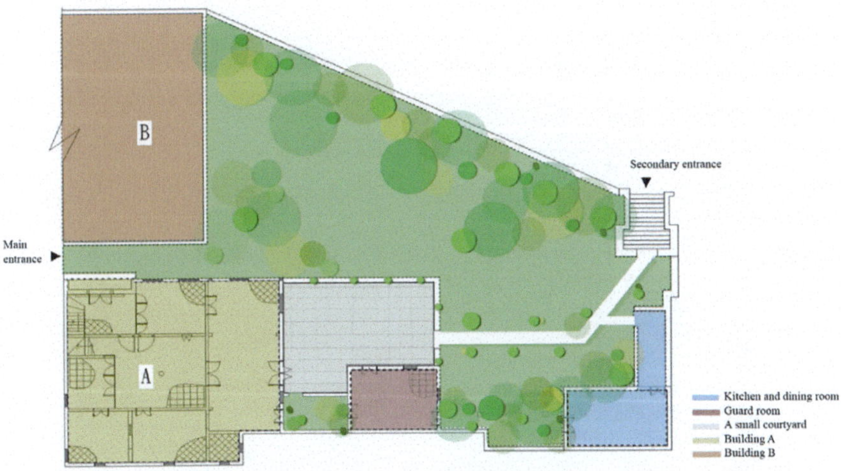

Fig. 15 Floor plan of Tang Shaoyi's former residence. (*Image source* redrawn according to *Immovable Cultural Relics of Zhuhai High-Tech Zone*, pp. 223)

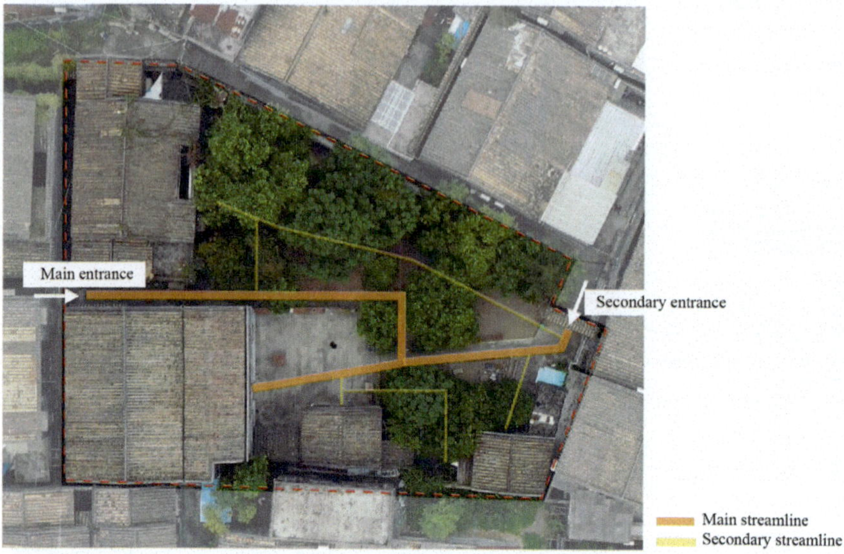

Fig. 16 Streamline map of Tang Shaoyi's former residence. (*Image source* redrawn according to *Immovable Cultural Relics of Zhuhai High-Tech Zone*)

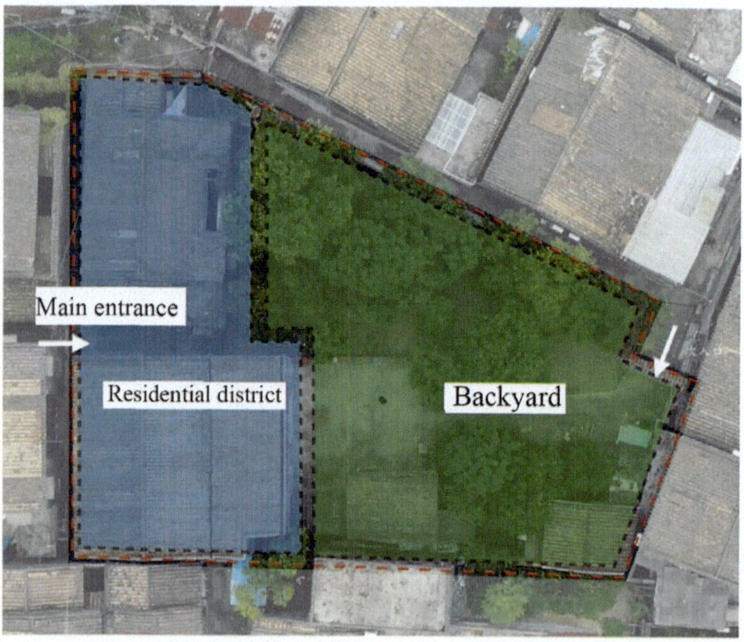

Fig. 17 The two major areas of Tang Shaoyi's former residence. (*Image source* redrawn according to *Immovable Cultural Relics of Zhuhai High-Tech Zone*)

Although the courtyard of Tang Shaoyi's former residence is small in area, the landscape is rich. It adopts a flexible spatial layout in the plan composition, so that the overall courtyard does not feel narrow. First of all, in the relatively small courtyard, in order to achieve the unity and openness of the landscape and avoid the fragmentation of the space, the courtyard adopts a main axis to organize the plan layout. The main axis leads to the main building, consisting of the kitchen, dining room, corridor, and the small courtyard of Building A, forming a main landscape belt. Building B is arranged next to the main axis and hidden in dense vegetation. This courtyard reasonably organizes the landscape space elements of the courtyard, and the whole presents order and a sense of space.

Secondly, if classified by function, the courtyard space is divided into two areas, A and B. Area A is mainly composed of Building A, supplemented by Garden A. This area is open and transparent, forming a front garden. The kitchen and dining room areas are smaller and clustered, with living and residence as the main function. Area B is mainly composed of Building B, leading to the courtyard via a path. This area is larger, occupying one-third of the entire area, with lush vegetation, quiet elegance, and high privacy, with leisure and entertainment as the main function.

2.2.3 Large-Scale Garden

In the development history of Lingnan gardens, there have been two large-scale gardens, which represent two peaks of Lingnan garden construction and have had a great influence on later generations. A relatively detailed record of early large-scale Lingnan gardens is the site of the Royal Garden of the Palace of the Nanyue King of the Western Han Dynasty (Figs. 18, 19 and 20), which is the earliest actual case in China that can reflect the large-scale Lingnan gardens at that time. Built by Zhao Tuo, the magistrate of the Qin Dynasty, and imitating palace gardens, the palace was built under Yuexiu Mountain in Guangzhou. In the palace, there were large stone-built water-receiving pools, stone channels, flat bridges, wells, and brick walkways (Huang Liping, 2024). According to archeological and documentary records, the area of the Nanyue King's Palace was about 13 hectares. This is a large-scale garden worthy of its name not only at that time but even today. Its appearance greatly advanced the starting time of large-scale Lingnan gardens.

During the late Tang Dynasty and the Five Dynasties and Ten Kingdoms period, the Lingnan region experienced rapid development, promoting the vigorous growth of Lingnan gardens. At this time, among the palace gardens, the most famous was the Xianhu Yaozhou, which was a large lake area of more than 500 zhang. Within the lake area, pavilions, terraces, and detached palaces were built. In the lake, there was a sandbar island, planted with flowers and medicinal herbs, which was the place for the emperor to refine elixirs, hence the name "Yaozhou" (Fig. 21), and it was one of the large-scale garden projects of the Southern Han dynasty.

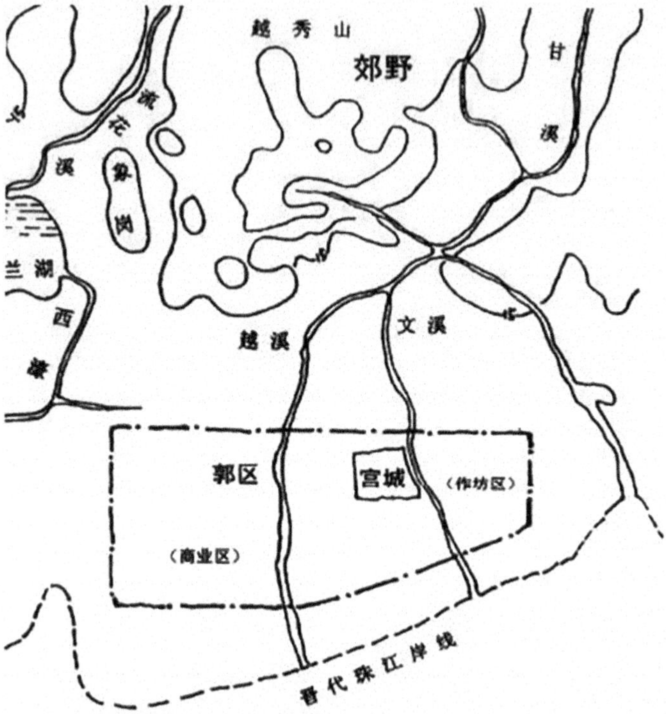

Fig. 18 Floor plan of the Imperial Garden site of the Nanyue King Palace Office in the Western Han Dynasty. (*Image source* Cao Jin's *Research on Lingnan Architecture in the Pre-Qin and Han Dynasties*)

During the Tang and Song Dynasties, transportation in areas such as central Guangdong was very developed, and the regional economy and culture developed rapidly. According to the *Guangdong Tongzhi* (廣東通志), regarding Lychee Garden, Qingyuan Wei's Garden, and Xinzhou Ten Immortal Garden, it can be discovered from relevant records that although they were all private gardens, judging from the general environment of Lingnan gardens at that time, the area can be regarded as a large garden. It is worth noting that Lingnan gardens at that time already had the characteristics of public gardens, even though the royal palace gardens were open to the public, providing places for ordinary people to engage in daily leisure activities, and also places for literati to chant poems and discuss learning. This "joy shared by all" open-mindedness laid the foundation for the development of public gardens in Lingnan in the later period.

In modern times, the earliest recorded instance is the Poontinqua's Garden (海山仙館), also known as Pan Garden, which was first built in 1830. It was a renowned overseas large-scale private garden. At that time, Guangdong's commodity economy was continuously improving, and the Thirteen Hong merchants of Guangzhou became new representatives of "Cantonese merchants." They relied on huge wealth

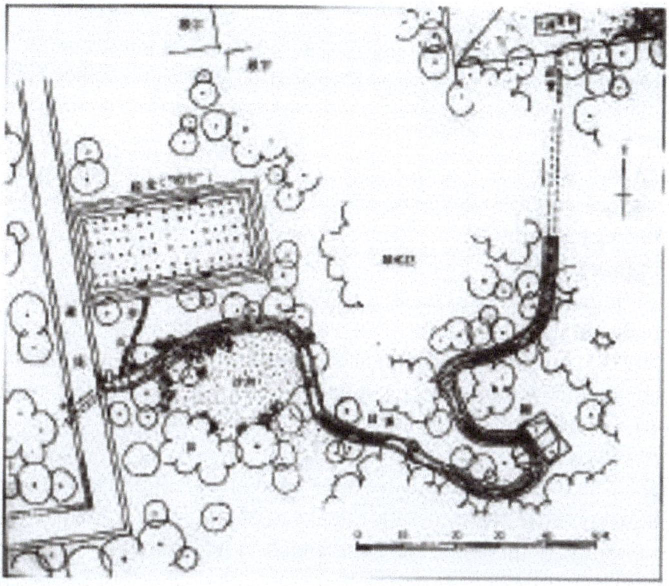

Fig. 19 Floor plan of the Imperial Garden site of the Nanyue King Palace Office in the Western Han Dynasty. (*Image source* Cao Jin's *Research on Lingnan Architecture in the Pre-Qin and Han Dynasties*)

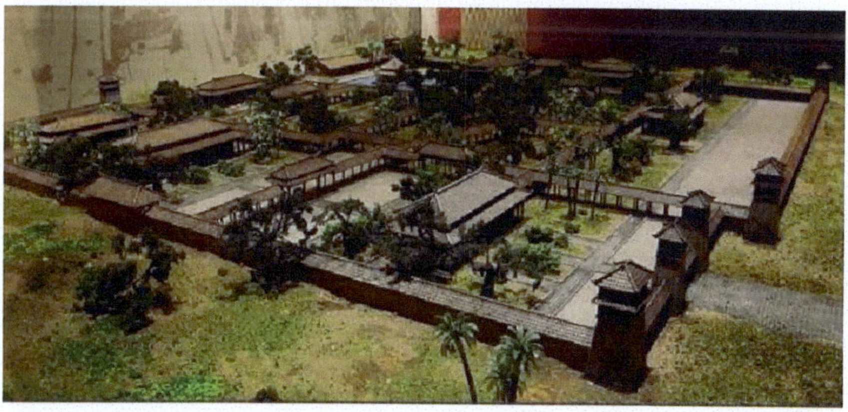

Fig. 20 Restoration model of the Palace and Garden of the Mausoleum of the Nanyue King. (*Image source* Museum of the Mausoleum of the Nanyue King)

to become the dominant force in the construction of large-scale Lingnan gardens. The *Continued Records of Panyu County* (番禺縣續志) once recorded that Poon-tinqua's Garden had a wide pond and a broad garden, making it the crown of Lingnan gardens. The pond in the garden was approximately 100 acres in size, and its water

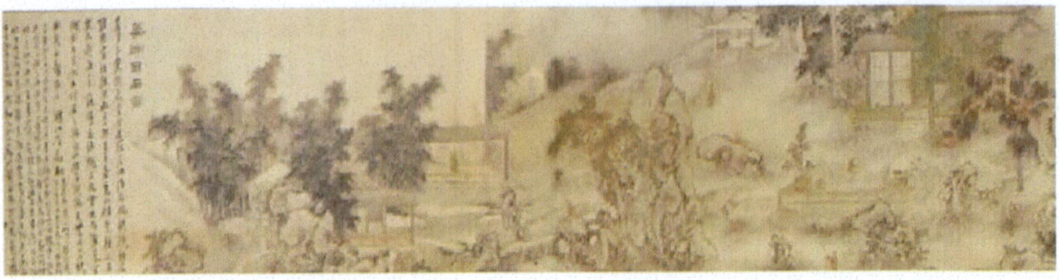

Fig. 21 Qing Dynasty, Su Liupeng's *Yaozhou Pinshi Map*. (*Image source Guangzhou Chengfang Chronicle Volume 6*)

connected directly to the Pearl River, sufficient for boating. It is representative of modern Lingnan large-scale gardens, not only inheriting the Lingnan garden style but also drawing on many Western Garden elements, especially in the interior decoration of the garden (Table 1).

The Lingnan region is economically developed and receptive to foreign cultures, possessing a strong sense of democratic openness. Most private gardens exhibit semi-open characteristics. In modern times, influenced by Western public gardens, public gardens in the Lingnan region arose, embarking on the path of modern evolution.

In 1921, Mr. Sun Yat-sen proposed the idea of "sharing happiness with the people" (与众乐乐), reflecting his ideology of people's livelihood and rights. Tang Shaoyi renamed his private garden "Gongle Garden" (共乐园), opening it completely to the villagers. During the same period, Mo Yongyu, the owner of Qixia Hall (栖霞仙馆) and a wealthy Hong Kong merchant, was also influenced by Western democratic thought, coinciding with the idea of "sharing happiness with the people." He shared his private garden with his clan or opened it to nearby villagers on a regular basis.

Table 1 Construction time and nature of some modern Lingnan large gardens

Garden name	Completion time	Location	Area	Garden type	Current use
Hai Shan Xian Guan	1830–1847	Lychee Bay, Western Suburbs of Guangzhou	Approximately 100 Mu	Merchant Garden	Museum
Liang Garden	1820–1850	Songfeng Road, Xianfeng Ancient Road, Foshan	Over 200 Mu	Scholar-Official Garden	Park
Qing Hui Garden	1888	Shunde District, Foshan	Approximately 19.5 Mu	Scholar-Official Garden	Park
Li Garden	1926–1936	Kaiping City, Guangdong Province	Approximately 17 Mu	Overseas Chinese Garden	Park

Source Compiled into a table based on *Lingnan Landscaping and Aesthetics*, *Lingnan Garden*, and *Southern Beauty*

During this period, the main creators of private gardens were mostly politicians, businessmen, and overseas Chinese, foreshadowing their later public transformation. When Gongle Garden became the first public garden in Tangjiawan, the Republic of China had only been established for 10 years, and many public gardens in China were just beginning, reflecting the progressiveness of Tangjiawan.

Large-scale gardens (≥10 acres, constituting 20% of total area), exemplified by Gongle Garden (共乐园) and Qixia Hall (栖霞仙馆), demonstrate distinctive wisdom in spatial organization, functional adaptation, and cultural synthesis. Gongle Garden, constructed by Tang Shaoyi in 1910 on mountainous terrain, employs a north–south axis to integrate its "Chinese Garden Quarter" and "Western Recreation Zone" [1, 20]. Leveraging elevation differentials, it creates a rhythmic "low court, high pavilion" spatial sequence. While its stone pagoda and resting pavilion perpetuate Lingnan traditions, the Stargazing Pavilion and tennis court introduce Western elements. Following its 1915 opening, added public facilities transformed the space from private garden to communal area. Qixia Hall centers on a Fasting Hall to create a hybrid "religious cultivation + Western leisure" program. Western geometric order manifests in its octagonal Guanyin Pond, Roman-columned clock tower, and Portuguese parquet flooring. Conversely, gray moldings on roofs and Cantonese bird-and-flower window lattices retain Lingnan architectural DNA. This ensemble embodies profound East–West craftsmanship and spiritual hybridization (Fig. 22).

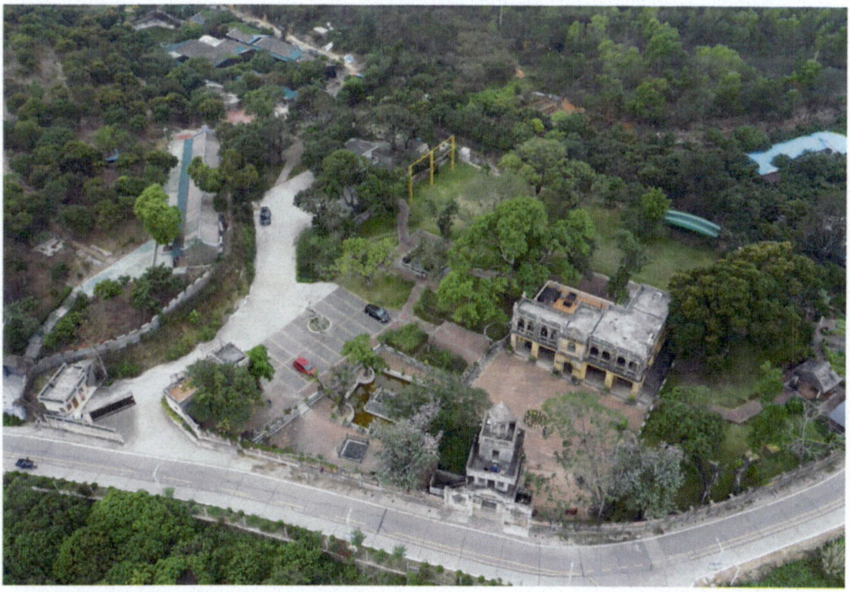

Fig. 22 Aerial photography of Qixia Hall. (*Image source* photographed by the authors)

3 Gardening Techniques of Modern Large-Scale Gardens

3.1 A Paradigm of Large Courtyards—Landscape Analysis of Gongle Garden

3.1.1 Site Selection

Early photos of Tangjiawan Town show that the entire area was largely composed of mountainous areas, farmland areas, and marine areas, with urban forms yet to develop (Fig. 23). And E'ling (Goose Ridge Mountain), where Gongle Garden is located, is on the edge of Tangjiawan Town, with the mountainous terrain higher in the west and southwest, and the mountainous height higher in the east and northeast. The overall trend slopes from southwest to northeast, and because the terrain at the foot of the mountain is relatively flat, most of the earliest buildings in the town were located around E'ling.

Gongle Garden, as a large garden, is built along the mountain, which can not only save farmland, leaving a large area of flat land for cultivation, but also has a better orientation. In winter, the northern mountains can block the southward movement of cold air, and in summer, the Southern Ocean brings refreshing sea breezes (Fig. 24).

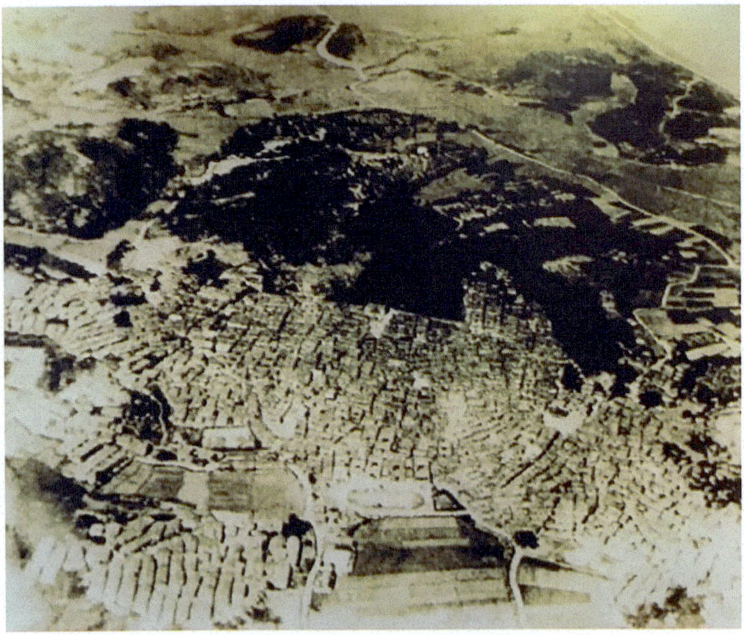

Fig. 23 Early aerial photograph of Tangjiawan. (*Image source* Luo Zuning's *Centennial Zhuhai Map*)

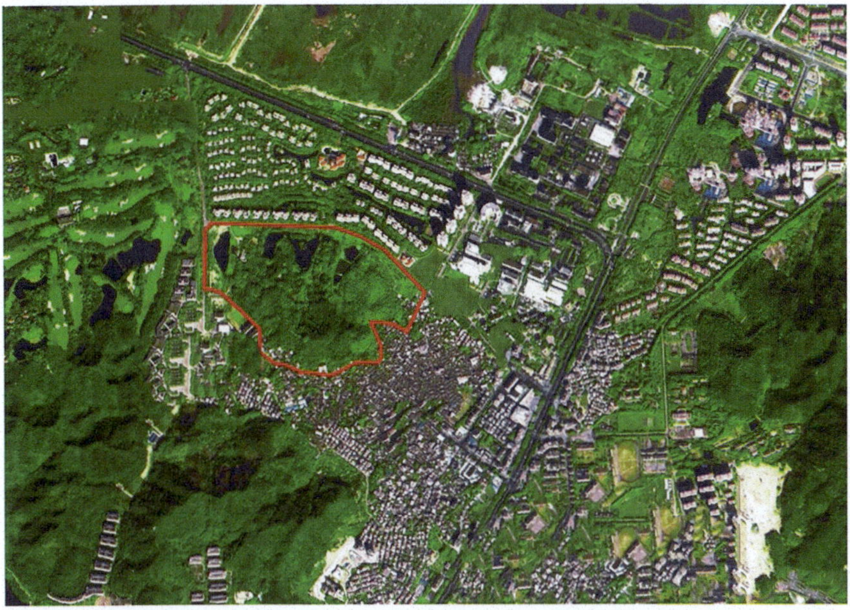

Fig. 24 Goose Ridge Mountain and Village. (*Image source* Redrawn from Google Maps)

Gongle Garden has a 100-year history since its construction in 1910. Due to political turmoil, the impact of wars, and later restoration and renovation, the area, layout, and roads within the park have changed. Even now, the entire view of Gongle Garden is not fully open to tourists. Therefore, the research scope of this chapter focuses on the core area of Gongle Garden: the area around the Stone Archway, Lotus Pond Nine Dragons Cun, Pastoral Villa, and Stargazing Pavilion (Fig. 25).

3.1.2 Grouped Layout

The Gongle Garden was an important place for Tang Shaoyi's life, work, and entertainment. In addition to simple material needs, there was also a spiritual pursuit of entrusting emotions to mountains and rivers. It was also a recreational place for entertaining guests (Fig. 26).

The planning of Gongle Garden generally embodies the thoughts and characteristics of traditional Chinese gardens, but influenced by phased construction and Western culture, Gongle Garden presents a traditional Chinese style and a modern Western style. First, divide Gongle Garden according to landscape function, which can be divided into two parts: the traditional building area and the Western-style recreational area (Fig. 27). The layout of the garden is based on the natural topography. The two areas transition naturally without separation. The buildings and small objects in the traditional building area are distributed widely and scattered, with lush

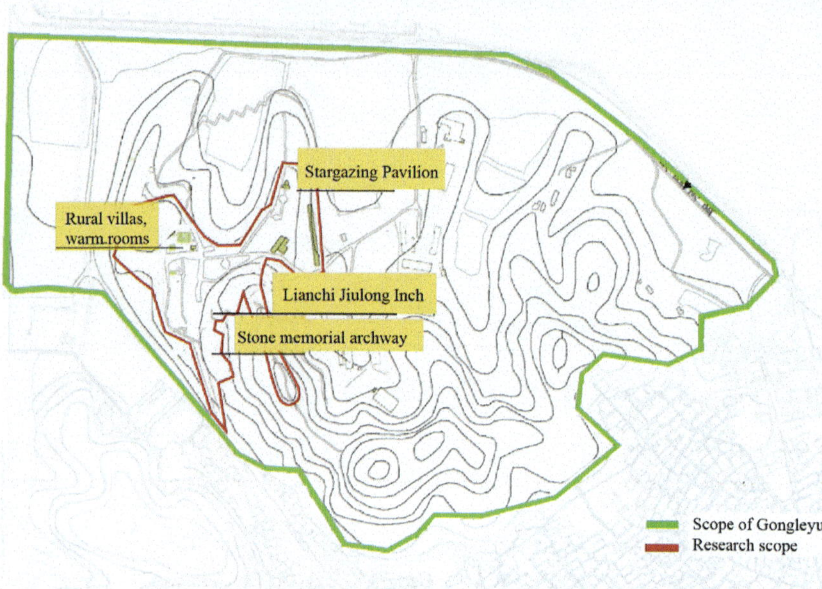

Fig. 25 The scope of the Gongle Garden. (*Image source* Redrawn according to the *Zhuhai Planning Research and Design Institute Plan Report*)

vegetation and towering ancient trees; the Western-style recreational area has fewer and more concentrated buildings, mostly based on geometric lawns, and the plants are planted in an orderly manner.

Based on the Chinese-style residential area and the Western-style recreational area, it is further subdivided. The Gongle Garden is divided into five groups according to the group landscape: Chinese-style garden area, Chinese-style building area, Western-style lawn area, Western-style building area, and mountain forest area (Fig. 28).

3.2 Tangjiawan Gardening Features

3.2.1 Tangjiawan Temple Courtyard Gardening Features

The modern ancestral halls, temples, and gardens in Tangjiawan are divided into spaces such as the front yard, central courtyard, patio, and side hall according to function and structure. The front yard has a larger area and is not only the main venue for clan festivals (such as the "flower basket grabbing" and dragon and lion dances held by the Tangjia San Miao during the "Land God's Birthday" on the second day of the second lunar month), but also closely connected to the lives of the villagers,

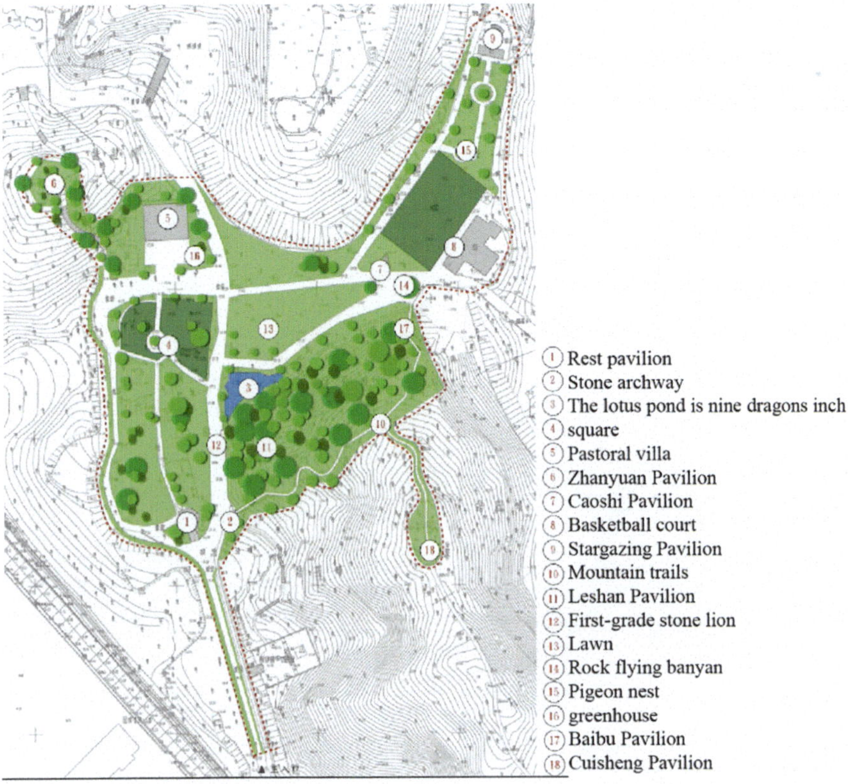

Fig. 26 The general plan of the Gongle Garden. (*Image source* Redrawn according to the planning data of Zhuhai High-Tech Zone)

sometimes used for labor, and carries cultural significance beyond the space itself (Fig. 29). The central courtyard and back courtyard are enclosed by buildings and are the core landscapes, combining etiquette and ornamental value. For example, the central courtyard of the Yuwo Tang Gong Ancestral Hall has a pond and wooden connecting bridge (Fig. 30), which is quite interesting, while the central courtyard of the temple emphasizes the etiquette function, often building pavilions or setting up altars (such as the Shengtang Temple, Jinhua Temple, and Qi'ao Village Wenchang Temple of the Tangjia San Miao) for sacrificial rituals; Larger ancestral halls and temples will have side halls combined with patios. The patios not only provide ventilation and drainage for the side halls but also add landscape elements to enrich the space (Fig. 31).

The ancestral temples and gardens take rituals, ethics, auspiciousness, and blessings as their core themes, with rich decorative subjects and profound meanings. The patterns and forms are diverse, including birds and animals (such as a rooster standing on a rock, implying "auspiciousness in the room," and dragons and phoenixes symbolizing the emergence of talented people), antique treasures (such as tripods and

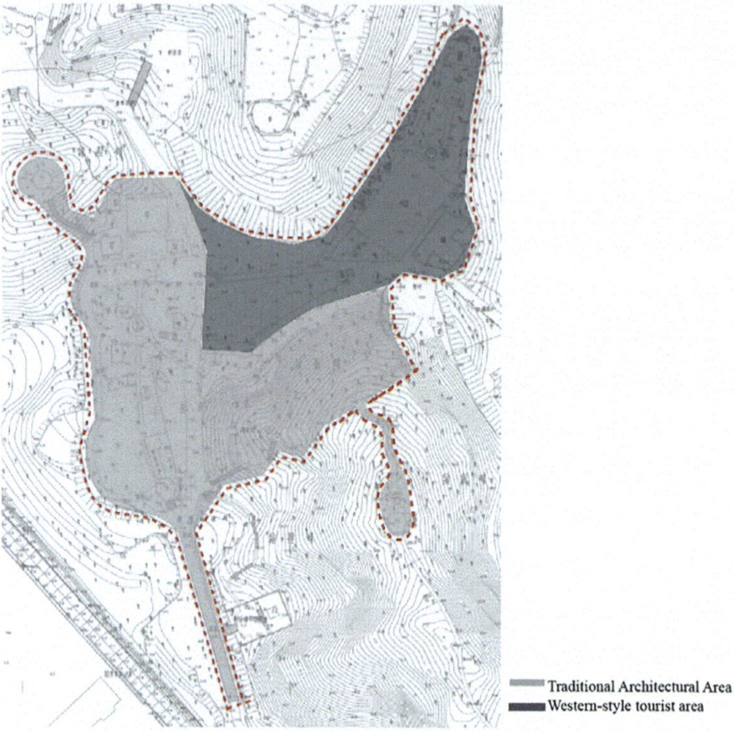

Fig. 27 There are two major divisions of the park. (*Image source* Redrawn according to the planning data of Zhuhai High-Tech Zone)

stationery, reflecting the learning of scholars), exquisite flowers and jade trees (such as banana trees implying "great career," peonies symbolizing wealth and nobility, and grapes and pomegranates conveying the wish for many children and grandchildren), landscape scenery (reflecting the harmonious coexistence of humans and nature), and figures and immortals (such as "Eight Immortals Crossing the Sea" and "Seven Sages of the Bamboo Grove"); the content of the themes is divided into Confucian ethics and morality (such as "Transmitting Scriptures" and "Oath of the Peach Garden," which impart ethical concepts), praying for blessings and well-being (such as "goldfish" homophonous with "gold and jade," implying talent and learning, and combined patterns of animals and plants conveying the wish for auspiciousness and longevity), and religious themes (such as "Eight Immortals Crossing the Sea" and bats symbolizing "ubiquitous blessings"), which not only demonstrates people's yearning for a better life, but also reflects the ideology and values of traditional society (Figs. 32 and 33).

The garden landscape is characterized by "Three Carvings and Two Sculptures" (mainly brick carving, stone carving, and stucco). Black decorative bands or colored paintings are commonly seen at the top of walls, which not only coordinate the colors

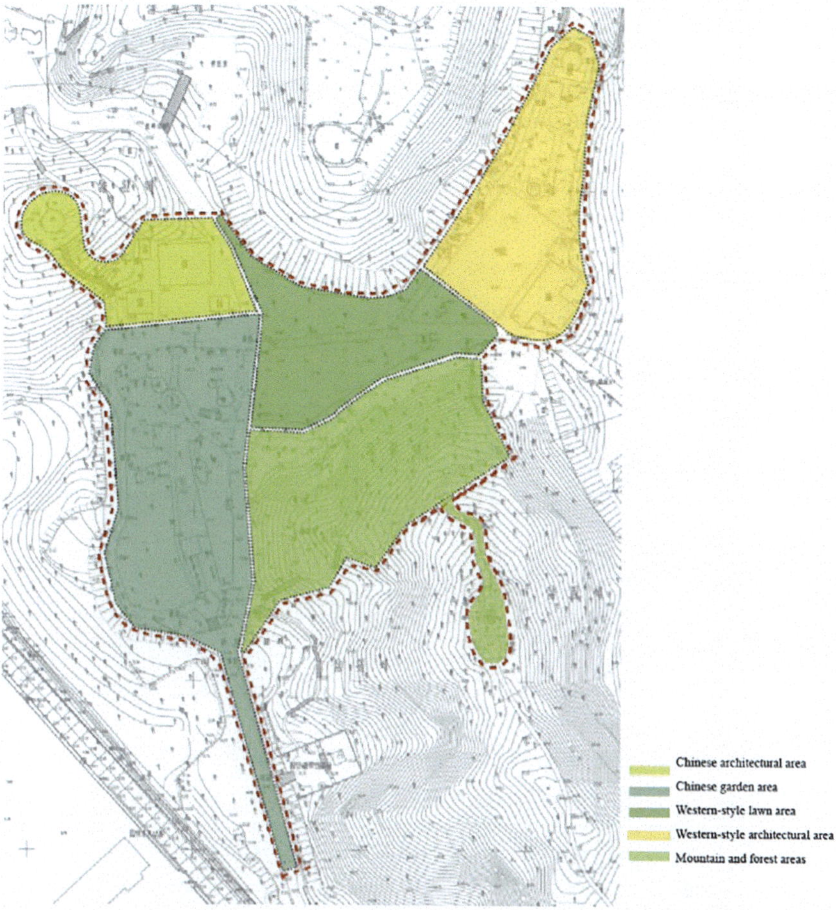

Fig. 28 There are five major group areas in the park. (*Image source* Redrawn according to the planning data of Zhuhai High-Tech Zone)

of purlins and blue brick walls, but also repair cracks and protect the walls. The murals on the gable of the main entrance are often painted in sections, with themes involving Lingnan flowers, fruits, figures, and landscapes; plinth piers are divided into two-section and three-section types, the former having wave-shaped carvings on the upper layer and figure legends or auspicious patterns carved on the lower layer, the latter having a brick-carved beveled surface decoration on the top, a transition layer in the middle, and the main painting (plinth pier body) expressing melons, fruits, figures, etc. through transparent carvings, with a combination of antique objects and flower patterns at the bottom. The craftsmanship integrates brick carving, stone carving, and stucco; roof ridges are divided into dragon boat ridges and antique ridges. The dragon boat ridge is shaped like a dragon boat and decorated with stucco, with boat supports and "Ao Fish Swallowing Ridge," and ridge eyes are set between the paintings and

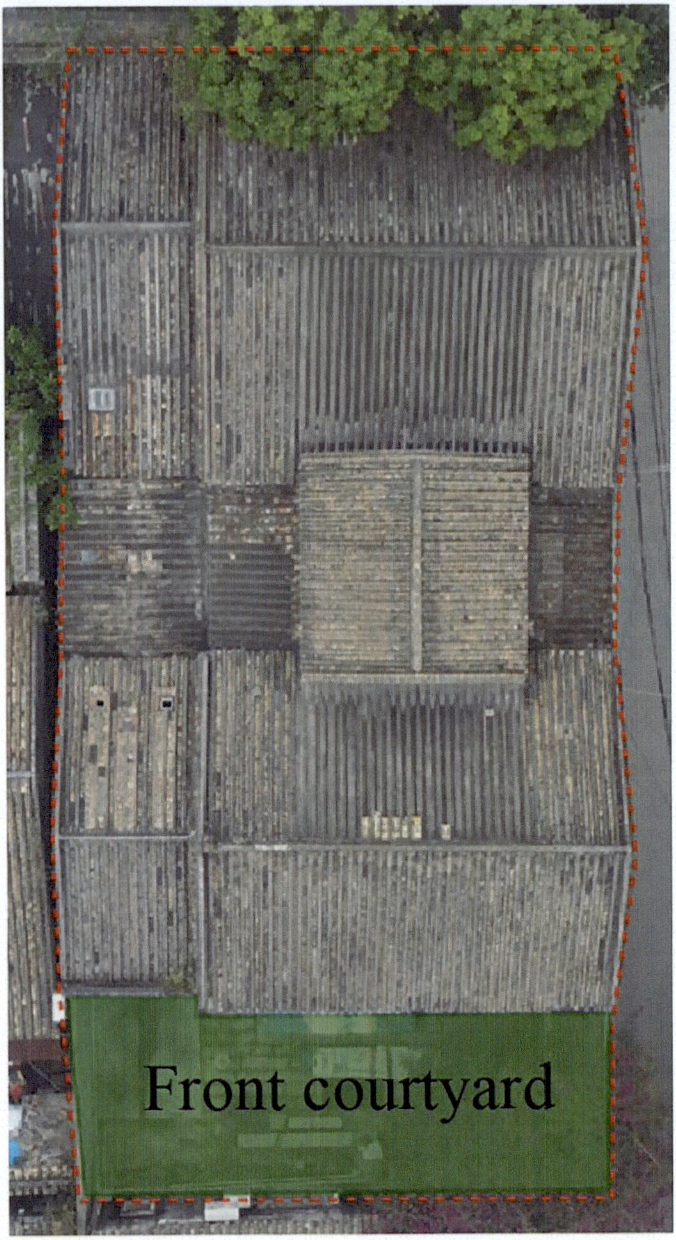

Fig. 29 Front courtyard of Juchuan Tang Gong Temple. (*Image source* Based on the renovation provided by the Cultural Center of Zhuhai High-Tech Zone)

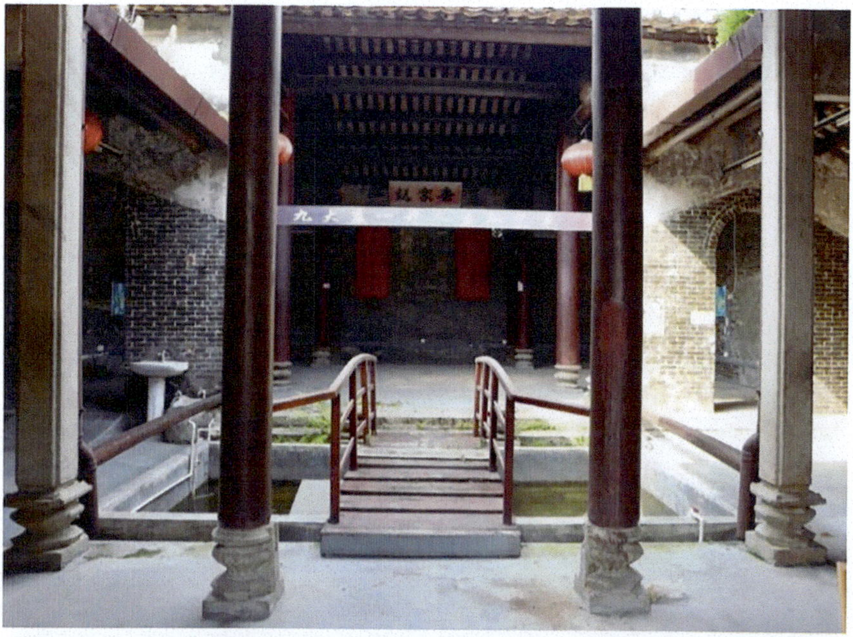

Fig. 30 Tangjiawan Yuwo Tang Gong Temple Courtyard landscape. (*Image source* by the author)

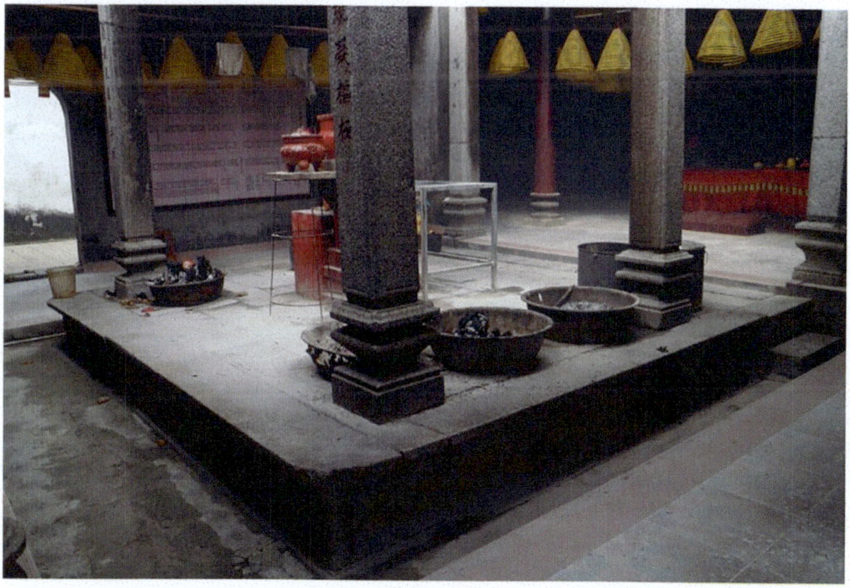

Fig. 31 Landscape of the Courtyard of Wenchang Temple in Qi'ao Village, Tangjiawan. (*Image source* by the author)

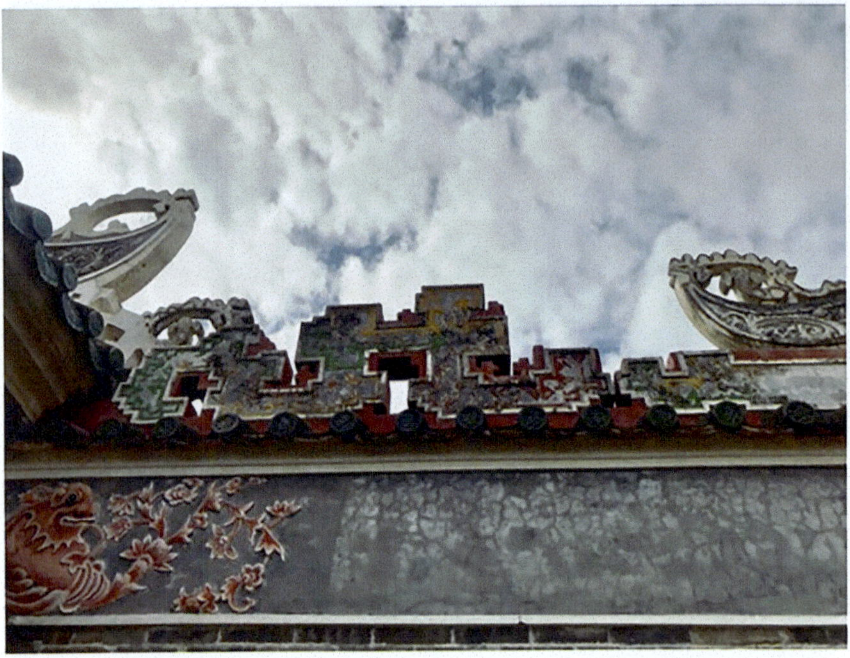

Fig. 32 The application of Tangjiawan Bogu pattern. (*Image source* by the author)

three-dimensional stucco is placed. The antique ridge is composed of a picture with stucco. There are two types of antique heads, simple (geometric Kui dragon patterns) and complex (carved vines, auspicious animals, etc., with bright colors), each with different styles (Fig. 34).

3.2.2 Landscaping Characteristics of Modern Residential Gardens in Tangjiawan

(1) Traditional Style and Combination of Chinese and Western-Style Courtyard Architecture

Modern residential garden architecture in Tangjiawan can be divided into two types: one is the traditional style of courtyard architecture, adopting the Lingnan traditional residential architectural style as the exterior, and the interior is also built according to the traditional style. Such as the Mo Family Mansion in Huitong and the former residence of Gu Yuan in Nazhou as representatives. The other is the "Western exterior and Chinese interior" style of courtyard architecture combining Chinese and Western styles, with the style of Western architecture as the appearance, while the interior of the house is still dominated by the traditional Lingnan residential architectural style. The former residence of Tang Shaoyi is a typical courtyard building of this type.

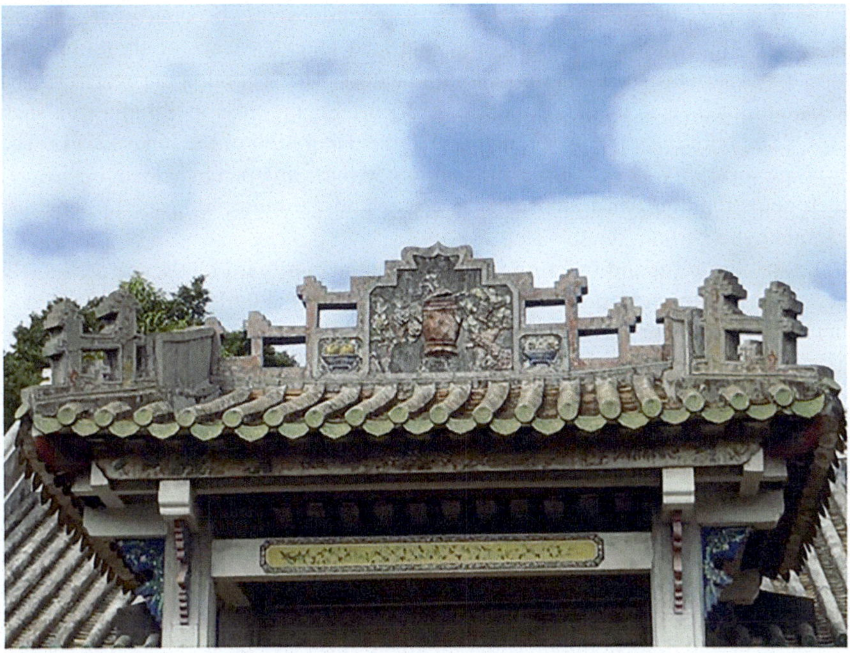

Fig. 33 The application of Tangjiawan Bogu pattern. (*Image source* by the author)

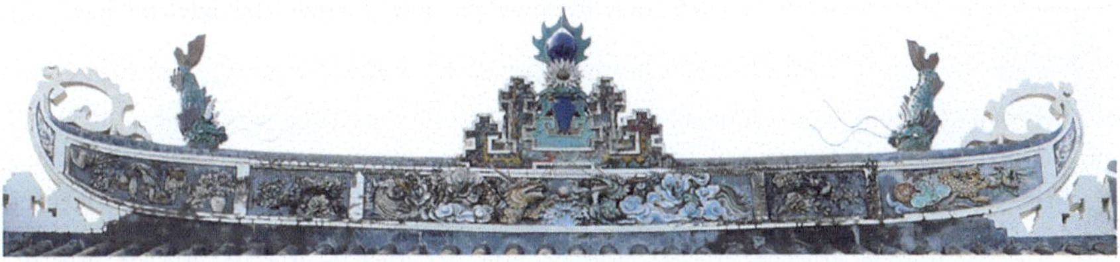

Fig. 34 Roof ridge decoration of the head gate of the Golden Flower Temple. (*Image source* by the author)

(2) Enclosed Courtyard Landscapes Meeting Different Functional Needs.

1) Enclosed Landscapes of Modern Residential Courtyards in Tangjiawan

Modern residential courtyards in Tangjiawan have different enclosed landscape interfaces. Different enclosing elements constitute these landscape interfaces, which have certain characteristic expressions. There are three characteristics of the enclosed landscape interface.

The first is centripetality. The degree of inwardness of the courtyard space is affected by the scale of the enclosing interface and the scale of the courtyard. These two scales are the most important indicators. The second is introversion. The enclosing interface elements of modern residential courtyards in Tangjiawan, such as walls and buildings, pay more attention to the decoration of the inner facade. The third is permeability. For example, the building will use an open facade for enclosure, which helps the courtyard landscape interface and the building space to penetrate each other and be integrated.

The specific form and function of the residential courtyard landscape are mostly determined by the enclosure of the building. Secondly, the building is also related to other enclosing factors. Therefore, the enclosure of the building is the basis for the existence of the courtyard landscape.

2) Landscape Space Division of Modern Residential Courtyards in Tangjiawan

The landscape space of modern residential courtyards in Tangjiawan meets the different functional needs of the garden owners, such as rest space, social space, and traffic space. First, modern residential courtyards in Tangjiawan mostly use ground paving to distinguish different spaces, making the traffic space clearly oriented and strengthening the dynamic connection between the landscape and the building. Secondly, landscape elements such as plants and water bodies in modern residential courtyards in Tangjiawan divide some relatively open spaces as interaction spaces. Some interaction spaces have a certain degree of privacy, serving as rest and private spaces.

3) The Relationship Between Architectural Space and Courtyard Space

The house sites in Tangjiawan are sometimes affected by topography, land area, and plot shape, resulting in irregularly shaped and non-spacious surplus land being exposed. However, the people of Tangjiawan use this surplus land to create gardens. It makes the architectural space and courtyard space permeate, connect, and intersperse each other, forming an integrated landscape of opposites, and realizing the conversion of different functional spaces in the residence and courtyard.

(3) Landscape Characteristics Based on Practicality

1) Natural Ecology

Since ancient times, China has had supreme worship of nature. This worship has influenced the construction of residential gardens, indirectly reflecting the Tangjiawan people's emotional view of nature. In order to create a living atmosphere with strong naturalness, natural objects are introduced into the landscape of modern residential gardens in Tangjiawan, and they are carefully transformed and utilized. Here, people cultivate their sentiments and enjoy the blessings of nature. Therefore, in the treatment of residential garden landscape space, Tangjiawan people choose a transparent space with natural ecology.

2) Spatial Permeability

Modern residential gardens in Tangjiawan have both "inner" and "outer" characteristics, such as the courtyard of Gu Yuan's former residence. The overall space has both introverted and external spaces. The introversion of the garden has the characteristics of an inner space, which is different from the infinitely expanding external space. The open top of the garden has the characteristics of external space. Therefore, the "inner" and "outer" characteristics in the garden reflect the ambiguity and permeability of the garden space, and this characteristic is precisely the highlight of the garden space. Modern residential gardens in Tangjiawan are not only internal public spaces but also outdoor open spaces, with great variability and adaptability in size.

3) Functional Practicality

The modern residential garden landscape in Tangjiawan has powerful functionality. First of all, it must meet people's living needs. The residential gardens in Tangjiawan are not large in size, and more consideration is given to how to make the best use of them. For this reason, they carry various life functions and daily behaviors of people. Secondly, there are aesthetic needs, such as planting flowers and plants and enjoying the scenery in the garden.

3.3 Axis Layout

First, modern large-scale gardens in Tangjiawan, based on Lingnan gardens, incorporate Western Garden landscape elements in the main buildings. The garden area consists of several garden spaces with different landscape characteristics. For example, the Qixia Hall garden space can be divided into two parts: building and green space. Gongle Garden is also divided into two parts: one part with buildings surrounding a geometric lawn as the main space, and the other part surrounded by natural mountains, forming a unique natural landscape. Compared with Lingnan garden design techniques, these diverse garden spaces are organized by their respective axes.

Second, modern large-scale gardens in Tangjiawan are different from Lingnan gardens. Lingnan gardens are generally laid out with mountains and rivers as the main body, supplemented by pavilions and other buildings, and planting flowers and trees, with architecture, mountains and rivers, flowers and trees as the elements, striving to achieve the effect of "although made by man, it seems to be naturally created." Modern large-scale gardens in Tangjiawan generally use large lawns, pools, and buildings as the main content, highlighting the effect of human craftsmanship (人力之美), and focus on buildings as the center, with a clear axial intersection layout (Figs. 35 and 36).

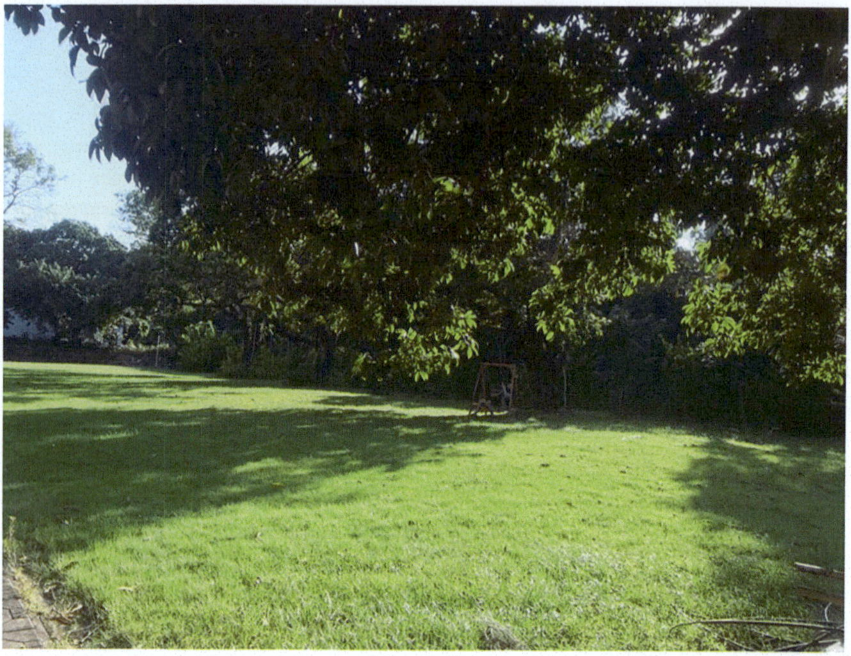

Fig. 35 A large Western-style lawn in a large garden. (*Image source* photographed by the authors)

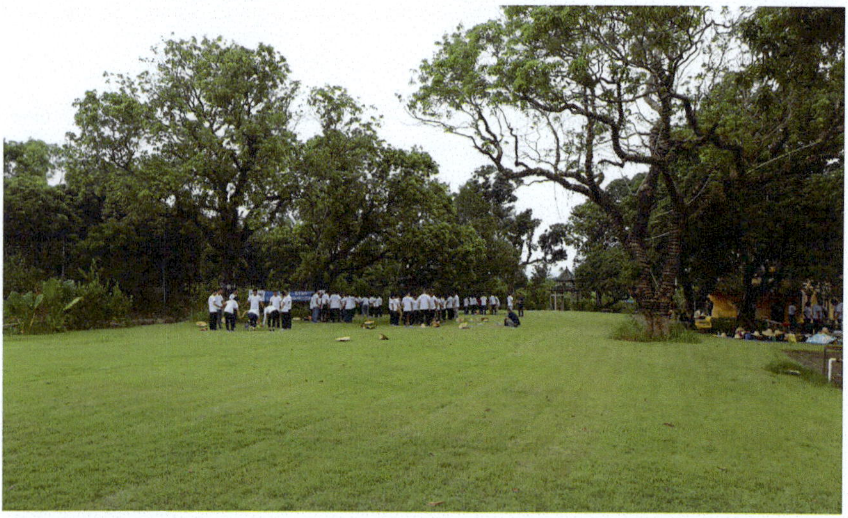

Fig. 36 Large Western-style lawn in a large garden. (*Image source* photographed by the authors)

Third, adaptive reinterpretation of axiality: While borrowing Western symmetry, these gardens avoid rigid formality. Western Gardens' axial symmetry presents an open form (Fig. 37), favoring the placement of features such as pools and plazas at road intersections. In contrast, the plan layout of large modern gardens in Tangjiawan does not strictly adhere to axial symmetry or control the garden. The main buildings, however, adopt Western Garden architectural techniques, employing axial symmetry and being constructed straightforwardly within the garden, unlike Chinese garden buildings which prefer to be hidden within the landscape. The plan layout of large modern gardens in Tangjiawan is relatively simple, lacking the twists, turns, and elaborately crafted spaces characteristic of Chinese gardens. Main structures often employ the "open-door-mountain-view" (开门见山) principle—exemplified by Qixia Hall's clock tower directly facing the main entrance. This approach contrasts with traditional gardens' "concealing-and-revealing" (藏景隐秀) aesthetics, establishing clear spatial rhythms through deliberate visual exposure.

3.4 *Cross-Cultural Reconstruction of Landscape Elements*

3.4.1 The East-Meets-West Interaction of Water and Stone

In water feature design, Tangjiawan gardens synthesize Lingnan and Western approaches: retaining the Lingnan preference for naturalistic boulder arrangements while introducing geometric Western pools—typically rectangular or polygonal basins with sculptural fountains. Exemplified by Qixia Hall's octagonal Guanyin Pond centered within Roman-tiered fountain columns, this fusion extends to innovative curved pool edges that blend artificial landforms with geometric rigor. Stacked marble formations continue Lingnan's "stone-as-landscape" (石景) tradition through bonsai-inspired rockeries and waterscapes. Yet Western elements permeate these features, as seen in Gongle Garden where sculptural fountains integrate with scholar's rocks—demonstrating experimental craftsmanship integration (Figs. 38 and 39).

The direct use of natural rocks for rock placement is also often seen in modern large-scale gardens in Tangjiawan. Because modern large-scale gardens in Tangjiawan are mainly artificial rockeries in the form of bonsai, they are often combined with water features or placed in the corners of the garden. At the same time, because artificial rockeries are rarely used in modern large-scale gardens in Tangjiawan, rockeries are rarely influenced by Western gardening. Only a few Western elements are integrated, such as the appearance of sculptural rockeries, which are also combined with fountains (Figs. 40 and 41).

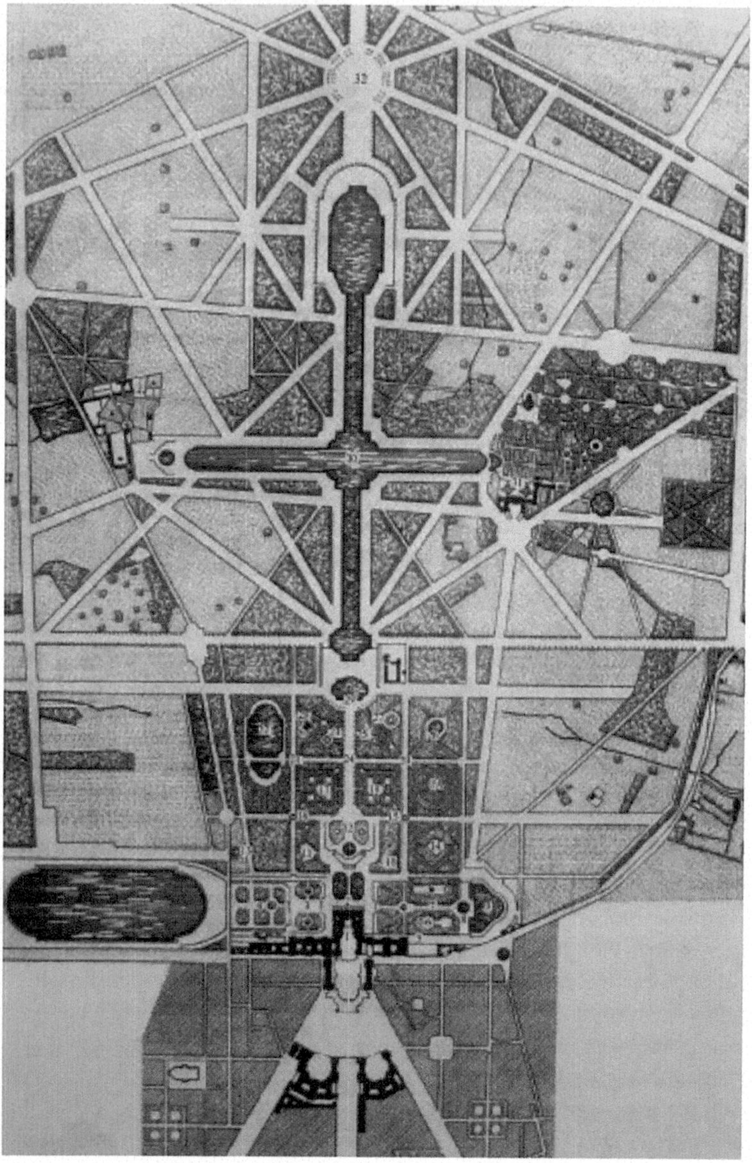

Fig. 37 Floor plan of the Palace of Versailles, Paris. (*Image source* Wang Xiangrong, *Theory and Practice of Western Modern Landscape Design*)

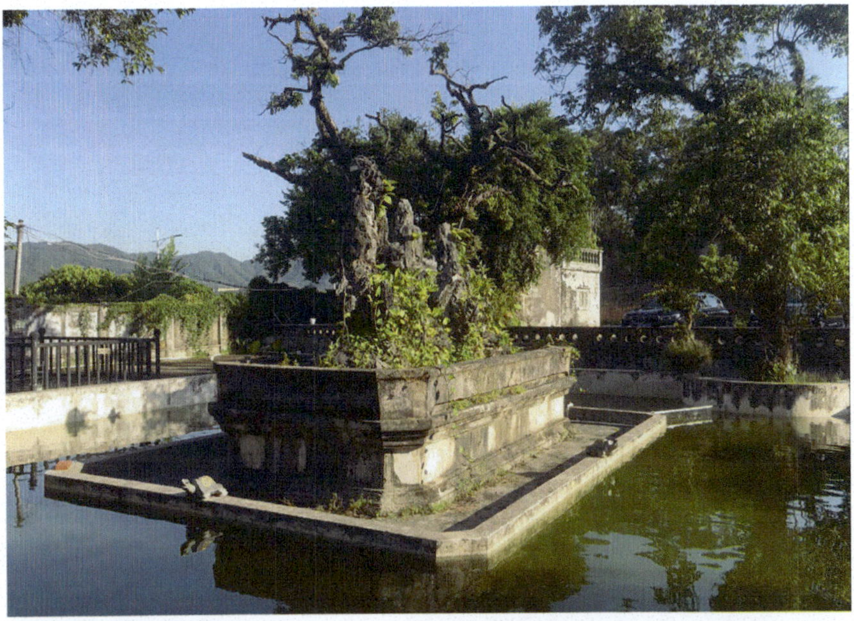

Fig. 38 Common Ground Mountain Stone. (*Image source* photographed by the authors)

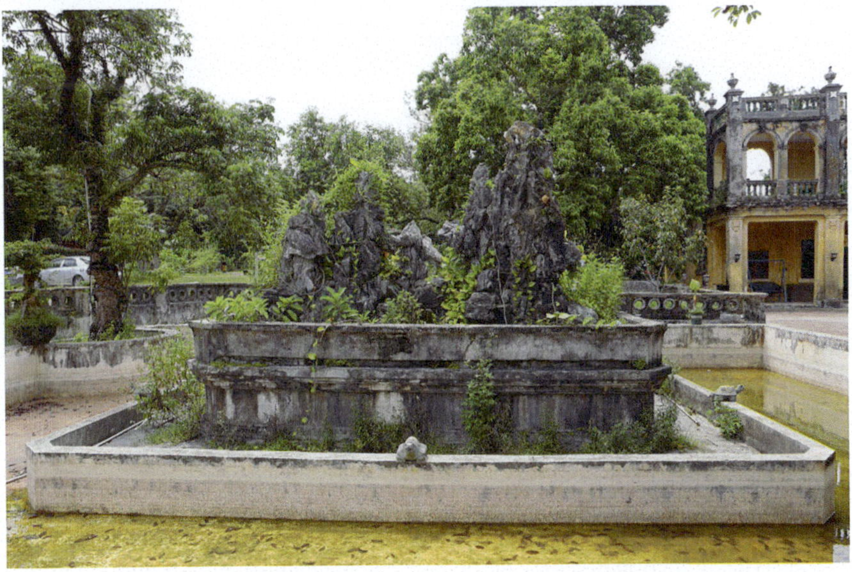

Fig. 39 Common Ground Mountain Stone. (*Image source* photographed by the authors)

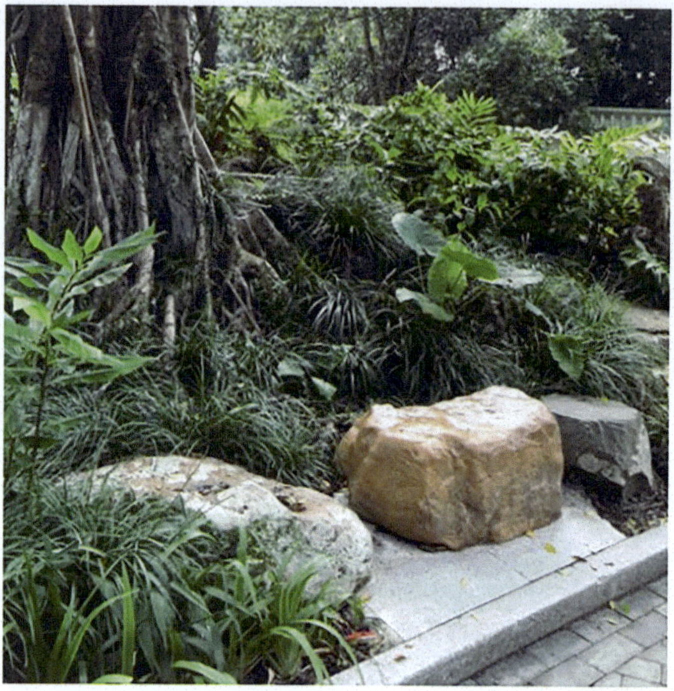

Fig. 40 Mountain Stone in Gongle Garden. (*Image source* photographed by the authors)

3.4.2 Western-Inspired Decorative Expressions

Tangjiawan gardens extensively incorporate Western decorative components—railings, vases, and balustrades—creating a distinctive ornamental language across roofs, balconies, staircases, and pool/flowerbed borders. The cast-iron railings at Qixia Hall Garden epitomize this cultural negotiation, merging Western scrollwork patterns with Lingnan's coiled lotus motifs. Strategically positioned as foreground ornamentation, these railings reinforce axial landscapes through deliberate vista framing (Figs. 42, 43 and 44).

3.4.3 Functional Evolution in Planting Design

Planting schemes embody a synthesis of local cultural heritage and introduce horticultural styles. Functionally, they perpetuate Lingnan's utilitarian tradition through lychee, longan, and other dual-purpose fruit trees offering economic and ornamental value. Simultaneously, Western ornamental lawns—occupying up to 50% of garden areas (e.g., Suxian Pavilion)—reinforce axial order with geometrically clipped shrubs. Species selection preserves symbolic native flora (lotus embodying purity, bamboo representing resilience) while introducing exotic palms to cultivate

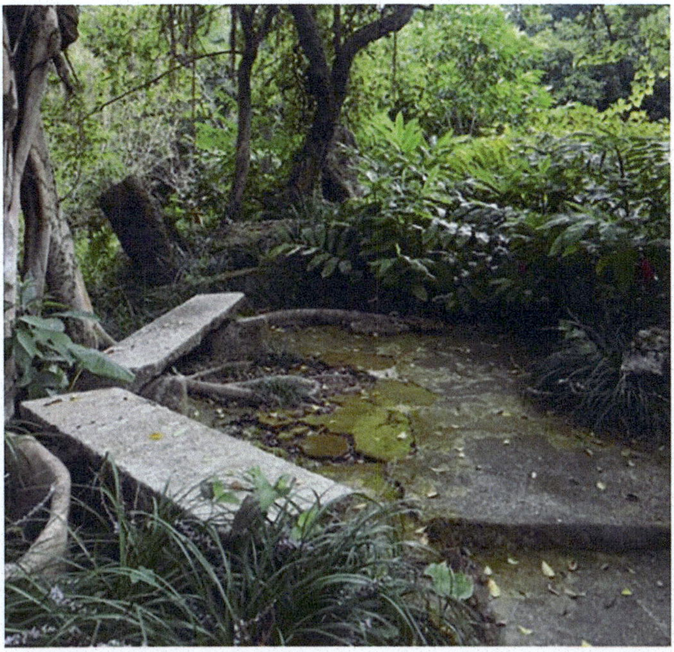

Fig. 41 Mountain Stone in Gongle Garden. (*Image source* photographed by the authors)

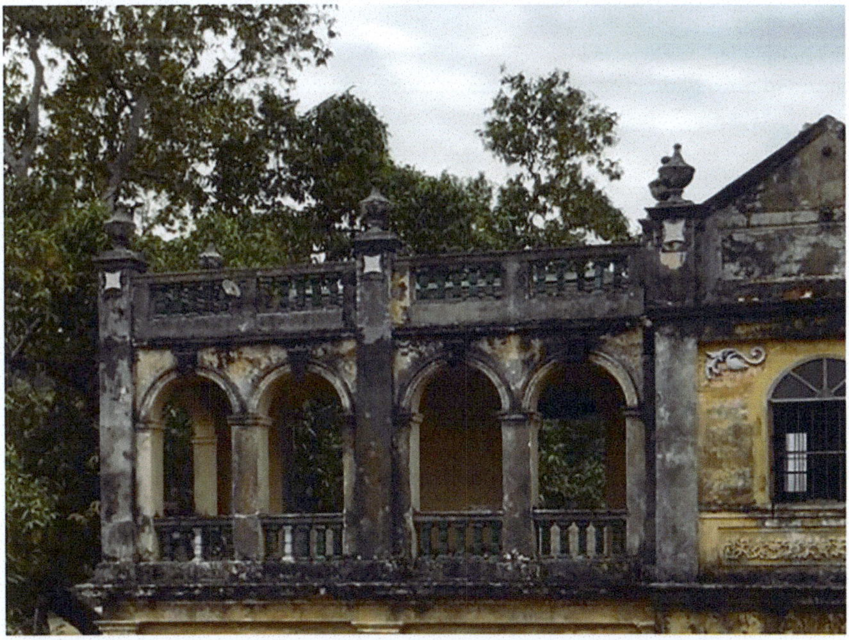

Fig. 42 Details of the decoration. (*Image source* photographed by the authors)

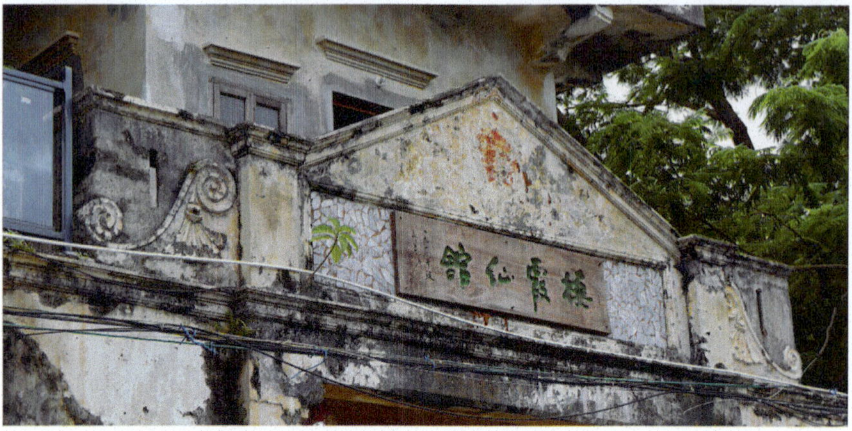

Fig. 43 Details of the decoration. (*Image source* photographed by the authors)

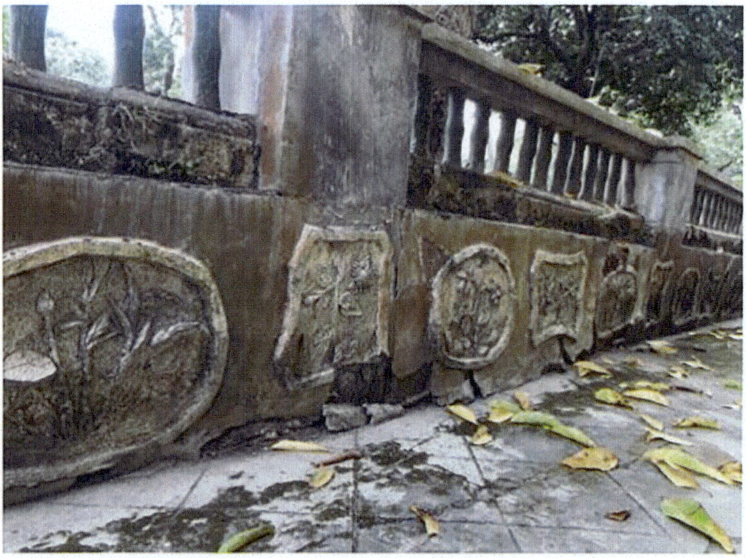

Fig. 44 Details of the decoration. (*Image source* photographed by the authors)

tropical ambiance. This layered juxtaposition creates distinctive phytoscapes. Moreover, freely arranged potted plants punctuate pathways and pool edges, softening rigid geometries and mediating Sino-Western aesthetic tensions through organic spontaneity.

3.4.4 Local Plant Vegetation Types and Configuration Pruning

(1) Types of Plants in Modern Large Gardens in Tangjiawan

Modern large gardens in Tangjiawan have a wide variety of plants. According to the classification of plant functions, they can be divided into the following five types:

Flowers and trees express noble character: This type of plant implies the garden owner's intention to express his ambition through objects. Such as lotus, bamboo, orchid, kapok, etc.

Flowers and trees create elegant and interesting scenes: When landscaping, the garden owner plants some plants that he likes and invites three or five friends to watch and chant together during the appropriate season. Such as hibiscus, cotton roses, azalea, etc.

Flowers and trees with auspicious meanings: Since ancient times, people in the Lingnan area believed in Feng Shui and placed auspicious meanings on plants. Such as pomegranate, pumpkin, longan, grape, etc.

Ornamental fruit trees: Tangjiawan has a diligent and pragmatic social atmosphere, and there are many fruit trees among the plants, which can not only be appreciated but also harvested. Ornamental fruit trees include lychee, mango, jackfruit, etc.

(2) Diverse Style Configuration

Tangjiawan has a complex mix of plant varieties. In terms of configuration, multiple different plant varieties are planted, or plantings are matched according to different spatial environments and planting areas to create rich spatial layers. Plant varieties are divided into two categories: one is native plants that reflect local culture and the characteristics of the tropics and subtropics, and the other is plants that absorb foreign varieties. Palm plants with tropical flair have long appeared in Lingnan gardens. According to "Guangdong Xinyu," fan palm is used for fans, oil palm for raincoats. Planting two palms at home yields many benefits. In modern large-scale gardens in Tangjiawan, palm plants are often paired with rockeries to create a unique landscape feature of modern large-scale gardens in Tangjiawan.

(3) Plant Pruning

A large number of lawns are used in the modern large-scale garden landscape of Tangjiawan, which are combined with Chinese pavilions to form lawn landscapes. In addition to regular lawns, potted plants are also introduced to decorate the garden. The position of potted plants is free, and the decoration is strong, adding a different kind of scenery to the garden. Shrubs arranged on both sides of roads, pools, and around flower beds are pruned into neat shapes using Western-style gardening techniques, which can well accentuate the trend of the main axis.

4 Factors Influencing the Modern Garden Landscape

The modern garden landscapes of Tangjiawan emerged through dynamic interplay between three forces: the natural environment, Lingnan's traditional horticultural DNA, and Western Garden culture. Within these spaces, Lingnan culture's characteristic pragmatism and adaptive capacity find full expression. Crucially, the region's unique geographical conditions and historical circumstances transformed these gardens into a crucible for Sino-Western horticultural experimentation—where traditions collided, negotiated, and ultimately synthesized.

4.1 Impact on the Natural Environment

Zhuhai is in the south of Guangdong Province and on the west bank of the Pearl River estuary [21]. Tangjiwan is in the northwest of the city, about 15 km away from the city center. Influenced by the topography, Tangjiawan is relatively independent, surrounded by mountains and forests in the west and south, and bordered by the estuary of the Pearl River and Qi'ao Island in the east, with a mangrove protection zone. Tangjiawan is located south of the Tropic of Cancer and has a southern subtropical maritime monsoon climate with warm winters and cool summers, an average annual temperature of 22.5°C, and an annual precipitation of 2,000–2,200 mm. The rainy and humid climate has prompted modern gardens to adopt special construction methods: enlarging the stone wall base, setting up hard paving and drainage holes, and utilizing high platforms and gable corridors to elevate the waterfront, which is not only conducive to storing water but also effective in repelling humidity, thus forming a unique composition of the garden features (Fig. 45).

Guangdong Province has a complex topography, with mountains, hills, plateaus, and plains interspersed with dense rivers, and the three rivers converge to form the Pearl River. As a typical delta river network area, Zhuhai has a well-developed water system, and Tangjiawan is located here. Its topography is high in the west and low in the east, tilting to the northeast, with the high hills and low hills of the Phoenix Mountain system accounting for 53.9% of the land area, and a variety of landforms such as alluvial plains and sandy beaches. Such topography and water systems provide natural conditions for gardening in Tangjiawan, shaping a unique garden style. The three-dimensional layout based on the mountains, the regular design that conforms to the flat land, and the landscape elements such as stacked rocks and rockeries modeled after the real mountains are all ingenious echoes and adaptive expressions of the local natural environment.

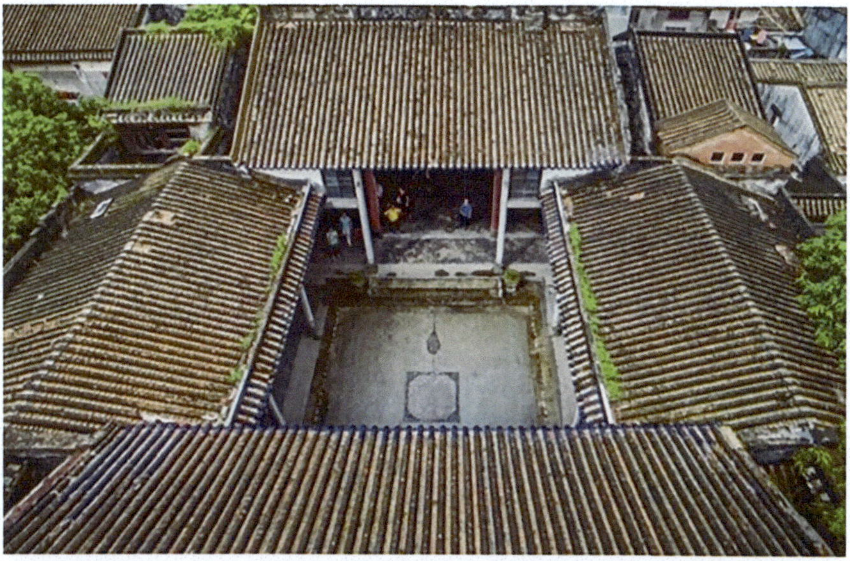

Fig. 45 Architectural layout of the courtyard of Ruizhi Ancestral Temple. (*Image source* Lu Qi, *The Layout and Spatial Characteristics of Lingnan Traditional Gardens*)

4.2 Landscape Technique of Mutual Penetration and Mutual Borrowing

Lingnan gardens are often closed and open to form a contrast, and through the elements of corridors, holes, leaky windows, etc., the scenery penetrates each other and borrows from each other, creating the marvelous effect of "there is a garden inside the garden, and there is a scenery outside the garden." This landscape technique is fully embodied in the modern garden of Tangjiawan: the use of plant sparseness and density, the contrast between the distant mountains and nearby water, and the curvature of the garden road to enhance the spatial hierarchy (Fig. 46); the introduction of scenery outside the garden through the construction of pavilions or the use of architecture and patio intermingling, to realize the borrowing of scenery, extending the spatial context (Fig. 47); and with the help of the window huan, screen, with the help of windows, screens, leakage walls to create a leakage landscape, blurring the boundary between reality and reality, combined with changes in light and shadow, adding a sense of subtlety and elegance; but also through the doorway, leakage windows and other scenic frames into a picturesque landscape, such as Ruizhi Ancestral Temple cave door framed plants and stone wells, giving the garden more interest (Fig. 48).

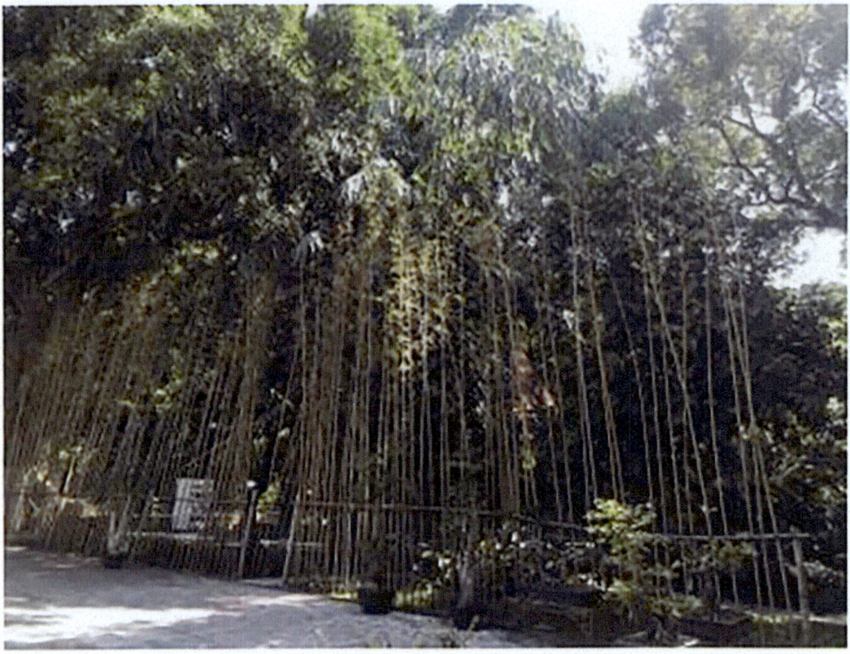

Fig. 46 Bamboo wall with the combination of reality and illusion. (*Image source* photographed by the authors)

4.3 Garden Decoration Under the Traditional Concept

The traditional decorative concept of Lingnan Garden has deeply influenced the modern garden of Tangjiawan, which is reflected in the shape, theme, and color. Plastic decoration through the architectural ridge, walls and other details of the carving, in the form of "three sculptures, three plastic" and other forms to enhance the artistic force, Tang Jiawan modern garden especially focus on the ridge decoration, to Bogu ridge, dragon boat ridge, etc. to show the Lingnan marine cultural characteristics (Fig. 49); the choice of subject matter, the Lingnan garden is often to auspicious birds and flowers, folk characters, natural landscape into the decoration, combined with the characteristics of the local water system, the water and grass pattern to set off the architectural silhouette, its "three carvings and three sculptures" more depictions of character stories and life scenes, which are widely used in the Tangjiwan ancestral temple garden; color, Lingnan garden to elegant green and gray as the base, with brightly colored details of the decorations, forming a unique

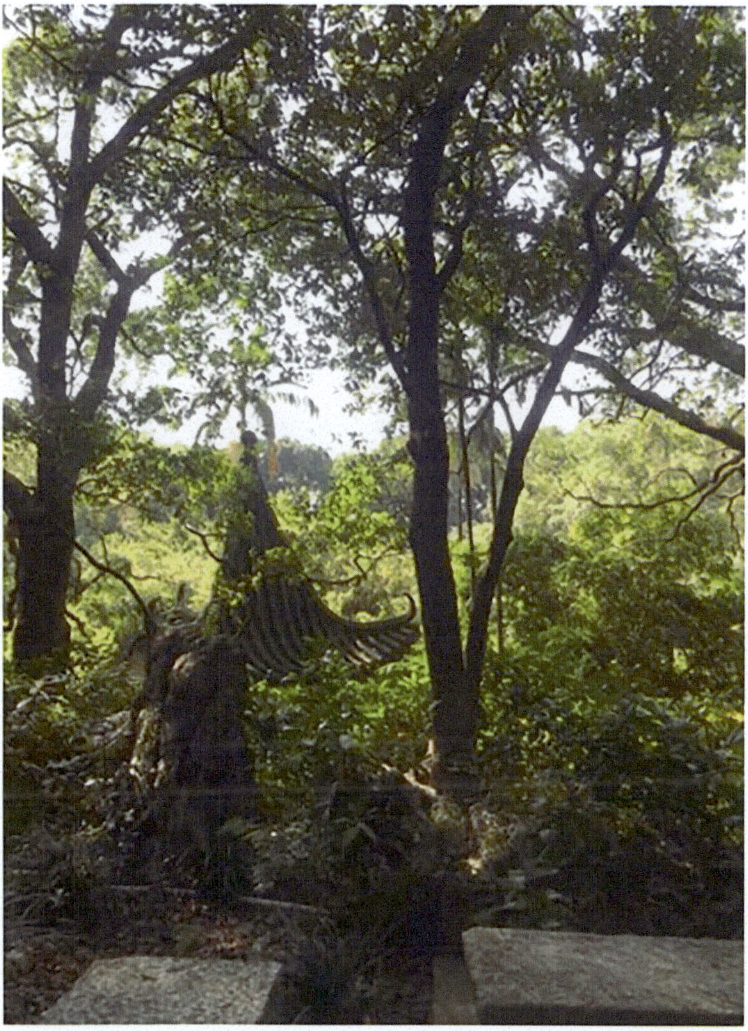

Fig. 47 "Infinite" garden view. (*Image source* photographed by the authors)

contrast, Tangjiwan The modern gardens of Tangjiawan are decorated with glazed and embedded porcelain roofs, and colorful landscapes are created by the use of tinted glass [22–27], showing the warm and cheerful qualities of the Lingnan culture (Fig. 50).

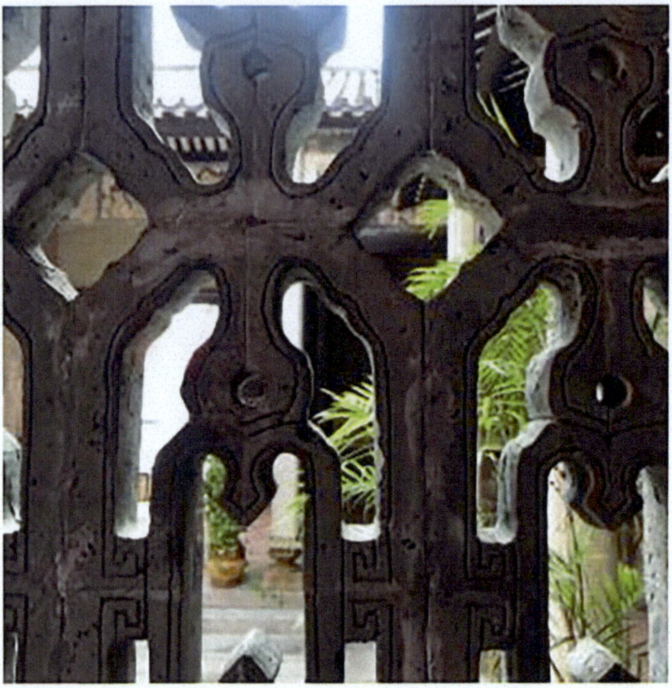

Fig. 48 Window Huan is in the Modern Garden of Tangjiawan. (*Image source* photographed by the authors)

4.4 Influence of Western Landscaping

While inheriting traditional Lingnan garden functions—leisure, social gatherings, and guest reception—Tangjiawan's modern gardens underwent significant transformation through Western influences. The introduction of novel Western recreational concepts expanded both functionality and spatial usage, exemplified by Gongle Garden's incorporation of formal flower parterres, dovecotes, astronomical observatories, and hard-surface tennis courts. Simultaneously, Western lifestyles introduced by the Xiangshan-Zhuhai Comprador Class catalyzed a paradigm shift from private enclosures to public spaces. Gardens like Qixia Hall and Gongle Garden opened to villagers, becoming communal hubs for floral appreciation and public events. This evolution realized the Confucian ideal of "hosts and guests sharing joy" (宾主同乐) through inclusive spatial democratization.

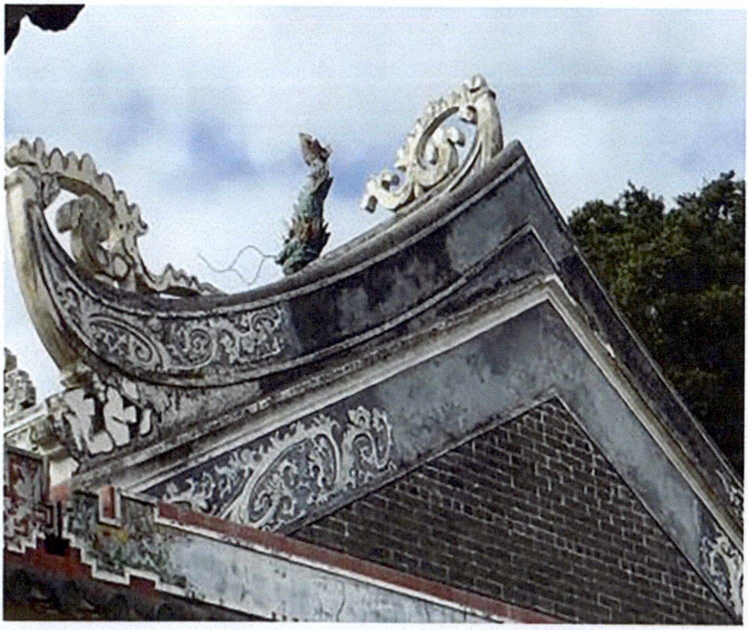

Fig. 49 Architectural plastic decoration in Tangjiawan Garden. (*Image source* photographed by the authors)

The symmetrical, regular, and rigorous layout style of Western gardens is borrowed and integrated into the modern garden of Tang Jiawan. Western gardens take the axis as the skeleton, and the spaces and elements are arranged along the axis, forming a strong sense of order (Fig. 51). This kind of layout influenced the modern garden of Tangjiawan, such as Qixia Hall adopting straight line garden road, creating diversified landscapes with a large lawn, tree groups, and flower beds; Guansing Pavilion complex of the Commonwealth Garden has flower beds as the axes, with left and right symmetrically arranged lawns, geometrical hedges and symmetrically arranged pseudo betel nuts, and the crossroads of roads with flower beds and architectural vignettes, echoing with arbors and shrubs. Flower beds and architectural vignettes are set up at the road intersections, echoing with trees and shrubs, and constructing a regular layout pattern (Fig. 52).

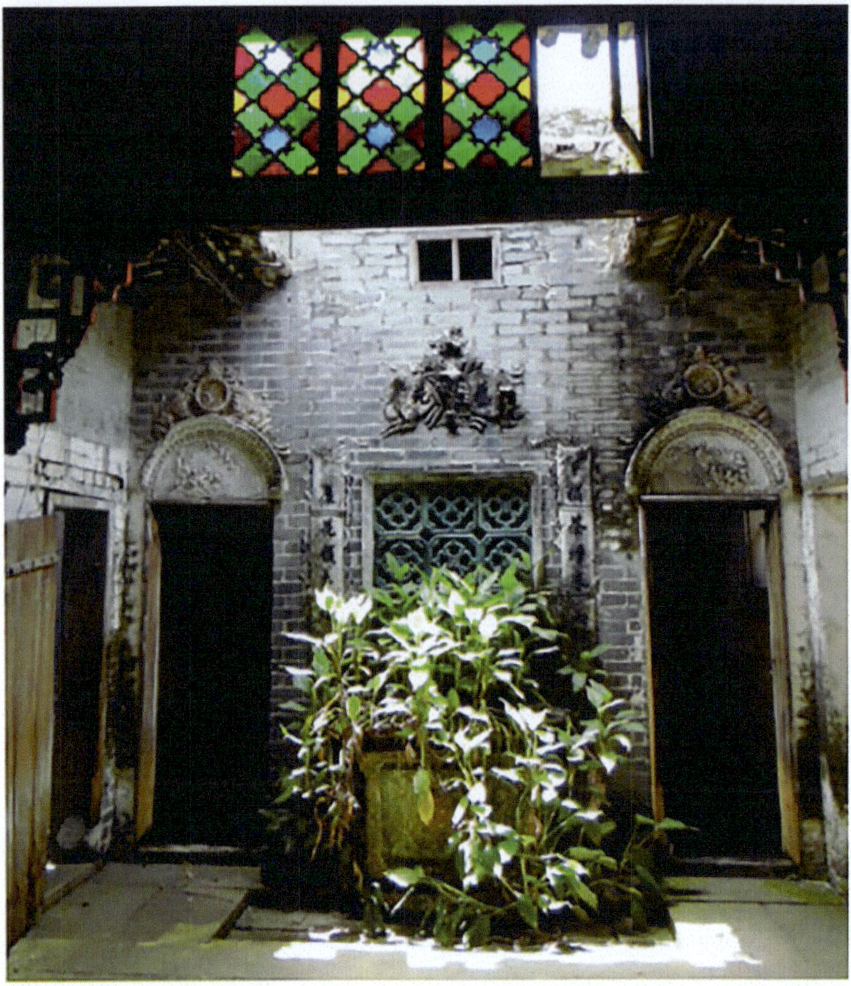

Fig. 50 Stained glass in the atrium of Yichu Ancestral Hall, Tangjiawan. (*Image source* photographed by the authors)

In terms of landscape vignettes, the combination of Western Garden elements and traditional features of Tangjiawan forms a unique landscape that combines Chinese and Western elements. Lawns as an important element of Western gardens, a large number of modern gardens in Tangjiawan, such as the Qixia Hall lawn area accounted for nearly half of the area, with the regular plant configuration; geometric pools in the

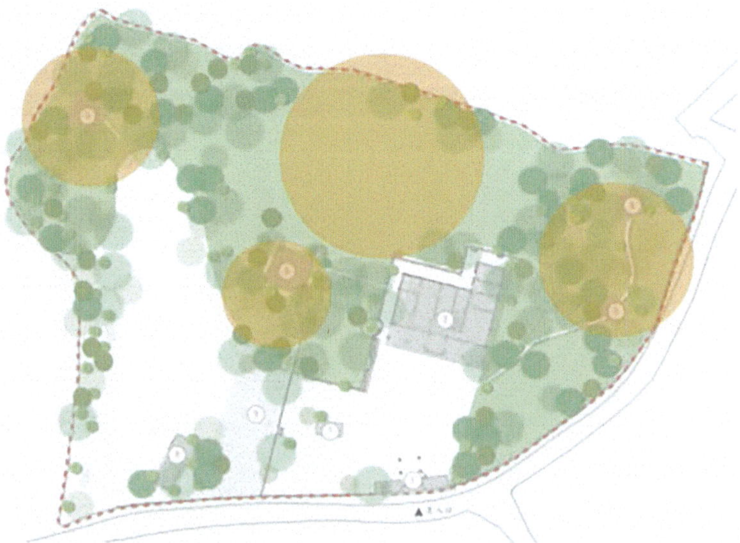

Fig. 51 Qixia Hall uses tree groups to divide the lawn. (*Image source* drawn by the authors)

Lingnan garden is common, the Western concept of the layout to further strengthen this aesthetic, so that it is better coordinated with the spatial relationship; Western sculpture and fountain combination of techniques, but also applied to the Tangjiawan garden, such as Qixia Hall Guanyin pool of sculpture, echoing the horizontal axis of the Hall, and the pool cleverly combined to create a pleasant vignette. The sculpture echoes the horizontal axis of the hall and is skillfully combined with the pool to create a pleasant landscape.

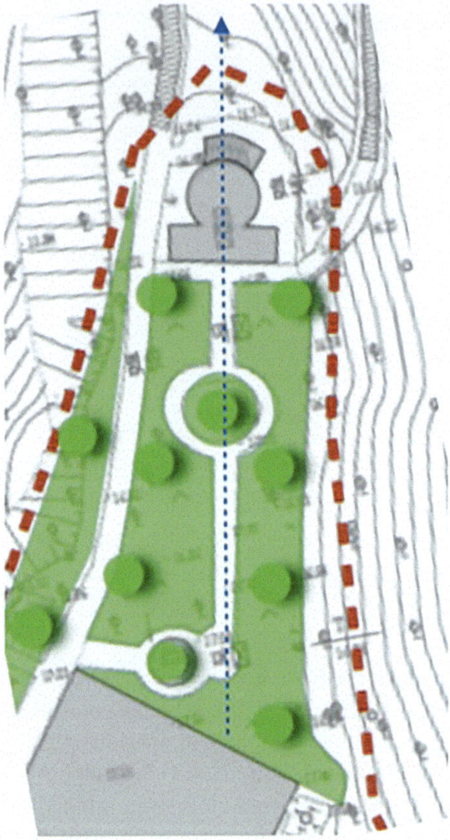

Fig. 52 Path of Stargazer Pavilion in Gongle Garden. (*Image source* redrawn according to the planning information of Zhuhai High-Tech Zone)

5 Conclusion

During the modern period in Lingnan, invasion and resistance, inheritance, and introduction coexisted, and the modern garden in Tangjiawan flourished by taking advantage of its geographical and humanistic location and became an important part of the history of gardening in modern Lingnan. However, the war and urbanization process caused a large number of gardens to fall into disrepair, and there is a lack of related research. This study finds that: in terms of landscape and architectural layout, influenced by the Lingnan gardens, the modern gardens in Tangjiawan present three types of architectural layouts, such as buildings around the garden, and the landscape space is divided into traditional, East-meets-West, and Western-style; in terms of the gardening techniques, stacked mountains and rocks are used in the large-scale gardens; the water management integrates Chinese and Western characteristics, with the regular geometrical pools as the main features, and the plantings are functional and ornamental, and the Western-style hedges and lawns are introduced; in terms of the landscape elements, both wood carvings and wood sculptures are retained, and the landscape elements are not only retained, but also used as a part of the modern gardens. On the landscape elements, not only retaining the traditional Lingnan crafts such as wood carving and gray molding but also integrating western architectural elements; on the use of functions, from closed to open, and the service object is more publicized. In conclusion, the diversified landscape characteristics of modern gardens in Tangjiawan are the result of the joint action of the natural environment, the development of the times, and the changes in aesthetics and lifestyles, showing the inclusiveness of the Lingnan culture and the innovative spirit of the people of Tangjiawan, and the present study also provides an important basis for its conservation and utilization.

This paper is the first to systematically explore the landscape characteristics, gardening techniques, and their influencing factors of the existing modern gardens in Tangjiawan; with the modern gardens in Tangjiawan as the research object, it fills in some of the blanks of the research system of Lingnan gardens and landscapes; and lays the research foundation for the protection, inheritance, and utilization of the modern gardens and landscapes of Tangjiawan, which is of important practical reference value. In the face of the rapid development of modern technology and materials, the modern garden has failed to effectively inherit regional characteristics and humanistic connotations. This study is expected to arouse attention to the gardening methods and landscape characteristics of modern gardens in Tangjiawan and to build a firm cultural foundation for the future development of the gardens and landscapes in Tangjiawan and Zhuhai. However, due to the constraints of anonymity of the original data and destruction of the site, it is difficult to guarantee the comprehensiveness of the research scope and the accuracy of the survey results, and the discussion of the modern garden landscape of Tangjiawan in this paper is inevitably deficient.

Funding This research received no external funding.

Institutional Review Board Statement Not applicable for studies not involving humans or animals.

Informed Consent Statement Not applicable for studies not involving humans.

Conflicts of Interest The authors declare that the research was conducted in the absence of any commercial or financial relationships that could be construed as a potential conflict of interest.

References

1. Song, J., Chen, Y., Zheng, L., & Tang, Q. (2025). The evolution of traditional ancient towns in South China: Tangjiawan Ancient Town in Zhuhai city. *Journal of Asian Architecture and Building Engineering, 24*(3), 2114–2132. https://doi.org/10.1080/13467581.2024.2349754
2. Song, J. (2022). Analysis of landscape characteristics of Lingnan private gardens from the perspective of landscape Gene: Tangjiawan Park in Zhuhai as an Example. *Journal of Environmental & Agricultural Studies*, *3*(3). https://doi.org/10.32996/jeas.2022.3.3.3
3. Sheng, N., & Tang, U. W. (2013). *Zhuhai Cities, 32*, 70–79. https://doi.org/10.1016/j.cities.2013.02.006

4. Deng, Q. S. (1985). The history of gardens. *Guangdong Landscape Architecture, 1*, 1–10.
5. Yan, J. S., & Chen, G. L. (2001). Creation of landscape environment in residential areas: Reflections on the environmental design of "Jiangnan Garden" community. *Chinese and Overseas Architecture, 4*, 17–18.
6. Peng, Y. G. (1986). *Analysis of Chinese classical gardens*. China Architecture & Building Press.
7. Changshi, X., & Bozhi, M. (1963). On Lingnan gardens. *Journal of Architecture, 3*, 11–14.
8. Yang, H. L. (2001). Characteristics of water management and landscaping in Guangzhou classical gardens. *Journal of Guangzhou University, 3*, 36–40.
9. Wu, Y. F. (2006). The art of rock modeling in Lingnan gardens. *China Construction Information, 13*, 22–25.
10. Zhou, L. (2011). *History of modern gardens in Guangdong*. China Architecture & Building Press.
11. Lu, Q. (2015). *Lingnan gardening and aesthetics*. China Architecture & Building Press.
12. Meng, Z. (2011). Focus on Guangdong gardens of three generations. *Guangdong Gardens, 3*, 4–8.
13. Yuanding, L., & Yanjun, W. (1981). Four gardens in central Guangdong. *Collection of research papers on the history of Chinese gardens* (Volume 1), 101–115.
14. Huang, H. (2013). On the historical value and protection of Lingnan ancient village buildings - taking Huitong Village in Zhuhai as an example. *Agricultural Archaeology, 3*, 216–220.
15. Yisan, R. (2015). Protection plan for Tangjiawan historical and cultural town. *Urban Planning*, 19.
16. Wang, Z., Zheng, R., Tang, J., Wang, S., & He, X. (2025). The aesthetic imagery of traditional garden door and window forms: A case study of the four major traditional gardens of Lingnan. *Buildings, 15*(4), 513. https://doi.org/10.3390/buildings15040513
17. Chen, C., & Zuo, J. (2018). The symmetrical beauty and its cultural connotation of Lingnan architecture. In *2018 International Conference on Management, Economics, Education, Arts and Humanities (MEEAH 2018)* (pp. 143–149). Atlantis Press. https://doi.org/10.2991/meeah-18.2018.25
18. Zheng, W. (2019). Analysis of gardening characteristics and style formation of Lingnan gardens. *Journal of Landscape Research, 11*(5), 130–134. https://doi.org/10.16785/jissnl943-989x.2019.5.028
19. Zhang, J., Xie, K., & Xu, Y. (2024). Study on the Plant Landscape Creation and Cultural Connotation of Lingnan Gardens. *OAJRC Environmental Science, 5*(2). https://doi.org/10.26855/oajrces.2023.12.002
20. Lai, H. M. (2007). *Zhongshan, Zhuhai, and Macau: Geographical and Historical Notes.*
21. Liu, Z., Ou, T., Chen, X., & Fu, C. (2019). Research on the value orientation and promotion path of rural revitalization in Zhuhai against the background of Guangdong-Hong Kong-Macao greater bay area. In *the 2nd International Conference on Contemporary Education, Social Sciences and Ecological Studies (CESSES 2019)* (pp. 838–842). Atlantis Press. https://doi.org/10.2991/cesses-19.2019.189
22. Yan, L., & Chen, Y. (2021). Stained glass in Lingnan historic buildings and its application in Manchuria windows. *Journal of Physics: Conference Series, 1885*(3), 032052. IOP Publishing.
23. Chen, Y. (2019). The development of decorative art in Lingnan house garden in Macau—Taking Lingnan oyster shell window and manchuria window as examples. In *3rd International Conference on Art Studies: Science, Experience, Education (ICASSEE 2019)* (pp. 479–484). Atlantis Press.
24. Wu, L. (2014). Discussion on the art of Lingnan Manchuria window. In *International Conference on Education, Language, Art and Intercultural Communication (ICELAIC-14)* (pp. 502–505). Atlantis Press.
25. Wang, X., & Du, Y. (2024). Research on Green glass of Cantonese colored windows based on color model. In *International Conference on Algorithm, Imaging Processing, and Machine Vision (AIPMV 2023)* (Vol. 12969, pp. 308–315). SPIE.

26. Du, Y., Wang, X., & Liu, J. (2022). Extraction and application standard of "Red" color model in Cantonese color windows. In *2022 4th International Conference on Frontiers Technology of Information and Computer (ICFTIC)* (pp. 712–717). IEEE.
27. Shuyuan, Z., Junling, Z., & Pohsun, W. (2023). Research on the decorative patterns of modern residential buildings in Macao from the perspective of social change. *Frontiers in Art Research, 5*(1).

Wharf Architecture in Macau: The Typological Transformation from Timber Structures to Reinforced Concrete

Liang Zheng, Yile Chen, and Jianyi Zheng

Abstract This chapter takes the 17 wharves along the Inner Harbour of the Macau Peninsula from Barra (媽閣) to Lam Mau Tong (林茂塘) as the objects of study, systematically revealing the coupled evolution of port–city relations, industrial organization, architectural types, and governance mechanisms. In the first half of the twentieth century, light industries such as firecracker production and passenger–cargo shipping routes shaped the Inner Harbour into a typical maritime cultural landscape. From around 1948 to 1953, the typology rapidly shifted from timber pile trestles to reinforced-concrete "box-shaped" structures. Starting in 1966, cold-chain facilities were embedded and the functional orientation transformed into a composite of fishery, refrigeration, wholesale, and catering uses. In the 1980s, the government promoted unified roofing and land concession reconstruction, standardizing 34 old-style wharves, institutionally nesting with the public passenger transport franchise (1953–1964), differentiated supervision of hazardous goods, and shoreline governance. In terms of scale, the majority had a frontage width of 15–33 m with dual berths as a typical configuration, and a unit berth width of 7–12 m. Based on these findings, this research proposes feasible conservation strategies such as compatibility restoration using washed granolithic plaster and masonry for "restoration in original style," reinstatement of platform drainage and fender components, and preservation of parapet inscriptions and Sino-Portuguese nameplates. It further suggests a combined governance framework integrating functional zoning, municipal facilities, and revitalization, to support the holistic conservation of the historical shoreline of the Inner Harbour and the reconstitution of its waterfront public character.

Keywords Macau inner harbour · Wharf architecture · Port–city relations · Shoreline governance · Industrial heritage revitalization

L. Zheng · Y. Chen (✉) · J. Zheng
Faculty of Humanities and Arts, Macau University of Science and Technology, Taipa, Macau, China
e-mail: chenyile@must.edu.mo

Y. Chen et al. (eds.), *City-Sea Symbiosis*, Springer Geography,
https://doi.org/10.1007/978-3-032-13725-8_8

1 Introduction

1.1 Research Background

The Inner Harbour of Macau is a typical arena where the port–shipping network of the Pearl River Estuary in modern times co-evolved with the urban fabric. The shoreline from Barra to Lam Mau Tong forms a cluster of wharves characterized by small-scale passenger and cargo transport and multilayered social networks. Based on prior field surveys of each numbered wharf (also known as Ponte) in the Inner Harbour, this study establishes a research approach centered on architectural mapping, historical investigation, and value assessment. At the levels of urban development, industrial organization, and maritime transportation, the Inner Harbour wharves served not only as passenger–cargo nodes between Macau and the western bank of the Pearl River in Guangdong, but were also key logistical nodes for Macau's fireworks and firecracker industry in the first half of the twentieth century. Beginning in the 1920s, the firecracker industry rapidly clustered and expanded in areas such as Taipa, generating sustained demand for loading, storage, and transshipment at the Inner Harbour. The rise and decline of this industry exerted long-term influence on functional allocation, route scheduling, and spatial usage of the wharves [1–4]. The firecracker industry once served as Macau's pillar industry [5], with a tightly coupled relationship between its production–transport–export chain and the Inner Harbour wharf system.

From the perspective of governance and construction systems, in the 1980s the government promoted the modernization of the Inner Harbour shoreline [6, 7], proposing that 34 old timber or simple-structure wharves along the approximately 1,400-m shoreline from Barra to Lam Mau Tong be gradually unified into standardized reinforced-concrete wharf buildings. In principle, reconstruction was to be self-funded by the original wharf owners, while the government implemented shoreline order and functional upgrading through land concession and regulatory control. Among them, a few wharves such as Wharf No. 21 were modernized relatively early, whereas others such as Wharf No. 8 and Wharf No. 11 did not submit unified reconstruction plans at that time due to either their prior use of advanced concrete construction systems or considerations of profitability and usage evaluation. During this stage, institutions such as maritime police inspection posts and fire stations were also embedded in the wharf system, indicating that the Inner Harbour wharves were gradually transforming from single-function passenger–cargo transfer nodes into governance spaces combining law enforcement, safety, community services, and commerce.

In the context of international academic discourse, the Inner Harbour wharves can be regarded as a typical component of maritime cultural landscapes—a concept emphasizing the holistic perception of maritime activities and onshore facilities and advocating interdisciplinary approaches to understanding the spatial and cultural unity of ports, navigation, settlements, and techniques. Classical theories of port–city relations interpret the cyclical migration of ports—coastline attachment, offshore relocation, and regeneration—from the dual dimensions of spatial form evolution and institutional reform [8, 9]. These theories help interpret the coupled mechanisms among morphology, function, and governance in old ports and historical shorelines like the Inner Harbour, providing methodological references for comparative studies of East Asian historical harbors and maritime cultural history under Portuguese administration. In terms of conservation practice and evaluation frameworks, Inner Harbour wharves are not only architectural heritage but also closely related to industrial heritage values. Internationally, the "Dublin Principles" issued jointly by ICOMOS (International Council on Monuments and Sites) and TICCIH (The International Committee for the Conservation of the Industrial Heritage), and the "*The Nizhny Tagil Charter for the Industrial Heritage*", propose principles of identification, documentation, minimal intervention, and adaptive reuse of industrial heritage, providing an operational framework for the conservation and renewal of waterfront industrial facilities and port-related remains [10].

In summary, the necessity of studying the wharf architecture of Macau's Inner Harbour lies in the highly coupled evolution of shoreline governance, industrial organization, and construction technology from the first half of the twentieth century to the 1980s. Through architectural history research on 17 case wharves in the Inner Harbour, this study provides solid micro-level data supporting an integrated empirical narrative across port, industry, architecture, and governance dimensions. Meanwhile, by incorporating international theories and conservation guidelines, the Inner Harbour wharves can be placed within a comparative framework of global port–city transformation and industrial heritage revitalization, offering practical technical approaches and policy recommendations for the holistic conservation of Macau's Historic Urban Area and the reconstruction of waterfront public spaces.

1.2 Research Object

This study takes the wharf buildings and platforms along the Inner Harbour shoreline of the Macau Peninsula as its core objects, focusing on 17 historical wharves distributed along Avenida de Demétrio Cinatti (爹美刁施拿地大馬路) and Rua da Prainha (火船頭街) that still retain identifiable superstructures or platform remnants. The selection criteria are as follows: (1) located along the main navigation channel of the Inner Harbour and directly adjacent to the urban street grid; (2) associated in urban history and oral history sources with freight transport, passenger transport, fisheries, warehousing, and commercial–industrial activities, reflecting typological evolution; and (3) historically or architecturally representative wharves. As shown in Fig. 1, this study covers 17 wharves including No. 8, 11, 20, 21, 22, 22A, 23, 25, 26, 27, 28, 29, 30, 31, 31A, 33, and 34. Spatially, the study area forms a linear sample belt along the shoreline from Wharf No. 8 at the southern end (located west of Rua de Belchior Carneiro) to Wharf No. 34 at the northern end (near the north end of Avenida da Marginal do Lam Mau).

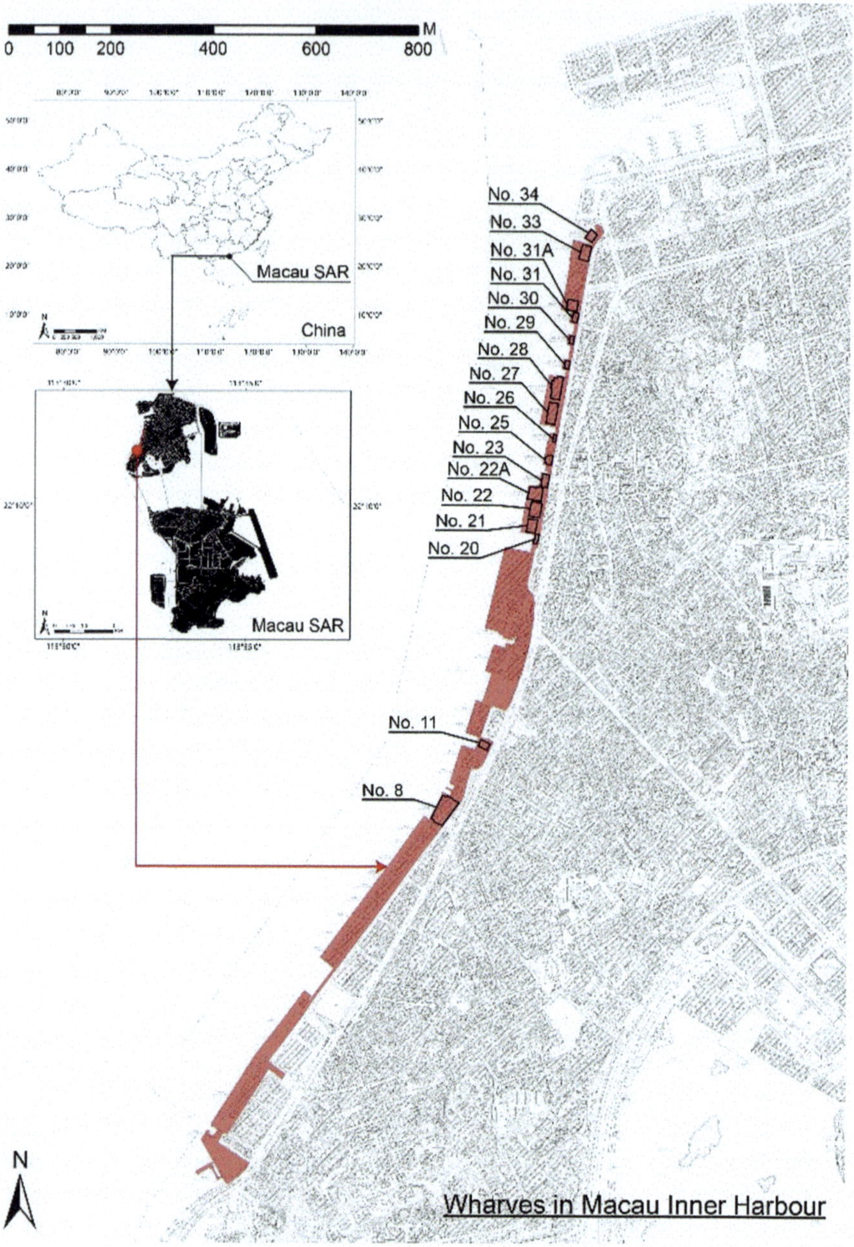

Fig. 1 Locations of Wharves in Macau Inner Harbour and research objects. (*Image source* Drawn by the authors)

The current conditions of the Inner Harbour wharves are shown in Fig. 2, with the temporal scope spanning from the twentieth century to the present. For example, Wharf No. 11, built in 1933, is an early representative; most wharves underwent reconstruction or functional transformation between the mid-twentieth century and the 1980s, exhibiting a chronological sequence of "old timber platform → reinforced-concrete superstructure → later adaptive reuse." The research content focuses on three scales: (1) Construction scale of individual buildings and platforms, such as width, number of births, and structural materials; (2) Street interface and land tenure/operating entity scale, including property rights and usage patterns; (3) Shoreline continuum scale combining landscape and social functions, including roles as shipping nodes, cold-chain/fishery facilities, and urban revitalization potential.

The basic information of the 17 wharves is shown in Table 1. Spatially, they form a linear sample belt from south to north, with Wharf No. 8 as the southern node (approximately 1 m from the South Sampanas Wharf (Cais de Sampanas Sul) (南舢舨碼頭) and about 3 m from No. 7A) and Wharf No. 34 as the northern node (approximately 20 m in width, with significant silt deposition). In terms of chronology and type, the sample covers a full structural spectrum from early constructions to modern times. Among them, Wharf No. 11, built in 1933, is an early dedicated cargo wharf (reinforced-concrete structure, 10 m wide, 2 berths, with Art Deco elements). Wharf No. 26 still preserves a timber structure (14 m wide, 2 berths). Wharves No. 22, 23, 25, 27, 30, and 31 are mostly modern reinforced-concrete wharves, with widths concentrated between 15 and 33 m and generally featuring 2 berths, presenting standardized dimensions and a continuous shoreline interface. In terms of functions and operational types, the sample includes historical freight/warehousing functions (e.g., Wharves No. 25, 29, and 30), fishery support and seafood retail functions (e.g., Wharves No. 31 and 31A), cold-chain-related industries (e.g., Wharf No. 33), as well as social public services (e.g., Wharf No. 20). In terms of detailed morphology and adjacency relationships, Wharf No. 28 is located between Wharves No. 27 and 29 (with a north–south spacing of approximately 7–25 m), and Wharves No. 31 and 31A are separated by about 2 m, forming a composite interface with the Mercado do Patane (沙梨頭街市綜合大樓).

1.3 Research Methods and Sources of Information

The research methodology of this study is oriented toward a comprehensive survey and assessment of the 17 wharves along the Macau Inner Harbour. Its aim is to establish a systematic interpretation of structural materials, typological features, locational context, and risk exposure based on verifiable on-site data and standardized drawings, combined with statistical information from authoritative agencies. The overall research design adopts a parallel multi-case approach using a uniform framework for data collection and analysis. Within a consistent spatiotemporal and scale framework, the study carries out on-site investigation and architectural surveying, verification and redrawing of historical drawings, documentary and archival research, integration of

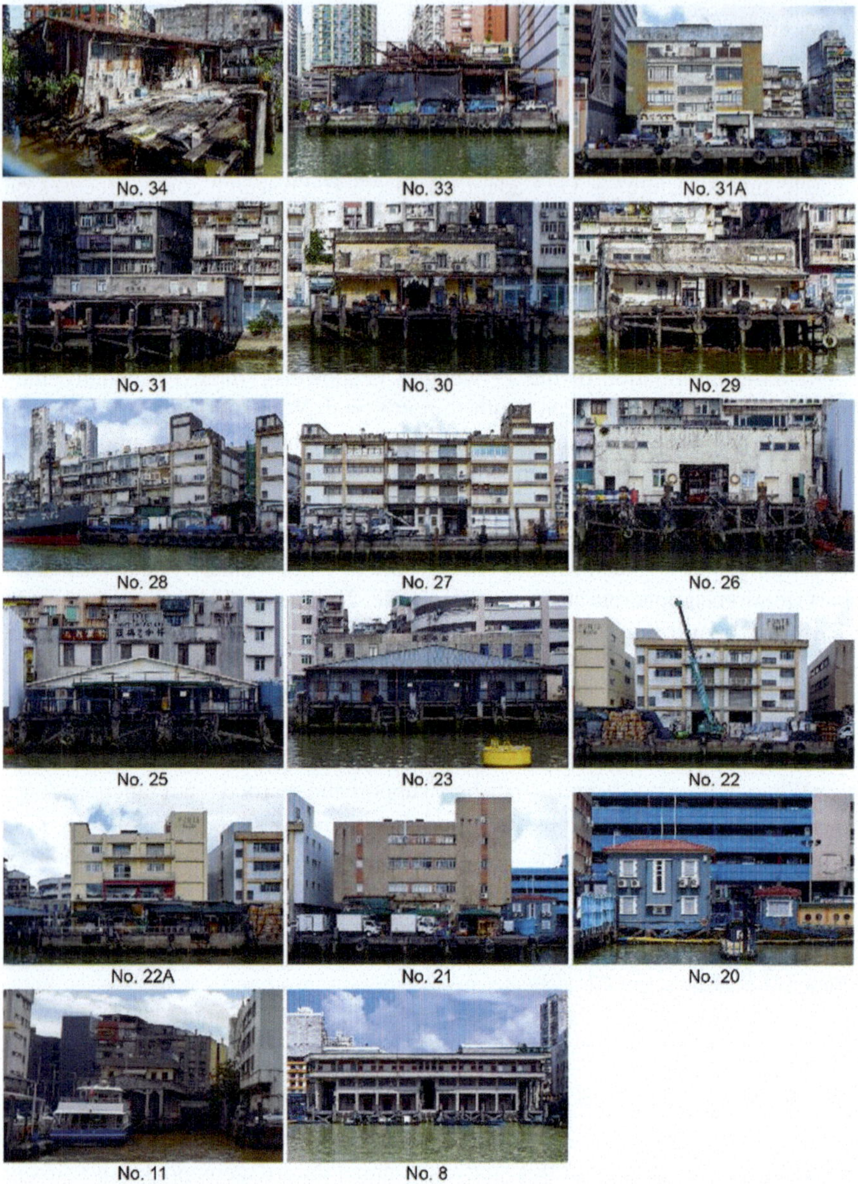

Fig. 2 Facades of inner Harbour Wharf buildings. (*Image source* Photographed by the authors)

Table 1 Basic information of the 17 inner Harbour Wharves in this study

Wharf no	Alias	Function	Description
8	Tai Fung Warehouse Wharf (大豐倉碼頭)	Formerly used for warehousing, temporary medical relief, shrimp peeling, and fish processing; currently occasionally used for cultural events, mostly vacant	Located west of Rua de Belchior Carneiro (比厘喇馬忌士街); about 1 m north of South Sampan Wharf (南舢舨碼頭), about 3 m from Wharf No. 7A; southern endpoint of the linear sample
11	Kwong Hing Tai/Cheng Cheong company firecracker factory Wharf (廣興泰/禎祥公司爆竹廠專用)	Formerly used by Kwong Hing Tai Firecracker Factory (廣興泰爆竹廠) and Cheng Cheong Passenger Service (禎祥客運); currently restricted to commercial service use	Built in 1933; reinforced-concrete structure; 10 m wide with 2 berths; located west of Rua da Prainha (火船頭街)
20	Former Maritime Police Inspection Post/ Community visit station (醫生局訪問所/火船頭訪問所/水警稽查總部/海傍水警署/內港水警訪問所)	Formerly long used as community support service building; currently vacant	Original site of the maritime police inspection post; historically functioned as a public service building; currently unused
21	Ping on Wharf (平安碼頭)	Formerly used for passenger–cargo transport and fuel transfer with firefighting function; currently seafood retail	Reinforced-concrete structure; 30 m wide with 2 berths; currently used for seafood retail; high revitalization potential
22	Tak on Wharf/Public Welfare Wharf (德安碼頭/公益碼頭)	Formerly Macau–Hong Kong/Macau–Jiangmen/ Macau–Qi'ao shipping routes, grain unloading, gambling boat berthing; currently for Macau–Hong Kong container transport	Reinforced-concrete structure; 33 m wide with 2 berths; adjacent to No. 22A and No. 21
22A	San Wo Fat Wharf (新和發碼頭)	Ground floor retail units with dining and car washing	Four ground-floor units (including Wo Fat Seafood Hotpot Restaurant (和發碼頭海鮮火鍋); adjacent to 22 and 23
23	Iao Tak Wharf (祐德碼頭)	Formerly office of Pico Food Company (比高食品) and others; currently vacant	Located opposite a parking building; formerly enterprise office; now idle; only 1.7 m from No. 22A

(continued)

Table 1 (continued)

Wharf no	Alias	Function	Description
25	U Wa Fat Kee Wharf (裕和發記碼頭)	Formerly passenger–cargo and oil transport between Macau–Hong Kong, Jiangmen, Shiqi; currently vacant	18 m wide with 2 berths; Inner Harbour pumping station added between No. 25 and No. 26 in 2017; surrounding area dominated by seafood stalls and fish trading
26	Pou Heng Wharf (寶興碼頭)	Formerly multipurpose wharf for shipping and fish loading; currently façade closed, used for fishing boat berthing	Timber structure; 14 m wide with 2 berths; adjacent to No. 27 and No. 25
27	Tai Shing Wharf (大誠碼頭)	Formerly Macau–Hong Kong shipping and knitted goods freight; currently in operation with fish stalls/ice storage/logistics	Good building quality, large area, flexible spatial layout, high potential for reuse
28	Hop Lei Godown Wharf (合利貨倉碼頭)	Formerly seafood transport wharf for Macau–Hong Kong shipping; rebuilt with cold storage; currently for fisheries and cold-chain warehousing	Located between No. 27 and No. 29; approx. 7–25 m north–south spacing
29	Chuen Lei Wharf (泉利碼頭)	Fishing boat berthing and commercial frontage from past to present	15 m wide with 2 berths; central partition wall; adjacent to No. 28 and No. 30
30	Hip Lei Wharf (協利碼頭)	Formerly used for Macau–Hong Kong freight; now for fishing boat berthing	Reinforced-concrete structure; approx. 19 m wide with 2 berths; three bays at the frontage
31	Iat Hing Wharf (義興碼頭)	Former major freight hub; now only for fishing boat berthing; registered operator Mei Kei Transport (美基貨運)	18 m wide with 2 berths; used exclusively for fishing boat berth; south of No. 30, north of No. 31A
31A	Kuan Kei Fish Stall (權記魚欄)	Formerly converted from Patane Health Station (沙梨頭衛生站) to fish stall/society office; now fresh fish wholesale and retail, also general freight	About 2 m from No. 31; adjacent to Mercado do Patane Complex Building (沙梨頭街市綜合大樓)

(continued)

Table 1 (continued)

Wharf no	Alias	Function	Description
33	Hap Tak Wharf (合德碼頭)	Formerly ice factory with cold storage; now still in use by Macau Ice Factory and Kam Wa Fish Stall (金華魚欄)	North of Wharf No. 34, south of Patane Market Complex; close shoreline continuity
34	Hoi Lei Wharf (海利碼頭)	Historically handled multiple freight routes; currently damaged and vacant	Approx. 20 m wide; intertidal siltation and environmental issues; formerly used for building materials, fuel, and freshwater fish trading

Source The author based on on-site surveys

visual materials, environmental and risk assessment, and value evaluation, ensuring comparability among the 17 case studies and objectivity of the conclusions. The research methods and process are described as follows:

(1) Data collection is centered on in situ observation and architectural surveying, focusing on recording bay width and depth, story height, component dimensions and joint construction methods, material and finish types, façade stylistic vocabulary, and current spatial usage. Standardized photography is used to document details and establish a numbered catalog system to ensure one-to-one correspondence between images and textual descriptions. The surveying output includes current-condition orthographic elevations, side elevations, sections, and necessary detailed construction drawings, which serve as base maps for subsequent historical reconstruction and comparative analysis.
(2) Verification and redrawing of historical drawings aim to reconstruct the original form and structural relationships of each wharf. Historical plans, elevations, sections, and site diagrams related to the wharves are collected and analyzed. Identifiable structural axes, dimensions, and component vocabularies are digitally redrawn at equal scale, with explicit annotation indicating reconstruction from historical sources. These redrawn maps are then overlaid at the same scale with current surveying drawings to identify major alterations such as extensions, partial demolitions, and finish replacements. This provides direct evidence for typological classification and value assessment.
(3) Documentary and archival investigation is used for external verification of locational functions, land-use attributes, port elements, and transport accessibility. Through authoritative sources such as port and maritime information from the Marine and Water Bureau, legislative texts and gazettes from the Printing Bureau, and public transport route data from the Transport Bureau, the study cross-checks the structural attributes, number of berths, functional license types, and surrounding transportation conditions of each wharf. When

necessary, historical functional transformations are chronologically sorted using a time-series approach.

2 Regional History and Port–Navigation System

2.1 Formation and Shoreline Evolution of the Inner Harbour

The Inner Harbour (内港) is located on the western coast of the Macau Peninsula. In earlier periods, the shoreline stretched approximately two kilometers from Barra (媽閣) in the south to Patane (沙梨頭) in the north, forming a typical estuarine inner bay under the combined effects of sediment deposition from the Pearl River Estuary and tidal dynamics. With the superimposed influence of long-term siltation and human activities, the officially numbered wharves now extend from No. 34 at Lam Mau Tong (林茂塘) to No. 1 at Barra Pier (媽閣塘), covering a total length of approximately 3.5 km. This indicates that the Inner Harbour shoreline underwent significant seaward extension and linearization in modern times. The shoreline advancement and wharf replacements in the North Bay area (内港北灣) directly stimulated the formation of adjacent urban districts such as Hia Van (下環) and Patane (沙梨頭), along with the concentration of commercial functions.

In the late nineteenth century, the Portuguese administration successively appointed engineers to propose Inner Harbour regulation and dredging schemes. In 1883, Adolfo Loureiro submitted the *Draft Project for the Improvement Works of the Port of Macau*, advocating dredging and embankment works to improve sedimentation and berthing conditions. Subsequently, in 1889, a reclamation contract was signed with the Isle Verde Cement Company (青洲英坭公司), and around 1890, the proposal of connecting Ilha Verde (青洲) by embankment for Inner Harbour regulation was put forward. The embankment and land reclamation along Ilha Verde altered tidal channels and sediment circulation pathways, marking a major turning point in the shoreline evolution of the Inner Harbour. In 1892, Ilha Verde was connected to the peninsula by an embankment, further changing the hydrodynamics of the North Bay and the trajectory of shoreline evolution.

From the early twentieth century to the 1920s, Inner Harbour underwent its first systematic phase of reclamation. The engineering approach adopted a combined "dredging and hydraulic fill" technique, directly consolidating dredged mud to form land. This process eliminated the irregular oyster beds and mudflats along the Inner Harbour, creating a straightened shoreline. Upon completion, three major sub-areas of the Inner Harbour were linked, leading to the emergence of early waterfront arcade commercial streets and warehouse districts, which in turn shifted the city's commercial center westward. Specifically, in the North Bay, reclamation between Wharf No. 23 and the western end of Avenida de Almeida Ribeiro (新馬路) was completed after 1925, and a new road—named Avenida de Demétrio Cinatti (爹美刁施拿地大馬路) after the then Harbor Authority Director—became the linear interface of the western shoreline and a wharf frontage. Correspondingly, the inland traffic axis, Avenida de Almeida Ribeiro (新馬路), was substantially completed in 1918 and continued to be developed in the 1920s, establishing the urban axis linking Nam Van (南灣) and the Inner Harbour.

With shoreline straightening and expansion, port financing and urban governance were crucial to infrastructure delivery. The substantial funds required for reclamation and port regulation were mainly sourced from colonial revenues such as the opium tax and gaming tax. Some Outer Harbour (外港) engineering projects were contracted to Dutch port engineering companies, while Inner Harbour works were coordinated and subcontracted by departments under the Land and Urban Construction Bureau (工務局). Given that the Inner Harbour was built on an alluvial foundation and reclaimed over oyster fields, limitations in early technology and fill quality meant that areas such as Ilha Verde–S. Lourenço (青洲—新橋—下環) became low-lying flood-prone zones subject to inundation during heavy rains. This geomorphological–engineering coupling effect profoundly influenced the subsequent pace of wharf infrastructure maintenance and renewal.

From the mid-twentieth century onward, as vessel types and cargo volumes increased, traditional Inner Harbour channel dimensions and berth capacity gradually failed to meet the requirements of modern large ships. In the 1990s, the Portuguese administration of Macau proposed the "Second Phase" dredging and reclamation plan for the Inner Harbour, following a technical approach of "channel dredging to improve water flow and reduce siltation, combined with localized reclamation to adjust the shoreline." This development coincided with the broader reclamation trends of the Pearl River Estuary, reflecting a rebalancing process within the port–city–estuarine system during regional transformation.

In summary, the evolution of the Inner Harbour shoreline can be understood along three parallel timelines: (1) A natural geomorphological line dominated by estuarine siltation and channel migration; (2) An engineering line of "port improvement–shoreline linearization–land reclamation" proposed continuously since the 1880s and intermittently implemented; (3) An urban development line shaped by the formation of city axes and waterfront districts, such as Avenida de Almeida Ribeiro (1918) and Avenida de Demétrio Cinatti in the 1930s. The interweaving of these three trajectories expanded the Inner Harbour shoreline from its original two kilometers to the present-day 3.5-km "combined harbor–continuous waterfront" system. This evolution, driven by local fiscal mechanisms and regional trade, catalyzed the typological transformation of wharves from timber piles and stone foundations to reinforced-concrete structures, leading to their standardized reconstruction.

2.2 *Port Administration Policies and the Concession System*

As a key interface between colonial urban governance and maritime order, the institutional logic of the Inner Harbour (內港) wharves was shaped by three major policy threads:

(1) Continuity of Administrative Authority. Since 1905, the port administration authority *Capitania dos Portos* was headquartered at the Moorish Barracks (Now the headquarters of the Marine and Water Bureau) at the foot of Barra Hill (媽閣山) and was responsible for supervising vessel traffic, patrolling waters, and issuing administrative permits in Inner Harbour. This institutional role continues today under the Marine and Water Bureau (Direcção dos Serviços de Assuntos Marítimos e de Água, DSAMA), reflecting the continuity and functional translation of Macau's port governance from the colonial period through the post-handover era.
(2) Route Concessions and Contract Operations. Taking public passenger transport between the Macau Peninsula, Taipa, and Coloane as an example, the establishment of the *Passenger Ferry Service between Macau, Taipa, and Coloane* (Carreira de Barcos de Passageiros entre Macau, Taipa e Coloane) in 1953 marked a milestone in institutionalized maritime transport. In 1963, the government formalized route operations under a concession (concession) model. However, in 1974, following the opening of the Macau–Taipa Bridge, improved land connectivity resulted in the termination of this waterborne passenger transport concession in the same year.
(3) Shoreline Governance Framework. In the 1980s, to standardize shoreline order and adapt to industrial restructuring, the Portuguese administration in Macau proposed the *Inner Harbour Wharf Reconstruction Plan* within a 1,400-m shoreline area from Barra (媽閣) to Lam Mau Tong (林茂塘). The plan aimed to unify 34 traditional wooden or simple-structured wharves into reinforced-concrete superstructure buildings. Reconstruction was to be self-funded by the original wharf owners, while additional land expansion was granted through a concession-of-land mechanism known as land concession, enabling unified superstructure construction.

Regarding the operational mechanism of concessions, during the wartime fuel price surge in 1943, the original six daily electric ferry services between Macau–Taipa–Coloane were reduced to four, rescheduled to depart at 8:00 a.m. and 3:00 p.m. This linkage between "energy–fare–service frequency" revealed the operational vulnerability of concession-based services under cost fluctuations. In July 1953, the franchise for the Macau–Taipa ferry service was awarded to Lei Chek Wai (李植槐) and Chan Wing-Hei (陳榮熙). They leased three electric ferries—*Kwong Cheng* (廣禎), *Kwong Chiu* (廣照), and *Kwong Kwong* (廣光)—under contract, with the first concession period running from November 1, 1953 to October 31, 1958. The contract specified that if the government renewed the concession, the original lessor must continue leasing the vessels. If the lessee withdrew early, 10% of voyage revenue would be deducted as compensation. The contract also defined clear divisions of responsibility for net profit sharing (30%, settled semi-annually), as well as insurance, crew, repairs, fuel, berthing, and utilities. Archival documents indicate that the company moved into Wharf No. 8 on October 30, 1953 and officially commenced service on November 1, illustrating a governance pathway of "concession–wharf occupation–commencement of service" linking administration and spatial allocation.

Parallel to the concession system was differentiated regulation over hazardous materials and industrial cargo. Finished fireworks produced in Macau could not be exported on mixed cargo vessels and had to be transported by designated ships. In the 1940s, the government established a Hazardous Materials Production Safety Committee, which required annual factory license renewal, and imposed multiple levels of review and supervision on hazardous goods importation. These differentiated regulations had direct implications for loading organization and storage layout at Inner Harbour wharves. They contributed to the adoption of a unified superstructure typology in the 1980s: "two lower floors as warehouse, two upper floors as offices," forming a coupled mechanism of "regulation–structure–function."

In terms of land and shoreline governance, during the colonial period, the institutional mechanism of "land concession (批地)–lease granting (批給)" became a key tool for port–shore spatial development. The 1980s reconstruction plan applied a combined strategy of "self-funded rebuilding + expansion land concession" along the 1,400-m shoreline. From a longer historical perspective, the *Draft Project for the Improvement Works of the Port of Macau* (1884), the *Plan of Port Requirements of Macau* (1912), the *Master Plan for Macau Artificial Port Engineering* (1922), and the *Prosperity Plan* (1946–1949) together established an institutional trajectory of "land reclamation–dredging–shoreline regulation–specialized wharf construction." As a result, the area of the Macau Peninsula expanded from approximately 2.8 km^2 to 5.3 km^2, forming a historical sequence of mutually reinforced "institutional supply–shoreline morphology–industrial space." Corresponding colonial land concession, regulations are documented in the *Regulation on Land Concession in the Colony of Macau* published in 1940, which demonstrates that administrative clauses of "land grant–renewal–land-use purpose" had long-standing legal continuity, providing institutional foundations for later Inner Harbour superstructure construction and functional transformation.

In summary, the policy–operation–spatial evolution of Inner Harbour wharves was the result of the intersection of four threads: port authority supervision, public transport concessions, differentiated regulation of hazardous cargo, and land concession with unified reconstruction. Concession contracts and government gazettes from 1953 to 1964 illustrate a governance path of "administrative licensing–contractual management–spatial allocation." Hazardous materials supervision and factory licensing formed in the 1940s incorporated "safety–logistics–storage" into a unified regulatory framework. The unified superstructure program supported by land concession in the 1980s completed a closed loop of "institution–engineering–typology."

2.3 Interaction Between the Inner Harbour and the Urban Economy

As both a functional interface and economic engine of modern Macau, the Inner Harbour (内港) long served as a key node mediating the circulation of people, goods, information, and capital between the city and the sea. Consequently, it maintained continuous spatial interaction with the commercial districts, industrial layout, and public services on the western side of the Macau Peninsula. Centered along Avenida de Demétrio Cinatti (爹美刁施拿地大馬路), waterfront wharves and arcade-style commercial buildings gradually formed a symbiotic relationship in the twentieth century. Hotels, restaurants, and retail businesses clustered densely around berthing points, giving rise to a waterfront economic corridor where "harbor frontage was street market," thereby transforming port operation flows directly into localized consumption and employment opportunities.

The evolution of transportation networks directly shaped daily urban life and consumption patterns. For example, the 1958 ferry schedule between Macau, Taipa (氹仔), and Coloane (路環) confirms that Inner Harbour ferry services once supported both intra-city commuting and tourism flows. With the advent of the bridge era, transportation dynamics were reshuffled. When the first Macau–Taipa Bridge opened in October 1974, it significantly reduced the former one-hour ferry advantage, leading to the gradual decline of passenger services at the Inner Harbour and their relocation to the Taipa Port system. The Old Taipa Ferry Terminal (舊氹仔碼頭), in particular, experienced structural decline in passenger usage after the construction of the Taipa Causeway in 1968 and the opening of the bridge in 1974.

The interaction between the Inner Harbour and daily urban supply was also embodied in the micro-circulation of small-scale trade and street commerce. From the mid to late twentieth century, flower growers from Wanzai (灣仔) and small-trade permit holders would gather at around 6:00 a.m. daily, reaching the vicinity of Wharf No. 20 via motorized sampans. They first registered at the Maritime Police Inspection Office (水警訪問所, located at Wharf No. 20), obtained landing permits, dispersed to markets from 7:30 a.m. for sales, and returned around 3:00 p.m. Government notices in the 1990s further specified that perishable goods such as vegetables and flowers were to be handled between 7:00 and 17:00 at the North Sampanas Wharf (北舢舨碼頭) and Coloane Slope (路環斜坡), and designated the North Sampan Wharf as the entry point for flower growers, with a daily quota of 450 persons in 2000. This "wharf–border checkpoint–market" chain effectively expanded peak-hour consumption and employment opportunities in nearby Inner Harbour communities. Contemporary management continues this micro-circulation in more institutionalized form: fishing vessels moored in the Inner Harbour anchorage must pre-register, maintain regulated berth spacing, and comply with safety and restricted operation rules, demonstrating how port regulation mediates community environments and the clustering of low-end economic activities.

In relation to the firecracker industry and light manufacturing, companies such as Kwong Hing Tai (廣興泰) and Cheng Cheong (禎祥) used the Inner Harbour in the first half of the twentieth century as the front-end node for worker transport and product shipment. Firecrackers, being classified as hazardous goods, had to be transported by dedicated vessels and could not be mixed with other cargo, thus intrinsically linking loading, storage, and urban fire safety. Around 1962, there were already eight firecracker factories operating between the Macau Peninsula and Taipa, including Po Seng (寶昇), Tung Cheong (同昌), Him Sun (謙信), Iat Long (益隆),

Kong Veng (光遠), Him Un (謙源), Kwong Hing Tai (廣興泰), and Hap Lei (合利), generating employment in upstream hazardous materials licensing and downstream packaging, loading, and transshipment services. To support export-oriented production, freight processes in the Inner Harbour became increasingly refined. For example, freight charges for carton-box exports were calculated based on the "three-party pricing system," with Macau dockworkers, shipping companies, and Hong Kong dockworkers dividing fees proportionally. Export shipments had to be inspected by the Economic Department (經濟廳), verified at the Inspection Office (訪問所, Wharf No. 20), and subjected to unpacking and sampling before loading. Tightened maritime regulation further reshaped enterprise behavior and spatial arrangements. After the new Maritime Police Inspection Headquarters (水警稽查總部) was established in 1951, the Inspection Office began to enforce unified control over export paper, document fees (人情紙), and wharf mooring charges. From 1953 onward, berthing vessels were charged fees regardless of operation status (2 patacas per day shift, 3 patacas per night shift). In the 1970s, import declaration and taxation were added, increasing institutional transaction costs and encouraging some industrial actors to shift operations to larger-scale regional port networks. By the 1970s, as mainland China resumed foreign trade and global supply chains were restructured, Macau's light industry exports transitioned from matches and firecrackers to textiles and garments. Kwong Hing Tai ceased operations in 1974, marking a structural shift in Inner Harbour industrial services from hazardous materials to light manufacturing and eventually to service-oriented industries.

With the consolidation of cross-border ferry routes and restructuring of port functions, the Wanzai Port (灣仔口岸) reopened on January 23, 2020 after renovation in 2016, restoring high-frequency ferry services between the Inner Harbour and Wanzai with intervals of up to one departure every 15 min and a capacity of approximately 280 passengers per vessel. Ticket prices were 20 MOP or 15 RMB for residents, reinforcing economic linkage between Inner Harbour commercial activities and the urban chain of the Greater Bay Area. Within Macau, public bus routes enhanced connectivity between "wharf frontage–commercial streets–community services." The Inner Harbour Ferry Terminal bus stop is served by routes 1, 2, 3A, 5, 10, 10A, 11, 18, 18B, 21A, 26, MT4, and N3, enabling mutual amplification between marine commuting and land-based transfers, promoting mixed-use development combining ground-floor commercial space with upper-level work or storage uses.

Looking ahead, the Inner Harbour has been incorporated into the "One River, Two Shores Cooperation Corridor" (一河兩岸合作軸帶) of the *Macao Special Administrative Region Urban Master Plan (2020–2040)*, which proposes the creation of waterfront commercial streets and riverside parks along the Inner Harbour, enhancement of flood resilience and slow-mobility systems, and the revitalization of waterfront public spaces to boost economic vitality and neighborhood livability.

3 Wharf Typology and Spatial Structure

3.1 Numbering System and Spatial Distribution

The Inner Harbour wharves adopt a numbering system based primarily on continuous numbers supplemented by alphabetic suffixes. Historically, numbering extended as far as Wharf No. 35 (Cooperação Agrícola Wharf, 協農碼頭), which was used for night soil transfer before disappearing after the installation of the city's sewerage system. The current numbering framework runs from No. 1 to No. 34 as the main sequence, with alphabetical insertions (e.g., 22A, 27A, 31A) added in certain gaps to reflect infill development arising from later reclamation projects, land concessions, or new functional insertions.

In terms of spatial configuration, the numbering proceeds progressively from south to north along the western coastline of the Macau Peninsula, passing through Barra (媽閣), São Lourenço (下環), Patane (沙梨頭), Ilha Verde (青洲), and Lam Mau Tong (林茂塘), corresponding to the historical patterns of shoreline straightening and reclamation. When the government proposed the "Unified Reconstruction" plan in the 1980s, it explicitly targeted "34 old-style wharves along the approximately 1,400-m shoreline from Barra to Lam Mau Tong," indicating a governance framework that integrated shoreline management with wharf numbering. Engineering discussions around 1985 also mentioned "straightening the curved shoreline between Wharves No. 9 and No. 17," highlighting the strong coupling between mid-section wharf numbering and shoreline reconfiguration.

The spatial distribution of the Inner Harbour wharves can be summarized into three segments: (1) Southern Segment (Wharves No. 1–12): Oriented along Rua das Lorchas (火船頭街), characterized by early waterfront buildings combining passenger and cargo functions with storage; (2) Middle Segment (Wharves No. 13–21): Historical cluster of passenger transport. Wharf No. 14 was once designated as an "additional passenger terminal," coordinating with the management radii of the South and North Sampan Wharves (南/北舢舨碼頭); and (3) Northern Segment (Wharves No. 22–34): Aligned with Avenida de Demétrio Cinatti (爹美刁施拿地大馬路) and dominated by corporate-named covered wharves such as Hap Lei (合利), Hoi Lei (海利), Hip Lei (協利), U Wa Fat Kee (裕和發記), and Pou Heng (寶興). These present a composite typology of warehouse–freight–office integration.

It is important to note that the alphabetic suffixes are not simple extensions of numbering but reflect cumulative institutional and engineering processes such as shoreline straightening, reclamation segmentation, functional insertion, and land concession renewals. For example, Wharves 22A and 27A often occupy "infill parcels" between adjacent main wharf numbers and exhibit transitional characteristics in façade vocabulary, platform structure, and berth configuration, making them valuable indicators of the evolutionary logic of the numbering system. Wharves 31 and 31A coexist with shoreline governance facilities such as civil defense broadcast poles and storm surge markers, showing their role as nodes within the contemporary resilience governance system.

In summary, the wharf numbering system of the Inner Harbour is not a simple locational sequence but a spatiotemporal index revealing an intertwined process of shoreline engineering history–institutional land concession–business networks–functional differentiation. The three-tier structure of South (Rua das Lorchas)–Middle (passenger concentration + sampan management)–North (covered freight/warehouse wharves) forms the backbone, anchored by the insertion of alphabetic wharves and local naming conventions.

3.2 *Platforms and Spatial Configuration*

In terms of scale and berth arrangement, the façade width (platform frontage) of covered wharves in the Inner Harbour exhibits a general increase from south to north. The southern section is dominated by small-scale units of 10–15 m in width; the middle section introduces transitional types around 30 m; and the northern section, where covered wharves are concentrated, ranges from 14 to 48 m, with widths around 20 m being the most common. Berth numbers are highly consistent, with most cases adopting a dual-berth configuration. The unit berth width generally ranges from 7 to 12 m, with a few freight-oriented cases reaching 15–16.5 m.

Differences in scale and berth size are closely correlated with vessel type, functional orientation, and historical period. Timber-structure wharves or fishing-oriented examples tend to have smaller unit berth widths, while reinforced-concrete framed wharves used for freight and storage have larger widths to accommodate loading demands. The "scale–berth–function" correspondence in the Inner Harbour can be summarized as follows: (1) Fishing and small cargo-oriented wharves: Smaller overall frontage width and unit berth width; typically timber structures or narrow-span concrete platforms; (2) Warehouse and freight-oriented covered wharves: Larger frontage and unit berth width; designed for dual berths and smoother vehicle-vessel circulation.

From a typological perspective, Inner Harbour wharf platforms evolved from pile-supported trestle structures to longitudinal shoreline platforms. Early wharves used tall timber piles inserted into the seabed, with platforms elevated above the water in a typical *stilted* configuration (Fig. 3). These were later rebuilt as continuous shoreline platforms with reinforced-concrete foundations and superstructures, significantly improving load capacity and durability (Fig. 4).

In terms of structural support systems, Wharf No. 11 exemplifies a reinforced-concrete multi-member inclined support system characterized by a "column–strut–ledger" structural logic (Fig. 5). The construction follows a sequential logic: main load-bearing columns are erected first; diagonal beams are then placed around them to form a supporting frame; finally, diagonal cantilever beams extend outward from the column tops to support the platform slab. This allows the platform to be suspended above the water as a multi-rod structural system that balances both mechanics and aesthetics. The use of rhomboid cross-braced concrete elements not only enhances

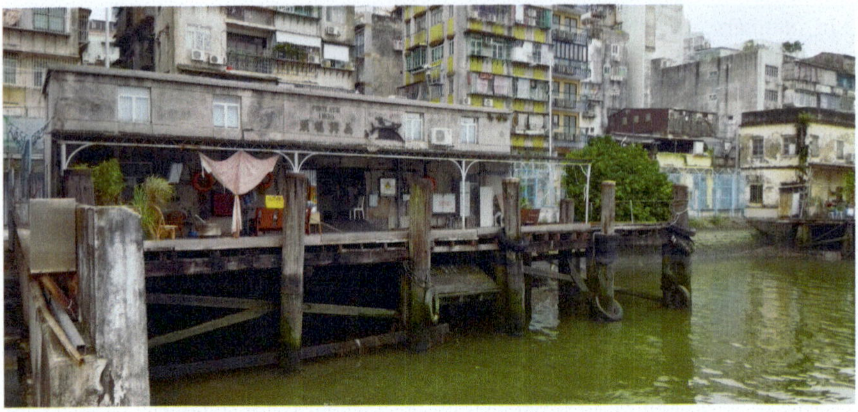

Fig. 3 Stilted form of Wharf no. 31. (*Image source* Photographed by the authors)

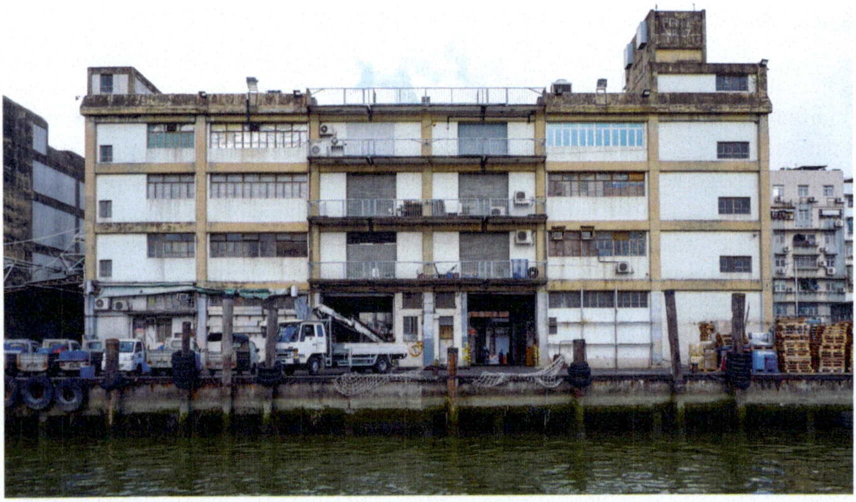

Fig. 4 Reinforced-concrete foundation of Wharf no. 27. (*Image source* Photographed by the authors)

the platform's spanning ability and lateral resistance, but also contributes a distinctive geometric rhythm to the waterfront façade.

In terms of safety construction and detailing, the waterfront protection system and boarding/disembarkation system exhibit clear functional differentiation. The fencing strategy distinguishes between waterfront zones and functional/office zones: (1) Waterfront sections generally employ square or rhomboidal concrete balustrades arranged at fixed intervals, prioritizing structural strength and drainage/ventilation; (2) Residential or office-related sections often use baluster-type railings (Fig. 6),

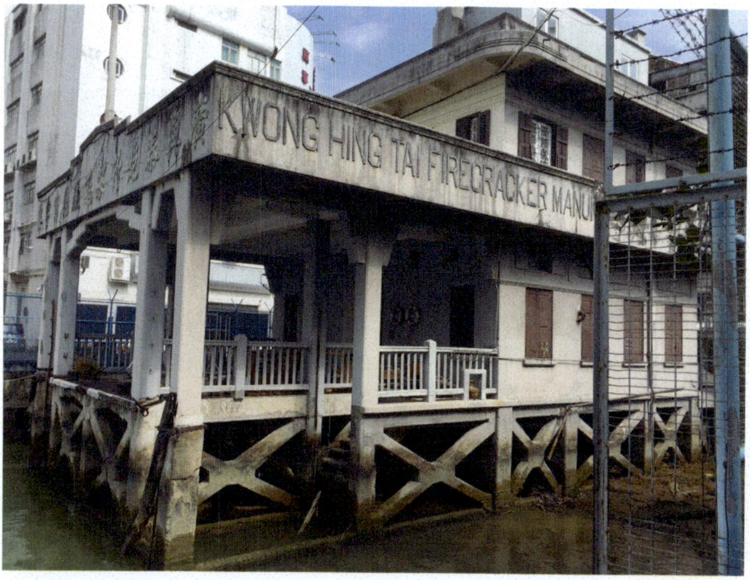

Fig. 5 Platform structure of Wharf no. 11. (*Image source* Photographed by the authors)

rectangular guardrails, or wrought iron fences, combining both viewing and barrier functions.

Stair systems are organized into three categories: (1) Direct access routes connecting the street to the wharf platform (Fig. 7); (2) Semi-open transitional stairways beneath platform arcades; (3) External corridor or roof service stairs.

The circulation of platform, entrance hall, and passenger waiting areas reflects an integrated spatial design based on axial symmetry, internal–external zoning, and all-weather operability. The platform of Wharf No. 8 combines both arcade-covered and open-air sections (Fig. 8), using a semi-open buffer zone to mediate passenger and cargo exchange. This reduces evacuation distance while providing rain protection. Internally, the space sequences from platform to entrance hall, to waiting/ticketing area, to office rooms, forming a linear chain of operational spaces.

The spatial structure of Inner Harbour wharf platforms can therefore be summarized as follows: (1) Structural System: reinforced-concrete columns and cross-bracing diagonals support platform slabs, controlling span and rigidity; (2) Decorative System: zoning with rhombic, square, or baluster-type parapets—simple on the waterfront side and refined on the functional/office side; and (3) Circulation System: a four-stage spatial organization comprising street access, transitional passage, semi-elevated buffer, and boarding interface ensures safe and orderly peak-period transfers and cargo handling.

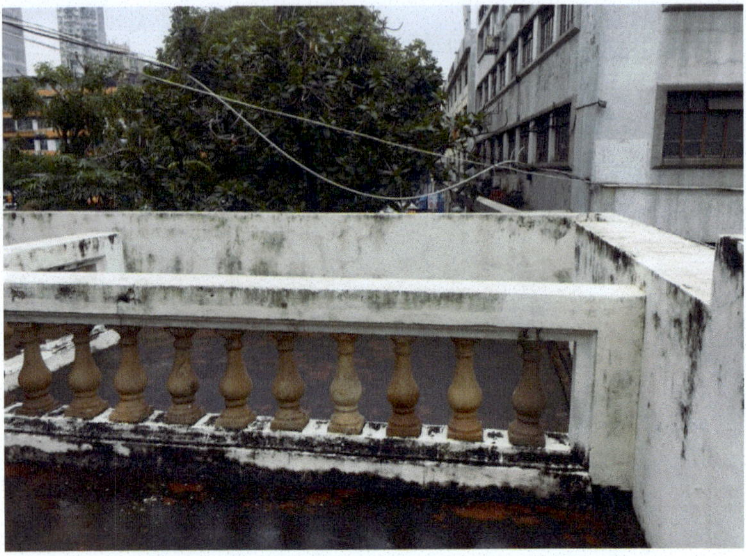

Fig. 6 Baluster-style railing on the roof of Wharf no. 11. (*Image source* Photographed by the authors)

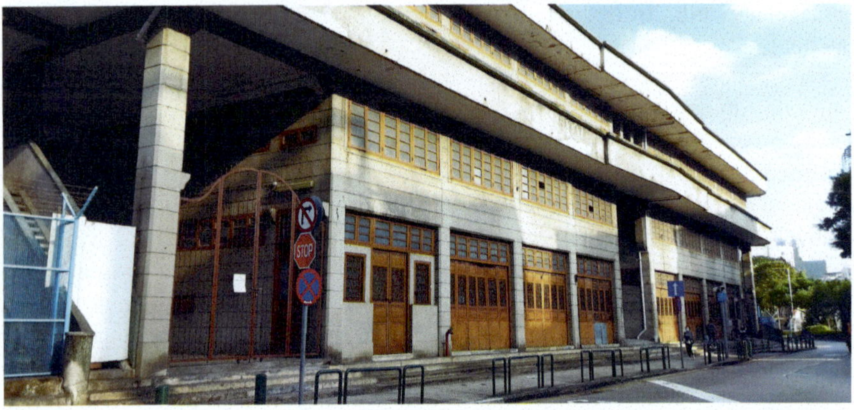

Fig. 7 Relationship between the building of Wharf no. 8 and the street. (*Image source* Photographed by the authors)

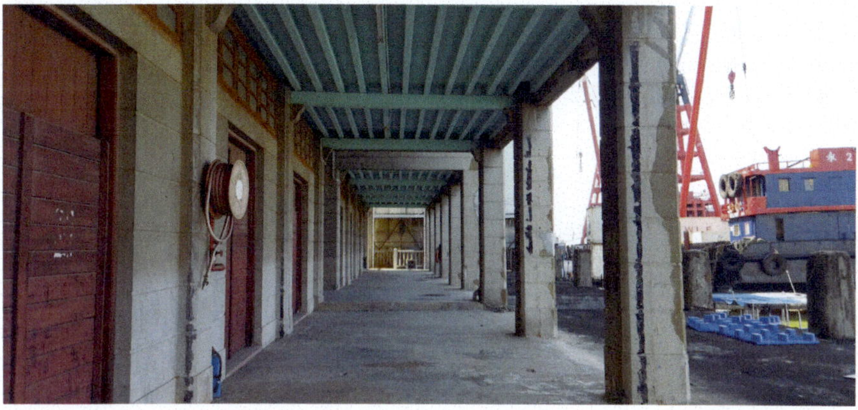

Fig. 8 Relationship between the building and platform of Wharf no. 8. (*Image source* Photographed by the authors)

4 Architectural Style, Materials, and Construction

4.1 Localization of Style

The Art Deco style of the Inner Harbour wharves is not a textbook imitation but rather a localized architectural expression shaped by the constraints of a humid, rainy, salt-laden coastal climate with strong solar exposure. Its most distinctive design vocabulary includes layered horizontal cornice bands, triangular pedimented parapets, cantilevered eaves, and functional rain canopies, often combined with washed granolithic finishes (水刷石, Fig. 9), differentiated baluster-type and diamond-pattern parapet railings, and Chinese–Portuguese bilingual inscriptions carved on lintels and parapet panels (Fig. 10). The result is a composite façade system in which structure, detail, and lettering together participate in architectural composition.

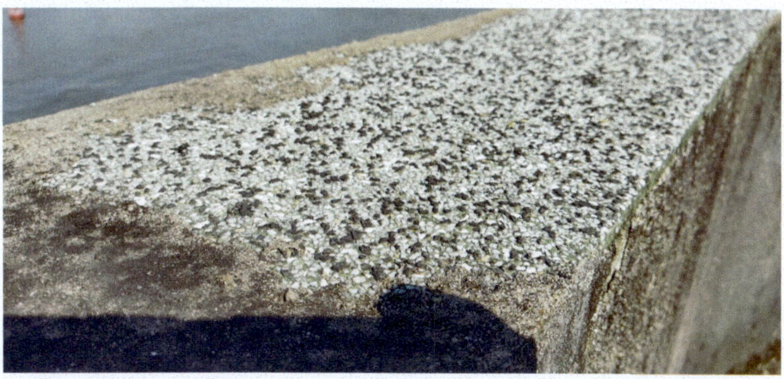

Fig. 9 Washed granolithic finish at Wharf no. 8. (*Image source* Photographed by the authors)

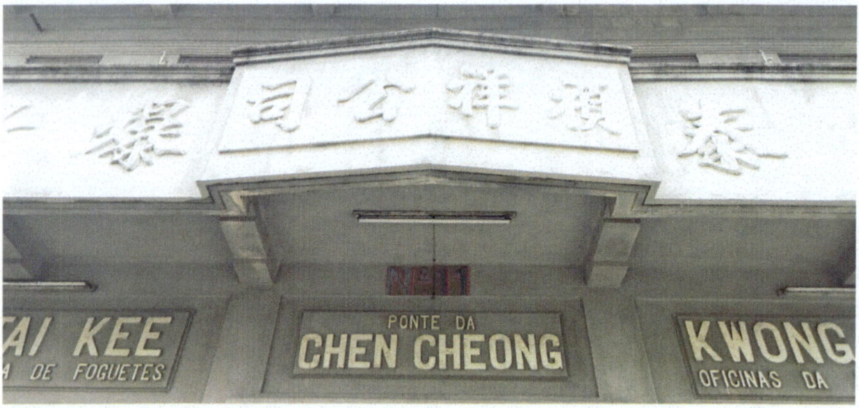

Fig. 10 Bilingual lintel inscription at Wharf no. 11. (*Image source* Photographed by the authors)

For example, the main street façade of Wharf No. 11 employs multiple horizontal cornice bands to mediate between the upper and lower storys. The cornice bands rise toward the central axis while tapering down toward both ends, emphasizing symmetry and hierarchy. The second-floor cantilevered slab wraps around the building, bending into a triangular form at the center. The parapet then retreats inward to define a terrace, crowned by layered horizontal moldings. A simplified Western-style pediment rises above, capped with a pointed finial detail for ceremonial emphasis (Fig. 11). This continuous composition of layered molding, projecting eaves, and parapet pediment embodies both the geometric discipline of Art Deco and a direct climatic response to heavy rainfall and strong sunlight along the Inner Harbour.

The localization of materials and surface texture is most evident in the extensive use of washed granolithic plaster. Wharves No. 8 and No. 11 apply different grain sizes and tones of washed stone finish to façades, stair balustrades, and parapet copings. Wall surfaces and parapet tops generally adopt modest gray–white tones for durability against pollution and abrasion (Fig. 12), while certain elements, such as the stair balustrades of Wharf No. 8, use a green-tinted washed stone to create controlled chromatic contrast (Fig. 13), enhancing visual guidance while retaining restrained regional character. This strategy of "same material, varied tone" couples' local craftsmanship with Art Deco geometric banding, allowing cornices and parapets to perform not only compositional roles but also waterproofing and weather protection functions.

Text and signage are also integrated into the façade grammar, forming a unique "readable ornamentation" characteristic of the Inner Harbour. On Wharf No. 11, the lintel is engraved with English corporate lettering such as *"CHAN TAI KEE CHEN CHEONG KWONG HING TAI,"* while parapet panels below are divided into segments inscribed with Chinese phrases such as "廣興泰" (Kwong Hing Tai), "禎祥公司" (Cheng Cheong Company), and Firecracker Factory. These combine corporate inscription with Art Deco geometry, embedding commercial identity into façade composition. In the northern segment, some wharves carry metal letter signage such

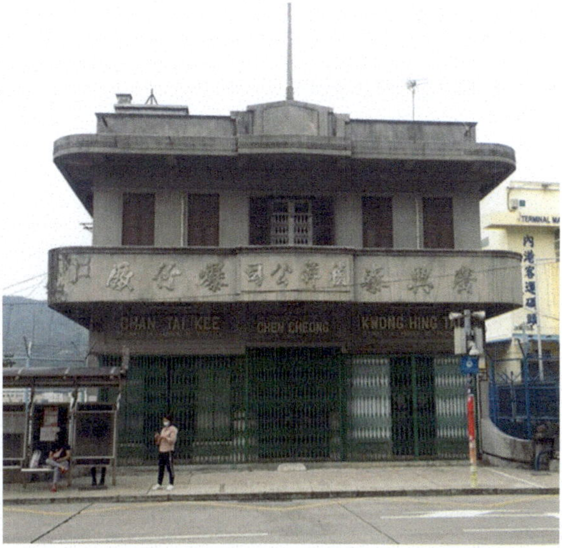

Fig. 11 Street façade of Wharf no. 11. (*Image source* Photographed by the authors)

Fig. 12 Washed granolithic finish at Wharf no. 11. (*Image source* Photographed by the authors)

as "PONTE Nº 31A" welded onto parapet corners for long-distance identification by vessels, turning signage into a formal façade element.

The treatment of cantilevered eaves and rain canopies demonstrates the combined constraints of local climate and operational use. The continuous second-floor cantilevered slab and rectangular rain sheds over doorways reflect Art Deco's horizontal linear emphasis while providing shelter for waiting passengers and loading/unloading during rain. For example, Wharf No. 25 features a cantilevered canopy

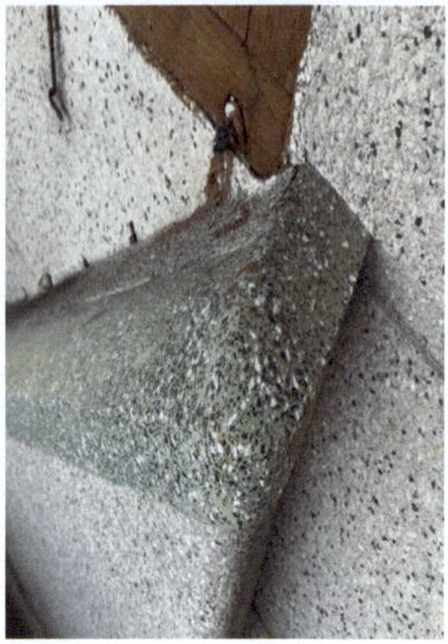

Fig. 13 Washed granolithic finish at Wharf no. 8 (staircase handrail). (*Image source* Photographed by the authors)

of approximately four meters square at its main entrance, integrated with a full-length horizontal lintel band. The result is a layered yet weather-resistant entrance composition (Fig. 14).

On the lateral and waterfront façades, Art Deco order is translated into structural and boarding rhythms. The west façade of Wharf No. 11 adopts stepped terraces

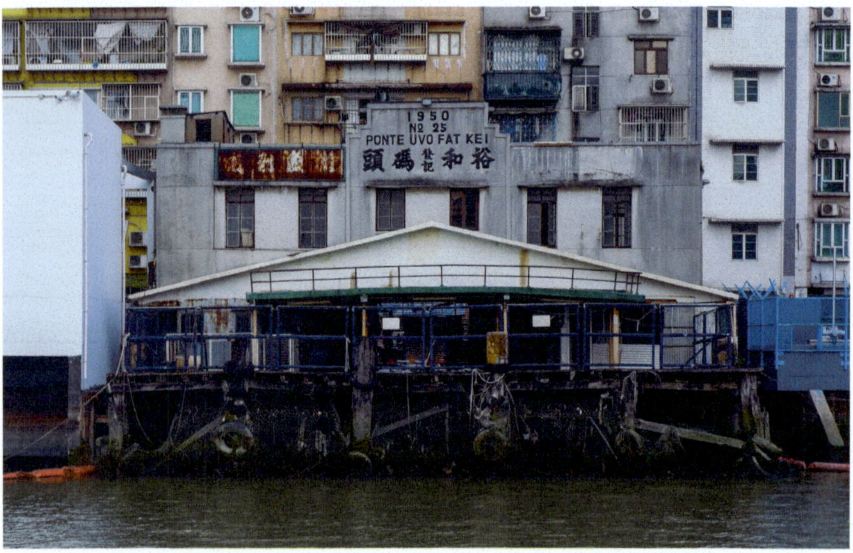

Fig. 14 Cantilevered rain canopy at Wharf no. 25. (*Image source* Photographed by the authors)

and a partially open ground floor, where concrete column grids, scissor-type diagonal beams (Fig. 15), and dual boarding staircases (Fig. 16) generate a sequence of regular voids and diagonal members. Structural components that were originally pragmatic acquire visual expression through Art Deco's geometric discipline, forming a localized paradigm in which structure becomes ornament.

In conclusion, the localization of Art Deco in the Inner Harbour is not a simple stacking of decorative parts, but a composite architectural expression shaped by material technique, climatic protection, industrial identity, and boarding regulations. Layered cornices and parapet pediments provide geometric order and emphasis; washed granolithic finishes and projecting eaves provide weather resilience; railings and boarding stairs translate safety logistics into façade rhythm; bilingual parapet inscriptions and lintel lettering embed industrial memory and commercial identity into the architectural skin.

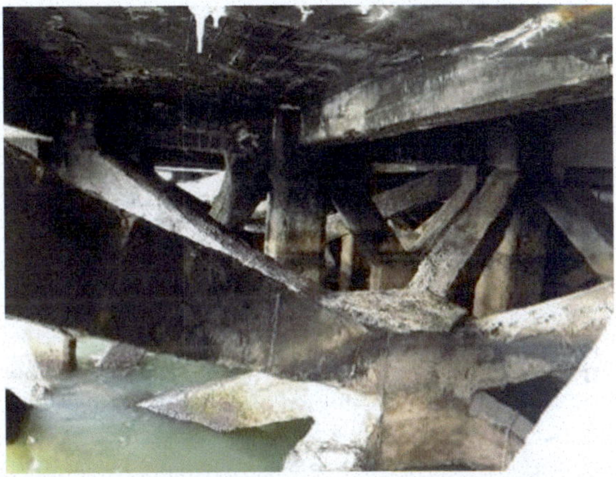

Fig. 15 Scissor-type diagonal beam at Wharf no. 8. (*Image source* Photographed by the authors)

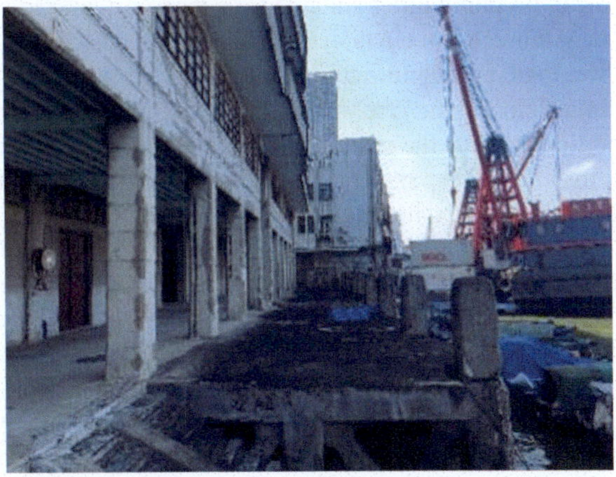

Fig. 16 Platform of Wharf no. 8. (*Image source* Photographed by the authors)

4.2 Façade Composition and Architectural Detailing

The façades of the Inner Harbour wharf buildings generally exhibit a transitional stylistic spectrum from Art Deco to Modernism. In terms of composition, they commonly adopt axial symmetry and a tripartite organization: a continuous hierarchical transition from base to wall body, to skyline through the use of door openings, stringcourses, parapets, and roof slab projections. In materials and details, they predominantly employ washed granolithic render (水刷石) and cement plaster finishes, supplemented by inscribed plaques, metal or concrete railings, louvered windows, and functional rain canopies. These elements together form façades that balance operational needs in loading/unloading, moisture protection, and sun-shading, while also maintaining clarity of identity and historical stylistic character in the urban setting.

After the 1940s, a group of modernist-oriented façades gradually took shape. The main street façades of Wharves No. 23 and No. 30 adopt vertical tripartite symmetry as their compositional framework. Above the main doorway, a rectangular relief panel frames inscriptions where "PONTE," "No.," and the Chinese name are presented in parallel. Construction dates such as "1948" are frequently marked along the lower edge of the parapet or within a gap below the cornice, thereby creating an informational sequence of "entrance opening–name plaque–stringcourse–parapet" (Figs. 17 and 18). Wharf No. 30 demonstrates exceptional clarity in dimensional articulation: the rectangular frame above the main entrance measures approximately 1.8 m × 1.2 m; a decorative stringcourse of 17 cm in height encircles the building between the first and second floors; window headers and sills project about 10 cm; and the façade is finished with a layered palette of light ochre paint with gray plaster trim and parapet capping, achieving a simple yet striking visual impression. In comparison, Wharf No. 23 distributes inscriptions such as "NO. 23 PONTE IAO TAC 1948" and "祐德碼頭" respectively across balcony parapets and lintels. Intervals of 40 cm between recessed horizontal lines visually stabilize the wall plane, producing a steady rhythmic order suitable for a transport-related façade.

Within the same modernist expression, Wharf No. 25 (25號) further emphasizes the alignment of information and structure. The central bay is articulated as a compositional focal point: the wall plane rises slightly above the height of the adjacent parapets, and above the main entrance a rectangular balcony doubling as a rain canopy is installed. The parapet is capped with washed granolithic coping, coordinated with grouped recessed horizontal grooves at a 40 cm module, forming a unified horizontal motif. Chinese and Portuguese bilingual titles—"裕和發記碼頭" and "PONTE N° 25 UVO FAT KEI"—are inscribed side by side, creating a highly recognizable façade focus (Fig. 19). The remaining three elevations are much simpler, with evenly spaced window openings prioritizing ventilation and lighting, establishing a clear hierarchy between an information-dense main façade and function-driven side and rear façades.

Among wharves rebuilt in the 1980s, Wharves No. 22 and No. 22A (22與22A號) exhibit a clearer structural rationalism and legibility in façade organization (Figs. 20 and 21). The east façade of Wharf No. 22A uses a 4 × 4 grid of beams and columns,

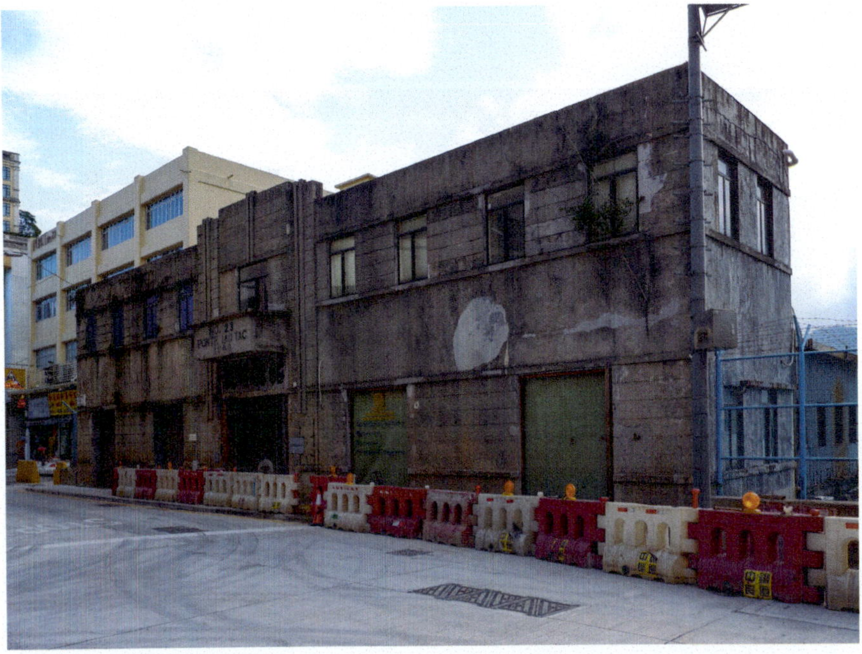

Fig. 17 Street façade of Wharf no. 23. (*Image source* Photographed by the authors)

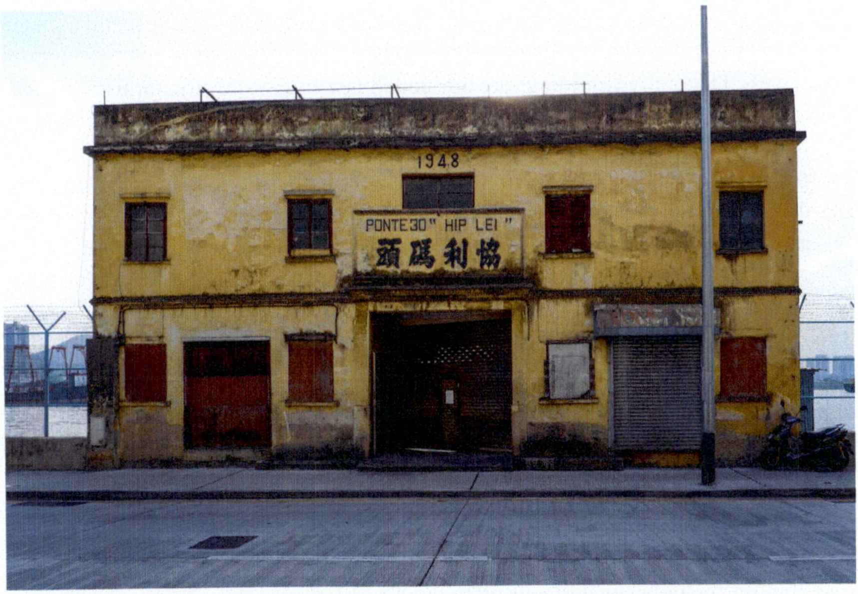

Fig. 18 Street façade of Wharf no. 30. (*Image source* Photographed by the authors)

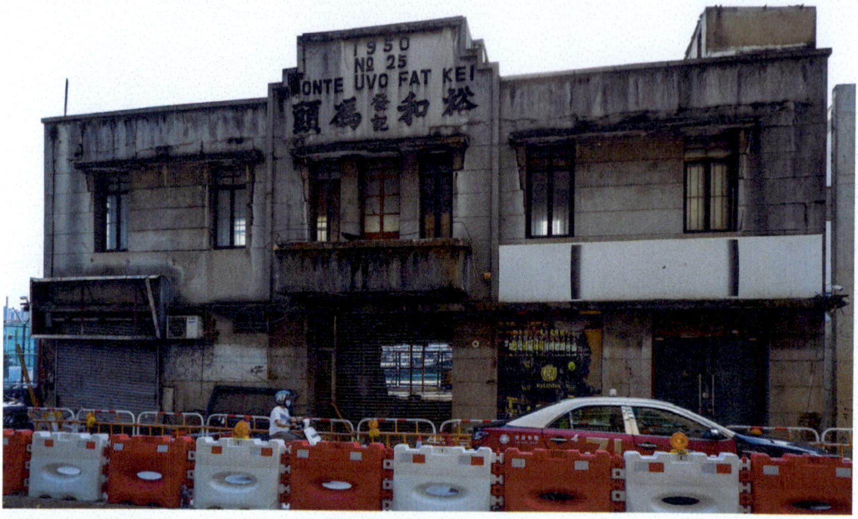

Fig. 19 Street façade of Wharf no. 25. (*Image source* Photographed by the authors)

with continuous strip windows inserted into the grid. Key nodes along the corners and waterfront side, including the stair tower and water tank, carry linear metal signage such as "PONTE Nº 22A", while two central bays include cantilevered balconies at upper levels. Decorative moldings are omitted in favor of a pure structural diagram of "frame–opening–balcony." Similarly, Wharf No. 22 retains elevated metal signage "PONTE Nº 22" on its west waterfront façade to ensure long-distance identification from the navigation channel. The platform uses rectilinear reinforced-concrete piles driven into the riverbed and continuous cantilevered platforms, with metal railings or retained timber fenders and tires as working façade details. Compared to earlier Art Deco examples, façades of this period emphasize modularity, function, and construction logic.

In segments where functional wharves coexist with industrial facilities, Wharves No. 33 and No. 34 present much more restrained exteriors. The east façade of Wharf No. 33 is mainly finished in cement-rendered plaster with a projecting concrete base (0.2 m) approximately 1 m high (Fig. 22). Symmetrical openings exist on both north and south sides, with enclosed balconies doubling as rain shelters. Chinese–Portuguese bilingual enterprise plaques are installed between the second and third floors, and cement lattice windows are used in sections for shading, introducing elements of Lingnan regional detailing. In comparison, Wharf No. 34 retains a large amount of timber waterfront platforming and sheet steel add-ons, expressing a rugged interface where façade and working platform are synchronized, preserving visible repair traces and time-layered maintenance logic. Here, detailing prioritizes

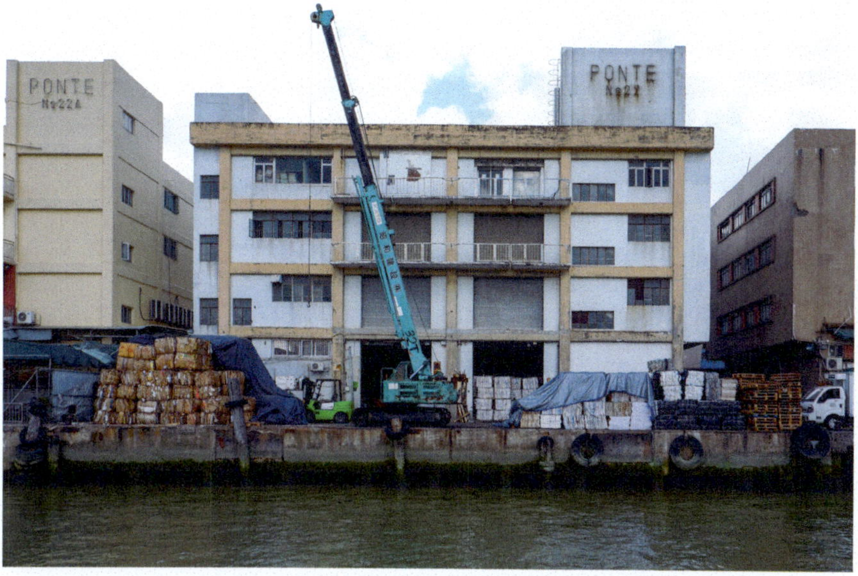

Fig. 20 Waterfront façade of Wharf no. 22. (*Image source* Photographed by the authors)

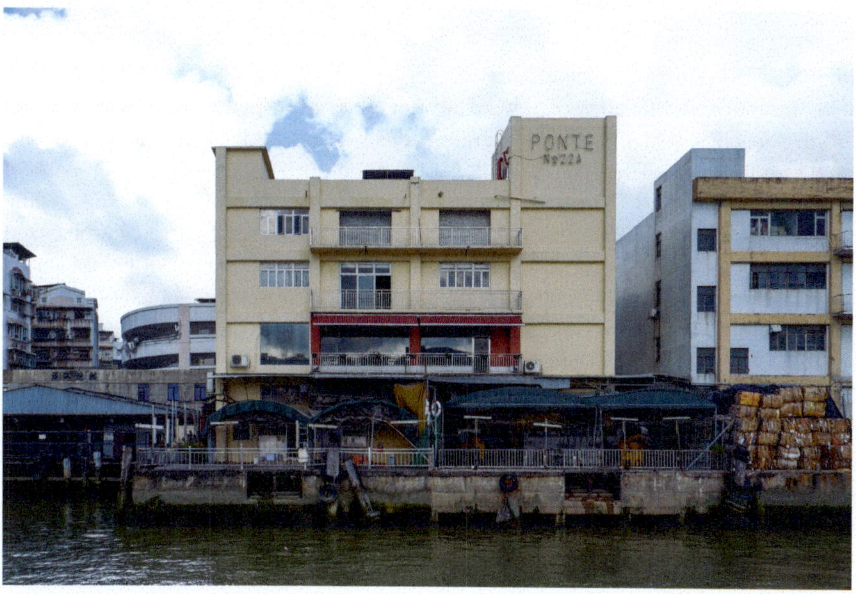

Fig. 21 Waterfront façade of Wharf no. 22A. (*Image source* Photographed by the authors)

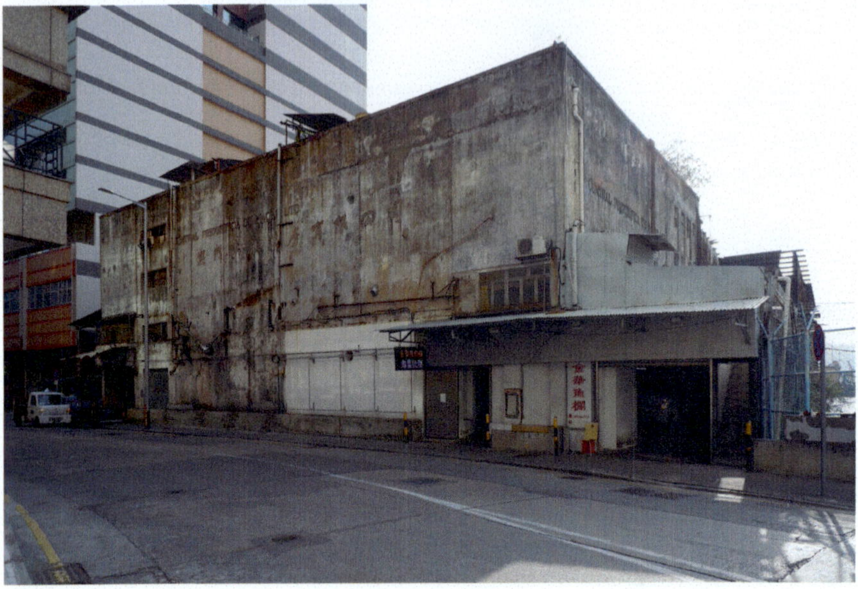

Fig. 22 Street façade of Wharf no. 33. (*Image source* Photographed by the authors)

durability, ventilation, and shading, with decorative intent giving way to functional pragmatism and maintenance efficiency.

In summary, the façades of Inner Harbour wharves are not decorative skin but working façades that integrate structure, construction logic, signage, and operational requirements. The linear pediments recessed horizontal grooves, and parapet inscriptions of the Art Deco period established a recognizable façade order, while the modular grids and cantilever balconies of the Modernist period emphasized frame articulation and functional clarity. In industrial wharf zones, restrained façades emphasize maintenance logic and durability. Based on this, future conservation and revitalization efforts should respect the compositional archetypes and material craftsmanship of each historical phase—for example, retaining parapet inscriptions and Chinese–Portuguese bilingual plaques in situ, repairing washed granolithic finishes using matching grain and tone, reconstructing parapet coping profiles, and maintaining the exposed authenticity of scissor bracing and fender structures along operational waterfront façades—so that the wharf façades retain historical legibility while extending service life.

4.3 *Materials and Construction Techniques*

The material spectrum and construction techniques of the Inner Harbour wharves demonstrate clear and readable evolutionary logic. Their fundamental construction pattern consists of reinforced-concrete floor slabs and roofs as the structural framework, blue brick or stone masonry with cement–lime mortar as wall and infill, washed granolithic plaster as the principal exterior finish, and wharf platforms constructed with either timber or reinforced-concrete piles in different historical periods, thus forming a hierarchical assembly of structure, masonry, finish, and platform systems. For example, Wharf No. 23 employs a rubble masonry foundation, reinforced-concrete roofing, and blue brick masonry walls laid in lime–sand mortar (Fig. 23). The façade is uniformly finished in gray washed granolithic plaster (Fig. 24), reflecting the Art Deco emphasis on surface treatment while meeting the durability requirements of a waterside environment.

The wall and finish process generally follows a standard sequence: masonry leveling → base plastering → washed-stone spray finishing → cornice/coping termination. Field surveys at Wharves No. 31 and 31A show traces of modern cement and lime–sand mortar in wall cracks, with façades consistently finished in gray-washed granolithic render. Recessed horizontal and vertical grooves at 40 cm intervals serve both as surface rhythm and as practical maintenance segmentation markers. In localized façade areas, green-glazed ceramic tiles (typically 240 × 115 mm) were added in later periods, corresponding with aluminum window replacements, revealing layers of material and component updates over time (Fig. 25).

In terms of doors, windows, and metal components, coexistence of original and later-added elements is a distinct feature of the Inner Harbour wharves. Earlier

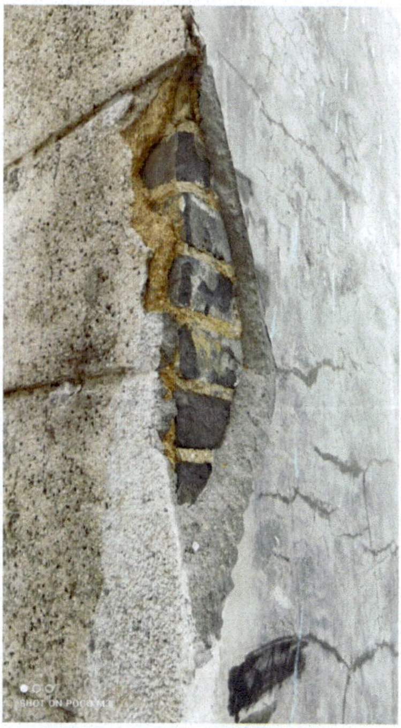

Fig. 23 Gray bricks masonry wall at Wharf no. 23. (*Image source* Photographed by the authors)

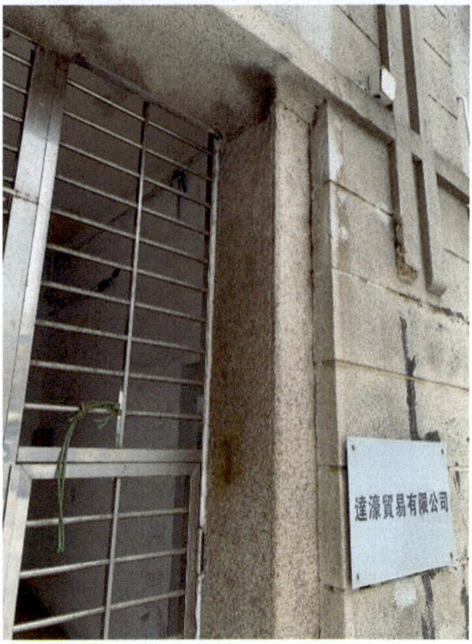

Fig. 24 Washed granolithic finish at Wharf no. 23. (*Image source* Photographed by the authors)

buildings commonly used timber-framed windows, wooden louver shutters, and iron grille gates, whereas later alterations introduced roller shutters and aluminum alloy windows. Along the waterfront frontage of Wharf No. 31, corrugated metal rain canopies are typically installed above entry points, projecting between 0.4 m and 1.5 m. Foldable iron gates of Macau type are common at doorways; each iron leaf measures about 0.2 m in width, with perforated ornamental patterns at the top for ventilation and daylight—this is a typical port-operational frontage component system (Fig. 26). Meanwhile, corrugated steel rain shades, aluminum windows, outdoor air-conditioning brackets, and other later metal additions, although visually inconsistent, serve practical needs in usage intensity and maintenance, reflecting an operation-first maintenance strategy.

Based on the above material and construction characteristics, the following targeted conservation and repair strategies are proposed: (1) Washed granolithic repair should follow a "restore as original" principle, matching aggregate grain size, color tone, and spray technique. Original coping lines and recessed grooves should be reinstated in situ. High-density coatings or putty leveling should be avoided, as they reduce wall breathability; (2) Masonry crack treatment should distinguish between load-bearing and non-load-bearing areas. Structural cracks should be repaired using grouting and structural joints, while non-structural cracks should be handled with partial chiseling and compatible mortar repair, avoiding rigid materials that may induce secondary damage in high-moisture environments; (3) Platform waterproofing and drainage should maintain the original system of brick channels, scuppers, drain holes, and cantilever slabs. Drain openings in parapets and semicircular gutters should be cleared. Minimal intervention methods should be applied when reinforcing

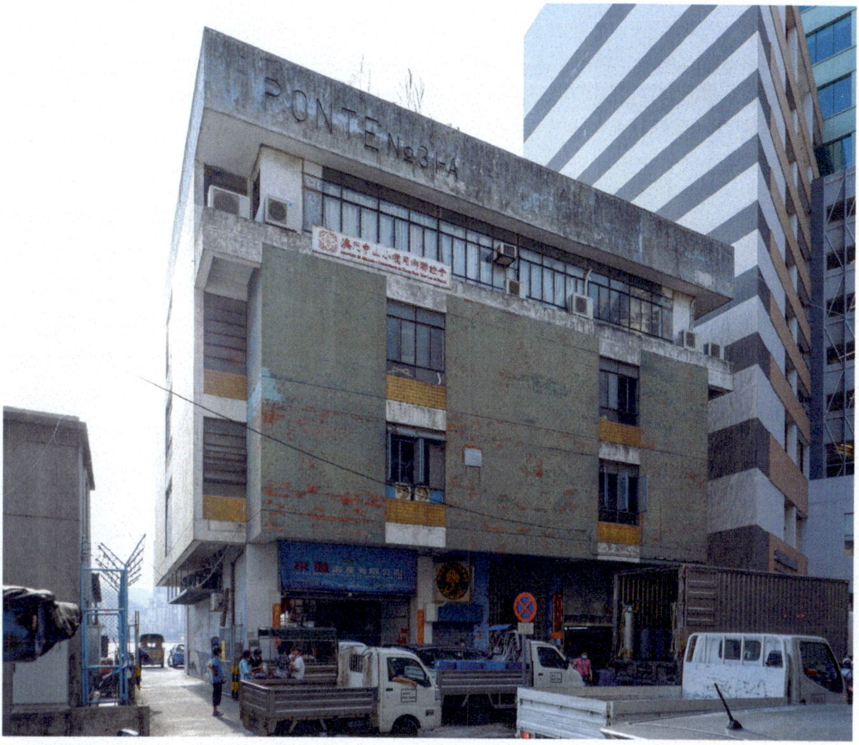

Fig. 25 Street façade of Wharf no. 31A. (*Image source* Photographed by the authors)

concrete pile caps and railings; (4) For doors, windows, and metal components, original features such as foldable iron gates should be preserved as priority heritage elements. When updating aluminum windows or rain canopies, their scale and rhythm should align with original façade patterns to minimize visual disruption.

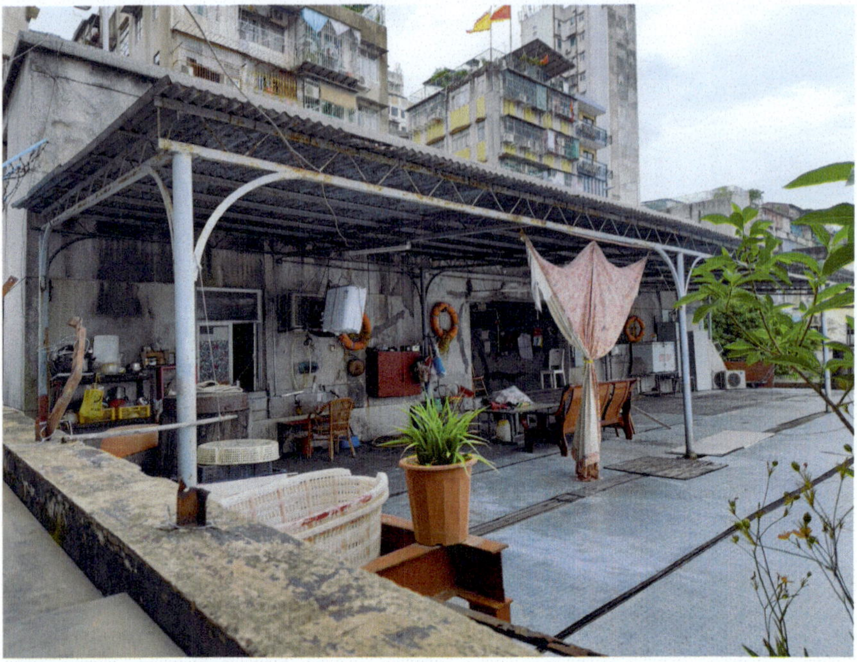

Fig. 26 Waterfront façade of Wharf no. 31. (*Image source* Photographed by the authors)

5 Conclusion

The core developmental logic of the Inner Harbour wharf system can be summarized as follows (Table 2): between 1948 and 1957, the dominant typology was the pile-supported timber platform, forming a transitional system from trestle-type to shoreline-parallel platforms. Between 1949 and 1950–1953, nine government-led reconstruction projects and private investments jointly introduced reinforced-concrete structural frames and Modernist "box-type" façades. In 1966, cold-chain facilities entered the harbor front, transforming wharf functions from single cargo handling to a composite system of fishery–refrigeration–wholesale–catering. After the 1980s, several wharves were reconstructed with four-story height limits or large-span floor plans, enhancing spatial adaptability. In recent years, due to storm surge risks and structural deterioration, wharves commonly face 0.5–1.0 m ground-floor flooding and rebar exposure hazards. Governance mechanisms are now shifting from individual operations toward an integrated model of "functional zoning + municipal facilities + adaptive revitalization."

In terms of chronological development, 1948 marks a critical starting point in Inner Harbour modernization. Wharf No. 8, described as the second large-scale reinforced-concrete wharf after Wharf No. 16, indicates that the introduction of modern structural systems followed postwar material redistribution and adjustments

Table 2 Typology of Inner Harbour Wharves

Type	Time period	Structure & materials	Typical dimensions	Façade & compositional features	Representative cases
Timber pile platform type (Transition from trestle type to shoreline type)	1948–1957	Timber piles + timber deck platform, with partial upper masonry or light-steel reinforcement	Platform approx. + 2.1 m above water; typical frontage 15–20 m	Corrugated steel sheets on waterfront side; timber purlins visible indoors	Wharf No. 29: platform elevation approx. + 2.1 m; Wharf No. 34: timber structure with 17 m frontage and 6 m depth; Wharf No. 26: timber wharf retaining old-style platform
Reinforced-concrete "Box-type"	From 1949; concentrated in 1950–1953	Reinforced-concrete frame with masonry infill	Frontage approx. 18–33 m; depth 17–18 m; 2–3 storys	Street façade divided into rectangular grids by beam–column frame; independent corner stair tower	Wharf No. 21: 25 m frontage, 17 m depth; Wharf No. 22: 28 m frontage, 18 m depth; Wharf No. 22A: 24 m frontage, 18 m depth
Cold-chain Warehouse Type	From 1966	Reinforced concrete/ hybrid structure with built-in cold storage machinery	Based on original Reinforced-concrete wharf massing (frontage approx. 25–33 m)	Functional openings and plain undecorated walls	Wharf No. 33: two-story cold storage approved in 1966; "wharf + ice house" composite
1980s New reconstruction type	1980s	New regulated reinforced-concrete frame, large spans	Larger floor area, flexible space, high adaptability	Setbacks and loading/ parking circulation better organized	Wharf No. 27: built in the 1980s, active and revivable structure
Fishery-oriented type	After 1990s zoning	Mixed timber/reinforced concrete; platform oriented to fishing boat berthing	Typical frontage 14–24 m	Wide loading openings and low platforms on waterfront side	Wharf No. 26: timber structure for fishing boats; Wharf No. 31: restricted to fishing vessel berthing

Source The author based on on-site surveys

in port governance. Between 1950 and 1953, Wharf No. 8 evolved from an initial proposal of a four-story complex with cinema and hotel into a three-story reinforced-concrete building, reflecting market contraction and investment retrenchment. During the same period, the Capitania dos Portos implemented reconstruction at Wharves No. 10, 14, 15, 16, 21, 23, 28, 29, and 30, institutionalizing the upgrade path from timber and stone railings to reinforced-concrete systems. The approval in 1966 for a two-story cold storage warehouse at Wharf No. 33 and the establishment of an ice factory company signaled the transformation of the Inner Harbour from a loading node to a cold-chain hub, reshaping the long-term spatial structure of adjacent fish stalls, wholesale operations, and seafood catering. The new Wharf No. 27 built in the 1980s adopted larger spans and flexible interior space, making it a preferred base structure for contemporary revitalization.

In structure and material, the pile-platform type typically features high timber piles supporting an elevated deck at approximately +2.1 m, exhibiting a stilted configuration. The reinforced-concrete box type is identified by "beam–column + ribbon window" street grids and corner staircase towers, with common frontage widths between 18 and 28 m, depths of 17–18 m, and 2–3 storys. Some wharves retain sections reinforced with timber, corrugated steel, or light-steel additions (e.g., Wharf No. 34), reflecting a "repair-and-continue" material renewal strategy. Notably, Wharves No. 29 and 26 preserve clear traces of trestle-to-shoreline evolution, providing important evidence for studying the technological shift from detached platforms to integrated shoreline structures.

In scale and composition, the minimum frontage width is 10 m at Wharf No. 11, the maximum is 33 m at Wharf No. 22, and the typical range falls between 18 and 28 m. Depth dimensions for reinforced-concrete box types are consistently 17–18 m, and two berths per wharf is the dominant configuration. Spatial gaps display a coexistence of dense adjacency and large separation: Wharf No. 11 has only 7–8 m to its neighbors, forming a continuous street–shore interface; Wharves 22A–23 are separated by just 1.7 m in a "stitched" configuration; whereas Wharf 28–29 (25 m apart) and Wharf 29–30 (33 m apart) illustrate a "spaced node" rhythm. These scale–composition–spacing relationships directly influence cargo movement patterns, fire evacuation routes, street-level commercial penetration, and spatial operability for adaptive reuse.

In function and governance, the case studies exhibit a layered transformation from freight/passenger transport → fishing berths → cold chain/wholesale → catering and leisure. Governance evolved from family- or company-run operations → concession-based port regulation → zoned control. Oral accounts from Wharf No. 30 indicate a conservative operational strategy based on maintenance without reconstruction and flexible mooring fees for fishermen. The Wharf 22–23–25 cluster demonstrates a short-chain "wharf–shop–cold-chain" commercial network. At Wharf No. 31A, clan associations and fish stall operations coexist within one building, embedding social networks directly into spatial organization along the Inner Harbour.

References

1. Zuev, D., Hannam, K., & Zhao, J. (2025). Industrial heritage tourism in Macau: reinventing the Iec Long firecracker factory. *Tourism Geographies*, 1–27.
2. Zhu, R., & Pinheiro, F. V. (2010). Preservation and sustainability of industrial heritage in the urban renewal: A case study of Iec long firecracker factory in Taipa, Macao. In *2010 international conference on e-product e-service and e-entertainment* (pp. 1–4). IEEE.
3. Huang, L., Huang, Y., Chen, Y., Lou, S., Chen, Y., & Jia, M. (2024). Application and development of firefighting technologies in industrial heritage: Experiences and Insights from Macau. *Buildings, 14*(9), 2699.
4. Zuev, D., Hannam, K., & Zhao, J. (2025). Industrial heritage tourism in Macau: reinventing the Iec Long firecracker factory. *Tourism Geographies*, 1–27. https://doi.org/10.1080/14616688.2025.2495179
5. Zhu, R., Wu, Y., Pinheiro, F. V., Zhehao, S. Z., & Chio, S. (2010). Preserving Macao industrial heritage for a sustainable urban future. In *Proceedings of the 8th ISAIA, Kitakyushu, Japan.*
6. Huang, L. (2022). Research on green renewal strategy based on LID adaptation toolkit selection—the example of the North Bay Section of Macau Inner Harbour terminal neighborhood. In *International conference on civil engineering* (pp. 258–267). Springer Nature. https://doi.org/10.1007/978-981-19-3983-9_23
7. Huang, Y., Chen, Y., Song, J., Zheng, L., Yang, S., Gao, Y., Li, R., & Huang, L. (2025). Explore the ultra-high density urban waterfront space form: An investigation of Macau Peninsula Pier District via Point of Interest (POI) and space syntax. *Buildings, 15*(10), 1735. https://doi.org/10.3390/buildings15101735
8. Westerdahl, C. (2011). *Conclusion: The maritime cultural landscape revisited. In The archaeology of maritime landscapes* (pp. 331–344). Springer.
9. Hoyle, B. S. (1989). The port—city interface: Trends, problems and examples. *Geoforum, 20*(4), 429–435.
10. Douet, J. (2016). *Industrial heritage re-tooled: The TICCIH guide to industrial heritage conservation.*

"Pátio" and "Beco" Architectural Forms: Chinese Living Spaces in Macau

Yile Chen and Jianyi Zheng

Abstract "Pátio" and "Beco" were important living spaces for Chinese in Macau after the mid-nineteenth century and were recorded in the official Macau street register, Macau Gazette and other documents, as well as laws and regulations. The article starts from four aspects: the interpretation and distribution of Macau "Pátio" and "Beco", the origin and cause of "Pátio" and "Beco", the spatial characteristics of traditional Chinese-style "Pátio" and "Beco", and the development and evolution of "Pátio" and "Beco" living form in the early twentieth century. This paper focuses on analyzing its spatial morphological characteristics from the outside shop and inside house block pattern, building plane pattern, type of building unit, ancillary structures, and architectural style features and finally puts forward some suggestions on the cultural value of the enclosure and its inheritance and protection.

Keywords Macau "Pátio" and "Beco" · Living space · Spatial form · Cultural value · Inheritance and protection

Since the opening of the port in 1557, the fusion and collision of Chinese and Western cultures have left Macau with unique urban morphological characteristics [1]. Residential space form is an important component of urban morphology. Urban morphology effectively summarizes and expresses the physical form and humanistic connotation of a city. Its analysis method can not only provide an effective theoretical tool for understanding the physical form and humanistic connotation of the city but also provide scientific methodological support for deepening and strengthening urban historical protection planning. "Pátio" (圍) and "Beco" (里) are the unique traditional Chinese residential space forms in Macau [2]. Compared with the urban texture constructed by the Portuguese medieval city "Largo" (前地) in Macau [3], "Pátio" and "Beco" retain the community texture of the traditional Chinese cities in the Guangzhou and Xiangshan areas and form a spatial structure with the former

Y. Chen (✉) · J. Zheng
Faculty of Humanities and Arts, Macau University of Science and Technology, Taipa, Macau, China
e-mail: chenyile@must.edu.mo

Y. Chen et al. (eds.), *City-Sea Symbiosis*, Springer Geography,
https://doi.org/10.1007/978-3-032-13725-8_9

Portuguese urban area as the main part and the Chinese community as the auxiliary part. Thus, the street system of different levels and scales of "Avenida (大馬路), Estrada (馬路), Rua (街), Calçada (斜巷), Rampa (斜坡), Caminho (路), Escada (石級), Travessa (巷), Beco, Pátio" in Macau is formed. "Pátio" and "Beco" are the lowest-level streets that constitute the urban texture. "Li" is also a basic residential unit in the cities and villages of Guangzhou Prefecture in the Ming and Qing Dynasties and sometimes refers to the streets and alleys in the residential area [4]. Macau's "Pátio" and "Beco" have obvious morphological inheritance relationships with the "Li" in the Cantonese region in terms of both residential and transportation functions. Macau's "Pátio" and "Beco" are a localized residential type of traditional Chinese residential culture in Macau, forming the main part of Macau's traditional Chinese dwellings.

1 Current Status of "Pátio" and "Beco"

1.1 Concept Interpretation

On July 31, 1963, the Land, Public Works, and Transport Bureau of the Portuguese Macau Government published Gazette No. 30/1963, approving Legislative Regulation No. 1600, "*General Regulations on Buildings in Macau Province, City, and District*" (Diploma Legislativo n. °1600 REGULAMENTO GERAL DA CONSTRUÇÃO URBANA) [5]. In Article 32, "Pátio" is defined as "several buildings on the same plot of land that can be connected to the street directly or through a crossroad". And in Article 400 of the Ordinance on Fines, it is clearly stated that "If the owner of a Pátio fails to keep the area intact and clean, he shall be fined 15 yuan". It can be seen that the understanding of "Pátio" at that time was based on the same plot of land, connected to the city's public roads, and became a branch of the street system. The owners of Pátio were obliged to maintain the cleanliness of the internal environment and were supervised by the government.

However, some scholars believe that "Pátio" and "Beco" contain three levels of connotation [6]. The first is the plot level. "Pátio" and "Beco" are a series of 4–5 m wide, 8–10 m deep, and closely arranged plots. The second is the building level. "Pátio" and "Beco" plots generally have 2–3-story residential buildings. The third is the street level. "Pátio" and "Beco" are a kind of lowest-level road with a width of 1.5–3 m that serves the "Pátio" and "Beco" plots. The buildings in the "Pátio" and "Beco" plots all have entrances and exits facing the enclosed roads. From the perspective of history and architectural typology, "Pátio" and "Beco" are urbanized, high-density, and relatively closed residential forms that originated from the traditional Cantonese rural residential model. The conceptual boundaries of "Pátio" and "Beco" lack clarity, leading to the frequent interchange of these names. The Macau Street website [7], managed by the Municipal Affairs Bureau (IAM) of Macau, has provided a description of these areas: "Pátio" and "Beco" are the smallest existing roads in Macau. In the past, most Macau residents lived in these "Pátio" and "Beco", placing tables and chairs in public spaces to eat, play games, and chat [8]. Neighbors had close contact and connections. This type of interaction is also the most important function of "Pátio" and "Beco" for modern Macau.

There are 140 existing places named after the Portuguese word "Pátio" and 122 named after the Portuguese word "Beco". There is no mixing of the Portuguese names "Pátio" and "Beco". There are many mixed names for the corresponding Chinese places based on Portuguese (Table 1). The places referred to by the Portuguese words "Pátio" and "Beco" do not correspond strictly, and there are many Chinese names for them, which shows some differences between Chinese and Portuguese cultures.

1.2 Existing Number and Distribution

According to the *Cadastro das Vias Publicas de Macau* and on-site surveys, there are currently 271 streets named "Pátio" or "Beco" (Fig. 1), including 145 "Pátio" and 126 "Beco" [9–24]. Most of these "Pátio" and "Beco" are located in the Macau Peninsula (251, accounting for 93.3%) and are relatively concentrated in the buffer zone around the Historic Center of Macau, that is, within the "Portuguese City" during the Portuguese era. They are distributed from Patane (沙梨頭) to the Mount Fortress and the Na Tcha Temple in a northwest-southeast direction; the middle section of the Rua da Tercena (果欄街) is distributed in a northeast-southwest direction, concentrated in the market area between the edge of the Complexo Municipal do Mercado de S. Domingos (營地街市), Baía do Mastro (爐石塘), and Rua dos Mercadores (營地大街), and some "Pátio" and "Beco" cross Avenida de Almeida Ribeiro (亞美打利庇盧大馬路) and extend to Rua da Felicidade (福隆新街) and Rua Do Gamboa (夜呣街); in the Praia do Manduco (下環) area, they are distributed discontinuously from Rua da Praia do Manduco (下環街) to Barra; and outside the city wall of "Portuguese City", they are mainly concentrated near Mong-Há (Fig. 2). These "Pátio" and "Beco" are very close to the residential space of Portuguese City distributed along the ridge in terms of location, but one is lower and the other is higher in vertical space. In the outlying islands, there are only a few "Pátio" and "Beco" in the old urban areas of Taipa and Coloane. The total length of "Pátio" is about 7292 m, and the average street length of each "Pátio" is 52.1 m; the total length of "Beco" is about 6507 m, and the average street length of each "Beco" is about 53.3 m.

Researchers found through on-site investigation that the original nine "Pátio" and "Beco" have disappeared, namely: Beco da Lavra (耕里/耕圍); Pátiodo Tabuado (板堆圍/板堆里); Pátio das Escondidas (內裡圍); Beco do Coronel Mesquita (美上校里); Beco dos Tingidores (染布里); Pátio da Boa Vista (竹室圍); Beco do Leite (黎地里); Pátio do Mirante (瞭望圍/安懷居); and Pátio da Greta (裝鏬圍). These "Pátio" and "Beco" still have road signs, but the buildings have been demolished, the plots have been merged and rebuilt, or they have become traffic shortcuts, streetside

Table 1 List of existing mixed usage of "Pátio" and "Beco" in Macau *Source* The author drew this map based on on-site surveys and street directory analysis

No.	Chinese name	Quantity	Description
1	圍	96	Portuguese name corresponding to "Pátio"
2	里	53	Portuguese name corresponding to "Beco"
3	社	1	Portuguese name corresponding to "Beco", such as Beco central (聚龍舊社)
4	台	10	Portuguese name corresponding to "Pátio"
5	巷	1	Portuguese name corresponding to "Beco", such as Beco da Rosa (玫瑰巷)
6	圍/里 (mixed use case)	84	Portuguese name corresponding to "Pátio"
7	圍/村 (mixed use case)	1	Portuguese name corresponding to "Pátio", such as Pátio do Espinho (茨林圍)
8	圍/坊 (mixed use case)	1	Portuguese name corresponding to "Pátio", such as Pátio do Jardim (花園圍/叢慶坊)
9	圍/街 (mixed use case)	1	Portuguese name corresponding to "Pátio", such as Pátio do Cotovelo (手肘圍/德隆新街)
10	圍/里/街 (mixed use case)	1	Portuguese name corresponding to "Pátio", such as Pátio da Felicidade (福隆圍/清和里/虱乸街)
11	圍/市 (mixed use case)	1	Portuguese name corresponding to "Pátio", such as Pátio do Mercado interior de Miguel Aires (美基內市場圍/味機市)
12	圍/巷 (mixed use case)	3	Portuguese name corresponding to "Pátio", such as Pátio do Ídolo (公仔圍/公仔巷); Pátio do Carpinteiro (木匠圍/木匠巷); Pátio das Esquinas (轉角圍/大鵬橫巷)
13	里/街 (mixed use case)	3	Portuguese name corresponding to "Beco", such as Beco da Melancia (西瓜里/果欄橫街); Beco da Pinga (擔桿里/草堆橫街); Beco do Tarrafeiro (沙欄仔里/沙欄仔橫街)
14	里/津/街 (mixed use case)	1	Portuguese name corresponding to "Beco", such as Beco da Rosa (玫瑰里/聚龍通津/騎樓街)
15	里/口 (mixed use case)	1	Portuguese name corresponding to "Beco", such as Beco dos Coulaus (高樓里/船澳口)
16	圍/里/巷 (mixed use case)	1	Portuguese name corresponding to "Beco", such as Beco do Cego (鹽埠里/鹽埠圍/椰子巷)

(continued)

Table 1 (continued)

No.	Chinese name	Quantity	Description
17	里/尾 (mixed use case)	2	Portuguese name corresponding to "Beco", such as Beco da Ostra (蠔里/果欄尾); Beco dos Cotovelo (手肘里/左堂欄尾)
18	里/巷 (mixed use case)	7	Portuguese name corresponding to "Beco"
19	里/仔 (mixed use case)	1	Portuguese name corresponding to "Beco", such as Beco do Gamboa (夜冚里/深巷仔)

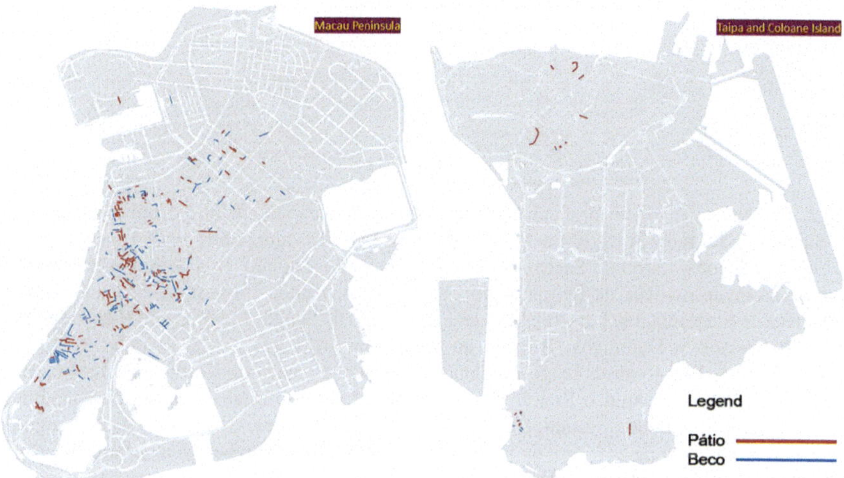

Fig. 1 Spatial distribution of "Pátio" and "Beco" in Macau. *Image source* drawn by the authors

recreational areas, or temporarily used as private parking lots. Therefore, the actual number of "Pátio" and "Beco" in Macau is 262.

2 The Origin and Cause of "Pátio" or "Beco"

2.1 The Docking of Tanka Boats and the Land-Based Migration of Tanka

Initially, the Chinese residents in the Inner Harbor of Macau were mainly Tanka people (疍家) [25–27]. Tanka people, also known as boat dwellers or boat people or the derogatory "Tankas", are a sinicized ethnic group in Southern China who traditionally lived on junk in coastal parts of Guangdong, Guangxi, Fujian, Hainan, Shanghai, and Zhejiang and along the Yangtze River, as well as in Hong Kong

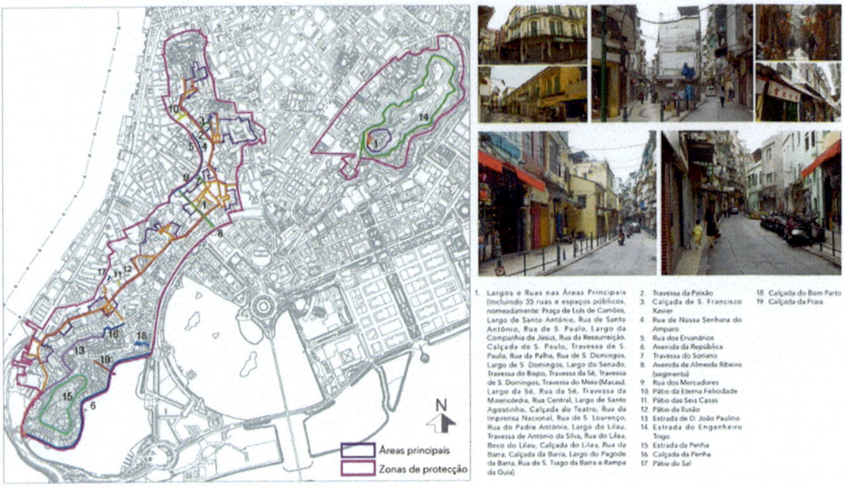

Fig. 2 The streets in the historic center of Macau. Number 1 in the picture details and translates to "Squares and Streets in the Main Areas (including 35 streets and public spaces, namely: Praga de Luís de Camóes, Largo de Santo António, Rua de Santo António, Rua de S. Paulo, Largo da Companhia de Jesus, Rua da Ressurreição Calçada de S. Paulo, Travessa de S. Paulo, Rua da Palha, Rua de S. Domingos, Largo de S. Domingos, Largo do Senado, Travessa do Bispo, Travessa da Sé, Travessade S. Domingos, Travessa do Meio (Macau), Largo da Sé, Rua da Sé, Travessa da Misericórdia, Rua Central, Largo de Santo Agostinho, Calcada do Teatro, Rua da Imprensa Nacional, Rua de S. Lourenco Rua do Padre António, Largo do Lilau, Travessa de António da Silva, Rua do Lilau, Beco do Lilau, Calçada do Lilau, Rua daBarra, Calgada da Barra, Largo do Pagode da Barra, Rua de S. Tiago da Barra e Rampada Guia)". Numbers 2 to 19 are the names of the streets. *Image source* the author adapted the drawing from the *Protection and Management Plan for the historic center of Macau*, and the photo was taken by the author

and Macau. The Qing government implemented management of the Tanka boats in Macau. For example, the Qing government uniformly registered all boats according to the "baojia system" and concentrated them in the Rua do Teatro Bay, where the Tanka people began to live together. Portuguese scholars pointed out that as the beach continued to silt up, boat houses began to gather on the dry mudflats, forming a group of "dry boat houses" and becoming part of the city (Fig. 3). This kind of residential prototype could still be seen in the Fai Chi Kei (筷子基) wooden house area in Macau in 1987 and even in the aerial photos of Tai O fishing village in Hong Kong in 1963. The inner harbor began to become the main area for Chinese to settle south of the Portuguese city wall in Macau.

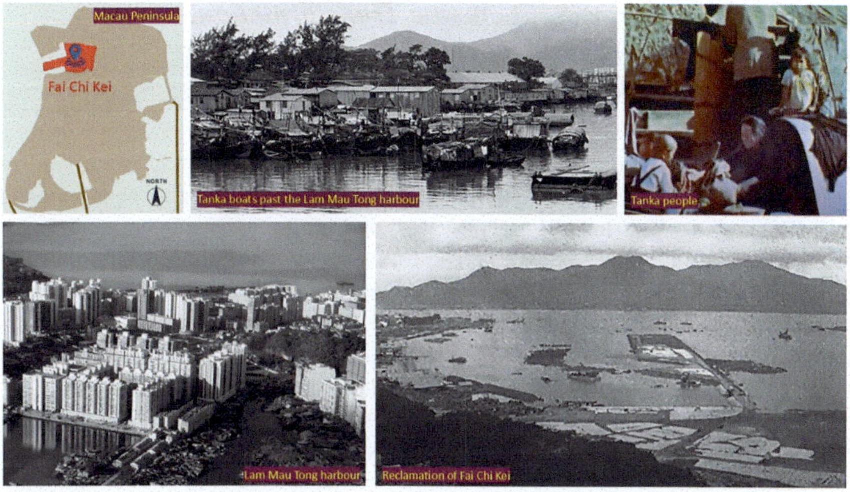

Fig. 3 The Tanka people and houseboats in the harbor past Fai Chi Kei. *Image source* author notes

2.2 The Disappearance of Chinese Villages Outside the "Portuguese City" and the Formation of Chinese Settlements Inside the "Portuguese City"

According to the *Cadastro das Vias Publicas de Macau*, in 1869, there were 212 "Pátio" or "Beco" in the Macau Peninsula (98 in the "Portuguese City" and 114 in Chinese villages outside the "Portuguese City"), of which 51 were in the Inner Harbor area; in 1878, there were 221 "Pátio" or "Beco" (139 in the "Portuguese City" and 82 outside the "Portuguese City"), of which 69 were in the Inner Harbor area; in 1896, there were 246 "Pátio" or "Beco" (160 in the "Portuguese City" and 86 outside the "Portuguese City"), of which 81 were in the Inner Harbor area; and in 1905, there were 210 "Pátio" or "Beco" (191 in the "Portuguese City" and 19 in the suburbs), of which 78 were in the Inner Harbor area. The change in quantity shows that the Chinese villages "Pátio" or "Beco" are gradually disappearing due to urbanization, while similar residential patterns are gradually increasing in the Chinese settlement areas in the Inner Harbor and the Macau Peninsula, that is, the urbanization of the Chinese villages "Pátio" or "Beco". If the statistics are based on the Inner Harbor and the Chinese villages outside the city, there were 165 Chinese settlements, "Pátio" or "Beco" in 1869, far more than the 47 in the Portuguese city. The huge difference in quantity is enough to show that "Pátio" or "Beco" are actually a continuation and promotion of the traditional residential model of Chinese villages.

2.3 *Population Growth, Economic Transformation, and Social Formation of the Chinese Population in Macau in Modern Times*

In modern times, with the interruption of international trade routes, the turmoil in the mainland's political situation, and the transformation of Macau's economy, the Chinese population has been rising in waves, and in the mid-nineteenth century it entered a period of explosive growth. The Chinese population in Macau has established absolute dominance since the mid-nineteenth century. Many Chinese people from the Pearl River Delta region needed more living space. Figures 4 and 5 show that from the known construction years of "Pátio" and "Beco", the construction of a large number of "Pátio" and "Beco" has a clear correlation with the growth of the Chinese population in Macau. The construction of many "Pátio" and "Beco" met the living space required by the growth of the Chinese population at that time.

Relevant data show that most Chinese immigrants to Macau in modern times are from Guangdong, especially from the Pearl River Delta. The primary occupations of these new immigrants are industry and commerce. In the context of the Portuguese-led urban planning and construction, these immigrants, who mainly come from the Pearl River Delta, can only build residential, production, and living spaces in the form of "Pátio" and "Beco" in the "marginal areas" of the city's "main planning" and construction. Macau's "Pátio" and "Beco" have thus become urban residential units derived from the traditional rural residential form of the Pearl River Delta.

As many Chinese immigrants have a growing demand for living space, Chinese businessmen in Australia have actively participated in real estate development centered on "Pátio" and "Beco", which not only meets their own needs for living but also allows them to rent and sell properties for profit. For example, famous Chinese

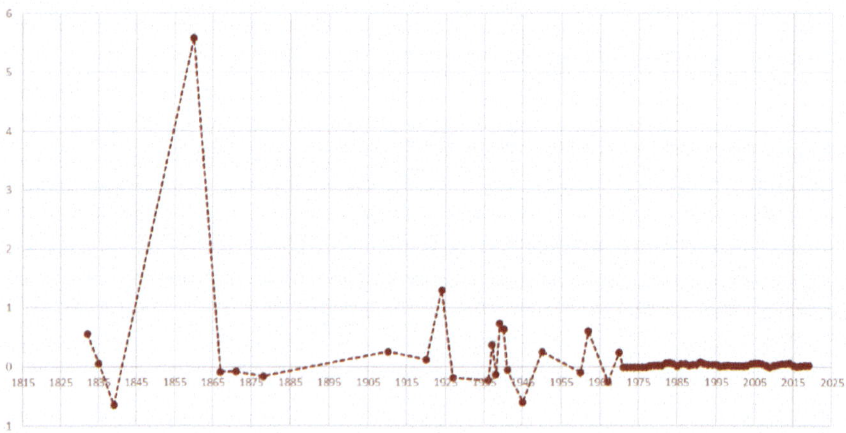

Fig. 4 Statistics on the population growth rate of Macau in modern times. *Image source* drawn by the author

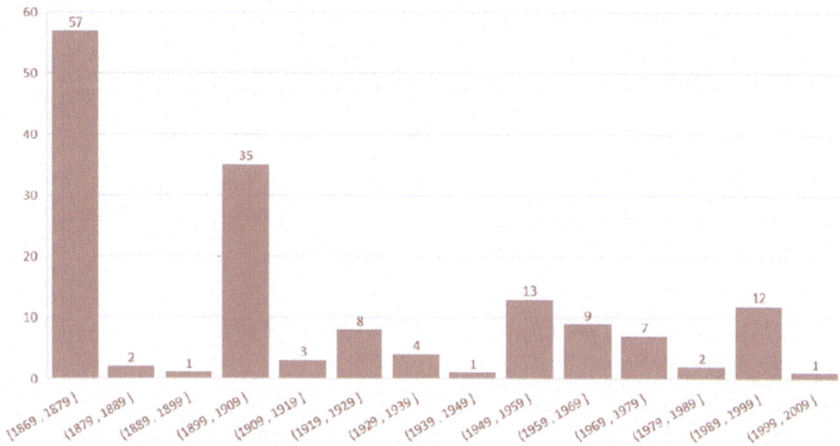

Fig. 5 Analysis of the construction period and quantity of the gates "Pátio" and "Beco" (1869–2009). *Image source* drawn by the author

merchants such as Vong Lok (王祿), Hó Lo Quai (何老桂), Chou Yau (曹有), and Lou Kau (盧九) owned a large number of properties in various parishes in Macau. Most of these properties were in the form of "Pátio" and "Beco" or within "Pátio" and "Beco", becoming the residence and production units for Chinese immigrants who had just entered Macau. In 1881, the Chinese merchant Jung Mun (容滿) owned a total of 198 properties, of which 39 properties were located in "Pátio" and "Beco".

It must be pointed out that in the process of promoting colonial expansion and leading urban construction, the Portuguese Macau government once recognized the Portuguese Macau government as a landlord and developed real estate "inside and outside the city" through the Chinese to achieve the purpose of land expansion and governance over the Chinese. The Chinese built and lived in the "Pátio" and "Beco" blocks, constructing residential units with clear ownership, self-sufficiency, safety and self-protection, and familiar neighbors, gathering the same geographical or business relationships, family groups, and social classes, as well as the same cultural customs and beliefs, etc., so that the Chinese community adjacent to the Portuguese was formed and its cultural customs were continued.

The entry of Chinese immigrants into Macau is characterized by stages and diversity. Chinese immigrants from different periods and different classes have had an overlapping impact on the construction of "Pátio" and "Beco" and the formation of their cultural characteristics. Specifically, in the early days, most of the "Pátio" and "Beco" were unit-type single-family houses, while in the later period, unit-type upper and lower two-family houses appeared, and the residential density continued to increase. The "Pátio" and "Beco" built in the early twentieth century, began to mix Portuguese architectural styles. The "Pátio" and "Beco" buildings of the wealthy Chinese are relatively large and beautifully decorated, such as Pátio da Eterna Felicidade (Beco da Enterna Celebração, 永福圍), Pátio das Seis Casas (六屋圍), and

Pátio da Ilusão (幻覺圍), while the "Pátio" and "Beco" buildings of the middle and lower classes are comparatively simple or even crude, such as the Pátio do Socorro (工匠圍), Pátio dos Cules (苦力圍), and Pátio do Mainato (洗衣匠圍). At the same time, the "Pátio" and "Beco" of different periods have also become Chinese family living spaces and production sites. For example, among the 78 "Pátio" and "Beco" streets in the Inner Harbour recorded in the *1905 Macau street directory*, 16 were named after Chinese surnames and 21 were named after industries or occupations.

3 Spatial Features of Traditional Chinese "Pátio" and "Beco"

3.1 The Street Layout of Outer Shops and Inner Houses Combined with Land Beliefs

The construction of Chinese-style "Pátio" and "Beco" residential areas is an important carrier of Chinese culture. The Chinese "Pátio" and "Beco" blocks and the newly reclaimed Portuguese blocks of Royal New Street are juxtaposed with each other, reflecting the unique Chinese business culture, that is, shops in front and houses in the back. The streets are surrounded by commercial shop lots, and the single buildings of the lots adopt the architectural space model of shops in front and houses in the back or shops below and houses above. "Pátio" and "Beco" as the main residential buildings are surrounded inside, filling the space inside the block. This approach not only meets the requirements of privacy and security of traditional residential buildings but also increases the density of street buildings (Fig. 6). Compared with the back-to-back model of Portuguese small blocks, "Pátio" and "Beco" blocks better reflect the balanced thinking of Chinese residence, life, and business.

The land temples that symbolize the traditional belief culture of the Chinese are also distributed in the "Pátio" and "Beco" units and streets formed during this period (Fig. 7). The "Pátio" and "Beco" residential units formed a pattern similar to the

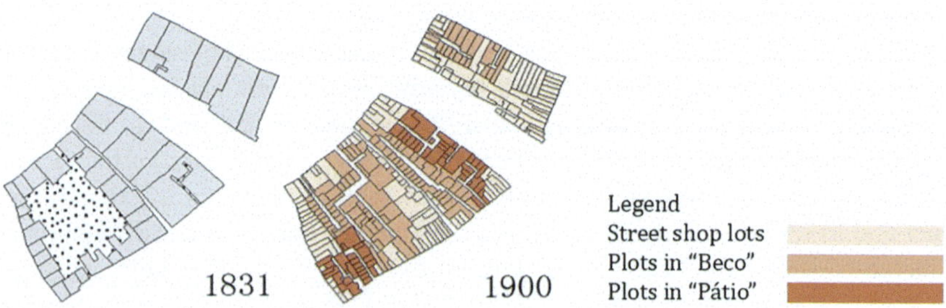

Fig. 6 The street pattern of the outer shop and inner house between Rua dos Mercadores and Rua de Camilo Pessanha. *Image source* drawn by the author

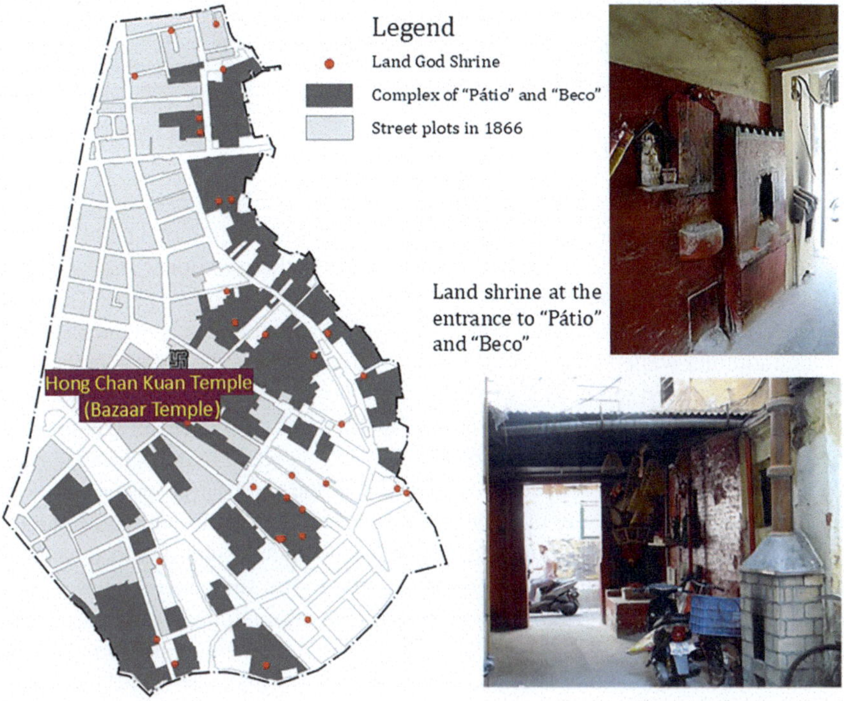

Fig. 7 Map of the location of land temples/shrines. *Image source* drawn by the author

residential areas of the Tang and Song dynasties. Each "Pátio" and "Beco" unit has a gate (also called a Pátio gate) at the entrance, and a land temple is set up next to the gate, forming a network of Chinese beliefs. At a higher level, there are temples to protect the community, such as Hong Chan Kuan Temple (Bazaar Temple), etc., which serve the area. Hong Chan Kuan Temple is the highest-level temple building in the Chinese community in the Inner Harbor area. It is an important symbol of the Chinese belief network and has become the center of Chinese belief in the Inner Harbor area and the center of Chinese business. The street morphology characteristics of "Pátio" and "Beco" are different from the Portuguese small block planning model. The construction of the Chinese "Pátio" and "Beco" streets and the Portuguese urban planning together constitute the collage-like street complex characteristics.

3.2 *Building Layout*

The traditional Chinese "Pátio" and "Beco" are the prototypes of Macau's "Pátio" and "Beco", and they are also the most distinctive "Pátio" and "Beco" that exist today. They reflect the origins and customs of Chinese life in terms of architectural space combinations, building unit types, and ancillary structures.

3.2.1 Architectural Space Combination

The "Pátio" and "Beco" buildings are modular residential forms, consisting of several building units arranged according to the shape of the plot. The internal building units and combination methods are actually very similar to the traditional Hakka walled houses in Guangdong. The Hakka walled house is divided into two parts: The front half is a combination of the main house and the horizontal house, and the back half is a semi-circular walled house. Depending on the size, the walled house in the back has one, two, or even more. The spatial form of the walled house has the following basic characteristics: (1) A simple arrangement of corridors + enclosure rooms. The enclosure rooms are closely arranged, with the entrance and main lighting surface of each room facing the corridor. The corridor is the public transportation of all enclosure rooms. (2) The size of each enclosure room is basically the same, and the first floor is the main one. (3) It is relatively independent and subordinate, located on the side or back of the main room or the horizontal room, and generally has a door.

The "Pátio" and "Beco" buildings in Macau (Fig. 8) continue the spatial combination model similar to the "walled house", combining the site conditions to build high-density residential units. Compared with the "Li" in Guangzhou, the "Pátio" and "Beco" buildings in Macau are more traditional residential units that form various combinations in response to the neighborhood environment and terrain.

3.2.2 Types of Building Units

The prototype of the Chinese "Pátio" and "Beco" building units is the bamboo tube house and bamboo pole house in Cantonese and Chaoshan folk houses. The span is about 4 m, and the width is calculated by the number of tile beams, which are generally 15–21 beams (each beam is about 27 cm, that is, 9 inches of wood), with a patio kitchen at the back. From the comparison of the floor plan units of Travessa das Venturas, Pátio das Seis Casas, and Beco da Felicidade, the floor plan layout of the "Pátio" and "Beco" building units is almost the same as that of the bamboo tube house (Figs. 9, 10, 11, and Fig. 12): The first floor is the living room, dining room, kitchen, and toilet, and the second floor is mainly used as bedrooms.

The double-bay combination of bamboo tube houses is called Mingziwu. The first bay is the living room, the other bay is the bedroom, and the back is the kitchen with a patio. Pátio da Eterna Felicidade is a typical "Pátio" and "Beco" building that is a combination of bamboo tube house units and Mingziwu units. The existing buildings in the "Pátio" and "Beco" are divided into two parts. The upper part is two Mingziwu units (Fig. 13) + 5 bamboo tube house units, and the lower part is five bamboo tube house units. In terms of quantity, "Pátio" and "Beco" buildings are mainly developed by bamboo tube house units in parallel, and Mingziwu units are rare. The "Pátio"

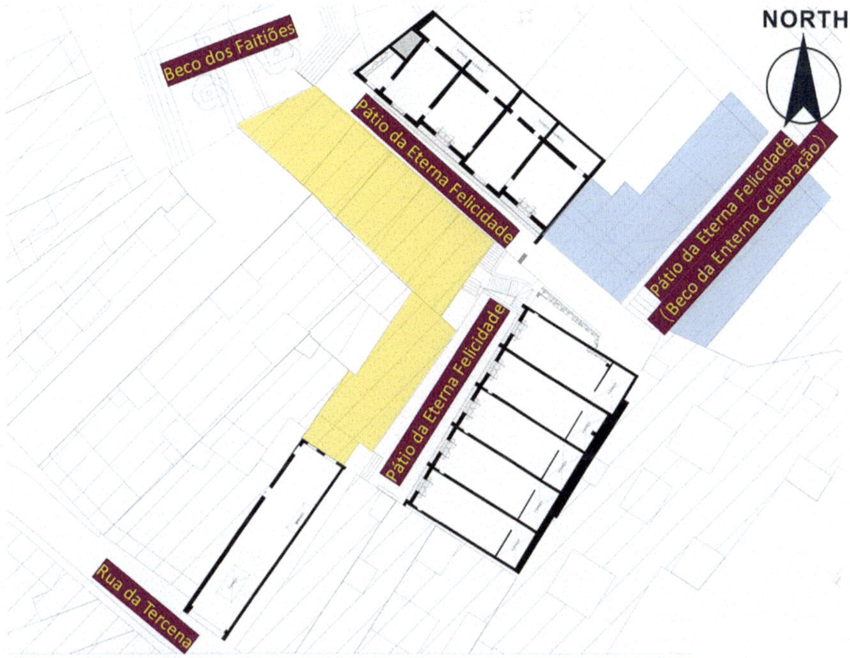

Fig. 8 Plan of Pátio da Eterna Felicidade, Macau. *Image source* drawn by the author

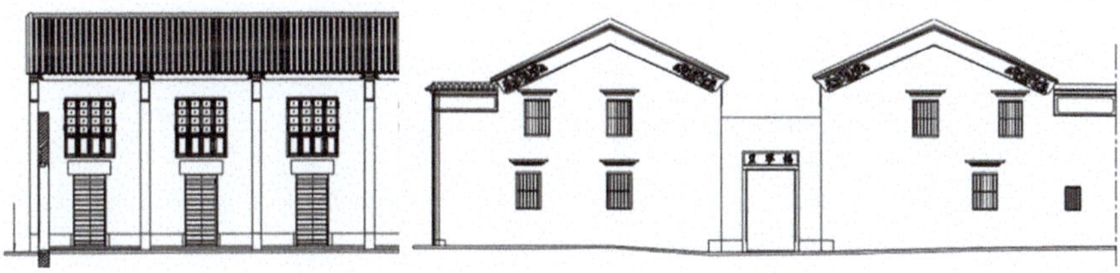

Fig. 9 Facade of Travessa das Venturas, Macau

and "Beco" buildings inherited the advantages and disadvantages of the combination of bamboo tube houses and Ming-shaped houses. Their advantages are economical use of land and compact combination, but the building density of these two types of dwellings is too high, ventilation and lighting are poor, and interference is great. In comparison, whether it is active "planning" or "passive" choice, building "Pátio" and "Beco" in the narrow geographical space of the Inner Harbor can make full use of the land, effectively increase the residential density, and better meet the actual life needs.

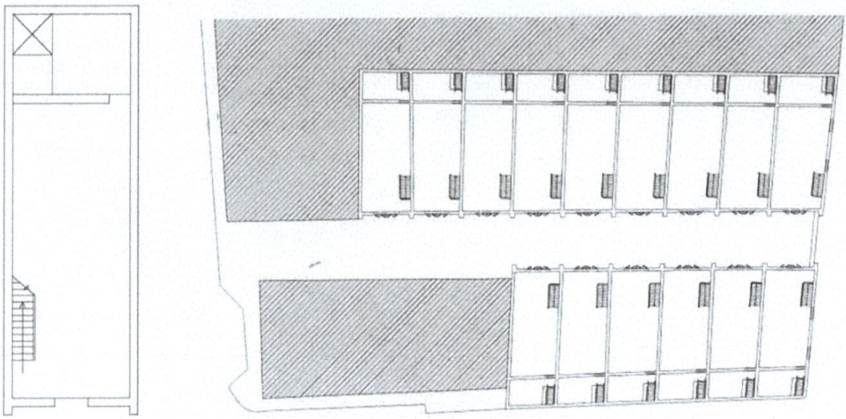

Fig. 10 Floor plan of Travessa das Venturas, Macau

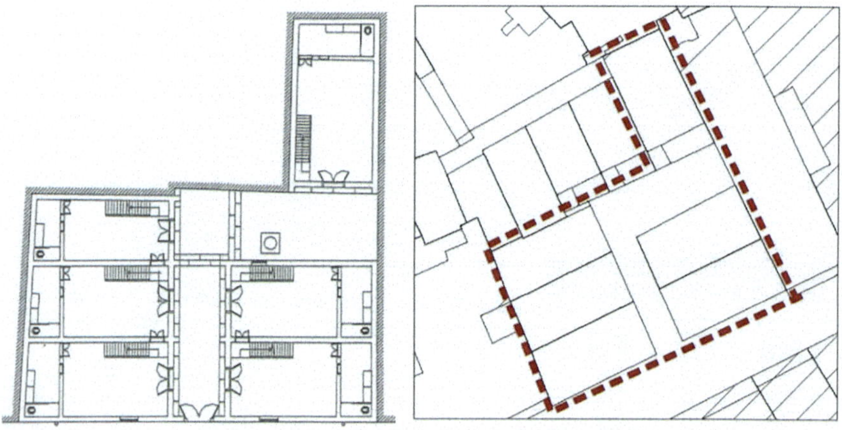

Fig. 11 Floor plan of Pátio das Seis Casas, Macau

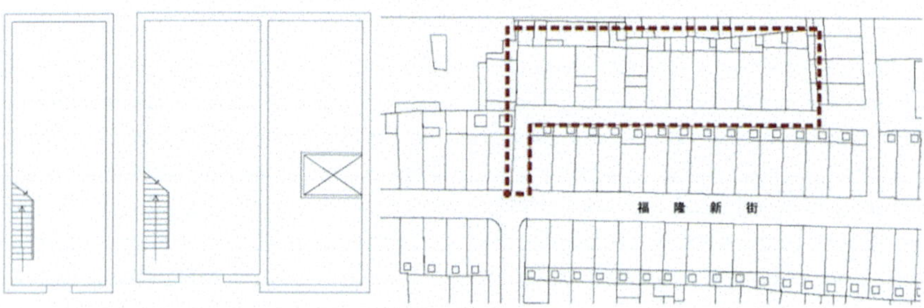

Fig. 12 Floor plan of Beco da Felicidade, Macau

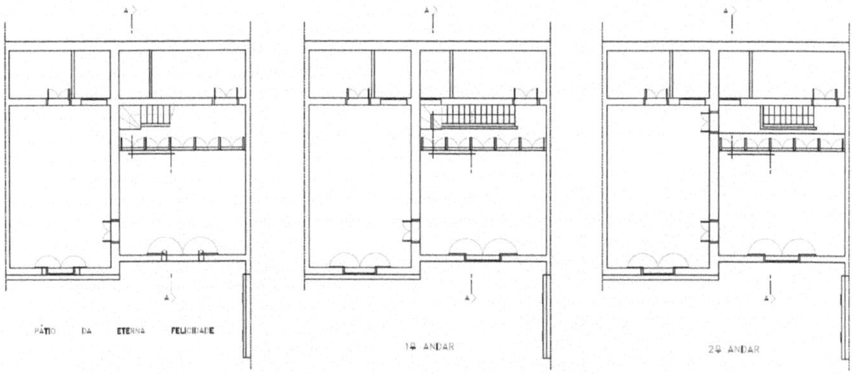

Fig. 13 Mingziwu residential buildings and Pátio da Eterna Felicidade units in Macau

3.2.3 Ancillary Structures

In addition to residential buildings, traditional Chinese "Pátio" and "Beco" generally have supporting ancillary structures, including gates, land god chapels, wells, etc. "Pátio" and "Beco" are very introverted and often have entrance gates at locations connected to the outside world to enhance defensiveness. At the same time, the entrance gate also plays an important decorative and identification role, defining the internal and external spaces and forming a relatively independent residential unit. The God of Land is very common in Chinese houses in Macau, especially in the Inner Harbor area. Traditional Chinese "Pátio" and "Beco" usually set it near the entrance. The God of Land can bless the residents to enter and leave safely. It is a religious need and spiritual symbol of the residents of traditional Chinese "Pátio" and "Beco", and it is also one of the important characteristics that distinguish it from other types of "Pátio" and "Beco" and other types of urban areas. On the other hand, before the installation of tap water in Macau, wells were the main source of drinking water for residents. Large "Pátio" and "Beco" usually have wells, such as Pátio das Seis Casas and Pátio do Espinho (Figs. 14 and 15). Wells can meet the daily needs of residents and are also important facilities for providing water for firefighting.

3.3 Main Architectural Features of "Pátio" and "Beco"

In general, there are few examples in Macau where the existing traditional "Pátio" and "Beco" architectural features are relatively well preserved. The notable features of the traditional Chinese "Pátio" and "Beco" are mainly reflected in the walls, doors and windows, roofs, and entrance gates.

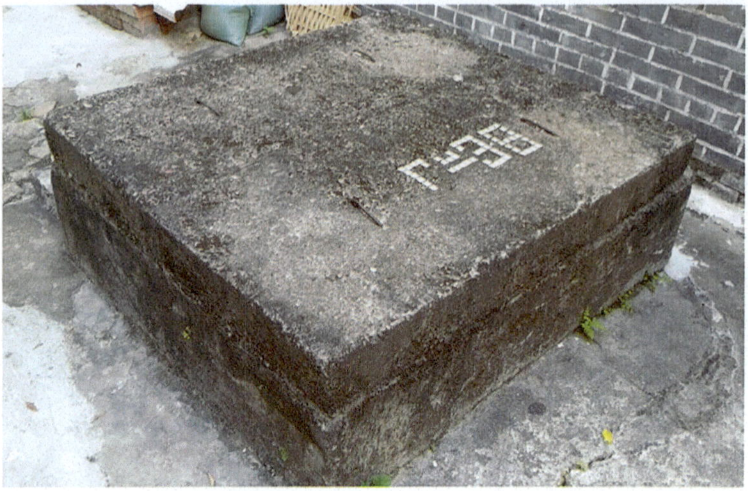

Fig. 14 Wells in the Pátio das Seis Casas. *Image source* photographed by the authors

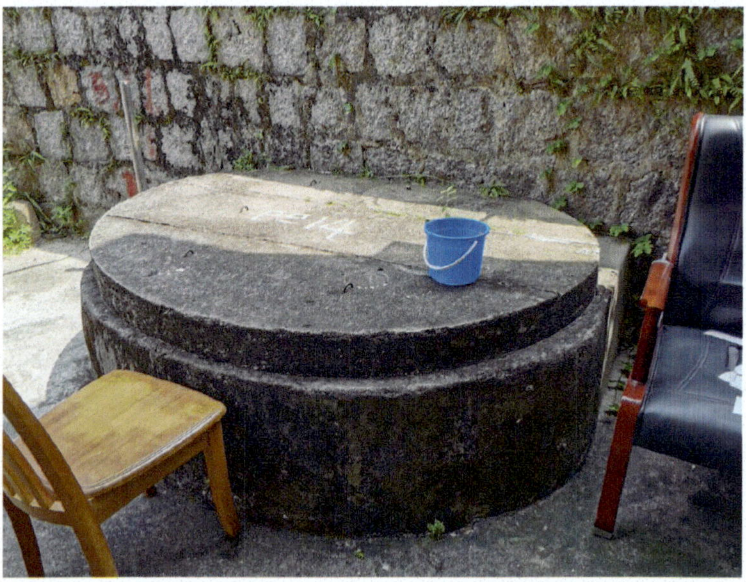

Fig. 15 Wells in the Pátio do Espinho. *Image source* photographed by the authors

3.3.1 Wall

There are two main types of walls in Macau's "Pátio" and "Beco". One is traditional blue brick masonry, with "strip masonry method" I-shaped masonry as the main method. Some "Pátio" and "Beco" walls use straight-edge masonry. For example, the "flemish garden wall bond" method is used in Pátio das Seis Casas (六屋圍/福六里), which is a relatively rare method. The other type of wall is mainly painted or water-brushed stone on the outer wall surface, and the wall structure is brick or stone masonry. The wall is the main load-bearing structure of the building. The walls of traditional Chinese "Pátio" and "Beco" have a front wall that protrudes about 0.5 m forward, with a lintel decoration on the wall. The lintel under the eaves is usually decorated with colorful paintings, emphasizing the vertical unit characteristics of the facade (Figs. 16 and 17). The Chinese "Pátio" and "Beco" have a bofeng at the junction of the mountain wall and the roof, and the surface is covered with black smoke, usually decorated with scroll patterns, but most of them are now damaged.

The outer walls of the Western-style "Pátio" and "Beco" are continuous, which further highlights the horizontal continuity of the entire "Pátio" and "Beco" building. The wall paint is also rich in color, including yellow, green, white, etc. The mountain walls of the Western-style "Pátio" and "Beco" are generally decorated with triangular pediment lines, reflecting the characteristics of Western classical elements.

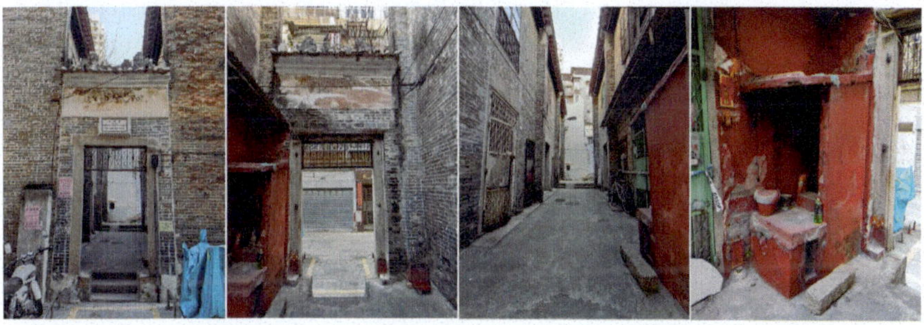

Fig. 16 The walls, eaves, doors and windows of the Pátio das Seis Casas

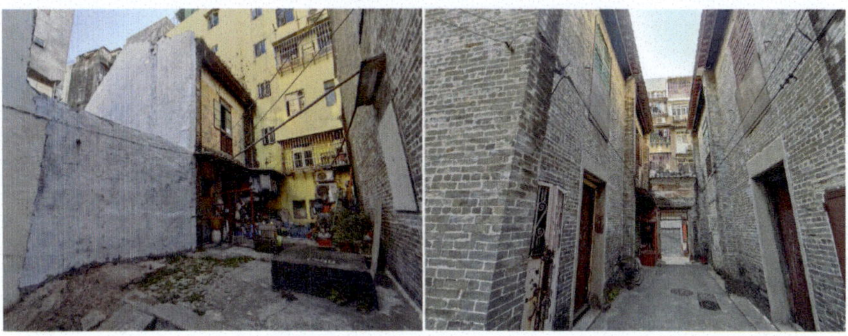

Fig. 17 The walls of the Pátio das Seis Casas. *Image source* photographed by the authors

3.3.2 Doors and Windows

The doors and windows of "Pátio" and "Beco" buildings mainly open to the inside of the "Pátio" and "Beco". There are no doors and few windows on the walls facing the outside streets. In traditional Chinese "Pátio" and "Beco" building units, the main door is in the most prominent position. The door opening is built with granite and is generally composed of three doors: The first is a low door (low-foot door); the second door is a sliding door, which is composed of several horizontal logs (Fig. 18); the third is a hardwood door, which is called "three-piece set" in the Cantonese area. It is a common small wooden component that has the functions of anti-theft, lighting, and ventilation [28]. There are still relatively well-preserved examples in Beco da Felicidade (福榮里), but most of the "Pátio" and "Beco" doors have been destroyed or replaced. Some "Pátio" and "Beco" are divided into upper and lower households, and a small door is opened next to the main door, which leads to the stairs directly to the first floor. The ground floor and the first floor are separated, and the entrances and exits are completely independent and do not interfere with each other.

The windows of "Pátio" and "Beco" buildings are another major feature. Traditional Chinese "Pátio" and "Beco" windows use double-leaf outward-opening sill

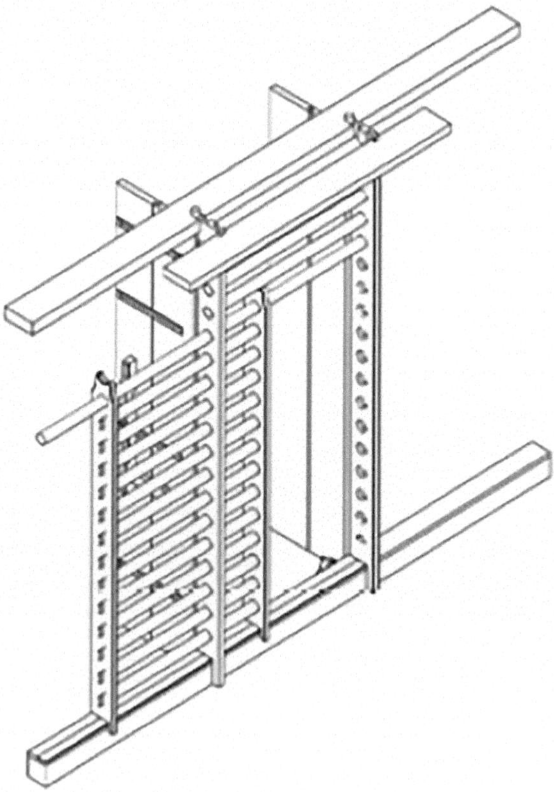

Fig. 18 Components of sliding door

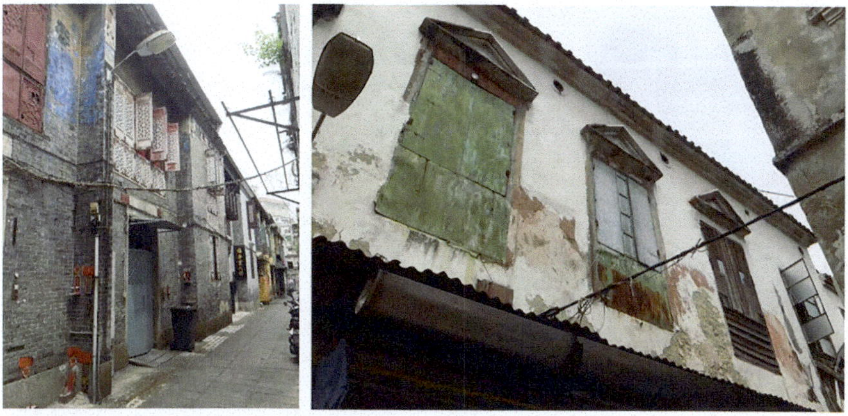

Fig. 19 Comparison of Chinese-style doors and windows (left) and Portuguese-style doors and windows. *Image source* photographed by the authors

windows with sill plates below the windows or are made up of outward-opening sashes and railings. The sill windows are mostly hexagonal ice cracks, which means "six sixes are endless", and the frames are inlaid with oyster shells or glass (Fig. 19). Some "Pátio" and "Beco" are influenced by Portuguese architecture and use shutters, or a combination of ice crack inner windows and shutter outer windows. Portuguese "Pátio" and "Beco" windows usually have window lintels above them, which can be in the shape of a straight line, triangle (Fig. 19).

3.3.3 Roof

The roof structure of "Pátio" and "Beco" buildings is mainly a double-slope roof with hard mountain and purlins (also known as a pyramid roof). The roof of Chinese "Pátio" and "Beco" generally does not have a ceiling inside, but Western "Pátio" and "Beco" or traditional Chinese "Pátio" and "Beco" influenced by them often use false ceilings on the roof, leaving ventilation holes at the four corners or around the four sides. The main roof material is tiles, which can be single layer or double layer. The former is to place rafters on the beams, lay flat tiles on top, with seven tiles on top and three exposed, and cover the flat tiles with tubular tiles, with mortar as bonding material; the latter is to lay two layers of tiles, and stop laying when the upper layer of tiles is 20–30 cm away from the ridge, in order to form an overhead layer between the upper and lower layers of tiles for heat insulation and ventilation. There are three main methods for eaves on the front facade: the traditional sloping roof in the Cantonese region, the stacked eaves, and the low guard wall. The traditional sloping roof in the Cantonese region is that the roof protrudes about 0.5 m from the wall, supported by end walls at both ends, and purlins are set between the end walls to support the upper protruding roof. The "three eaves" are used under the

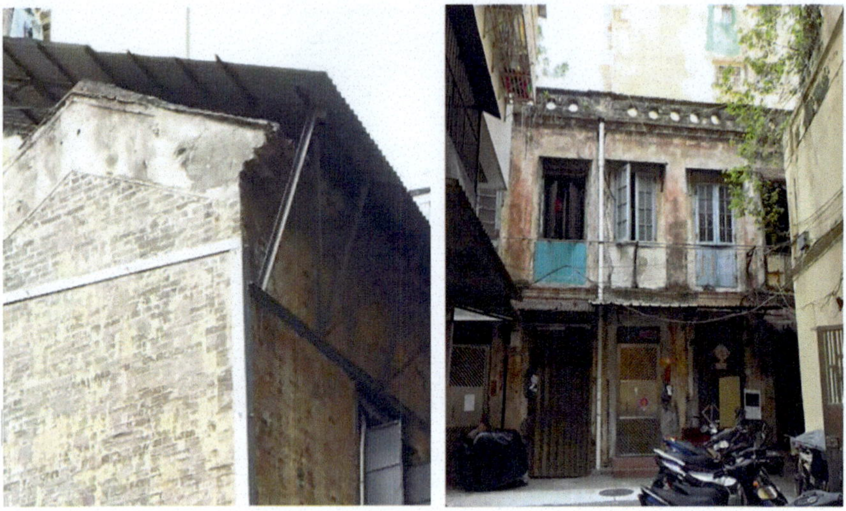

Fig. 20 The overlapping overhangs of the traditional Chinese "Pátio" and "Beco" roof (left) and the low guardrail of the Portuguese "Pátio" and "Beco" (right). *Image source*: photographed by the authors

eaves of the roof, namely the eaves board, the chicken breast fly, and the drip tile. Its main function is to drain rainwater from the roof, thereby protecting the walls of the building.

The method of stacking the cornice is relatively simple and is commonly used in "Pátio" and "Beco" buildings. However, because the distance of the stacking cornice is generally only about 0.3 m (Fig. 20), the lower wall surface is easily stained. The outside of the stacking cornice is whitewashed to form a concave curve. The Western-style "Pátio" and "Beco" buildings in Macau mostly use low retaining walls (also known as parapets) on their front facades, which originated from Western architecture. Visually, the parapet is an upward extension of the main facade wall, which blocks the pyramid roof behind it, making the vertical space of the entire building stronger (Fig. 20). To increase the decorative effect of the parapet, Portuguese parapets generally use round holes as decorative elements. The eaves do not extend beyond the outer wall, and an open ditch is set behind the parapet to guide rainwater into the drainage pipe.

3.3.4 "Pátio" and "Beco" Gates

"Pátio" and "Beco" doors are the entrance doors of traditional Chinese "Pátio" and "Beco". They play the role of living, management, and defense. They are often the decorative focus of "Pátio" and "Beco". They are divided into two types: linear type (Fig. 21) and gatehouse type (Fig. 22). The linear type is generally a door opening on the entrance wall of the enclosure, with granite as the door frame, and the name of the enclosure or village is usually engraved on the lintel. For example, the lintel of Pátio do Socorro (工匠圍/居安里) is engraved with "居安里". Some linear gatehouses combine Portuguese elements in detail, such as Portuguese moldings, low guardrails with round holes. What is even more special is that the gatehouse of Beco dos Vendilhões (桂和里) adopts Bogu ridge and Western-style decoration, forming a unique style that combines Chinese and Western styles; the gatehouse style, that is, the upper part of the wall gate is raised, and the roof and ridge decoration are added. When it is processed, it is often combined with the form of the wall to form an overall gatehouse style. The gatehouses of "Pátio" and "Beco" in Macau are both single-eaved gatehouses, that is, the roof is added to the wall. The gatehouses of Pátio das Seis Casas (六屋圍/福六里) and Pátio do Sal (鳳仙圍) are the more typical single-eaved gatehouse styles that exist today, but they are slightly different. The roof of the Pátio das Seis Casas (六屋圍/福六里) gatehouse adopts the stacked eaves method, which is relatively small, and the ridge adopts the Bogu ridge; the Pátio do Sal (鳳仙圍) gatehouse is more exquisite, with end walls on both sides, overhanging eaves, and green glazed tiles on the roof. The Pátio da Ilusão (幻覺圍/高樓里) gatehouse is the highest-standard gatehouse in Macau. It has two floors on the outside, end walls on both sides, large overhanging eaves at the front and back, and a depth of about 2 m. The brick-built arched doorway, shutters on the top, and eaves under the eaves, the whole gatehouse is majestic and exquisite. In comparison, the gatehouses of Pátio dos Cules (苦力圍/聚龍里) and Pátio do Mainato (洗衣匠圍/青雲里) embody the mixed characteristics of Chinese and Portuguese.

4 The Evolution of "Pátio" and "Beco" Residential Forms in the Early Twentieth Century

During this period, "Pátio" and "Beco" gradually integrated into the urban space of Macau, and the Chinese gradually incorporated the style features and internal structure of traditional Portuguese architecture into the newly built "Pátio" and "Beco". Simultaneously, under the influence of foreign architectural culture, some "Pátio" and "Beco" also displayed neoclassical architectural styles. Portuguese "Pátio" and "Beco" are mainly reflected in the architectural layout and facade style. The most typical ones are Pátio da Claridade (光復圍/光復里), Pátio da Águia (順成圍/順成里), Beco do Lilau (龍頭里), Pátio do Serralheiro (馮家圍), etc. Unlike traditional Chinese "Pátio" and "Beco", which use end walls to separate each parallel building unit, Portuguese "Pátio" and "Beco" form a continuous whole on the facade. The roof uses overlapping eaves, or the facade extends upward to form a low wall. The top of the side wall is decorated with triangular patterns, and the exterior wall material is mainly painted. There are many types of floor plans, and most of them have two households above and below each unit, with independent entrances. The first-floor cantilevered balcony of Pátio da Claridade forms a unique styling feature.

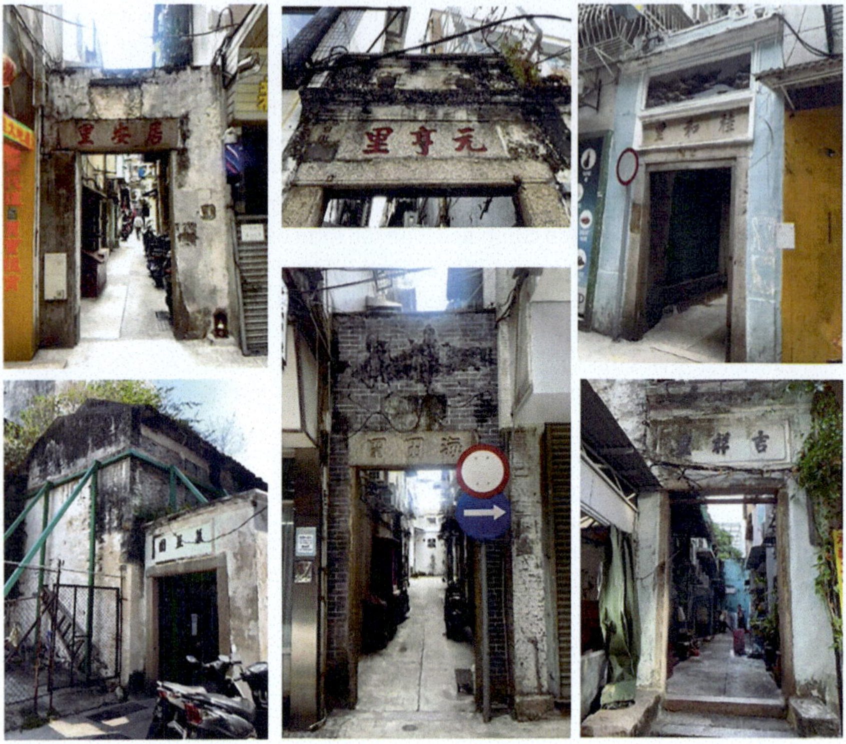

Fig. 21 Linear "Pátio" and "Beco" gates. *Image source* photographed by the authors

The characteristics of neoclassical "Pátio" and "Beco" are mainly reflected in the use of classical columns and balconies on the front facade and large windows on the building facades. The side facades are mainly solid walls, with neatly arranged windows. The front and side facades are one virtual and one real, with a strong contrast. A typical example is Pátio do Padre Narciso (Figs. 23 and 24).

It is worth noting that in the 1950–1970s, early modernist residential architecture began to influence Macau. During this period, some modern residential areas similar to "Pátio" and "Beco" were built. These "Pátio" and "Beco" were mainly early modernist style residential buildings named "Pátio" and "Beco" built after World War II. Early modernist "Pátio" and "Beco" buildings are mainly arranged in neat strips with a large distance between the front and back. The building is 4–5 stories high, with stairs running through each floor, and two households on each floor. The facade uses simple horizontal or vertical lines and blinds as the main composition elements. Typical ones include Pátio do Comandante Mata e Oliveira, Pátio da Barra, Beco da Barra, Pátio da Sé, etc.

The abundance of modernist "Pátio" and "Beco" suggests that Macau's urban development process has transformed the traditional "Pátio" and "Beco" into modern residential areas due to population growth and space constraints. The naming of

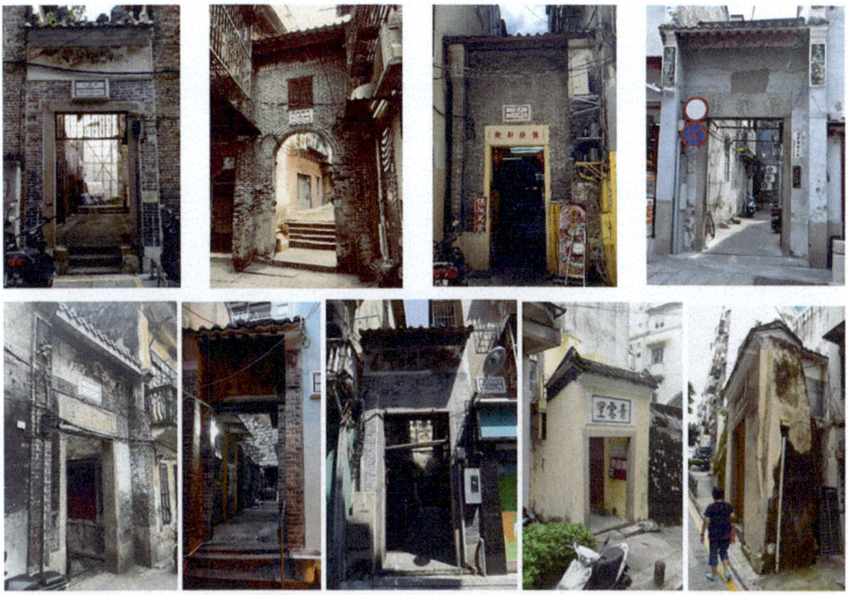

Fig. 22 The gatehouse-style "Pátio" and "Beco" gates. *Image source* photographed by the authors

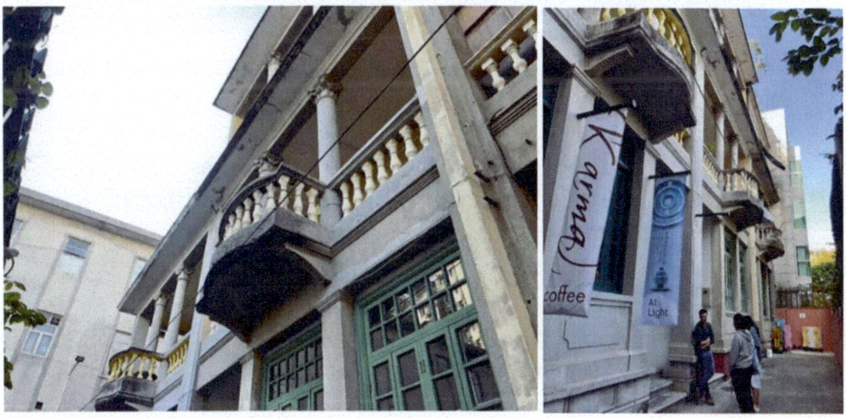

Fig. 23 Classical elements on the main facade of Pátio do Padre Narciso (partial). *Image source* photographed by the authors

"Pátio" and "Beco" also mirrors the "memory" and residential space inherited from traditional "Pátio" and "Beco" elements in modern urban development. Unfortunately, this attempt is limited in terms of its development intensity. The continuous growth of Macau's population and the substantial increase in housing demand have resulted in the demolition and rebuilding of most modernist "Pátio" and "Beco" structures, with no new construction of these structures occurring since the 1980s.

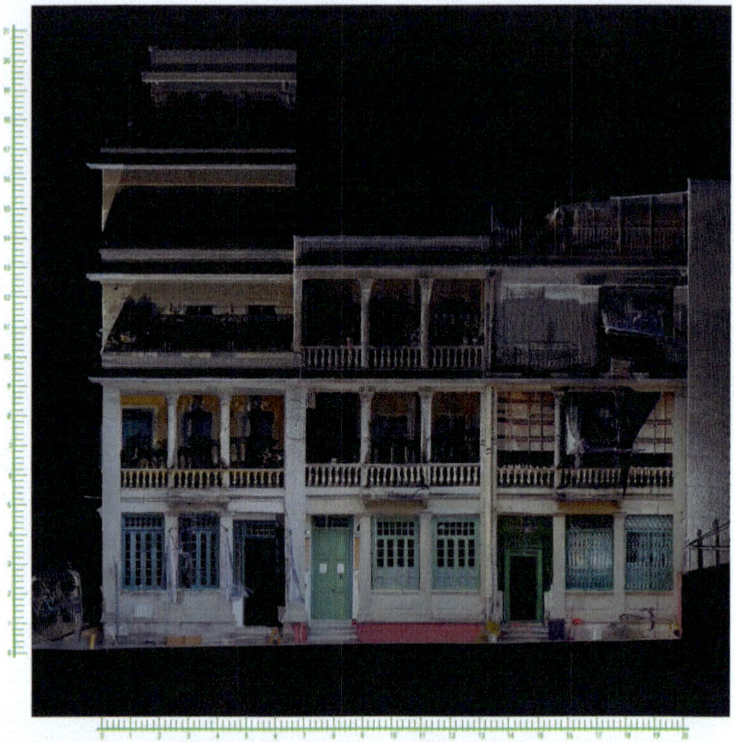

Fig. 24 Classical elements on the main facade of Pátio do Padre Narciso (partial). *Image source* photographed by the authors

5 Discussion: Heritage Protection of "Pátio" and "Beco"

"Pátio" and "Beco" have important value in the history of Macau's urban development. They are an important part of Macau's urban texture and are also the spatial carrier and ethnic grid for the life of Macau's Chinese and the integration of Chinese and Western cultures. When compared to the "Pátio" and "Beco" streets that have vanished in the surrounding areas, Macau's 262 preserved "Pátio" and "Beco" streets stand out as the largest in scale and possess unique characteristics. In particular, the coexistence and integration of Chinese and Western "Pátio" and "Beco" characteristics show the world the unique historical and cultural value of Macau's traditional dwellings. Despite the large-scale urban construction, Macau's "Pátio" and "Beco" remain closely integrated with the life, beliefs, and customs of the surrounding neighborhoods. They have organically connected with the city's street system, preserving it for over 450 years since the city's founding, a rare occurrence in Chinese cities. The Ming and Qing Dynasties introduced the "Pátio"- and "Beco"-like architectural forms to Macau. In modern times, these architectural forms made their way to Shanghai City. In modern times, the "Pátio"- and "Beco"-like architectural forms spread to Shanghai City, Wuhan City, and other places as the Heungsan people moved north to conduct business. Chinese residential culture also merits further study. The "scene" of the city is usually filled with the light and shadow of history. By observing and comparing the "Pátio"- and "Beco"-like residential forms in Portuguese cities and Macau, it can provide an important entry point for a deeper understanding of the characteristics of the intersection of Chinese and Western cultures in Macau.

5.1 Problems in the Protection of "Pátio" and "Beco" in Macau

"Pátio" and "Beco" are Macau's unique urban texture and traditional residential buildings, representing Macau's architectural history and urban memory over the past 400 years since its opening. From the perspective of cultural heritage protection and urban memory preservation, the current protection of Macau's "Pátio" and "Beco" presents the following challenges:

5.1.1 Accelerate the Inclusion of Heritage Items in the List for Protection

During the Portuguese Macau government period, traditional Chinese buildings represented by "Pátio" and "Beco" were not effectively protected and repaired, and most of the "Pátio" and "Beco" were destroyed or rebuilt. After Macau's return to the motherland, the economy developed rapidly, but even in the favorable environment of the successful application for the historic city, the protection of "Pátio" and "Beco" buildings has not attracted enough attention. In addition, Macau is prone to wind and flood disasters, and many "Pátio" and "Beco" buildings have been ruined or collapsed (Fig. 25), especially the roofs, doors and windows, decorations, etc., which urgently need to be included in the Macau Cultural Heritage List as soon as possible through cultural heritage protection mechanisms, legal supervision, technical support and financial support, and effectively protected by classification, phase, and area.

On May 13, 2020, commissioned by the Cultural Affairs Bureau of the Macau Special Administrative Region Government, the Institute of Social and Cultural Studies of the Macau University of Science and Technology conducted a survey and study on the Macau "Pátio" and "Beco" from the perspective of the protection of "Pátio" and "Beco" and its existing ancient buildings. By the end of December 2020, the survey and record of "Pátio" and "Beco" in Macau had been completed. Until the end of 2020, the Municipal Affairs Bureau of Macau commissioned a team from the University of Saint Joseph to prepare a Conceptual Masterplan for the Public Domain of the Southern Inner Harbour, focusing on the public spaces in the area from Barra to Avenida de Almeida Ribeiro, such as the release of space in the Praia do Manduco

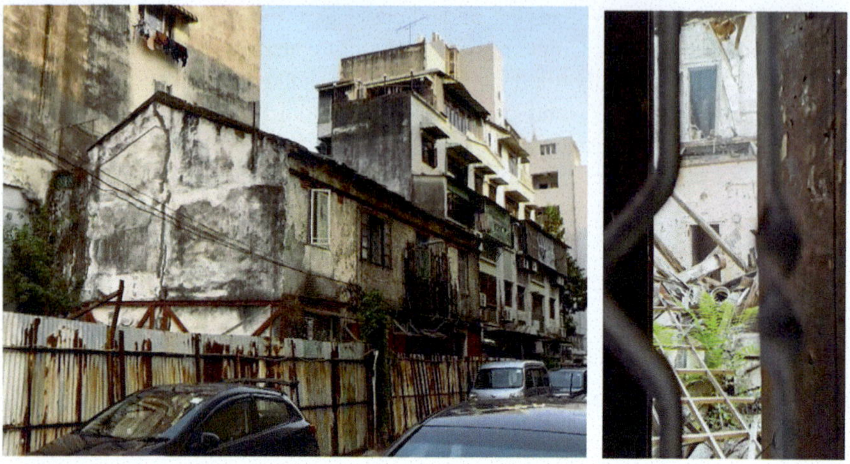

Fig. 25 An example of external damage and internal collapse of "Pátio" and "Beco". *Image source* photographed by the authors

(Pátio da Ilusão, Pátio do Bonzo, Pátio das Seis Casas, and Pátio do Sal), proposed the concept of a "heritage trail", and attempted to connect communities through the street attributes of the enclosures to activate the vitality of the area. A community engagement workshop for the preliminary draft was held in December 2021, and other parts of the project are still in progress. In October 2021, the Cultural Affairs Bureau initiated the assessment procedure for the third batch of 12 real estate in Macau in accordance with Law No. 11/2013, "*Cultural Heritage Protection Law*", completed the draft administrative regulation "*Assessment of the Third Batch of Real Estate*" and announced 12 new real estate to be assessed through Administrative Regulation No. 37/2021, including two sections of the ruins of the wall of St. Paul's College located in Pátio do Espinho and Beco do Craveiro respectively. The current assessed real estate category is "monument".

5.1.2 "Pátio" and "Beco" Are Gradually Deserted, and the Living Scene Is about to Disappear

Due to the deterioration of infrastructure and the surrounding environment, most of the original residents of "Pátio" and "Beco" have moved away, resulting in the decline and loss of vitality in the "Pátio" and "Beco" communities. The traditional living scene of "Pátio" and "Beco" has vanished and is no longer reproducible. Most "Pátio" and "Beco" have become cheap residences for non-local residents, further exacerbating the disorderly development and continuous decline inside and outside "Pátio" and "Beco".

5.1.3 The Cultural Value of "Pátio" and "Beco" Is Not Explored, Protected, and Revitalized

Traditional "Pátio" and "Beco" are rich in historical stories and cultural traditions that reflect the life and production of the early Chinese in Macau. They also serve as an important symbol of Macau's cultural positioning, where Chinese culture dominates and multiple cultures coexist. For a considerable amount of time, a lack of thematic historical and cultural research on "Pátio" and "Beco" has prevented the effective exploration of their deep cultural value. The practice of protecting Portuguese residential culture similar to Casas-Museu da Taipa has not been replicated in Chinese "Pátio" and "Beco" buildings; the relationship between "Pátio" and "Beco" and community revitalization and tourism development is still lacking professional planning and social participation.

5.1.4 Strengthen and Improve the Technical and Professional Nature of the Restoration Process

The well-preserved "Pátio" and "Beco" buildings are generally older, and the technical, craftsmanship, and professional issues in the restoration process are worth noting. For example, the brick joints of the recently rebuilt wall of Pátio da Eterna Felicidade (Beco da Enterna Celebração) are too thick (Fig. 26), and cracks appeared in the brick joints just after the wall was built. The new wall was not built according to the original nine-straight bricks and one-tie bricks method (if the brick wall is all straight bricks and no tie bricks, it means that there is no male heir in the family in Cantonese culture). Some brick walls have been machine-polished (Fig. 26), right, with traces of polishing), which destroyed the protective layer on the brick surface and may have accelerated the weathering of old bricks; salty bricks have yellow and black surfaces due to acidification.

In the future, we can try to repair and improve the living ecology and sanitary environment of "Pátio" and "Beco" from the perspective of municipal construction and cultural tourism and explore the paths and mechanisms for residents to participate in the revitalization of "Pátio" and "Beco". At the same time, it is necessary to strengthen the coordination with the relevant departments of the Macau SAR government's urban renewal plan, make unified arrangements for the protection, revitalization, and renewal of "Pátio" and "Beco" in a timely manner, and jointly formulate micro-modification methods for the maintenance and renewal of "Pátio" and "Beco" with multiple departments, combine cultural and creative tourism, build the cultural tourism IP of "Pátio" and "Beco" and revitalize the "Pátio" and "Beco" space. Finally, a unified special plan for the protection and sustainable development of Macau's "Pátio" and "Beco" will be formulated to accelerate the training of local architectural heritage restoration technical personnel in Macau and improve the level and quality of "Pátio" and "Beco" architectural protection.

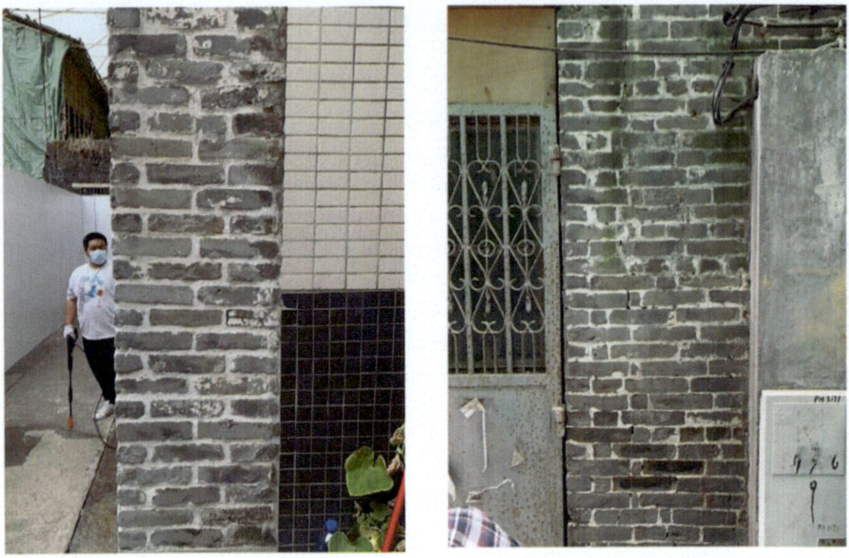

Fig. 26 Photos of the exterior wall of the Eternal Felicidade (Enter Celebração) after restoration using the nine-by-one method. *Image source* photographed by the authors

5.2 Comparison with Other Places: Pearl River Delta, Shanghai, Lisbon

5.2.1 The "Li" and "Wei" of China's Pearl River Delta

In Chinese history, "Li" and "Fang" are not only social management units but also the basic residential units and urban space planning units of ancient cities. "Li" refers to the passage around a small block, "Fang" refers to the alleys subdivided within "Li", and "streets" are above "Li" and "Fang". According to the *Guangzhou City and Fang Chronicles*, there were 57 main streets and alleys named "Li" in Guangzhou and its surrounding areas during the Qing Dynasty. At present, many "Li" are still preserved in the traditional street names of Guangzhou, such as Huaningli (華寧里), Zhuojiali (擢甲里), Yingenli (迎恩里).

Since the Ming Dynasty, China has generally implemented the Lijia system (community-tithing system, 里甲制), which is an administrative organization system based on the four-level household registration system of "hu-jia-li-xian" (戶—甲—里—縣). "Li" has thus become a basic unit of social management based on household registration and taxation. In the Qing Dynasty, Guangdong generally used the Tujia system (圖甲制) as a local management system for household registration and tax collection. "The capital governs the fort, the fort governs the map, the map governs the Jia, the Jia governs the household, and the household governs the claw", forming a tax collection and management hierarchy of "capital-fort-tu-jia-hu-zhua"

(都—堡—圖—甲—戶—爪). The "Li" place names still found in many towns of the Pearl River Delta region are remnants of the Lijia system from the Ming and Qing Dynasties.

In addition, the "Li" in the Ming and Qing dynasties was different from the "Li" in the ancient capital's Lifang system (里坊制). In the Lifang system, residents primarily utilized the "Li" as a relatively simple form of spatial organization. However, the "Li" in the Pearl River Delta towns during the Ming and Qing dynasties was both a social management unit and a spatial organization form and gradually formed a street and alley structure and texture pattern associated with the Lijia system and the Tujia system during the development process. The "Li" in the Ming and Qing dynasties was mainly responsible for counting households and managing taxes, and there were no strict regulations on its spatial pattern. Lixiang (里巷) originally belonged to specific residential land, and the residences were usually connected to each other to realize the relevant taxation and anti-theft functions. With the continuous changes in the rural management system and the complex organizational changes, such as the multiple rounds of household division and transfer, the structural style and spatial hierarchy of the "Li" have also changed. With the large number of "Li" left in the Pearl River Delta today, its spatial form has lost its clear administrative correspondence with the Lijia system and the Tujia system.

In general, the spatial structure and basic form of the "Li" in the Pearl River Delta towns (such as Guangzhou city, Panyu, Foshan) can be described as varied in form and structure. Donghuali (東華里) in Foshan and Huagaili (華蓋里) in Daliang, Shunde are large in scale and complex in spatial organization, with two or even three levels of Lixiang. The most important Lixiang are often named "Li", or street, while the secondary passages are named alleys, cross alleys, etc. Its layout and road design feature styles like fishbone, T-street, cross street, comb, and radial layouts, which can be open or semi-closed; the complexity of its shape mainly comes from how many levels of Lixiang there are. Sometimes Qingyun lane (青雲巷), which originally belonged to the interior of the house, will become a secondary Lixiang as the property rights are split or transferred.

As for the form of "Li", "Li" in the Pearl River Delta region usually consists of two "Li" gates, a "Li" lane (often named after "Li"), and two rows of buildings along the lane. Larger and more complex "Li" is formed by multi-level "Li" lanes and basic types. The orientation of the buildings in "Li" is either facing the "Li" lane or parallel to the "Li" lane. The former uses the front gate as the entrance, and the latter uses the side courtyard gate as the entrance. The main hall mostly faces east or south, and the "Li" wall is often combined with the exterior wall of the building.

According to records, the origins of place names containing "Wei" in Guangdong are mainly related to "Dike Wei" and Hakka "Weiwu" (Hakka walled village). Place names related to "Dike Wei" are mainly distributed in the coastal areas of Guangdong. Place names containing "Wei" in areas outside the coast are mainly distributed in mountainous and hilly areas with relatively high elevations and steep slopes. They are relatively far from rivers and are less affected by floods. They are mostly related to the Hakka building "Weiwu".

"Wei" is a very typical place name in the Pearl River Delta. People build the so-called Wei on the banks of rivers and lakes, creating strip-shaped water conservancy facilities that resemble dams. The construction of "Wei" is to prevent floods and drain water on land. Generally, "Wei" is constructed parallel to the river's surface, with its inner side being low-lying; in some cases, it is even lower than the river itself.

In ancient times, the Pearl River Delta was not land. The ancients built a circle of dikes around the islands at the mouth of the Pearl River. When the seawater rose, the river water was pushed back, and at the same time, the silt from the riverbed rolled into the dikes. Day after day, year after year, the silt in the dikes was higher than the river surface, forming land, which was then turned into fertile fields. Then, a larger circle of dikes was built outside, and the land area continued to expand. This repeated process increased the land carrying capacity of the Pearl River Delta region. Therefore, "Wei" was expanded from a linear place name to a surface place name or a point place name.

Similar practices of "Pátio" and "Beco" were also found in Guangzhou residential areas during the Republic of China period, which are called "concave"-shaped ground planes. This practice is very similar to introducing internal roads (alleys) into the original larger plots in British residential areas and further subdividing them into smaller plots. This method of plot filling can achieve two purposes: One is to increase the building density and build more houses; the other is to increase the number of plots inside the street when the building density is saturated to accommodate more independent residents.

5.2.2 Shanghai's Cantonese-Style Alleys

Canton Road is where the Cantonese people lived in modern times and made their livings in Shanghai. By the end of the nineteenth century, after decades of evolution, a type of single-bay, two-story brick-and-wood structure began to appear in the Cantonese settlements in Shanghai. Because it was different from the old Shikumen (石庫門) buildings and its appearance was similar to that of local Guangdong dwellings, it was called "Guangdong-style lanes".

In the 1850s, the port trade of Shanghai developed rapidly and replaced Guangzhou as the center of China's domestic and foreign trade. Many Guangdong merchants, represented by Xiangshan people, began to migrate their businesses to Shanghai and promoted the organized and large-scale immigration of thousands of Guangdong workers to Shanghai, setting off the first wave of immigration to Shanghai in modern times. After arriving in Shanghai, Guangdong people initially rented houses and lived in scattered places; they then gradually built houses and lived together, forming a residential area with typical Guangdong characteristics. According to statistics, the largest number of Guangdong people who came to Shanghai were from Xiangshan (Zhongshan), followed by Chaoshan people, Foshan people, and Guangzhou people. The early Cantonese houses were also similar to the residential houses in the above-mentioned areas.

During the Westernization Movement, many enterprises supervised by Westernization officials and run by merchants emerged in Shanghai, among which shipyards were the most representative. These newly built shipyards needed many skilled workers familiar with shipbuilding technology, so many workers were recruited from Guangzhou and other areas with more developed shipping industries. Cantonese shipping workers mainly lived in the houses of the dock factories built along the Huangpu River in the Hongkou area. These residential buildings were made of simple and cheap materials, had no patio, and had no decorations on the facades. They were more like factory dormitories, which were the prototype of Cantonese-style lane buildings.

Cantonese-style lanes are cheap, simple in shape, and practical, mainly to meet the housing needs of Cantonese workers, hawkers, and ordinary employees. Around 1900, Cantonese-style lanes first emerged in industrial areas in eastern Shanghai, such as Hongkou and Yangshupu. Typical Cantonese-style lanes appeared in Range Road (now Wujin Road), Qiujiang Road, Wuchang Road, and Wusong Road, which are adjacent to factories and have cheap land prices. Thereafter, Cantonese-style lanes also appeared in Huangpu, Nanshi, Changning, and Putuo District. Since these lanes were built at a relatively close time, the architectural style and craftsmanship are generally consistent, which is quite different from the new Cantonese-style lanes that appeared during the Republic of China (A.D. 1912–1949), so they are called "old Cantonese-style".

The entrance of the old Cantonese-style lane is usually a street-crossing building, which is arranged in a row, with single bays adjacent to each other. Each unit does not have an entrance courtyard but opens directly to the lane, and the main entrance generally faces south. The floor plan has removed the Shikumen and the front patio, and the height of the two floors is about 3.3 m on the ground floor and about 3 m on the second floor. There is a single-story kitchen at the back. The windows on the ground floor of the south facade are generally large, and the doors and windows fill the entire bay, and the second floor is slightly cantilevered; there are also cases where the ground floor does not fill the entire bay, and the second floor has one or two small windows per bay. The building materials vary depending on the floor. The ground floor is mostly brick-paved and plastered, and the walls are compacted and have a large load-bearing capacity; the walls of the second floor are mostly made of wooden boards, some of which are covered with plaster, and some of the wooden strips are directly exposed. The old Cantonese-style lane houses are not heavily decorated. Except for some brick and stone carvings at the main entrance of the lane leading to the outside road, the facade is almost undecorated. The existing representative old Cantonese-style lanes include Baitaitou (八埭頭) in Yangpu District, Hong'anli (鴻安里) in Hongkou District, and Jiujiangli (九江里) and Rongshouli (榮壽里) in Huangpu District.

The Guangdong elements in the process of modern transformation of Shanghai residential architecture promoted the evolution of the layout, materials, exterior decoration, and functions of the later Shikumen houses and directly generated a new type of residential architecture, the "Guangdong-style lane". This concept of focusing on cost, durability, and space saving has become a major feature of modern Shanghai urban residential construction.

5.2.3 "Bêco" in Lisbon, Portugal

The classification of the street and lane system in Macau follows the Portuguese practice. The names "Pátio" and "Beco" are derived from Portuguese and are the lowest-level roads in the street and lane system in Macau. In Lisbon, Portugal, Bêco and Pateo have always been important components of urban space. As early as the fourteenth and fifteenth centuries, the concepts of Bêco and Pateo as urban micro-regions were gradually established. Today, "bêco" often refers to short and narrow streets or dead ends with no exits. The urban area of Lisbon currently includes four larger areas, namely the Old Town, the University Town, the Belém District, and the Expo District. According to statistics, there are currently 153 named Bêcos in Lisbon, especially concentrated in the Alfama district with a medieval style (Fig. 27) and the Bairro Alto district with rugged mountain roads. From a spatial perspective, the existing Bêco buildings show a certain degree of agglomeration. Unfortunately, due to the pandemic, it is impossible to conduct an on-site survey and collect relatively complete information on the names, addresses, types, construction years, number of bays, number of floors, number of units, etc. of the existing old streets in Lisbon. It is worth noting that "patio" is not a type of street in the Portuguese street system [29]. Patio as a street type is likely to be a characteristic of Macau.

5.2.4 Similar Residential Patterns: "Ilhas" in the Porto, Portugal

Through the analysis of similar "Pátio" and "Beco" in different cities mentioned above, similar "Pátio" and "Beco" have similarities in land intensiveness and residential density. From the perspective of block planning, they can increase the intensity of land development and adapt to population growth and the residential needs of middle- and low-class residents. In order to fully improve the utilization efficiency of street block land, "Pátio" and "Beco" use small alleys perpendicular to the external streets to penetrate into the larger street block. The "Pátio" and "Beco" plots can be flexibly arranged along both sides of the small alleys, occupying the remaining space inside the large block and forming more internal plots.

For example, the high-density residential model of "Ilhas" in Porto is a common type of working-class housing in Portugal from the nineteenth to the twentieth century [30]. It is located behind the large street-facing buildings in the front and arranged along narrow passages. In the second half of the nineteenth century, Porto was experiencing a climate of industrial euphoria that attracted rural populations to the city, coming from Minho, Trás-os-Montes, and Alto Douro, and Beira Alta, fleeing the rural crisis that was being experienced there. The search for cheap accommodation then made these clusters of shack buildings, with a single entrance, an attractive business, mainly operated by small landowners who, having little capital, saw in the Ilhas the guarantee of a quick recovery of the invested capital and, in the short term, significant profits. Since the 1869 Municipal Code of Conduct limited municipal oversight to what was visible from the street (which in this case were the fronts of a common house, usually that of the developer) and the Ilhas ended up being built

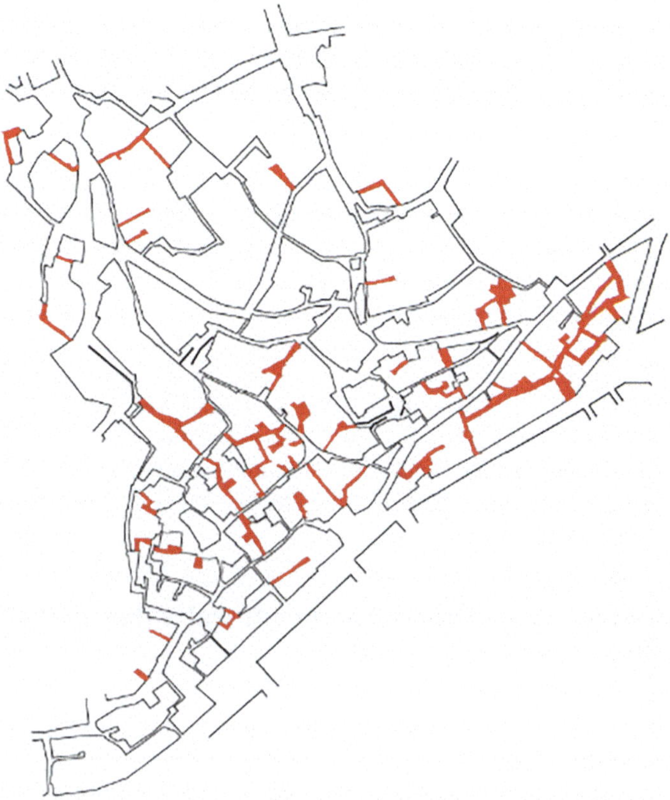

Fig. 27 Bêco in the Alfama district of Lisbon, Portugal. *Image source* References [29]

inside the blocks, they were not subject to municipal control and were not visible to the public. At first glance, the city did not seem to be in such poverty. Between 1878 and 1890, 5100 homes were built on the Ilhas (half of those that existed in 1900), where, according to Ricardo Jorge, a third of the city's population lived in 1899. According to a survey carried out by the Porto City Hall in 1939, there were 1152 Ilhas in the city at the time, housing 45,291 inhabitants, that is, at that time, 17% of its total population. Following an attempt to clean up the city to prevent epidemic outbreaks, from the 1940s onwards, municipal authorities began to gradually demolish the Ilhas of Porto, relocating families to large housing complexes far from the city center. Fifty years later, the Ilhas were still not completely eradicated from Porto and Greater Porto. Many remain firmly made of stone and mortar and are trying to renew themselves to counter the cold and impersonal spirit that increasingly defines life in social housing. The results of the socioeconomic study carried out in 2001 by the Porto City Hall revealed that there were 1182 Ilhas in the city, of which 1109 were inhabited. In 2011, the data update revealed that the number of inhabited Ilhas had decreased to 930.

This type of residential unit is similar to the internal filling model of the British burgage plot, which is the Portuguese model of burgage. From the perspective of urban morphology, the combination of "Pátio" and "Beco" and shophouses in Macau may have been influenced by the Ilhas residential model. This phenomenon of filling in the interior of the block is similar to the subdivision and infill redevelopment of the tail of the Burgage plot in the old town of Alnwick in the UK. The large increase in internal plots has led to new public roads inside the plots, called "right-of-way channels". This internal channel often has no significance in urban transportation and is also called a "pseudo-street system".

The organization of "Pátio" and "Beco" forms a relatively independent land group, and the "Pátio" and "Beco" roads become semi-private and semi-public spaces within the group. Although the spatial forms of "Pátio" and "Beco" have many forms such as "I", "L", and "U", the root of these forms is to adapt to population growth and improve land use efficiency and to achieve high-density filling or land reorganization. It can be said that this dynamic factor has led to similar residential patterns in Macau, Guangzhou, Portuguese cities, British cities, etc.

6 Conclusion: The Historical and Cultural Value of "Pátio" and "Beco" in Macau

Macau's "Pátio" and "Beco" are in the "deepest part" of the city. The "Pátio" and "Beco" serve as the residential form and spatial carrier of the convergence of Chinese and Western cultures in Macau. They represent a traditional Chinese architectural style that has a rich history in Macau. They are the material carriers of traditional Chinese residential life. They embody the construction ideas and skills of the Chinese in Macau. They are also the material form and carrier of Macau's "exchange and cooperation base with Chinese culture as the mainstream and multiple cultures coexisting" [31], proposed in the *Outline Planning of the Guangdong-Hong Kong-Macau Greater Bay Area*. They serve as crucial resources for studying the lives and contributions of Chinese and Portuguese residents in Macau since modern times, as well as the history of Macau's urban development and architecture. They are also an important base for carrying out local historical and cultural education in Macau.

"Pátio" and "Beco" are an indispensable part of the urban fabric of Macau that coexists with the Portuguese city (i.e., the Historic Centre of Macau). Most of "Pátio" and "Beco" are in the "subordinate" area of the World Heritage, the Historic Centre of Macau, and have an important impact and value on the protection of the World Heritage Historic City. From the perspective of urban development and architecture, "Pátio" and "Beco" in the old city of Macau Peninsula are currently facing the contradictions and challenges of urban renewal. How to fully protect "Pátio" and "Beco" and their buildings, revitalize their physical space, and preserve their cultural value is a major issue worthy of attention and consideration in the urban renewal plan.

"Pátio" and "Beco" were common residential types in the Pearl River Delta during the Ming and Qing Dynasties. Although Macau was small in size, Chinese immigrants from the Pearl River Delta were like "marginal people" in the city and could only live in this intensive traditional residential form on the edge of the main urban area. During the implementation of the Portuguese-style street and lane hierarchy in Macau, this traditional residential form was retained and connected with a Portuguese street and lane name system with similar semantics. This street and lane construction, naming characteristics, and its remains may be the result of cultural intersection and compromise formed by the early mixed residence of Chinese and Portuguese and have unique significance and value in the cultural exchange between China and the West and the history of Macau's urban development.

In modern times, the Portuguese Macau government has formulated four renovation plans, including the *Overall Renovation Plan for the Market Area (1900)*, the *Avenida de Almeida Ribeiro Plan (1903)*, the *Avenida de Almeida Ribeiro-Nam Van Plan (1927)*, and the *Macau New Port Plan (1935)*. The purpose is to demolish the "Pátio" and "Beco" residential areas in the inner harbor area to conform to the Portuguese grid-style blocks, facilitate transportation layout, and meet the needs of urban development. But in the end, only some of the "Pátio" and "Beco" on both sides of Xinmalu were demolished, and most of the "Pátio" and "Beco" have been preserved to this day.

In modern times, Shanghai was one of the main commercial gathering places for Xiangshan people. Many Xiangshan people became rich in Shanghai through business and brought the intensive, simple, private, and safe living mode of "Pátio" and "Beco" to Shanghai, which became the main living mode of local Xiangshan people and even Cantonese people, called Guangdong-style lanes. Therefore, in addition to the Pearl River Delta, the lanes in Shanghai are also closely related to the "Pátio" and "Beco" in Macau. Although the correlation and degree of the "Pátio" and "Beco" in the above-mentioned regions with the "Pátio" and "Beco" in Macau still need further research, the "Pátio" and "Beco" in Macau are influenced by the architectural culture of the Pearl River Delta and Portugal. There is no doubt that the living mode like "Pátio" and "Beco" has cross-cultural and cross-regional characteristics.

Funding This research was jointly supported by the Cultural Affairs Bureau of Macau, Project "Research on the Recording of 'Pátio' and 'Beco' in Macau and Their Cultural Value" (0002/DEPROJ-PCT-122/2020), and the National Social Science Foundation of China, Project "Research on the Interaction and Influence of Eastern and Western Architectural Culture during the Period of Guangzhou's Open Trade" (21VJXT011).

Institutional Review Board Statement Not applicable for studies not involving humans or animals.

Informed Consent Statement Not applicable for studies not involving humans.

Conflicts of Interest The authors declare that the research was conducted in the absence of any commercial or financial relationships that could be construed as a potential conflict of interest.

References

1. Porter, J. (2018). Macau: The imaginary city. *Routledge, New York.*. https://doi.org/10.4324/9780429499128
2. Chen, Y., Zheng, L., & Yan, L. (2024). Research on the intelligent generation of the spatial form of the Island city historic district based on parameterization: Taking Macau Taipa Village as an example. *Journal of Asian Architecture and Building Engineering, 23*(3), 1094–1125. https://doi.org/10.1080/13467581.2023.2257274
3. Xie, Q., Ursini, F. A., & Samo, G. (2023). Urbanonyms in Macau. *Names, 71*(1), 29–43. https://doi.org/10.5195/names.2023.2421
4. Li, M., & Xie, J. (2023). Social and spatial governance: The history of enclosed neighborhoods in urban China. *Journal of Urban History, 49*(4), 723–744. https://doi.org/10.1177/00961442211040460
5. Macau Government Printing Bureau: "Legislative Regulation No. 1600: General Regulations on Urban Buildings", 1963, page 13, (unofficial published text, for reference only, reference material provided by the Macau Legal Affairs Bureau). https://images.io.gov.mo/bo/i/63/30/dil-1600-63-cn.pdf
6. Wang, W. J., & Cheong, C. K. (2011). *Regenerating pátio: Study of Macau's historical urban fabric.* Culture Institute of Macau. http://hdl.handle.net/10722/169144
7. Municipal Affairs Bureau (IAM) of Macau, Macau Street Website: "Lifestyle in the "Pátio" and "Beco", December 17, 2018., https://Macaustreets.iam.gov.mo/c/culturaltraildetail/detail.aspx?id=19
8. Li, J., Lin, F., & Chen, Y. (2025). Spatio-temporal patterns of the construction and development of urban open spaces in Macau from 188 2020. *Journal of Asian Architecture and Building Engineering*, 1–23. https://doi.org/10.1080/13467581.2025.2458807
9. 1847 Macau Place Names (from the Portuguese Overseas Historical Archives).
10. 1869 Macau Street Statistics (Relatorio da commissão nomenda para classificar as ruas da cidade).
11. 1874 Macau City Public Streets List (Nomenclatura das Praças, Largos, Ruas etc da Cidade de Macau).
12. 1878 Macau Streets (Mappa das ruas, travessas, becos, etc.)
13. 1896 Macau Streets (Mappa das vias publicas).
14. 1905 Macau Street List (Cadastro das Vias Publicas de Macau).
15. 1925 Macau Street List (Cadastro das Vias e Outros Lugares Publicos da Cidade de Macau).
16. 1933 Macau Place Name Street List (Vias Publicas e outros sítios da Cidade de Macau); List of Street Names in Macau in 1933.
17. List of Macau Place Names and Street Names in 1934 (Vias Publicas e Outros Sítios da Cidade de Macau).
18. List of Macau Place Names and Street Names in 1935 (Vias Publicas e Outros Sítios da Cidade de Macau).
19. List of Macau Place Names and Street Names in 1936 (Vias Publicas e Outros Sítios da Cidade de Macau).
20. List of Macau Place Names and Street Names in 1937 (Vias Publicas e Outros Sítios da Cidade de Macau).
21. List of Macau Place Names and Street Names in 1941 (Vias Publicas e Outros Sítios da Cidade de Macau).
22. List of Macau City Streets in 1944 (Cadastro das Vias e Outros Lugares Publicos da Cidade de Macau).
23. List of Macau Place Names and Street Names in 1952 (Vias Publicas e Outros 1955 Street Names List of Macao (Vias Públicas and Outros Sítios da Cidade de Macau) 1957 Street List of Macao (Cadastro das Vias Públicas and Outros Lugares da Cidade de Macau); 1957 Public Roads of Macao (Vias Públicas da Cidade de Macau) 1993 Street and Other Places List of Macao (Cadastro das Vias Públicas and Outros Lugares da Cidade de Macau).

24. Street and Other Places List of Macao (Updated - 1997) (Cadastro das Vias Públicas and Outros Lugares da Cidade de Macau).
25. Chen, Y. (2022). Research on the protection of Macau water-faring community (Dan Jia) from the perspective of urban heritage. In: Mendonça, P., Cortiços, N.D. *(Eds.)* Proceedings of the 7th international conference on architecture, materials and construction. ICAMC 2021. Lecture Notes in Civil Engineering, Vol 226. Springer. https://doi.org/https://doi.org/10.1007/978-3-030-94514-5_49.
26. Wongis, K. K., & Wei, C. G. (2014). Macao–cultural interaction and literary representations.
27. Antony, R. (2019). The pirates of Macau in historical perspective. *Review of Culture, 60,* 7–31.
28. Guangzhou Housing and Urban-Rural Development Committee. (2018). *South China University of Technology architectural design institute, "Guangzhou historical building renovation plan"*. Guangzhou Municipal People's Government Publishing House.
29. Proença, S. D. S. B. (2014). *A diversidade da rua na cidade de Lisboa: morfologia e morfogénese.* (Doctoral dissertation,. Universidade de Lisboa (Portugal).
30. Teixeira, A. T. (2018). As Ilhas do Porto e as tipologias de habitação dos séculos XIX e XX. 5%, 1–17. http://revista5.arquitetonica.com/ojs/index.php/revista5/article/view/16
31. Office of the Chief Executive of the Macau Special Administrative Region: Chief Execu-tive's Notice No. 40/2017 "Framework Agreement on Deepening Guangdong-Hong Kong-Macau Cooperation and Promoting the Development of the Greater Bay Area", July 26, 2017.

Evaluation and Optimization of Modern Urban Space in the Xiangshan Area

The third part focuses on the present of traditional urban space, using evaluation methods such as AVC, spatial syntax, GIS, POI, and other big urban data analysis methods to evaluate and reflect.

Evaluation of Carrying Capacity for the Revitalization of Traditional Villages: Beishan Village in Zhuhai

Wangyunshu Zhang

Abstract The acceleration of urbanization has brought many challenges to traditional villages, and the revitalization of traditional villages has become inevitable in this context. At present, there are several traditional villages in Zhuhai that have completed cultural and creative transformation, among which Beishan Village is an influential case, and the results of its transformation are widely discussed. Based on the AVC theory (attraction, vitality, and capacity), this study uses AHP analytic hierarchy process to evaluate the protection scope of Beishan Core Planning and Protection. Through field research and collation of the current situation data of local cultural elements such as landscape, folk customs, lifestyle, traditional skills, and population number, the classification and analysis of traditional spatial forms, life forms, and ecological environment provide reference for the selection of evaluation indicators. At the same time, the case of Beishan cultural and creative transformation was analyzed, and the transformation method was summarized. Finally, the evaluation system of the three forces of local culture AVC was constructed, and the comprehensive weighted scores of the indicators were obtained by analytic hierarchy process as the evaluation basis, and the corresponding strategies were proposed for the indicators with low scores.

Keywords AVC · Evaluation research · Continuity · Zhuhai Beishan Village · Local culture

1 Research Background

In recent years, the cultural and creative industries in China have been developing with each passing day, and more traditional villages have chosen to carry out village renewal and transformation through new approaches that integrate cultural creativity. The integration of cultural creativity with traditional cultural resources not only

W. Zhang (✉)
Jiangsu Branch, CAUPD (Beijing) Planning & Design Consultants Ltd, Nanjing, China
e-mail: 1435815956@qq.com

Y. Chen et al. (eds.), *City-Sea Symbiosis*, Springer Geography,
https://doi.org/10.1007/978-3-032-13725-8_10

preserves the historical and cultural connotations of villages but also provides conditions for the rational development and utilization of traditional cultural resources within their original spaces, enabling them to exert their economic and cultural vitality while inheriting culture. Among them, Guangdong Province has actively responded to this trend and successively issued policies related to the transformation of traditional villages, such as *Regulations on the Protection of Famous Historical and Cultural Cities, Towns and Villages in Guangdong Province and Regulations on the Protection of Traditional Style Buildings in Guangzhou*, encouraging multi-stakeholder social cooperation and providing solid financial and policy guarantees for the revitalization and transformation of traditional villages.

Currently, several traditional villages in Zhuhai have been renovated with cultural and creative elements and completed. Among them, Beishan Village in Zhuhai, as a traditional village with unique historical and cultural value, holds significant representativeness and certain influence in the research on the carrying capacity evaluation of traditional village revitalization. In the process of renovating famous historical and cultural villages, there are some inevitable contradictions between protection and revitalization-utilization. Consequently, the effectiveness of the cultural and creative renovation of Beishan Village has attracted extensive attention from society and academic circles, and its subsequent development paths and experiences merit in-depth exploration.

2 Base Information on Beishan Village

2.1 Location Information and Current Status

Beishan Village is in the east of Nanping Town, Xiangzhou District, Zhuhai City (Fig. 1), with a geographical area of approximately 5 km^2. To its north lies Zhuhai Avenue, an urban expressway. The village derives its name from Beishan Mountain, which is situated within the village. Historically, there was a river winding along the front of the village to its north; nestled against mountains and facing water, the village is surrounded by mountains and rivers, where vitality converges.

Beishan Village was first established in the first year of Jiaxi in the Song Dynasty (1237). It boasts a profound historical and cultural heritage, and after hundreds of years of development, it has formed a unique Lingnan architectural style and folk culture. In 2009, with the approval of the People's Government of Guangdong Province, Beishan Village was listed as one of the second batch of Famous Historical and Cultural Villages in Guangdong Province by the Department of Housing and Urban-Rural Development of Guangdong Province and the Department of Culture of Guangdong Province. The main research scope of this chapter is the core protection area in the planning of Beishan Village, with an area of 8.97 ha (Fig. 2).

Beishan Village has preserved the overall layout and style of clan villages in the Pearl River Estuary region during the Qing Dynasty. The village buildings, oriented

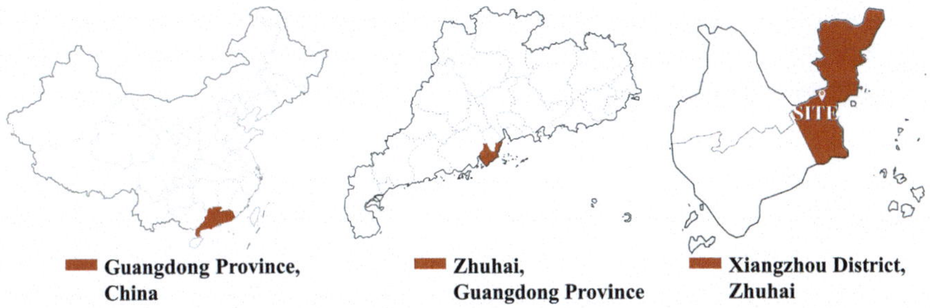

Fig. 1 Location of Beishan Village. *Source* created by the author

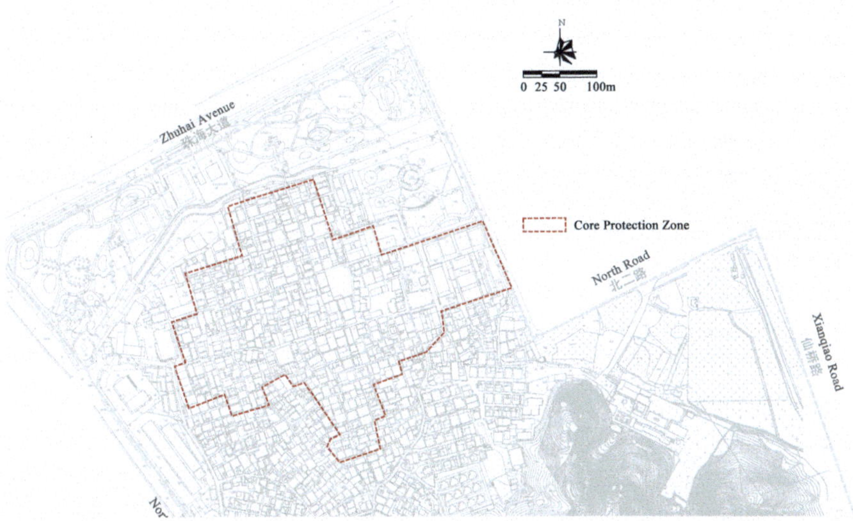

Fig. 2 Beishan core protection zone. *Source* redrawn by the author from: *Protection Plan for Beishan Village, Zhuhai City—A Famous Historical and Cultural Village in Guangdong Province (2020–2027)*)

from south to north, are relatively concentrated with a complete and clear form, presenting an intensive and irregular pattern. The gaps between the front and rear buildings in the village are small, and they are arranged in north-south rows like a comb. Between each row of buildings, there is an alley about 2 m wide, which serves as the main passageway for the residents of Beishan Village.

The core protection area of Beishan Village is the region where the historical style is most intact, historical buildings are most concentrated, and the preservation status is best, with a complete and clear form. The external spaces of the settlement are mainly alleys and courtyards, and there are some shops on both sides of some alleys that allow vehicle access. The total area of the core protection area is 8.97 ha, among which the existing cultural relic buildings, historical buildings, traditional-style buildings, and historical environmental elements in the core area cover an area of 2.23 ha, accounting for 25% [1].

2.2 Historical Value

The existing traditional Lingnan architectural complex in Beishan includes eight immovable cultural relics built during the Qing Dynasty. The core protection area of Beishan is approximately 8.97 ha, within which the eight immovable cultural relics built in the Qing Dynasty occupy an area of about 6796 m^2, accounting for approximately 7.58% of the core protection area. Historical buildings cover an area of about 8503 m^2, accounting for roughly 9.48% of the core protection area, while buildings with traditional styles occupy about 13,790 m^2, making up approximately 15.37% of the core protection area [2].

The distribution of buildings at various protection levels within Beishan's core protection area is shown below (Fig. 3). Among them, Chengchuan Yang Ancestral Hall and Baoxia Yang Ancestral Hall have been reused as art exhibition venues and art centers through integration with the cultural and creative industries, attracting many visitors.

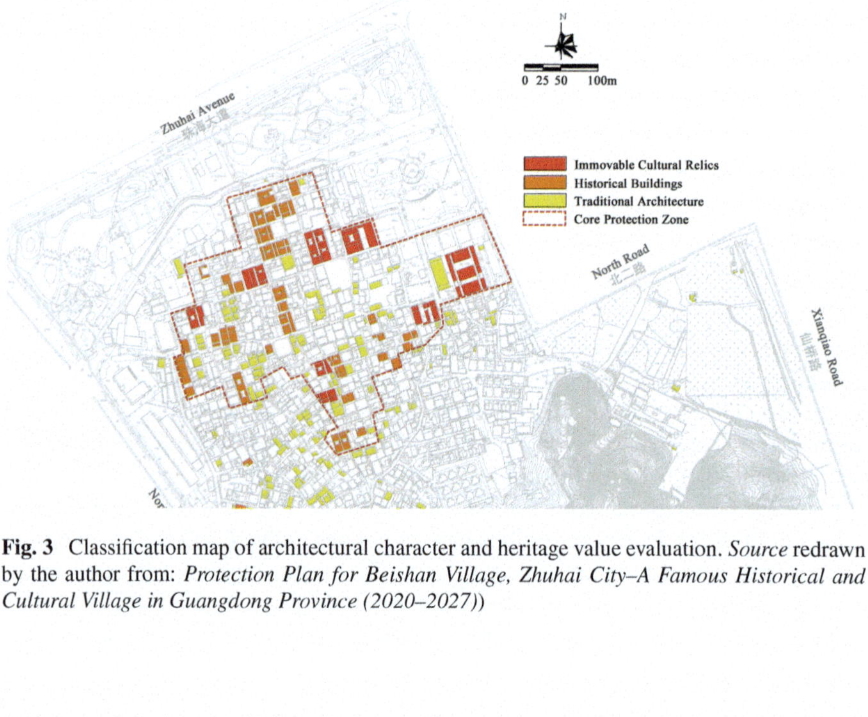

Fig. 3 Classification map of architectural character and heritage value evaluation. *Source* redrawn by the author from: *Protection Plan for Beishan Village, Zhuhai City–A Famous Historical and Cultural Village in Guangdong Province (2020–2027)*)

2.3 *Population Size*

According to statistics from the village committee, in 2006, the permanent population of Beishan was approximately 17,000, consisting of 520 households, with a floating population of nearly 9000. The population of Beishan Community reached around 30,000 in 2019, among which the registered permanent population was 2648. Statistics in 2021 showed that the permanent population exceeded 18,000, with a floating population of about 14,400. Over 70% of the population has relatives in Hong Kong, Macau, and overseas, with approximately 4000 overseas compatriots [3]. The Yang clan accounts for about 90% of the total permanent population. The relatively diverse population structure of Beishan has laid a solid foundation for its modern cultural and creative transformation.

3 Spatial Form Analysis

3.1 *Public Space*

The village's public spaces are classified into three types: points, lines, and planes, based on the criteria of the village's spatial form and composition. The analysis of the classification is as follows.

3.1.1 Point-Like Spatial Form

The coherent core nodes scattered across Beishan Village collectively form the village's point-shaped public spaces. These nodes, characterized by diverse functions and accessibility, can be categorized into the following types:

(1) Nodes with distinct themes and historical significance. Nodes with distinct themes include Container Square, ancestral halls, and their attached square courtyards (Fig. 4), while functionally significant historical spaces encompass point-shaped areas such as ancestral halls used for rituals in Beishan's history.
(2) Nodes with a strong sense of place. Examples include village entrances/exits, areas beneath ancient trees, and spots beside old wells (Fig. 5).
(3) Fully open nodes within lanes outside buildings, such as the long stone benches outside residential houses where villagers engage in daily leisure and conversations (Fig. 6).

The leisure square near Beishan Vintage Warehouse features numerous container-based restaurants where visitors can dine and chat. Additionally, there are many tree-plot seats under the trees, making it a popular spot where both visitors and villagers gather to rest and converse.

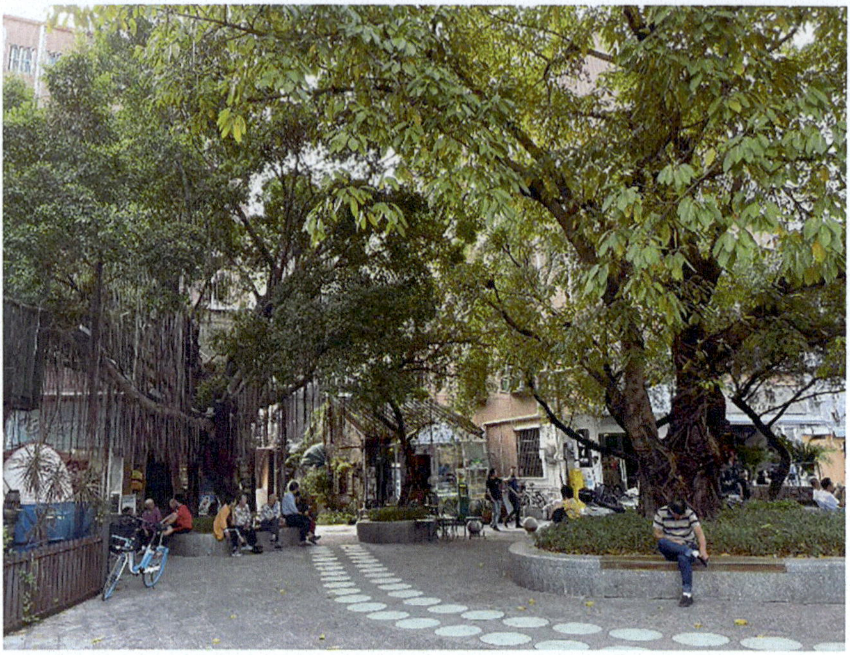

Fig. 4 Public space of the Container Plaza. *Source* photographed by the author

3.1.2 Linear Public Space

The linear public spaces in Beishan Village serve the following main functions: (1) transportation and (2) connecting the ancient architecture within the village.

These various linear public spaces together form an orderly network of streets and lanes in the village. All the street and lane spaces systematically link the node spaces within the village, contributing collectively to ensuring smooth traffic flow, fire prevention, and ventilation in the village. The distribution map of Beishan's main current spaces is shown in Fig. 7.

3.1.3 Areal Public Space

The planar public spaces in Beishan Village are network-shaped spaces formed by the integration of various important functional node spaces, linear spaces such as streets, lanes, and water systems within the village. The status of planar public spaces within Beishan's core protection area is shown in Fig. 8.

The main functions of the planar public spaces in the village include villagers' daily social interactions, recreational activities, and large-scale religious and belief activities. For instance, the square in front of Yang Clan Ancestral Hall serves

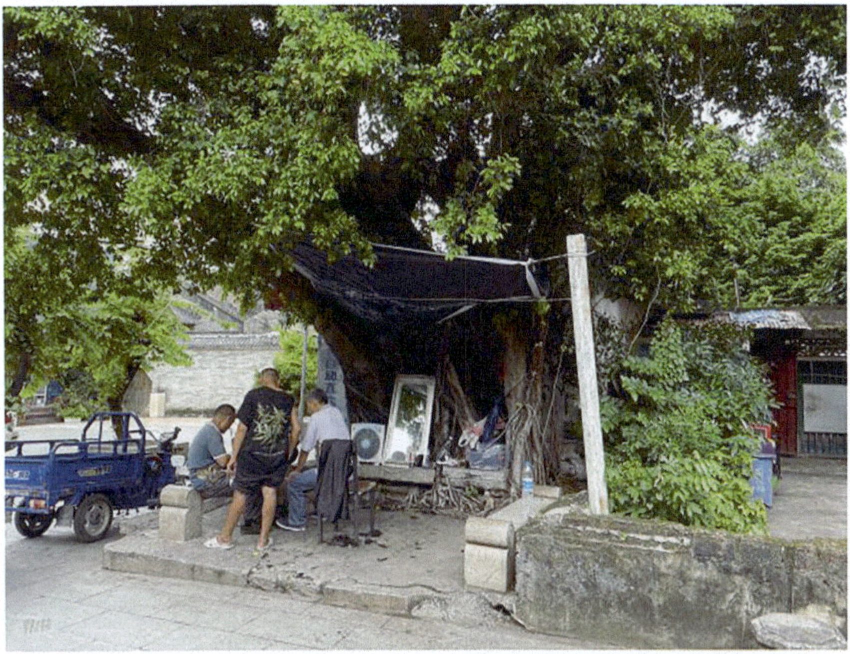

Fig. 5 Spontaneously formed recreation space under ancient trees. *Source* photographed by the author

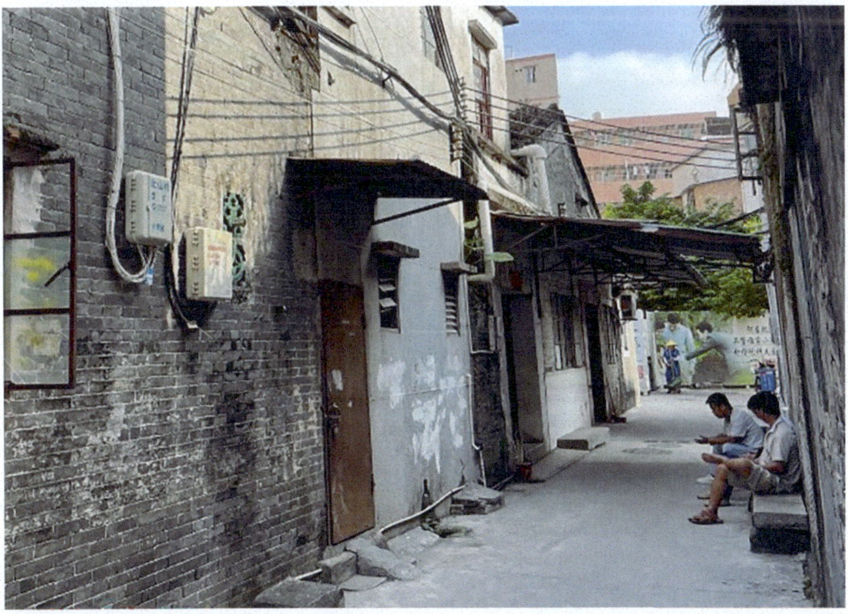

Fig. 6 Long stone benches in alleys for villagers' daily leisure. *Source* photographed by the author

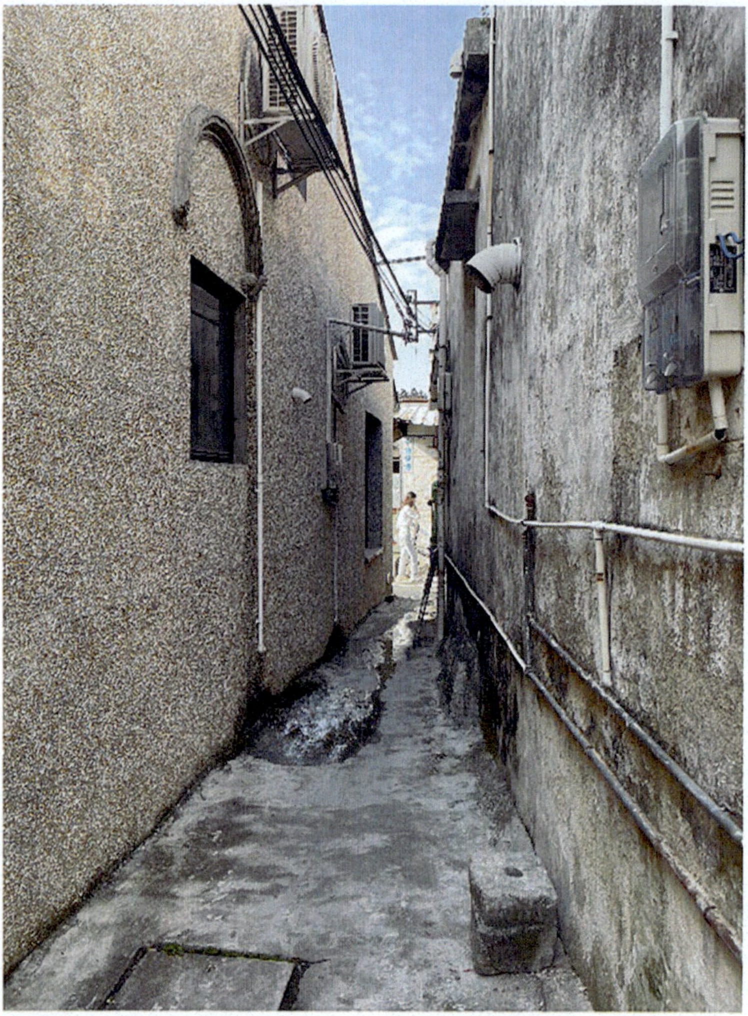

Fig. 7 The current situation of linear public spaces. *Source* photographed by the author

as a significant public space for clan gatherings, festival celebrations, and folk performances (Fig. 9).

3.2 Traditional Streets and Lanes

Beishan Village's historical streets and lanes, as shown in Fig. 10, are mostly traditional stone slab roads. With few pedestrians passing through, their preservation

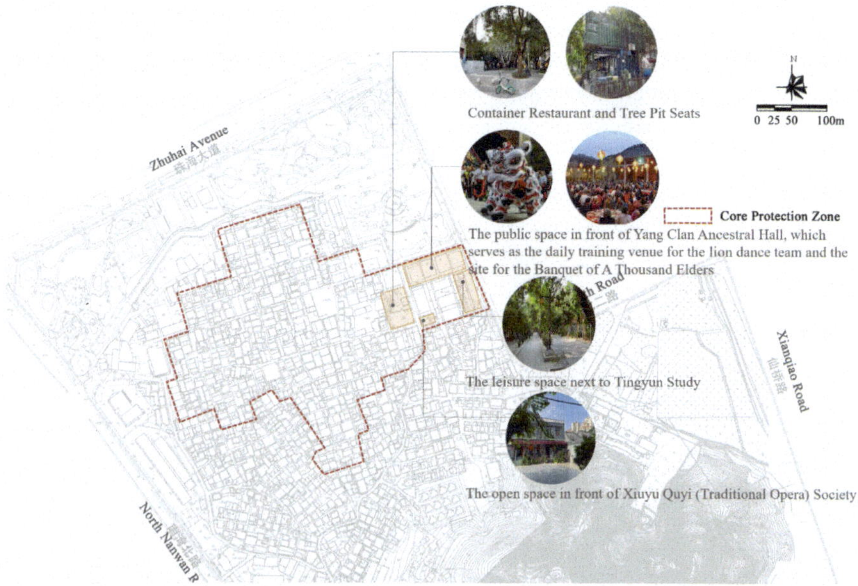

Fig. 8 Current status of point-like public space. *Source* redrawn by the author from: *Protection Plan for Beishan Village, Zhuhai City—A Famous Historical and Cultural Village in Guangdong Province (2020–2027)*)

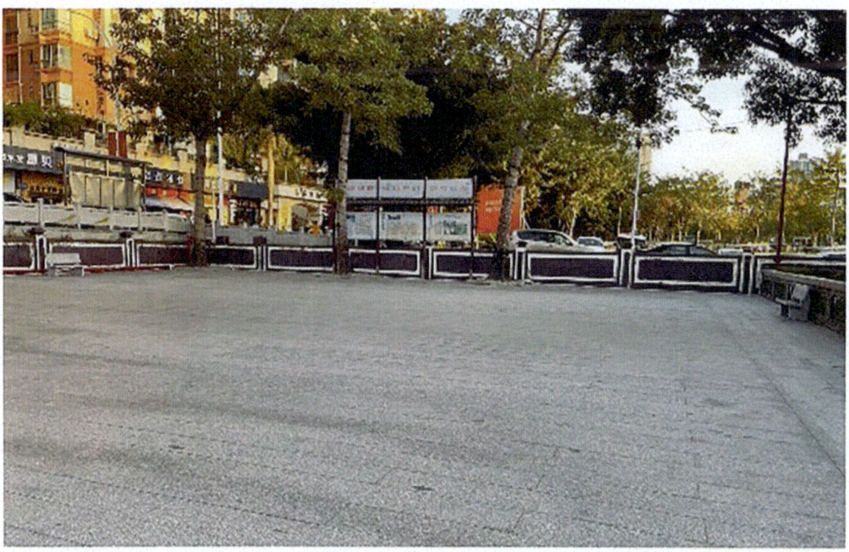

Fig. 9 Plaza in front of Yangshi Ancestral Hall. *Source* photographed by the author

status is poor, as illustrated in Fig. 11. The users of such street and lane spaces are mainly residents, and both sides of the streets and lanes are mostly residential houses.

Beishan Main Street, as the main road connecting the east and west of Beishan Village, sees a large volume of pedestrian flow, as shown in Fig. 12. The historical

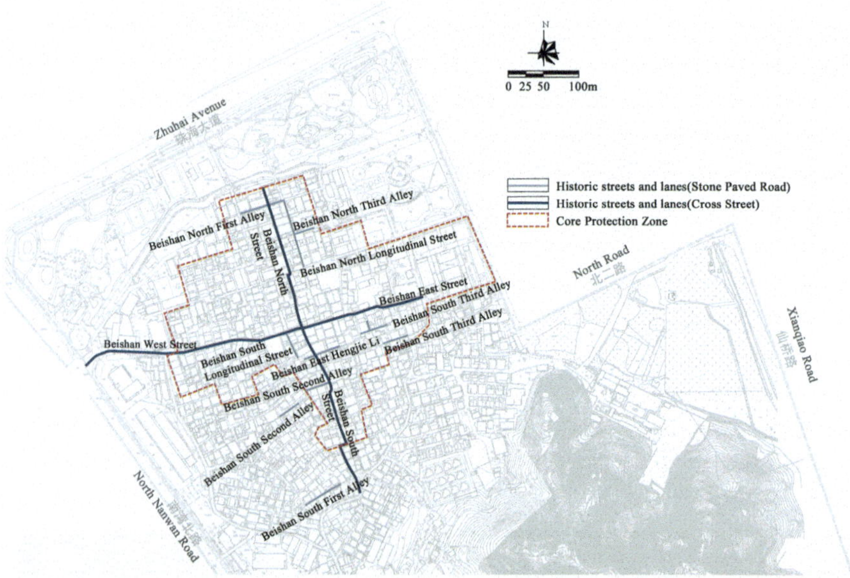

Fig. 10 Map of historic lane pattern features and current status of historical alleys. *Source redrawn by the author from: Protection Plan for Beishan Village, Zhuhai City—A Famous Historical and Cultural Village in Guangdong Province (2020–2027)*

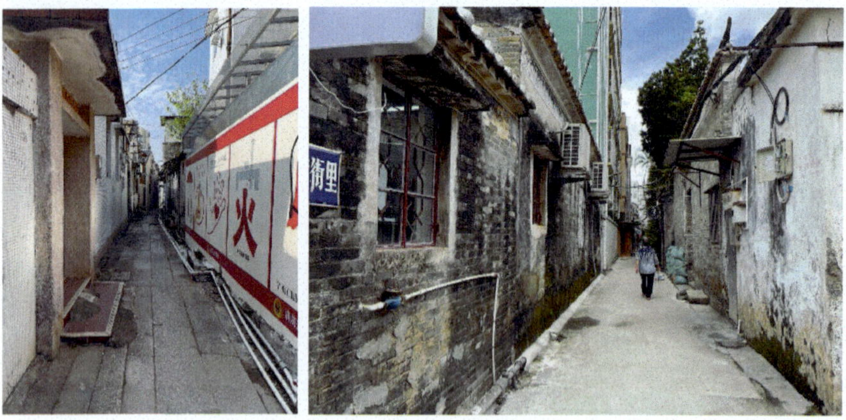

Fig. 11 Alley space of Beishan South Street and East Hengjieli Street. *Source* photographed by the author

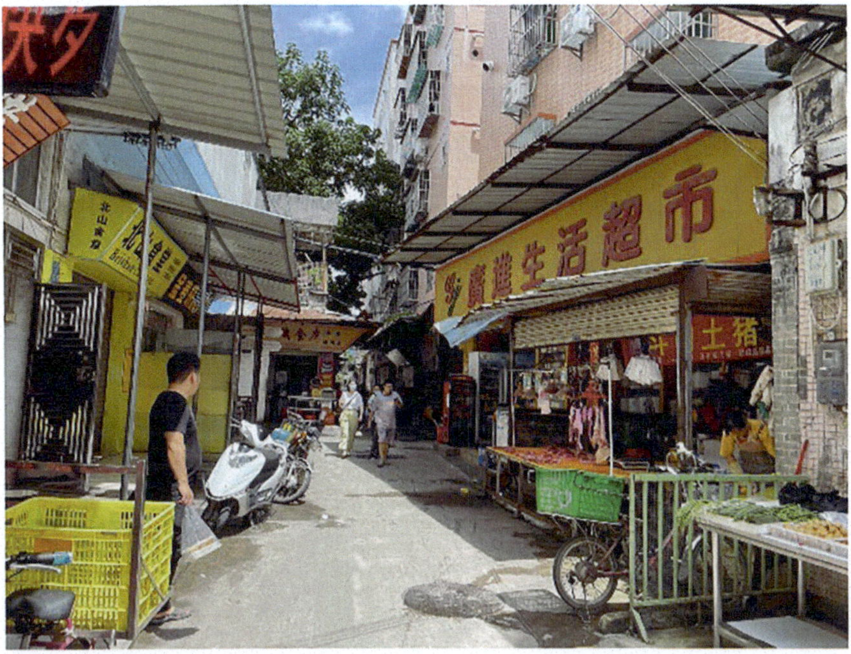

Fig. 12 Beishan Zheng Street. *Source* photographed by the author

Cross Street is a major thoroughfare linking the four directions (east, west, south, and north) within the village. With many shops on both sides, it is mainly used by vendors and residents.

One of the important measures to preserve the village's memories is to protect and inherit the traditional street and lane spaces. The planning for Beishan also mentions corresponding measures: taking traditional streets and lanes as links to connect the ancient buildings within Beishan Village, forming a cultural experience path. This approach extends the exhibition function of the museum to the outdoors, allowing visitors to immerse themselves in Beishan's culture.

3.3 Traditional Architectures

There are several cultural relics protection units at the Guangdong provincial level in Beishan Village (Fig. 13), such as Yang Clan Ancestral Hall (Fig. 14), Baoxia Yang Ancestral Hall, and Beishan Yidi Temple (Fig. 15).

The immovable cultural relic buildings are in good condition after renovation, and their current usage status is shown in Fig. 16. Yang Clan Ancestral Hall covers an area of 8838 square meters, with a width of five bays and a depth of three bays, featuring a gable roof. There is a Tingyun Study on its east side. As the oldest

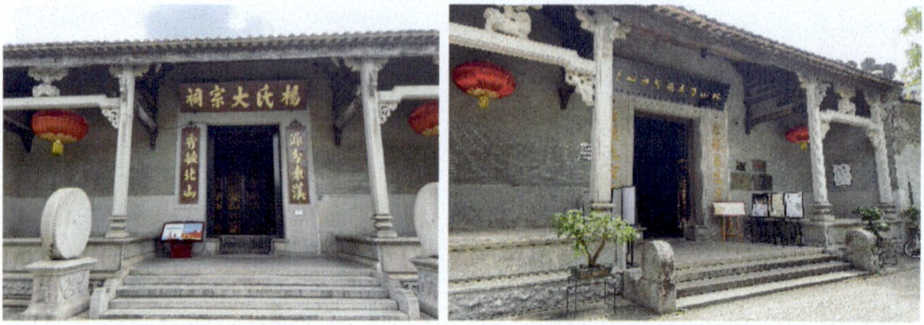

Fig. 13 Main facade of Yangshi Ancestral Hall (left) and Baoxia Yang Ancestral Hall (right). *Source* photographed by the author

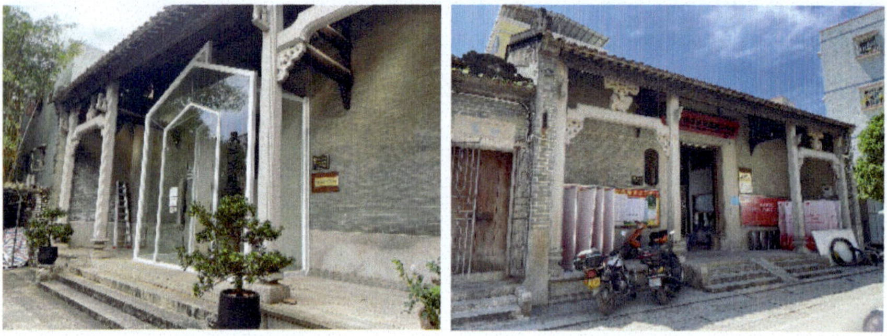

Fig. 14 Main facade of Chengchuan Yang Ancestral Hall (left) and Dongchi Yang Ancestral Hall. *Source* photographed by the author

ancestral hall in Beishan Village, it is also the largest folk ancestral hall in Zhuhai. Founded in the seventh year of Tongzhi in the Qing Dynasty (1868), the hall adopts an axisymmetric architectural layout and a beam-lifting structure. Adorned with exquisite stone carvings, wood carvings, and lime carvings, along with blue tiles and red ridges, it is a distinctive Lingnan ancient building that embodies the artistic aesthetics and technical standards formed during the Qing Dynasty.

3.4 Lifestyle Analysis

3.4.1 Folk Activities

Since the founding of Beishan Village, the genealogy has been compiled eight times. This compilation has connected clan members scattered around the world, fostering a sense of belonging among them. Every year during the Spring Festival, the Beishan Village Committee hosts a banquet at Yang Clan Ancestral Hall, inviting elderly

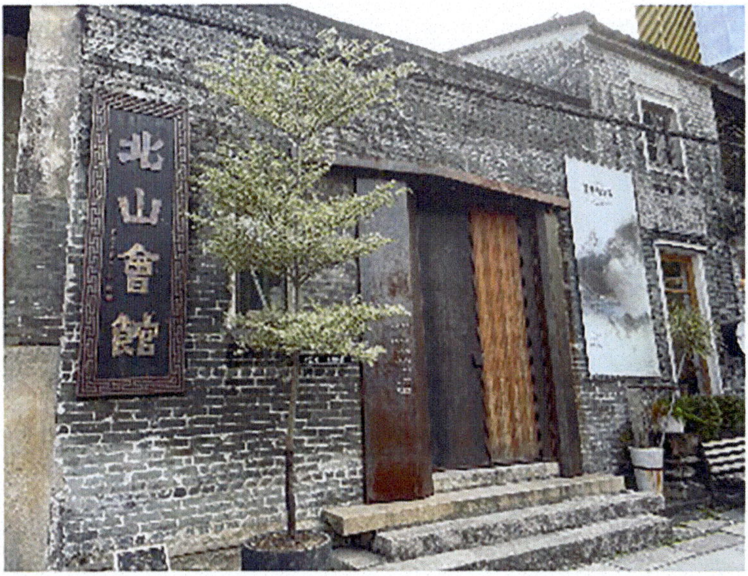

Fig. 15 Main facade of Beishan Yidi Temple. *Source* photographed by the author

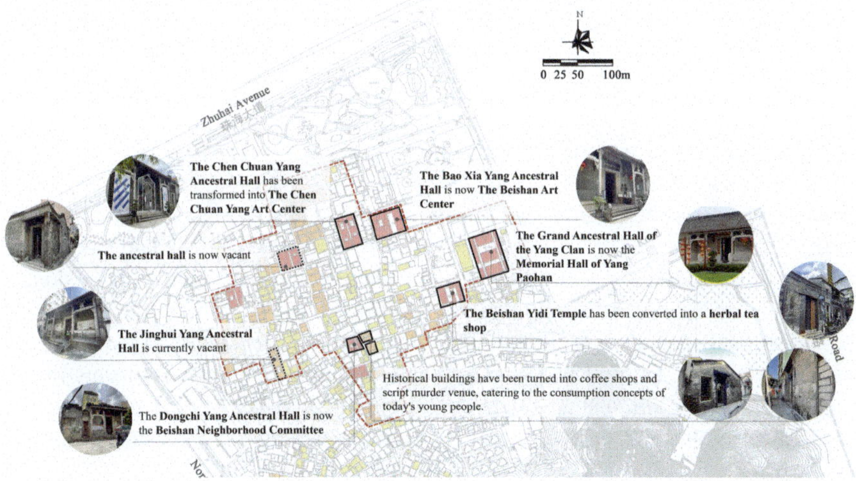

Fig. 16 Map of building distribution and current usage status. *Source* redrawn by the author from: *Protection Plan for Beishan Village, Zhuhai City—A Famous Historical and Cultural Village in Guangdong Province (2020–2027)*

clan members from the village as well as those residing in Hong Kong, Macao, and overseas [4]. Known as the "Beishan Thousand Elders Banquet," this grand clan event has been held without interruption since 1999, embodying a quiet adherence to clan traditions. As a major celebration in Beishan, it has significantly strengthened the bonds among clan members.

The Beishan Shengli Lion Dance Troupe, founded during the Republic of China period and passed down to the present day, performs exquisite traditional lion dances at major festivals and events. At the entrance of Yang Clan Ancestral Hall, there is an exhibition area dedicated to displaying the training and performance props of the troupe, as shown in Fig. 17. Villagers who join the troupe have fixed weekly training sessions. Both Xiuyu Quyi Society and the Beishan Lion Dance Troupe fully support villagers and visitors in learning lion dancing, offering free lion dance classes through Quyi (traditional folk arts) activities.

Field research and interviews reveal that both the Beishan Thousand Elders Banquet and lion dance performances are held inside or in the front square of Yang Clan Ancestral Hall.

The active participation of Beishan's resident organizations has contributed significantly to the better preservation and inheritance of local cultural and entertainment activities such as lion dancing and Quyi (traditional folk arts). Through the cultural and creative transformation of the entire village and the influx of cultural and creative brands, villagers have recognized the cultural value of Beishan Village. The transformation of the village's business forms and the guidance of artists have strengthened the cultural confidence and sense of identity among Beishan villagers.

3.4.2 Local Characteristic Shops

Among the existing tea pastry shops in Beishan Village, there is a distinctive one named "Beishan Tea Pastry" outside Baoxia Yang Ancestral Hall. Tea pastries, originally made as snacks for sacrificial rituals, have now integrated into people's daily lives. "Beishan Tea Pastry" is an authentic traditional food store, but there are relatively few local specialty snacks and intangible cultural heritage (ICH) foods in the village. Before the cultural and creative transformation, the business forms in the village were rather simplistic. Currently, the village boasts a variety of distinctive shops, such as those offering cultural experiences, handicraft workshops, and art galleries.

3.5 Ecological Environment Analysis

3.5.1 Ancient and Famous Trees

The natural leisure landscapes and leisure agriculture within the village both fall under the category of industrial culture [5]. In Beishan Village, ancient trees, old

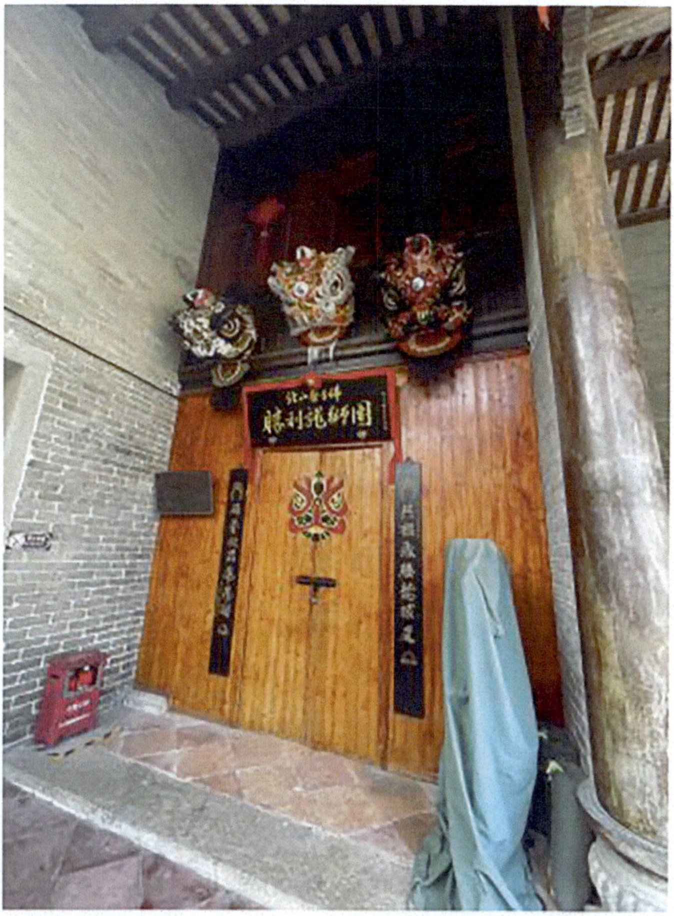

Fig. 17 Beishan Lion Dance Team training center. *Source* photographed by the author

wells, village texture, and natural scenery constitute abundant elements that designers can utilize; they also form integral components of villagers' traditional life scenes. For instance, the ancient trees behind Yang Clan Ancestral Hall and those adjacent to Beishan Yidi Temple have become significant site symbols, where villagers spontaneously gather for communication and recreation. These ancient trees have thus become an inseparable part of villagers' daily lives.

Beishan Village is home to two ancient and famous trees and five important trees that are listed and registered in Zhuhai. Field investigations reveal that most of these ancient and famous trees have suffered long-term damage without timely repair.

The existing ancient trees and historical river inlets in Beishan are located in areas as illustrated in Fig. 18.

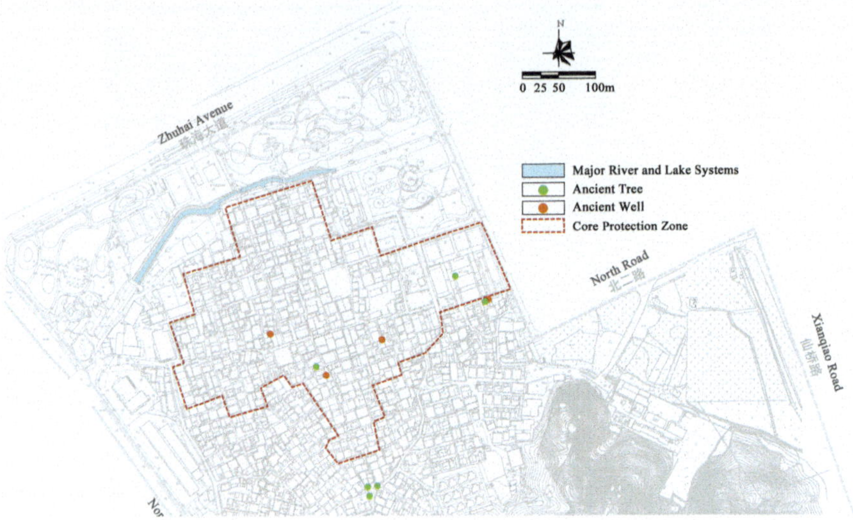

Fig. 18 Map of locations of existing ancient trees, ancient wells and historical river courses in Beishan. *Source* redrawn by the author from: *Protection Plan for Beishan Village, Zhuhai City—A Famous Historical and Cultural Village in Guangdong Province (2020–2027)*

3.5.2 Water Systems and Ponds

As indicated in the distribution map of the ancient well water system, wellheads are primarily distributed along Beishan Main Street, Beishan West Street, and Beishan East Street. It is speculated that in the past, Beishan Main Street gradually evolved into a major cross thoroughfare running through the east, west, north, and south of the village to facilitate water access for villagers. According to data, many villagers still used ancient wells to draw water even as late as 2003, when all residents of Beishan Village relocated to Xiuyuyuan (as shown in Fig. 19).

4 Characteristics of Cultural and Creative Transformation

4.1 Cultural Characteristics Analysis

4.1.1 Preservation of Material Culture

Beishan first carried out a comprehensive transformation of the ancient village, adhering to the concept and principle of "Restoring the old as the old." It renovated the ancient buildings and facades within the village and endowed the vacant ancient buildings with new functions. Meanwhile, the public space environment was comprehensively upgraded, with the addition of several landscape nodes. Not only

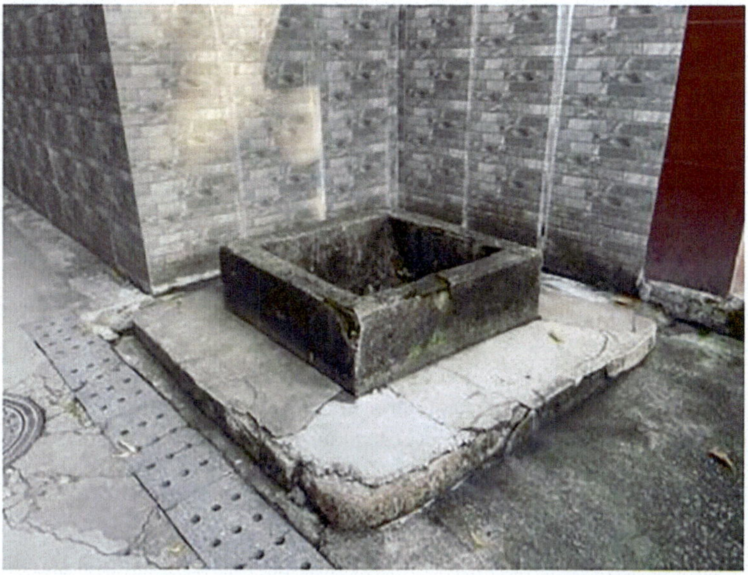

Fig. 19 Water well. *Source* photographed by the author

has the overall architectural style of Beishan Village been preserved, but the restored spaces such as vacant ancient houses have also provided room for the growth of the cultural and creative industry.

4.1.2 Diversification of Industrial Development

In terms of industrial development, the elements of cultural and creative industries have endowed the village with greater development potential based on the original basic elements of the village. It has driven the development of the original agriculture and agricultural product processing industry in a "point-to-face" manner, which in turn has increased the income of new villagers, original residents, and the village collective, enabling Beishan Village to be reborn under the general trend of integrating culture and tourism.

Beishan Village first reshaped its style and features to endow idle buildings with new functions for industrial development, while attracting more cultural and creative workers to settle in Beishan. It has made full use of the natural resources in Beishan Village and transformed them into cultural and creative products through cultural and creative means. These measures have not only attracted many industries to settle in Beishan but also brought many tourists to the village, which in turn has increased tourism income and consolidated the development foundation of Beishan Village.

4.1.3 Construction of Endogenous Forces

According to Secretary Guan of the Beishan Village Committee, many villagers in the village still adhere to cultural and entertainment activities such as the Beishan Lion Dance Troupe and Quyi Society, providing free teaching activities for residents and tourists. These villagers, who silently persist in and selflessly inherit traditional culture, play an indispensable role in the process of cultural and creative transformation of the entire Beishan Village.

The spontaneous participation of local villagers is of far-reaching significance to the cultural and creative development of the village. As inheritors of the original culture, the indigenous villagers have a deeper understanding of cultural inheritance such as the Beishan lion dance, Quyi Society, and ancient building restoration. They have taken on the most important role in reshaping the native culture and played an irreplaceable role in the inheritance and display of traditional culture. Secondly, the indigenous villagers have jointly promoted the sustainable economic development of the village by providing original ecological agricultural products and spontaneously operating local food restaurants. The natural resources in the village and the daily labor of the villagers have become original materials that help cultural and creative workers create more "down-to-earth" literary and artistic works. Currently, since most local villagers have moved out, only a few remain in the village, so there is still great potential to tap into this internal force.

At present, the business forms in the village are relatively rich. Buildings of different protection levels have been rationally utilized, and the villagers' active participation in business operations and entrepreneurship has significantly improved their economic conditions.

4.1.4 Guidance of External Forces

The arrival of external artists has significantly enhanced Beishan villagers' cultural identity and their courage to embrace new things. With the continuous influx of cultural and creative industries, diverse cultural and creative activities such as music festivals and art exhibitions have driven the rapid development of tourism in Beishan Village. Meanwhile, villagers spontaneously provide agricultural products, traditional foods, homestays, and other tourism commodities and services. Under the synergy of new and old industries, the village has generally formed a diversified business development model, with per capita income increasing significantly and the village economy being activated and developed.

4.2 Case Study on the Cultural and Creative Transformation of Beishan

4.2.1 Renovation and Reuse of Old Building

Enterprises such as "Old Things Warehouse" and "Little Days" have continuously settled in Beishan. The successive influx of mature and distinctive cultural and creative enterprises has not only enriched the cultural and creative connotations of Beishan but also diversified the local business forms (Fig. 20).

Beishan Theater, built during the Republic of China period, was converted into a commune auditorium during the Cultural Revolution and can accommodate over 1000 people. In addition to hosting concerts, it also holds fashion shows, educational lectures, art exhibitions, and auctions at the Beishan Guild Hall, among other events. Revenues from the theater are used to support local artists, fund cultural relic restoration, and maintain the normal operation of children's care funds. Located within Beishan Theater, Beishan Old Things Warehouse displays and sells various old objects from the last century, evoking a sense of time travel among visitors.

"Little Days," a cultural and creative store converted from Caibo Temple—the earliest bank in Beishan—has greatly catered to the popular culture and consumption trends of young people today, gaining favor among young consumers and significantly expanding the audience base. Despite its success in appealing to young consumer groups, when considered as a traditional village needing to promote its culture, the cultural and creative products that can reflect local characteristics are lacking.

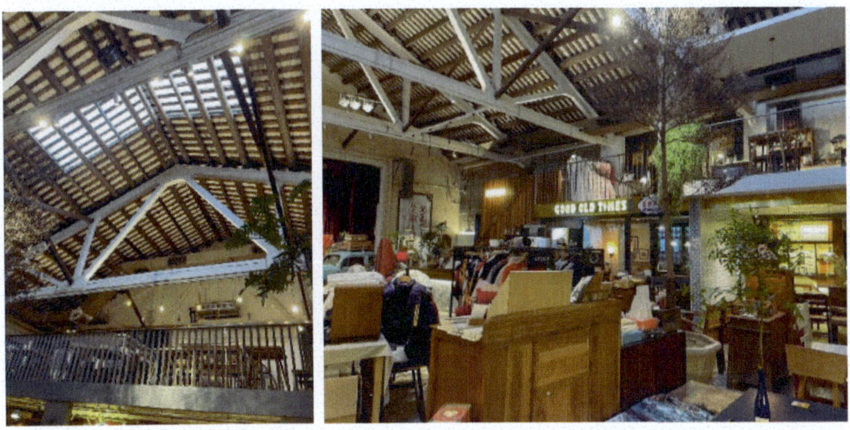

Fig. 20 Beishan old objects warehouse: **a** Roof structure, **b** Indoor space. *Source* photographed by the author

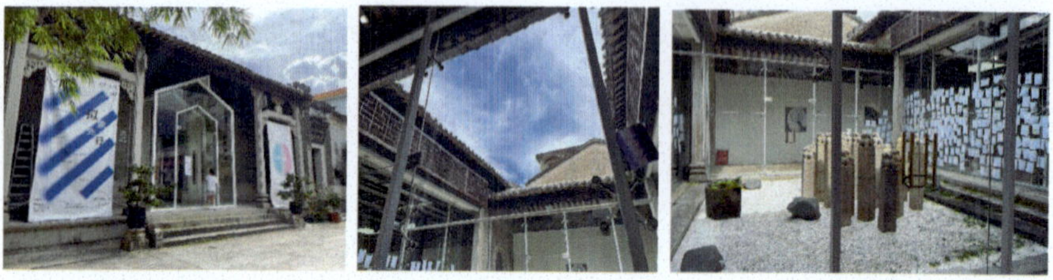

(a) (b) (c)

Fig. 21 Chengchuan Yang Ancestral Hall: **a** Main facade; **b** Indoor space; **c** Indoor space. *Source* photographed by the author

4.2.2 Reuse of Immovable Cultural Relics

Chengchuan Yang Ancestral Hall, a cultural relics protection unit in Zhuhai, is now also known as Chengchuan Art Space (Fig. 21). This ancient building has been converted into a modern art exhibition space. The artist exhibitions held here have attracted many visitors. By providing painting tools and paper for visitors, it enables them to participate in art displays, enhancing the interactivity between the venue and visitors. In terms of architecture, Chengchuan Yang Ancestral Hall has achieved a combination of traditional and new buildings. Culturally, using the public ancestral hall as a pop-up art store also represents the integration of old and new art cultures. Instead of adopting a "fenced-in" protection approach for the old building, its mode of use has been transformed—from a space with a single original function to a proactive "viewable" space. This allows visitors to enter the building to appreciate artworks and engage in artistic creation, thereby achieving the goal of cultural continuity.

4.2.3 Renovation of Street Landscapes

Beishan Main Street, as the main lane running east-west through the core protection area of Beishan Village, has its width controlled at 5 m through planning and renovation. Beishan South Street and Beishan North Street, the main lanes traversing the core protection area from north to south, should also have their widths restricted to 5 m. These lanes must meet the basic design requirements for fire truck access while restricting motor vehicle traffic daily. Lanes such as Beishan West Second Lane and Beishan North Second Lane serve as primary pedestrian thoroughfares (Fig. 22). In addition to the historical street and lane system, rectification measures are applied to incompatible building facades. Dilapidated and damaged walls, as well as street surfaces, are repaired, with the latter restored using materials such as stone slabs or blue bricks, striving to recreate the original appearance of the street and lane spaces.

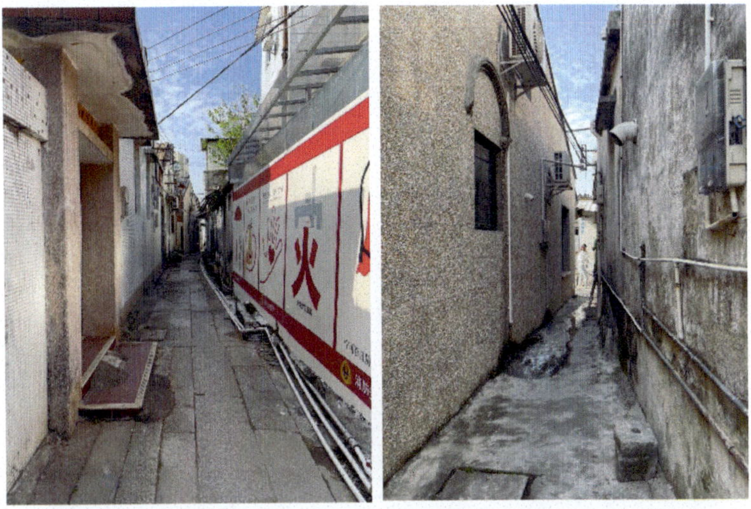

(a) (b)

Fig. 22 Traditional alley space: **a** Southern section of Nanzong Street Lane, Beishan; **b** Northern section of Nanzong Street Lane, Beishan. *Source* photographed by the author

4.2.4 Renovation of Ecological Spaces

With the cultural and creative transformation of Beishan, the ecological environment has also been improved. Measures such as the listing and protection of ancient trees and the renovation of historical river courses have significantly enhanced Beishan's landscape quality. The tourism landscape nodes in Beishan Village are conducive to transforming these natural resources into tourism products.

Field research shows that the water system located in the northern part of the village as depicted in the planning map no longer exists (Fig. 23). Through environmental improvement, various ornamental plants are now planted all over the steps. Thanks to the cultural and creative transformation of Beishan, the overall ecological environment in the village has been reasonably improved, thus greatly enhancing the landscape quality of Beishan's environment.

4.2.5 Construction Volume of Villages

According to the Protection and Utilization Plan for Beishan Village, Nanping Town, Zhuhai (hereinafter referred to as the Protection Plan) compiled in 2002, there were 80 well-preserved traditional-style buildings in Beishan at that time. Data from 2006 showed that there were 418 remaining traditional buildings in Beishan. From 2006 onward, 156 traditional buildings in Beishan Village were demolished and rebuilt over 3 years (Yang Ting, Feng Jiang, October 2016). In June 2012, another 169

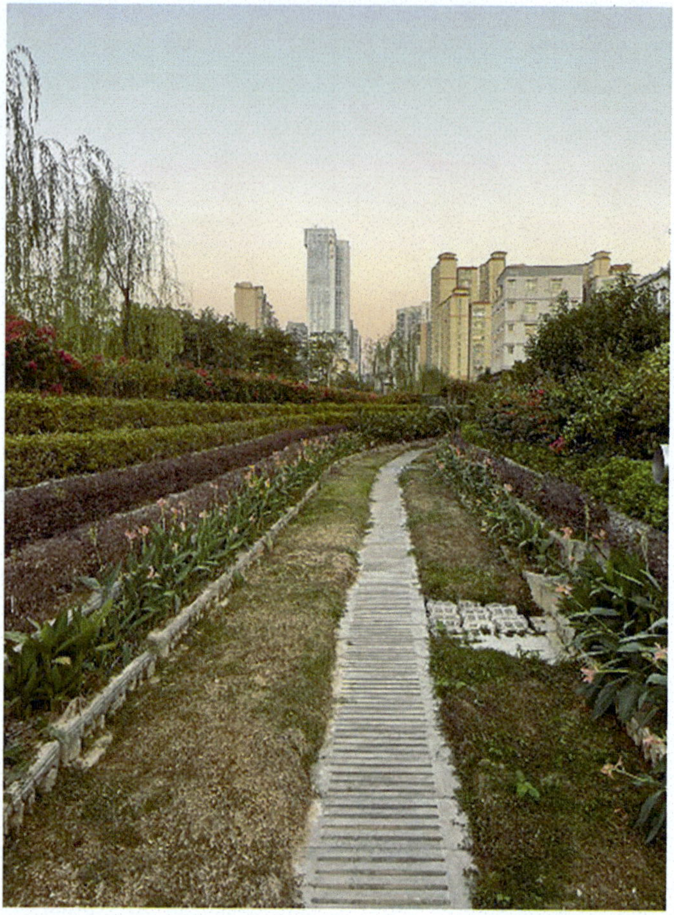

Fig. 23 Current status of key river and lake systems. *Source* photographed by the author

traditional buildings in Beishan Village were demolished, just before the "rectification of illegal construction and land use." By then, only 93 traditional buildings remained, accounting for 22% of the 418 traditional buildings counted in 2006. Survey data from the end of 2014 indicated that only 32 of these traditional-style buildings were left (Yang Ting, Feng Jiang, October 2016). Between 2002 and 2014, the demolition rate of traditional-style buildings in Beishan Village reached as high as 48% (Fig. 24). In fact, many traditional buildings not confirmed by the Protection Plan were destroyed, resulting in an overall demolition rate of 78%, both traditional buildings and the streetscape suffered severe damage (Fig. 25).

As visually shown in Fig. 24, many traditional buildings in Beishan Village were demolished after the cultural and creative transformation. Field investigations reveal that after all these traditional buildings were demolished, they were replaced by catering establishments such as Japanese and Thai restaurants. Both the food sold

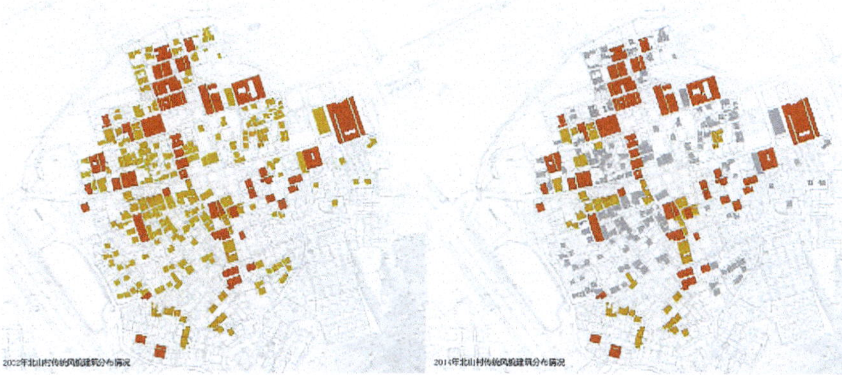

Fig. 24 Comparison of distribution of traditional-style buildings in Beishan Village between 2002 and 2014. *Source* redrawn by the author from *Property Law, Rural Residential Land Policy, and the 12 Years of Beishan Village*, Yang Ting, Feng Jiang, 2016

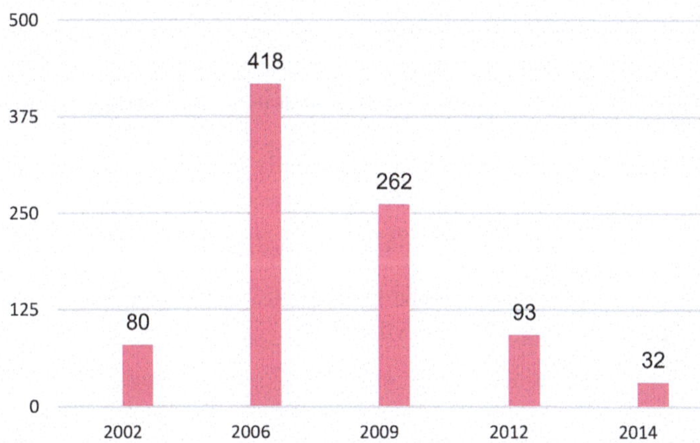

Fig. 25 Annual changes in the number of traditional-style buildings in Beishan Village. *Source* created by the author

and the architecture itself have caused the unique characteristics exclusive to Beishan to vanish without a trace. It can thus be seen that the changes brought by cultural and creative industries to Beishan have both positive and negative aspects. Many shops, to attract attention, demolished traditional buildings and rebuilt them into structures that match the style of the catering services they offer (Fig. 26).

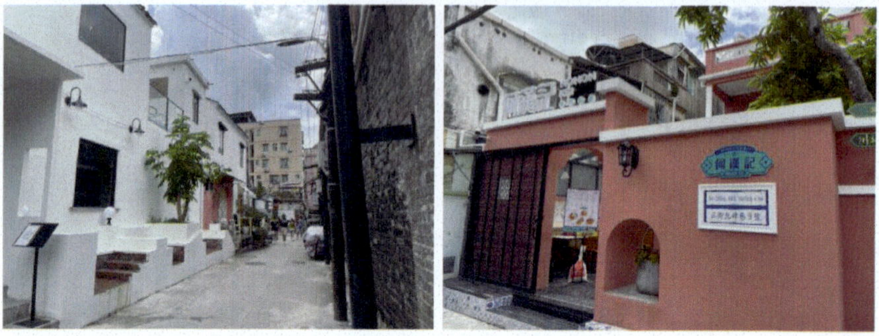

Fig. 26 Current status after demolition of traditional buildings: **a** Café, **b** Restaurant. *Source* created by the author

4.2.6 Population Return Migration

The changes in the population of Beishan Village over the years are shown in Fig. 27. Nearly 1500 indigenous residents of Beishan Village chose to relocate to "Xiuyuyuan" after the completion of the third-phase housing project in 2006, leaving the original old village houses facing the choice of being rented out or renovated and rebuilt.

As can be seen from Fig. 27, Beishan currently has many floating populations residing in the village, and most of the indigenous residents have moved out. However, the number of registered permanent residents has shown a gradual upward trend in recent years. It is speculated that this is due to the influx of many industries; the favorable development of these industries has attracted villagers to return to start businesses, thus revitalizing Beishan Village.

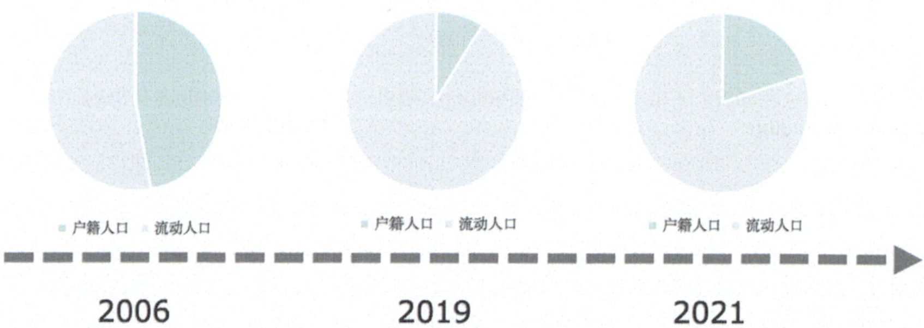

Fig. 27 Population change in Beishan Village. The light blue in the figure represents the registered population, and the gray represents the floating population. *Source* created by the author

5 Construction of the AVC Evaluation System

5.1 *AVC Theory*

AVC refers to Attraction, Vitality, and Capacity. Attraction refers to the elements that conform to tourists' aesthetic preferences and satisfy their entertainment and leisure needs; Vitality denotes the possibility and potential of a tourist destination for long-term prosperity; and Capacity refers to the capacity of the ecological, resource, and environmental systems (Wang Yuncai, Wang Shuhua, December 2007).

Since the AVC theoretical evaluation involves factors such as tourism and economy, with relatively comprehensive evaluation methods and results, it is adopted to assess the comprehensive development status and potential of Beishan Village after cultural and creative transformation.

5.1.1 Basis for Construction

Before constructing the AVC comprehensive evaluation system for Beishan Village after its cultural and creative transformation, reference is made to the evaluation and selection criteria specified in the Convention Concerning the Protection of the World Cultural and Natural Heritage adopted by the United Nations Educational, Scientific and Cultural Organization (UNESCO) in 1972. As a vital element of cultural heritage, historic and cultural cities hold certain referential value in the selection of evaluation indicators (Table 1). With the characteristics of historic and cultural villages and towns as the core, 15 specific indicators are selected for reference from the perspectives of both tangible and intangible cultural heritage.

Reference is made to national standards and regulations such as *Classification, Investigation and Evaluation of Tourism Resources (GB/T18972-2003)* to provide a basis for constructing the evaluation system. Combined with local cultural characteristics, evaluation indicators are selected and confirmed, thereby establishing the evaluation system.

The AVC evaluation indicators in this study are based on theoretical research from relevant literature on landscape evaluation of tourist areas. Through on-site investigation and observation, summary, consultation with relevant academic experts and local staff, and considering the actual situation of Beishan, the evaluation indicators applicable to the AVC evaluation of Beishan's local culture are sorted out.

After the indicators are established, relevant professional experts and staff of Beishan Neighborhood Committee are invited to score the evaluation indicators to determine the weight score of each indicator. Scores for the indicator layer are collected from users through online and offline questionnaires, and the final score is obtained by calculating the weighted scores combined with the weights.

Table 1 Evaluation system for the protection of historic and cultural villages and towns

Layer A	Layer B	Layer C	Layer D
Protection of historic and cultural towns and villages	Tangible cultural heritage B1	Natural environment C1	D1: Degree of harmony between settlements and natural environment
			D2: Integrity of natural ecological environment
			D3: Aesthetic value of natural ecological environment
		Spatial form C2	D4: Integrity of the overall form and style
			D5: Integrity of the street–alley spatial pattern
			D6: Aesthetic quality of spatial form and style
		Architectural features C3	D7: Authenticity of the preservation of cultural relics and historic sites
			D8: Integrity of the protection of vernacular architecture
			D9: Cultural value of architectural art
	Intangible cultural heritage B2	Historical influence C4	D10: Long-standing history of villages and towns
			D11: Influence of historical events and figures
			D12: Distinctiveness of the historical functions of the town–village
		Folk culture C5	D13: Uniqueness of traditional cultural folk customs
			D14: Preservation degree of folk customs, arts and crafts
			D15: Preservation degree of rare folk heritage items

Source Redrawn by the author from: Study on the Evaluation System and Methods for the Protection of Historic and Cultural Villages and Towns: A Case Study of China's First Batch of Famous Historic and Cultural Towns (Villages) Zhao Yong, Zhang Jie, Li Na, and Liang Li (2006)

5.1.2 Establishment of Evaluation Indicators

The specific process of determining the evaluation indicators for Attraction, Vitality, and Capacity through consultations and review of relevant literature is as follows:

(1) Attraction

Liu Binyi divided Attraction into four aspects: landscape uniqueness, landscape diversity, traffic accessibility and comfort, tourism destination popularity, and scenic tourism destination image [6]. Considering the diverse characteristics of rural landscape Attraction [7–9], he comprehensively considered factors such as natural and humanistic landscape resources, rural economic development, and ecological environment quality, selected corresponding indicators comprehensively, and conducted a comprehensive analysis of the indicator system after obtaining the evaluation results [10]. In summary, he identified natural conditions, location, settlement, and humanistic conditions as important factors influencing the comprehensive evaluation system of rural landscapes.

The landscape criterion within Attraction generally involves landscape beauty, harmony, uniqueness, and symbolism. In Beishan Village, this criterion is specifically reflected in existing ancient and famous trees, ancient ponds, water systems, and other aspects. The recreational criterion within Attraction usually refers to environmental comfort and tourist participation. In Beishan Village, this criterion corresponds specifically to existing traditional streets and lanes, public spaces, etc. The historical value criterion within Attraction is generally related to popularity and cultural connotations. In Beishan Village, this criterion is specifically embodied in the value and popularity of Beishan's spatial places. Therefore, the final selected Attraction indicators are shown in Table 2.

(2) Vitality

Vitality refers to the potential for a village's growth and development, with an emphasis on the optimal utilization of resources to drive the village's economic development [11, 12]. Wang Shengyue and Yan Shuhong (April 1994) argue that Vitality encompasses aspects such as the material environment, cultural inheritance, tourism demand, and resource management.

Table 2 Indicators of attraction

Recreational nature	Comfort level of public spaces
	Interestingness of traditional artistic culture
	Interactivity of viewing and appreciation
Landscape quality	Degree of landscape elegance and harmonization
	Symbolic value
Historical value	Degree of recognition
	Degree of humanistic and literary connotations

Source Author's statistics

The criterion layer of cultural inheritance within Vitality includes social economy and residents' livelihoods. In this study, it specifically refers to Beishan's traditional folk customs, celebrities and historical figures, living habits, the number of traditional and contemporary businesses, and the occupancy rate of industries. Combined with the theme and content of this study, the index layer of Vitality is selected, as shown in Table 3.

(3) Capacity

Liu Binyi and Wei Liya (2005) proposed that space and psychological carrying capacity are the main influencing factors affecting Capacity.

In this study, the spatial carrying capacity criterion within carrying capacity specifically refers to the passenger flow and consumption capacity of Beishan, the construction volume of the village, as well as the quantity and distribution of villagers. Psychological carrying capacity refers to satisfaction with participatory experiences and environmental perception.

By integrating the research content of this paper, the indicators of carrying capacity are finally determined as shown in Table 4.

In summary, the following table concludes that the evaluation indicators of the continuity of local culture in Beishan are derived (Appendix Table 17), and the relevant factors are summarized according to the literature theory combined with the characteristics of local culture. By consulting the opinions of five experts, the evaluation structure for the continuity of local culture in Beishan after its cultural and creative transformation is determined to include 1 first-level indicator, 3 second-level indicators, 6 third-level indicators, and 18 fourth-level indicators, as shown in Table 5.

Table 3 Indicators of vitality

Protection of local culture	Continuity of cultural inheritance
	Embodiment of cultural preservation
Socio-economic aspect	Rate of industrial settlement
	Branding of villages or business households

Source Author's statistics

Table 4 Indicators of capacity

Spatial carrying capacity	Visitor flow and consumption capacity
	Volume of village construction
	Number of permanent residents in villages
Psychological carrying capacity	Satisfaction with participatory experience
	Satisfaction with environmental perception

Source Author's statistics

Table 5 Evaluation indicators

Overall objective layer (A)	Sub-objective layer (B)	Criterion layer (C)	Indicator layer (D)
AVC evaluation of local cultural continuity in the cultural and creative transformation of villages and towns	Attraction (B1)	Recreational nature (C1)	Comfort level of public spaces (D1)
			Interestingness of traditional artistic culture (D2)
			Interactivity of viewing and appreciation (D3)
		Landscape quality (C2)	Degree of landscape elegance and harmonization (D4)
			Symbolic value (D5)
		Historical value (C3)	Degree of recognition (D6)
			Degree of humanistic and literary connotations (D7)
			Regional characteristics of architecture (D8)
	Vitality (B2)	Protection of local culture (C4)	Continuity of cultural inheritance (D9)
			Embodiment of cultural preservation (D10)
		Socio-economic aspect (C5)	Rate of industrial settlement (D11)
			Branding of villages or business households (D12)
	Capacity (B3)	Spatial carrying capacity (C6)	Visitor flow and consumption capacity (D13)
			Volume of village construction (D14)
			Number of permanent residents in villages (D15)
		Psychological carrying capacity (C7)	Satisfaction with participatory experience (D16)
			Satisfaction with environmental perception (D17)

Source Author's statistics

5.2 Expert Questionnaire Study

This study mainly adopts the Analytic Hierarchy Process (AHP) to construct a judgment matrix by inviting experts to score the importance of indicators through pairwise comparison. The weight of each individual indicator is calculated using the sum-product method [13, 14]. To ensure the accuracy of the calculation, a consistency test is required. The CI value is the consistency index. If CI and the average random consistency index (RI) satisfy $CR = CI / RI < 0.1$, the calculation result is valid; otherwise, the data needs to be re-adjusted until $CR < 0.1$ is met [15].

$$\lambda_{\max} = \sum_{i=1}^{n} \frac{[AW]_i}{nw_i} \tag{1}$$

The formula for calculating the maximum eigenvalue of a judgment matrix is shown in Formula (1), where $\lambda_{\max}$ represents the maximum eigenvalue of matrix A; n denotes the order of matrix A, meaning that matrix A is an $n \times n$ square matrix and can also be interpreted as the number of factors or indicators under consideration; A refers to an $n \times n$ square matrix, which is the judgment matrix formed by pairwise comparisons of various factors; W is an n-dimensional column vector obtained through certain methods (e.g., the eigenvector method), where the element ω_i in the vector represents the weight of the corresponding factor and satisfies $\sum_{i=1}^{n} \omega_i = 1$; $[AW]_i$ indicates the i element of the new vector AW derived from the multiplication of matrix A and vector W; and nw_i represents n multiplied by the i element w_i of the weight vector W.

$$CI = \frac{\lambda_{\max} - n}{n - 1} \tag{2}$$

The formula for the consistency index is shown in Formula (2), where CI denotes the consistency index, which is used to measure the degree of deviation of the judgment matrix from complete consistency. The closer CI is to 0, the better the consistency; $\lambda_{\max}$ represents the maximum eigenvalue of the judgment matrix; and n denotes the order of the judgment matrix, which can also be understood as the number of factors or indicators considered.

$$CR = \frac{CI}{RI} \tag{3}$$

The formula for the consistency ratio is shown in Formula (3), where CR denotes the consistency ratio. If $CR < 0.1$, the consistency of the judgment matrix is considered acceptable; if $CR \geq 0.1$, the judgment matrix needs to be adjusted. CI represents the consistency index; RI refers to the random consistency index, which is the average value calculated from many randomly generated positive reciprocal matrices. It is related to the order n of the judgment matrix, with different orders corresponding to different RI values, which can be found in relevant Table 6.

Table 6 Average random consistency index (RI)

Order	1	2	3	4	5	6	7	8	9	10	11	12	13	14	15
RI value	0	0	0.52	0.89	1.12	1.26	1.36	1.41	1.46	0.49	0.52	1.54	1.56	1.58	1.59

In a judgment matrix, Aij refers to the degree of importance of indicator i compared with indicator j. When i = j, the two indicators are identical, meaning they are equally important, which is recorded as 1. Every element is greater than 0 and Aij * Aji = 1 [15].

Judgment matrices are constructed based on the established evaluation indicator model, and expert questionnaires are distributed. By inviting 1 professional teacher, 2 village Party secretaries of Beishan Village, and 2 staff members to compare and score the importance of indicators at each level pairwise, the weights of the indicator system are obtained. To ensure the scientific validity of the data, a consistency test is required.

The evaluation scores and consistency test results from one of the experts are shown in Tables 7, 8, 9, 10, 11, 12, 13, and 14:

λ_{max}: 3.0037; Consistency ratio of the judgment matrix B1-B3: CR = 0.0036<0.1.
λ_{max}: 3.0183; Consistency ratio of the judgment matrix C1-D3: CR = 0.0176<0.1.
λ_{max}: 2.0000; Consistency ratio of the judgment matrix C2-D5: CR = 0.0000<0.1.

Table 7 Judgment matrix A1-B3

AVC evaluation of the impact of cultural and creative transformation on the continuity of local culture (A1)	Attraction (B1)	Vitality (B2)	Capacity (B3)	w_i
Attraction (B1)	1.0000	2.0000	5.0000	0.5816
Vitality (B2)	0.5000	1.0000	3.0000	0.3090
Capacity (B3)	0.2000	0.3333	1.0000	0.1095

Source Author's statistics

Table 8 Judgment matrix C1-D3

Recreational nature (C1)	Comfort level of public spaces (D1)	Interestingness of traditional artistic culture (D2)	Interactivity of viewing and appreciation (D3)	w_i
Comfort level of public spaces (D1)	1.0000	3.0000	4.0000	0.6250
Interestingness of traditional artistic culture (D2)	0.3333	1.0000	2.0000	0.2385
Interactivity of viewing and appreciation (D3)	0.2500	0.5000	1.0000	0.1365

Source Author's statistics

Table 9 Judgment matrix C2-D5

Landscape quality (C2)	Degree of landscape elegance and harmonization (D4)	Symbolic value (D5)	w_i
Degree of landscape elegance and harmonization (D4)	1.0000	4.0000	0.8000
Symbolic value (D5)	0.2500	1.0000	0.2000

Source Author's statistics

Table 10 Judgment matrix C3-D8

Historical value (C3)	Degree of recognition (D6)	Degree of humanistic and literary connotations (D7)	Regional characteristics of architecture (D8)	w_i
Degree of recognition (D6)	1.0000	0.3333	2.0000	0.2493
Degree of humanistic and literary connotations (D7)	3.0000	1.0000	3.0000	0.5936
Regional characteristics of architecture (D8)	0.5000	0.3333	1.0000	0.1571

Source Author's statistics

Table 11 Judgment matrix C4-D10

Protection of local culture (C4)	Continuity of cultural inheritance (D9)	Embodiment of cultural preservation (D10)	w_i
Continuity of cultural inheritance (D9)	1.0000	0.3333	0.2500
Embodiment of cultural preservation (D10)	3.0000	1.0000	0.7500

Source Author's statistics

Table 12 Judgment matrix C5-D12

Socio-economic aspect (C5)	Rate of industrial settlement (D11)	Branding of villages or business households (D12)	w_i
Rate of industrial settlement (D11)	1.0000	2.0000	0.6667
Branding of villages or business households (D12)	0.5000	1.0000	0.3333

Source Author's statistics

Table 13 Judgment matrix C6-D15

Spatial carrying capacity (C6)	Visitor flow and consumption capacity (D13)	Volume of village construction (D14)	Number of permanent residents in villages (D15)	w_i
Visitor flow and consumption capacity (D13)	1.0000	2.0000	0.2500	0.2081
Volume of village construction (D14)	0.5000	1.0000	0.2500	0.1311
Number of permanent residents in villages (D15)	4.0000	4.0000	1.0000	0.6608

Source Author's statistics

Table 14 Judgment matrix C7-D17

Psychological carrying capacity (C7)	Satisfaction with participatory experience (D16)	Satisfaction with environmental perception (D17)	w_i
Satisfaction with participatory experience (D16)	1.0000	5.0000	0.8333
Satisfaction with environmental perception (D17)	0.2000	1.0000	0.1667

Source Author's statistics

λ_{max}: 3.0536; Consistency ratio of the judgment matrix C3-D8: $CR = 0.0516 < 0.1$.

λ_{max}: 2.0000; Consistency ratio of the judgment matrix C4-D10: $CR = 0.0000 < 0.1$.

λ_{max}: 2.0000; Consistency ratio of the judgment matrix C5-D12: $CR = 0.0000 < 0.1$.

λ_{max}: 3.0536; Consistency ratio of the judgment matrix C6-D15: $CR = 0.0516 < 0.1$.

λ_{max}: 2.0000; Consistency ratio of the judgment matrix C7-D17: $CR = 0.000 < 0.1$. The final weight values derived from the scores given by experts, sorted in descending order, are shown in Table 15.

5.3 Formulation of Tourist Questionnaires

The author quantified the established evaluation indicators using data obtained from the distributed questionnaires. The basis for constructing the judgment matrix is

Table 15 Weight values of indicators

Indicator layer (D)	Weight value
Degree of humanistic and literary connotations	0.1575
Embodiment of cultural preservation	0.1132
Continuity of cultural inheritance	0.1132
Satisfaction with participatory experience	0.0876
Rate of industrial settlement	0.0754
Number of permanent residents in villages	0.0664
Regional characteristics of architecture	0.0625
Comfort level of public spaces	0.0494
Interestingness of traditional artistic culture	0.0487
Degree of recognition	0.0481
Number of cultural and creative stores	0.0377
Interactivity of viewing and appreciation	0.0271
Traditional architectural style	0.0237
Volume of village construction	0.0237
Satisfaction with environmental perception	0.023
Visitor flow and consumption capacity	0.0206
Degree of landscape elegance and harmonization	0.0094
Preservation of traditional folk activities	0.0085
Preservation of ancient trees, wells, and streets	0.0043

Source Author's statistics

the scoring results from 100 tourists who have visited Beishan, collected through the questionnaires designed by the author (Appendix 1). The final comprehensive score was calculated by combining the weights of the evaluation indicators with the questionnaire scores through weighted score calculation.

In accordance with the content of the weight indicators, the user-oriented questionnaire includes 18 indicators in total. Each indicator is assigned a score of 5, 4, 3, 2, or 1, corresponding respectively to satisfaction evaluations of "very satisfied," "quite satisfied," "average," "dissatisfied," and "very dissatisfied."

This questionnaire was distributed through a combination of online and offline channels, with the distribution period from November 1 to November 15, 2022. A total of 120 questionnaires were distributed, and 100 valid ones were recovered. The questionnaires were randomly distributed to tourists and villagers who had visited Beishan, both on-site and online. When distributing questionnaires on-site, explanations and interviews were provided to respondents for any unclear questions. For online questionnaires, considering respondents who had never visited Beishan Village, pictures and text descriptions were attached to inform them of the current situation of the site, aiming to ensure the validity and authenticity of the data obtained from the questionnaires.

The average scores and weighted scores obtained from the questionnaires are shown in Table 16.

Table 16 Average scores and weighted scores obtained from questionnaires

Type	Specific details	Average score	Weighted score
Recreational nature	Comfort level of public spaces	4.2286	0.2089
	Interestingness of traditional artistic culture	3.8571	0.1878
	Interactivity of viewing and appreciation	3.6286	0.0983
Landscape quality	Symbolism in landscape expression	3.8000	0.0163
	Symbolic value	4.1429	0.0982
Historical value	Degree of recognition	3.4857	0.1677
	Humanistic and literary dimensions	4.0571	0.6390
	Architectural locality	4.2571	0.2661
Protection of local culture	Degree of preservation and reuse of ancient/traditional buildings	4.3714	0.4948
	Experientiality of folkways and folk customs	3.6571	0.4140
	Sense of psychological identification with local culture	3.7429	0.4237
	Quality of traditional shops	3.7714	0.4269
Socio-economic aspect	Diversity of cultural and creative shops	4.2857	0.3231
	Branding of villages or business households	3.9714	0.2208
Spatial carrying capacity	Visitor flow and consumption capacity	3.6857	0.0759
	Volume of village construction	3.9714	0.0941
	Number of permanent residents in villages	3.5143	0.2333
Psychological carrying capacity	Satisfaction with participatory experience	3.7714	0.3304
	Satisfaction with environmental perception	3.9429	0.0907

Source Author's statistics

5.4 Weight Analysis of AVC

Based on the scoring results derived from the AVC evaluation system for Beishan's local culture, as shown below, the weight of Attraction (A) is 52%, Vitality (V) is 37.05%, and Capacity (C) is 10.95% (Fig. 28). Among the AVC three dimensions, Attraction accounts for the largest proportion, while Capacity accounts for the smallest. Therefore, after the cultural and creative transformation, Attraction has the most significant impact on the continuity of the village's local culture, followed by Vitality, and finally Capacity.

The weighted score of Attraction is 1.6823, that of Vitality is 0.9839, and that of Capacity is 0.8245. The line chart of the weights and weighted scores of the AVC is shown in Fig. 29.

It can be visually observed from the line chart that the weighted scores change with the weights. Since the trend lines of the weight values and weighted scores tend to converge, an additional trend line representing the ratio of weighted scores to weights is included for reference.

As shown in Fig. 30, there is a significant gap between the weighted scores of the AVC three dimensions. The trend of the ratios is completely different from that of the weighted scores and weights, presenting a result where the score for Carrying Capacity is the highest, followed by that for Attraction, and finally that for Vitality. This also indicates that, regarding the current development status of Beishan's local culture, the current development status of Attraction is inferior to that of Carrying Capacity, and it is roughly on par with that of Vitality. From this, it can be concluded that the current development status of Beishan Village's Carrying Capacity is relatively good.

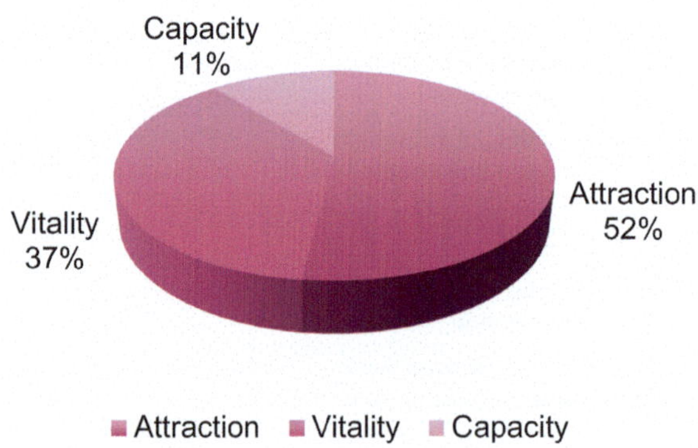

Fig. 28 AVC evaluation weight—proportion in Beishan Village. *Source* created by the author

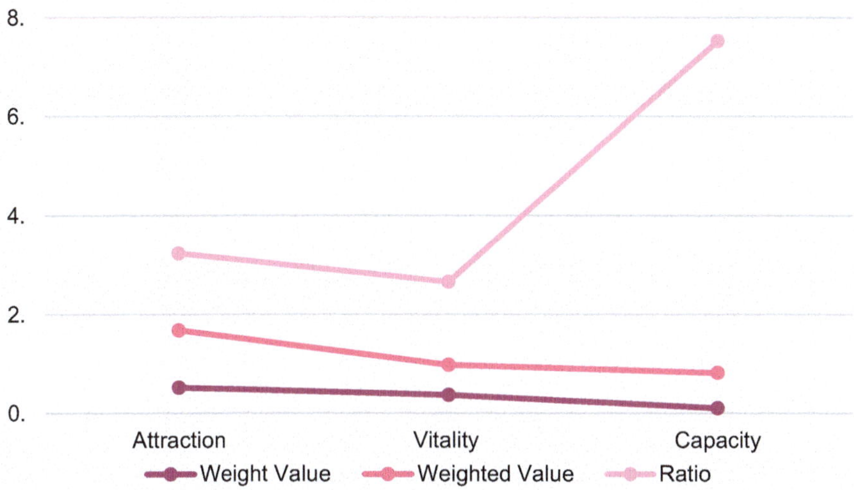

Fig. 29 AVC weight value, weighted value, and ratio. *Source* created by the author

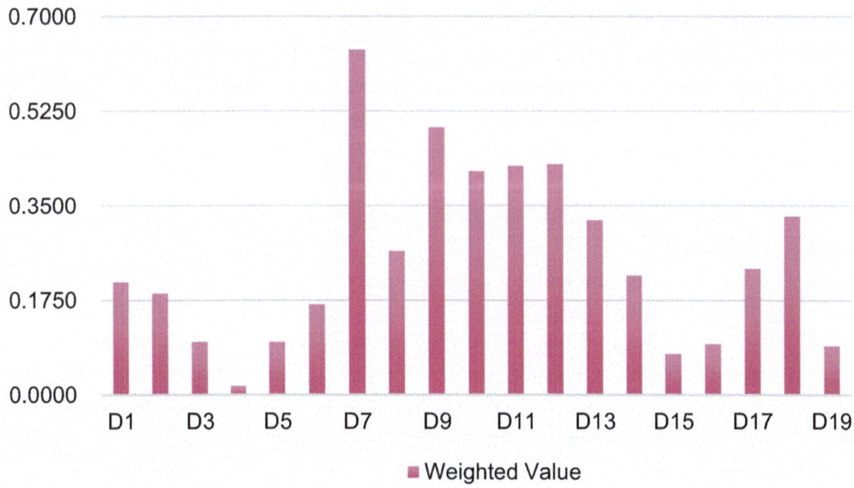

Fig. 30 Weighted scores of each indicator. *Source* created by the author

5.5 *Analysis of Weighted Scores of AVC*

5.5.1 Analysis of Scores for Attraction Indicators

In the AVC evaluation of Beishan Village's local culture, the score for Attraction is 1.6823. Attraction consists of three factor layers and eight indicator layers.

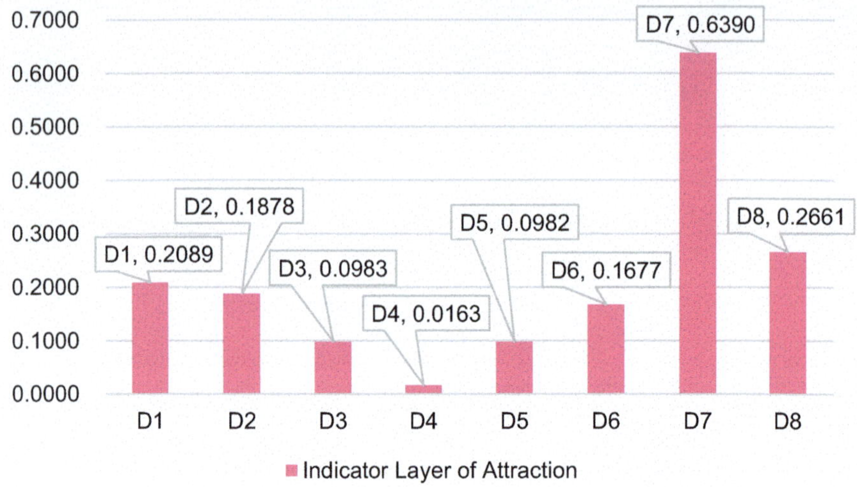

Fig. 31 Weighted scores of attraction indicators. *Source* created by the author

Among these, one indicator is rated as "excellent" (i.e., scoring above 0.4), and three indicators are rated as "poor" (i.e., scoring below 0.1).

It can be clearly observed from Fig. 31 that three indicators—"Degree of Landscape Elegance and Harmonization," "Symbolic Value," and "Interactivity of Viewing and Appreciation"—have relatively low scores. Tourists generally perceive that the expression of local culture fails to bring a sense of novelty, and descriptions such as "too conventional" and "too broad" have become their basic evaluations of Beishan's landscape.

Questionnaire interviews reveal that tourists consider the sense of interaction and participation to be low, mostly due to the weak interactivity and participation in traditional culture. An analysis of the reasons suggests that Beishan's promotion of local traditional art and culture, such as lion dance and folk art, is insufficient. Although there are free teaching and experience sessions offered by the lion dance team and folk-art team, the lack of systematic management has caused many tourists to miss the opportunity for personal experience.

5.5.2 Analysis of Scores for Vitality Indicators

In the AVC evaluation of Beishan Village's local culture, the score for Vitality is 0.9839. Vitality consists of two factor layers and six indicator layers. Among these, four indicators are rated as "excellent" (i.e., scoring above 0.4), and the overall score is relatively high.

The three indicators—"Experientiality of Folkways and Folk Customs," "Diversity of Cultural and Creative Shops," and "Branding of Villages or Business Households"—scored relatively low (Fig. 32). Analysis indicates that this is due to the

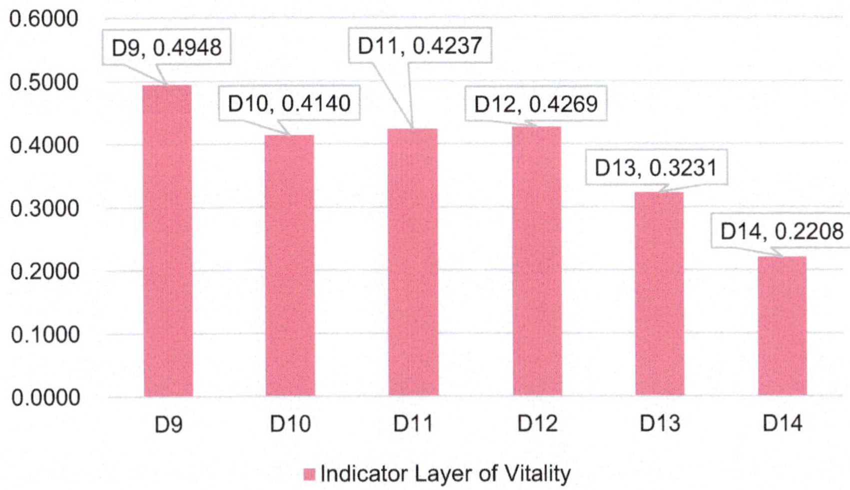

Fig. 32 Weighted scores of vitality indicators. *Source* created by the author

absence of a fixed venue in the village for villagers to engage in daily training for lion dance and other cultural and recreational activities. Naturally, tourists also have no place to experience such activities, which is a major factor contributing to the low score in "Sense of Psychological Identification with Local Culture."

5.5.3 Analysis of Scores for Capacity Indicators

In the AVC evaluation of Beishan Village's local culture, the score for Capacity is 0.8245. Capacity consists of two factor layers and five indicator layers. Among these, one indicator is rated as "good" (i.e., with a score ranging from 0.3 to 0.4), and three indicators are rated as "poor" (i.e., scoring below 0.1). Tourists generally gave low satisfaction scores regarding the "Visitor Flow and Consumption Capacity," "Volume of Village Construction," and "Satisfaction with Environmental Perception" (Fig. 33).

Analysis suggests that with the massive influx of cultural and creative industries, some self-employed villagers have blindly followed the currently popular industries, lacking integration with local culture. Meanwhile, to align the style of traditional-looking buildings with the operated shops, many traditional-style buildings have been demolished, resulting in a significant loss of local culture. Due to the lack of standardized management, many villagers randomly discard garbage during their operations, causing environmental squalor, which in turn leads to low user ratings for environmental perception.

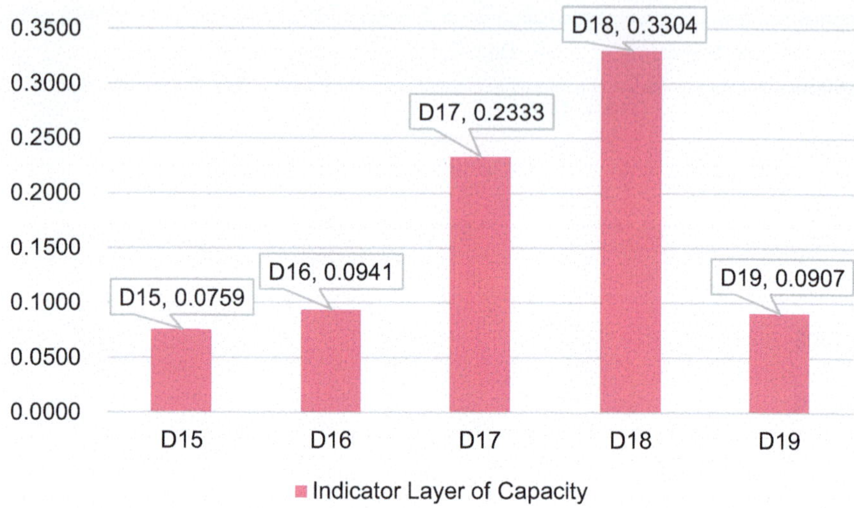

Fig. 33 Weighted scores of capacity indicators. *Source* created by the author

5.6 *Development Suggestions for AVC*

By referring to relevant literature on tourism landscape evaluation research, it is known that the comprehensive development potential of a tourist area and the driving force behind the long-term development of its landscapes are determined by the balance of the AVC development of the tourist area (Liu Yurong, 2015). Therefore, based on the weighted score results obtained earlier, the development level of local culture in Beishan Village can be tested using the following formula. With the overall framework of the AVC evaluation system for Beishan's local culture as the standard, the standard deviation of AVC is calculated as follows:

$$S = \frac{1}{n}\sqrt[3]{\left[(A - M)^2 + (V - M)^2 + (C - M)^2\right]}$$

$$\text{Among them, } M = \frac{1}{3}(A + V + C) \tag{4}$$

From the calculation using Formula (4), it can be determined that the balance degree of the AVC development of Zhuhai Beishan's local culture is $S = 0.5062 < 1$. The value of S is inversely proportional to the balance degree of the three-dimensional development, indicating that Beishan has a relatively high balance degree of AVC development and strong development potential.

As can be seen from the weight value table in the evaluation system, the weighted scores of the AVC change with the weight values. Among them, Attraction (A) accounts for a relatively large proportion of the total weight of the A.V.C, which indicates that Attraction contributes significantly to the entire evaluation system, while both Vitality (V) and Capacity (C) still need to be strengthened. Based on the current development status of Attraction, Vitality, and Capacity, the following strategies are proposed.

5.6.1 Strategies for Enhancing Attraction

Environmental Renovation. Carry out an overall renovation of the village environment, standardize the construction environment of ongoing construction projects, implement standardized management for villagers' shops and stalls that litter, formulate relevant regulations, and promote environmental protection among villagers.

For Beishan Village, it is necessary to extract unique cultural characteristics. For example, cultural elements such as lion dance, folk art, village texture, and legendary stories can be integrated to create cultural and creative products or landscapes, enrich expression forms, and develop characteristic rural cultural and creative designs that align with local culture.

Enhancing Interactivity. Interactivity is a necessary condition for both natural landscapes and venues and projects for tourists' recreation. It is suggested to set up a Beishan Village Activity Center to provide a fixed venue for villagers to inherit culture. For instance, Beishan's folk-art society can offer folk-art experience classes, the lion dance team can conduct performance rehearsals and daily training in the training room, etc. Additionally, folk custom experience classrooms should be established to provide tourists with all-round interactive experiences, such as costume changing, handcraft making, and VR simulation, to attract people and truly engage them in participation.

5.6.2 Strategies for Enhancing Vitality

Enrich the Number of Cultural and Creative Stores. The existing cultural and creative stores offer few local characteristic cultural and creative products that stand out. It is suggested to design products based on local cultural features, for example, using cultural elements such as Beishan's lion dance, folk art, and myths and legends to create items like wool felt crafts and pendants, thereby enriching the diversity of products.

Enhance Folk Customs Experience. Folk customs are important products of local culture. First, a fixed venue is needed to allow tourists and villagers to carry out cultural practice and experience activities. It is also feasible to celebrate folk customs at fixed festivals, for example, commemorating lion dance culture on the anniversary of the establishment of Beishan Lion Dance Team and creating a "Lion Dance Experience Month" to engage the whole people in cultural experience. Such grand activities will naturally attract public attention, and more people will pay attention to lion dance culture.

5.6.3 Strategies for Enhancing Capacity

Enhancing Environmental Quality. Firstly, criticize and educate tourists for their uncivilized behaviors and villagers for littering during business operations. Strengthen the management of tourist landscapes and innovate traditional greening methods, such as adding landscape art installations.

Village Branding. Branding is an effective means to truly protect rural culture. Under the protection of a brand, the uniqueness of village culture can be better showcased. The primary task of branding is to find a positioning; determine the main image of rural tourism; give full play to the uniqueness, ethnicity, artistry, and tradition of local culture; and meet tourists' expectations for Beishan's image. For example, Longjing Village in Zhejiang Province is well known for its brand image of "the first village of tea town" and has become a representative of rural tourism featuring tea culture. For Beishan, it can leverage the red resource of revolutionary martyr Yang Pao'an or representative characteristic cultural resources such as Beishan lion dance culture and position the main image of the brand based on the uniqueness of these resources.

6 Conclusion and Discussion

6.1 Conclusion

This chapter first obtained a large amount of factual data through field investigations and interviews, selected key words from them through induction and collation, and then selected indicators based on the AVC theory combined with literature materials. Using the analytic hierarchy process (AHP), an evaluation system for local culture applicable to villages undergoing cultural and creative transformation in Zhuhai was formulated. Subsequently, a specific analysis was conducted by combining this evaluation system with the actual case of Beishan Village in Zhuhai. It was concluded that the weighted score of Attraction among the AVC in Beishan Village accounts for a relatively large proportion, and corresponding countermeasures were proposed based on the status of the A.V.C.

At the initial stage of the research, the specific definition of local culture in this study was established through literature review and field research interviews: all material and non-material elements constituting traditional villages are part of local culture.

Several aspects such as the reuse of old buildings, industrial diversification, and ecological space renovation are the interventions of the arrival of cultural and creative transformation on local culture. The continuity between cultural and creative transformation and local culture is reflected in three aspects: historical origin, natural environment, and humanistic factors.

The conclusions derived from the indicator evaluation are as follows.

In terms of both importance and performance among Beishan Village's AVC dimensions, the order is Attraction (A) > Vitality (V) > Capacity (C), and the development of the AVC three dimensions is relatively balanced. Through the ratio of weighted scores to weights, it is found that among the AVC three dimensions, Capacity currently has the best development status, followed by Attraction, and finally Vitality, which still has significant room for improvement.

Based on the evaluation results, cultural and creative transformation undoubtedly promotes the continuity of local culture. Measures such as the reuse of ancient buildings, cultural experiences, and environmental renovation are the best protection and inheritance of local culture. The opening of cultural and creative stores, art exhibitions, and the Beishan Music Festival have attracted many visitors to Beishan Village, bringing possibilities for its development. However, there are also some negative impacts. With the continuous influx of cultural and creative industries, villagers have spontaneously engaged in commercial activities, demolishing many traditional-style buildings, which is quite detrimental to cultural inheritance.

In the scoring of Attraction indicators, it is evident that the inadequate improvement measures for ecological spaces have led to low scores in "Degree of Landscape Elegance and Harmonization," "Symbolic Value," and "Interactivity of Viewing and Appreciation." This indicates that the expression forms of local culture fail to bring a sense of novelty to tourists, and "being too popular and broad," has become tourists' basic evaluation of Beishan's landscape.

The overall weighted score of Vitality is relatively good, but due to the low questionnaire scores for Vitality, the ratio of weight value to weighted score is the lowest among the three AVC dimensions, which also suggests that the current development status of Vitality is poor. Among them, the weighted scores of two indicators related to the village's cultural and creative transformation—"Experientiality of Folkways and Folk Customs," "Diversity of Cultural and Creative Shops," and "Branding of Villages or Business Households"—are relatively low. This also shows that the products offered by characteristic stores in the village are too popular and lack innovation, which is a major reason why operators find it difficult to achieve branding.

Capacity has the lowest weight among the AVC three dimensions, but its current development status is the best. Among its indicators, "village construction scale" and "tourist flow and consumption capacity" have the highest scores. It is obvious that as Beishan Village gradually takes shape as an "Internet-famous cultural and creative village" with cultural creativity at its core, the goal of branding has achieved initial results, making Beishan Village gradually become a major tourist destination among surrounding villages. The arrival of cultural and creative transformation has brought economic vitality and development possibilities to Beishan Village.

6.2 Discussion

With the introduction and support of national policies, many villages have joined the ranks of cultural and creative transformation. This study addresses such issues as the status of local culture in villages after cultural and creative transformation, how local culture relies on cultural and creative industries for existence, and tourists' evaluation of local culture after cultural and creative transformation.

From the questionnaire survey results, the current development status of Vitality is poor, which can be explained by the following reasons:

(1) Although there is a wealth of traditional cultural and entertainment activities, there is no systematic promotion, nor are there fixed venues for villagers to train and for tourists to experience, making it difficult to inherit the culture.
(2) The number of traditional shops is small, while other shops in the village are too conventional and lack the embodiment of local characteristic culture.
(3) Beishan's own image is not distinct enough, and a strong brand theme atmosphere and impression have not yet been formed, making it impossible to enhance consumers' cognition and memory.
(4) Cultural and creative shops have not been well integrated with local culture, and the phenomenon of homogenization is serious.

6.3 Summary

This chapter expands the research field of the AVC theory; analyzes the commonalities between culturally and creatively transformed villages and the AVC theory in terms of society, economy, and environment; establishes an AVC comprehensive evaluation system for local culture in culturally and creatively transformed villages; and proposes corresponding improvement countermeasures.

Regarding the shortcomings of this paper, the AHP (analytic hierarchy process) and Delphi method used in this study rely on experts' personal perceptions for indicator selection and scoring, which are highly subjective. The scoring results vary from person to person, so the obtained results are not fixed and may have a certain gap with the objective results in the real world. What the author can do is minimize the gap with the objective results by selecting multiple experts to score the indicators, thereby improving the authority of the results.

Achieving rural revitalization through the development of cultural and creative industries essentially involves a dynamic process of grasping regional characteristics, agglomerating development, and realizing development effectiveness and regional Attraction. The influence of cultural and creative transformation is evident in the transformation of an ordinary traditional village into a popular cultural and creative village sought after by numerous artists and swarmed by young consumer groups.

This study evaluates the local culture of Beishan using the AVC three dimensions and proposes corresponding countermeasures. Affected by the mutual restriction of the three dimensions and the subjectivity of evaluation indicators, how to conduct a more scientific and objective evaluation and how to achieve the all-round development of the three dimensions of villages still require further discussion and research.

Acknowledgements Institutional Review Board Statement Not applicable for studies not involving humans or animals.

Informed Consent Statement Not applicable for studies not involving humans.

Conflicts of Interest The authors declare that the research was conducted in the absence of any commercial or financial relationships that could be construed as a potential conflict of interest.

Data Availability Statement The datasets used and analyzed during the current study are available from the corresponding author on reasonable request.

Funding This research received no external funding.

Appendix

(See Tables 17 and 18).

Table 17 Expert indicator scoring form

Overall objective layer (A)	Sub-objective layer (B)	Criterion layer (C)	Indicator layer (D)
AVC evaluation of local cultural Continuity in the cultural and creative transformation of villages and towns	Attraction (B1)	Recreational nature (C1)	Comfort level of public spaces (D1)
			Interestingness of traditional artistic culture (D2)
			Interactivity of viewing and appreciation (D3)
		Landscape quality (C2)	Degree of landscape elegance and harmonization (D4)
			Symbolic value (D5)
		Historical value (C3)	Degree of recognition (D6)
			Degree of humanistic and literary connotations (D7)
			Regional characteristics of architecture (D8)
	Vitality (B2)	Protection of local culture (C4)	Continuity of cultural inheritance (D9)
			Embodiment of cultural preservation (D10)
		Socio-economic aspect (C5)	Rate of industrial settlement (D11)
			Branding of villages or business households (D12)
	Capacity (B3)	Spatial carrying capacity (C6)	Visitor flow and consumption capacity (D13)
			Volume of village construction (D14)
			Number of permanent residents in villages (D15)
		Psychological carrying capacity (C7)	Satisfaction with participatory experience (D16)
			Satisfaction with environmental perception (D17)

Source Author's statistics

Table 18 Questionnaire content

Criterion layer	Indicator layer	Scoring
Recreational nature	Comfort level of public spaces	5 4 3 2 1
	Interestingness of traditional artistic culture	5 4 3 2 1
	Interactivity of viewing and appreciation	5 4 3 2 1
Landscape quality	Symbolism in landscape expression	5 4 3 2 1
	Symbolic value	5 4 3 2 1
Historical value	Degree of recognition	5 4 3 2 1
	Humanistic and literary dimensions	5 4 3 2 1
	Architectural locality	5 4 3 2 1
Protection of local culture	Degree of preservation and reuse of ancient/traditional buildings	5 4 3 2 1
	Experientiality of folkways and folk customs	5 4 3 2 1
	Sense of psychological identification with local culture	5 4 3 2 1
	Quality of traditional shops	5 4 3 2 1
Socio-economic aspect	Diversity of cultural and creative shops	5 4 3 2 1
	Branding of villages or business households	5 4 3 2 1
Spatial carrying capacity	Visitor flow and consumption capacity	5 4 3 2 1
	Volume of village construction	5 4 3 2 1
	Number of permanent residents in villages	5 4 3 2 1
Psychological carrying capacity	Satisfaction with participatory experience	5 4 3 2 1
	Satisfaction with environmental perception	5 4 3 2 1

Source Author's statistics

References

1. Yang, T., & Feng, J. (2016 October). Property law, homestead policy, and twelve years of heritage protection in Beishan Village, Zhuhai. *Urban Architecture*, 104–107. (In Chinese).
2. Zhuhai Xiangzhou District Government. (2020). Notice from the people's government of Xiangzhou District, Zhuhai City, on the announcement of the protection plan for Beishan Village, Zhuhai City, a famous historical and Cultural Village in Guangdong Province (2020–2027). http://www.zhxz.gov.cn/xxgk/tzgg/content/post_2639969.html
3. Yu, Q., & Zhou, X. (2020). Research on the development of cultural resources in urban ancient villages from the perspective of creative tourism: A case study of Beishan Community in Zhuhai. *Journal of Northwest Adult Education College, 6*, 62–67. (In Chinese).
4. Mai, Y., Yang, C., You Kexin, X., & Jiaqi, H. X. (2021). System dynamics of spatial production in "cultural creativity +" historical blocks: A case study of Beishan Community in Zhuhai. *Geographical Research, 40*(02), 446–461. (In Chinese).
5. Zhang, F. (2008). Research on the development of local cultural industry in Taiwan. *Fujian Forum (Humanities and Social Sciences)*, 49–51. (In Chinese).
6. Liu, B. (2003, May). Quantitative model for AVC evaluation of scenic areas and tourism areas: A case study of the master plan of Xuanwu lake scenic area. *ChineseGarden*, 61–64. (In Chinese).

7. Du, Y. (2021, February). The construction of evaluation system of rural tourism development of intangible cultural heritage from the perspective of tourism landscape. In *6th international conference on economics, management, law and education (EMLE 2020)* (pp. 234–239). Atlantis Press. https://doi.org/10.2991/aebmr.k.210210.037
8. Liu, B. (2023). Analysis and assessment of human settlement, inhabitation, and travel activity. In *The trialism and application of human settlement, inhabitation and travel environment studies: Applications in water-net region* (pp. 141–183). Springer Nature Singapore. https://doi.org/10.1007/978-981-19-9143-1_5
9. Binyi, L. I. U. (2022). Towards landscape interaction: Inheritance and development of landscape perception and visual evaluation. *Landscape Architecture, 29*(9), 12–17. https://doi.org/10.14085/j.fjyl.2022.09.0012.06
10. Liu, B. (2023). The origin of trialism of human settlement, inhabitation, and travel environment. In *The trialism and application of human settlement, inhabitation and travel environment studies: Applications in water-net region* (pp. 1–30). Springer Nature Singapore. https://doi.org/10.1007/978-981-19-9143-1_1
11. Liu, S., Ge, J., Bai, M., Yao, M., He, L., & Chen, M. (2022). Toward classification-based sustainable revitalization: Assessing the vitality of traditional villages. *Land Use Policy, 116*, 106060. https://doi.org/10.1016/j.landusepol.2022.106060
12. Yang, Z., Liu, T., Shen, N., Guan, M., Zheng, Y., & Jiang, H. (2022). Resource and environment constraints and promotion strategies of rural vitality: An empirical analysis of rural revitalization model towns. *Frontiers in Environmental Science, 10*, 956644. https://doi.org/10.3389/fenvs.2022.956644
13. Munier, N., & Hontoria, E. (2021). *Uses and Limitations of the AHP Method.* https://doi.org/10.1007/978-3-030-60392-2
14. Podvezko, V. (2009). Application of AHP technique. *Journal of Business Economics and Management, 2*, 181–189.
15. Wang, Q. (2021, March). *A media convergence evaluation index system based on AHP* (pp. 263–266). Digital Communication World.

Potential Assessment of Traditional Residential Area Revitalization: "Pátio" and "Beco" in Macau Peninsula

Hongyu Bao and Junxin Song

Abstract The success of applying for World Heritage in Macau's historic district has brought new opportunities and challenges to the protection of Macau's cultural heritage. Among them, Patio and Beco, which used to represent the daily social and cultural life of residents, have once attracted people's attention, but they are still not fully activated and utilized. Therefore, how to activate and develop Patio and Beco has become the focus of many studies. According to this research background, this paper puts forward the following research questions: firstly, from the perspective of the current situation and the overall system of the city, what are the characteristics of the space of Patio and Beco and what are the influence potentials of these characteristics on its subsequent activation and development? Secondly, how to comprehensively evaluate the activation potential of these spatial potentials in combination with previous studies on cultural values? This paper attempts to analyze the current characteristics of this low-level street by combining space syntax and GIS, including the accessibility and openness characteristics of each Patio and Beco street at different scales and the harmony characteristics of the block related to human cognition. These characteristics are calculated by the integration degree, traversal degree, and comprehensibility degree in space syntax. Finally, the quantitative results are combined with the cultural value evaluation results made by predecessors, and the activation potential indicators of Patio and Beco streets in Macau Peninsula are obtained. On the one hand, the significance of the indicators provides the corresponding planning reference for the orderly activation of Patio and Beco, On the other hand, it also provides the corresponding method reference for the orderly protection of cultural heritage in Patio and Beco and even more cities. Through the above research, the findings of this paper can be summarized into two main points: Firstly, in terms of spatial potential, the accessibility and openness of Patio and Beco at different scales are lower than the average level of Macau city, which indicates that they are inaccessible and private spaces on the whole. However, compared with the urban scale,

H. Bao
Politecnico di Milano, Mantova, MN, Italy

J. Song (✉)
Institute for Research on Portuguese Speaking Countries, City University of Macau, Taipa, Macau, China
e-mail: a23092100150@cityu.edu.mo

Y. Chen et al. (eds.), *City-Sea Symbiosis*, Springer Geography,
https://doi.org/10.1007/978-3-032-13725-8_11

people are more likely to reach or cross these low-level streets from the perspective of pedestrian scale activities, and the accessibility status of Patio and Beco pedestrian scale is significantly increased, even close to the average level of Macau streets. However, from the perspective of openness, Patio and Beco streets always remain introverted. From the distribution law, the accessibility of Patio and Beco streets has obvious external dependence, and its easy access is more because the external street connected with it is easier to reach. Although their openness is also influenced by the external streets, it is more related to their own forms. In addition, the block harmony related to human spatial cognition shows that most of the blocks in Patio and Beco are easy to understand, which is conducive to people's activities in the block, but the harmony in Freguesia de São Lourenço (St. Lawrence Parish) is mostly low, which is not conducive to the activities of external people in this area. Secondly, from the perspective of activation potential, this paper obtains the activation potential scale of Patio and Beco from the perspective of cultural value and spatial potential. The higher the activation potential is, the more priority the Patio and Beco should be given and suitable for development as public commercial space or tourism cultural space, while the lower the activation potential is, the later it should be considered and suitable for use as local residential area.

Keywords Patio and Beco · Space syntax · GIS · Potential activation

1 Introduction

The success of Macau's application for heritage status has played an important role in promoting the city's culture and urban development. The local government and residents are also actively concerned about and participating in the protection of urban cultural heritage. But at the same time, Macau's urban development also faces many challenges: for example, the large-scale development of buildings for gambling and commerce and the conflicts between the tourism industry and residential communities. Among these, many traditional daily living spaces in Macau, such as the "Pátio" and "Beco" and lanes of the "Pátio" and "Beco" have not been well integrated into the overall urban development during their revitalization and protection efforts. Instead, they appear as independent and redundant parts attached to the city's interior, forming a stark contrast with the bustling and prosperous external spaces. With the increasing attention from the Macau SAR government and the community to Chinese cultural heritage, research on "Pátio" and "Beco" have also gradually increased. However, as an important component of Macau's unique urban fabric, the "Pátio" and "Beco" are still subject to various forms of damage and are gradually declining. For example, the visual impact of external high-rise buildings on the alley spaces, changes in the street profile ratios, the decay and collapse of historical buildings, and the loss and aging of the population within the community. These phenomena indicate that amid the rapid development of Macau's urban society and economy, the alley spaces have not been integrated well into this process. Therefore, the current situation faced by the alley streets is not optimistic. And the more important question is, considering this current situation, how should these alley streets be revitalized and developed to bring more social, cultural, and economic benefits to Macau? From the current state of revitalization, only a small portion of the Macau alley streets have been repaired and renovated, and the progress is slow. Meanwhile, most of the remaining streets have not been well revitalized or reused, often hidden deep within the city and receiving little attention.

There are numerous "Pátio" and "Beco" in the Macau Peninsula, and Gong Kai believes that it is very difficult to carry out specific work on so many streets and "Pátio" and "Beco" in a short period of time [1]. Therefore, it is very important to conduct preliminary assessments and classifications of the alleys, which helps in the orderly revitalization of these areas. This paper combines theoretical analysis of space syntax with visualization techniques of geographic information systems to achieve the following two objectives: (1) From the perspective of space syntax theory, explore the spatial characteristics of the traditional hierarchy of the "Pátio" and "Beco" street in Macau within the overall modern urban system of Macau. (2) Integrating cultural value and space syntax perspectives, propose reference indicators for revitalization potential and conduct an objective quantitative assessment of all "Pátio" and "Beco" on the Macau Peninsula.

2 Materials and Methods

2.1 Research Area: "Pátio" and "Beco" in Macau Peninsula

Since the "Pátio" and "Beco" streets primarily exist in the Macau Peninsula, this paper focuses on the current Macau Peninsula as the research scope, taking all the existing old and new "Pátio" and "Beco" streets on the peninsula as the research subjects. After field research, it was found that there are currently 241 street spaces named with "Pátio" and "Beco" on the Macau Peninsula (Fig. 1).

2.2 Analytical Methods

The distribution of streets and "Pátio" and "Beco" within the "Pátio" and "Beco": This paper proposes using objective quantitative analysis through spatial syntax to describe the current vitality of the "Pátio" and "Beco" and combines geographic information systems to provide auxiliary visualization and correlation analysis for the syntactic model. This paper also considers using urban big data and field survey methods to reduce model errors, enhance model interpretability, and assist in analyzing the spatial characteristics of the "Pátio" and "Beco". Finally, the various data will be aggregated and organized.

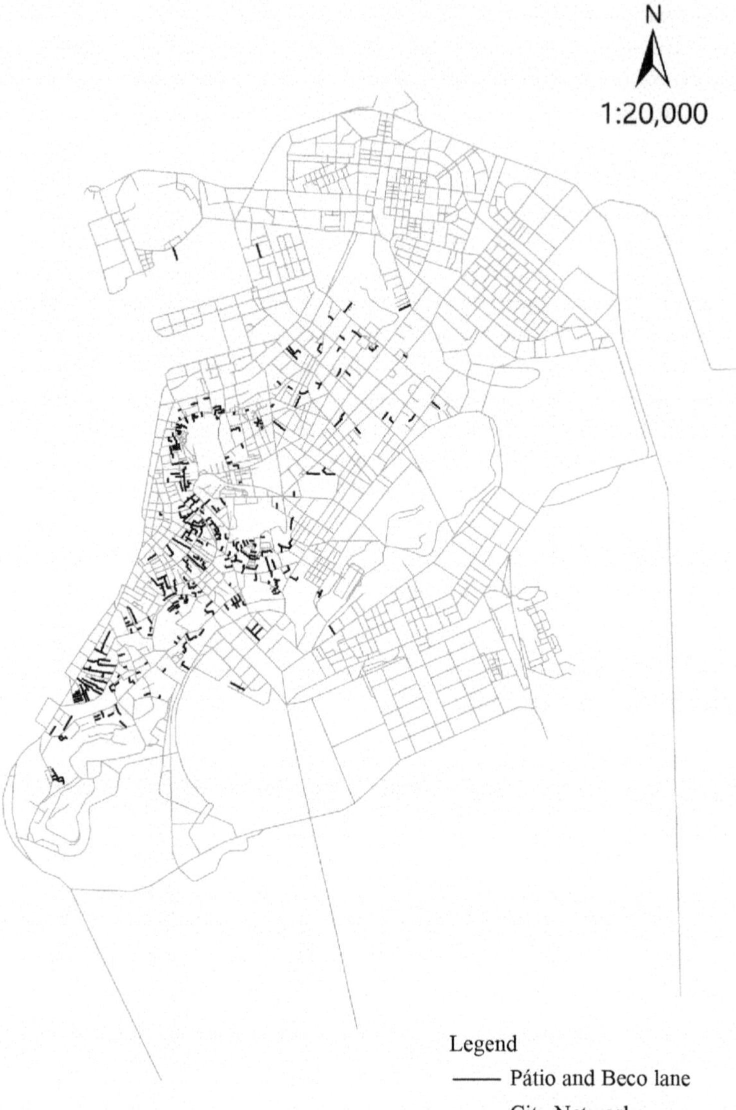

Fig. 1 Distribution of "Pátio" and "Beco". (*Image source* drawn by the author)

2.2.1 Space Syntax

Space syntax is an urban and architectural theory co-developed by Bill Hillier and Julienne Hanson since the 1970s. The main idea of this theory is that the overall structure of space can influence various aspects of a city by affecting the flow of people, such as land use functions and street density [2]. This paper applies space syntax theory, focusing on integration, permeability, and intelligibility, to the study of the enclosed streets. It reinterprets these three parameters as accessibility, openness, and block harmony, and categorizes and analyzes each enclosed street based on these quantitative results. At the same time, these quantitative results are also important influencing factors on the activation potential of each "Pátio" and "Beco", affecting their spatial potential. In terms of actual calculation methods, this paper uses DepthmapX software to perform calculations at various scales on the street network of the Macau Peninsula.

2.2.2 Geographic Information System (GIS)

Geographic information systems originated in the 1960s. They can support "the collection, management, processing, analysis, modeling, and display of spatial data, facilitating the resolution of complex planning and management issues". This paper mainly uses ArcGIS software to model the street network of the Macau Peninsula and conducts corresponding auxiliary visualization and correlation analysis of spatial syntax parameters.

2.2.3 Fieldwork

This study also conducted field research on all "Pátio" and "Beco" and their surroundings in the Macau Peninsula. From the perspective of objectives, one is to understand the specific locations and current conditions of the various enclaves on the Macau Peninsula, and the other is to specifically grasp the distribution of the urban street network in the Macau Peninsula area, improve the accuracy of spatial syntax modeling, verify the model, and eliminate discrepancies with reality.

2.3 Research Process

The specific analysis will be conducted from the following three perspectives:

(1) First, an evaluation and analysis of the urban scale accessibility and public openness of Patio and Beco are conducted. This step aims to assess the position of the Patio and Beco, which serves as a lower-grade street, within the overall system of the current Macau Peninsula region, and to understand its spatial characteristics and how these characteristics impact on its current state. At the same time, by dividing the averages, a classification analysis of the overall scale accessibility and openness of each walled space is conducted.
(2) Next, the walkability and public openness of the "Pátio" and "Beco"s will be classified and analyzed. This step involves analyzing the local spatial characteristics of the "Pátio" and "Beco"s on the Macau Peninsula at a pedestrian

scale and exploring how these characteristics affect their current state. Furthermore, the results of the local analysis also need to be compared with the results of the overall analysis to determine whether the walkability and openness of these "Pátio" and "Beco"s will change significantly with changes in scale. This analytical approach can better capture the scale characteristics of the "Pátio" and "Beco"s.

(3) Finally, the evaluation and analysis of the harmony of the blocks within the "Pátio" and "Beco" are conducted. The object of the intelligibility calculation is a local block, which explores whether the small-scale direct perception and movement of people can correctly understand the large-scale morphological characteristics of the area. This is not only related to people's spatial cognition but also to the harmony between the streets. This paper explores the harmony characteristics of the blocks where the enclosed streets are located and conducts a classification analysis of the group areas where each "Pátio" and "Beco" is situated based on the high and low standards of intelligibility. At the same time, it also explores the significance of these characteristics for the current state of the neighborhood and its future revitalization and development.

3 Morphological Characteristics of "Pátio" and "Beco" at the Urban Scale

3.1 Reachability Analysis

Integration is an important parameter in space syntax. At the global scale of urban computation, it reflects the ease of access from all street segments to that space. And this level of difficulty is completely different from the Euclidean distance weighting in the actual sense. The spatial syntax segment model in this paper is cost-weighted based on topology and corner angles. Hillier believes that angle analysis is highly related to human wayfinding cognition patterns [3]. In other words, from the perspective of space syntax, the global scale integration of the "Pátio" and "Beco" reflects the degree of accessibility from a cognitive perspective within the entire network of the Macau Peninsula.

After GIS mapping and on-site field research, the urban road network data of the Macau Peninsula was obtained and imported into the DepthmapX software for calculation. The integration degree of the streets and "Pátio" and "Beco" in Macau and the integration degree of the overall road network of Macau was extracted and drawn into a histogram for analysis (Fig. 2, Table 1, Fig. 3, and Table 2).

From an average perspective, the average value of the global integration degree of Pátio and Beco in the Macau Peninsula (790.88) is significantly lower than the average level of the street network in the Macau Peninsula (893.35). This indicates that as part of the urban streets in Macau, Pátio and Beco belong to the parts that are more difficult to reach in the overall street network structure of Macau. Meanwhile, the relevant manual of space syntax also points out that the average value of integration defines

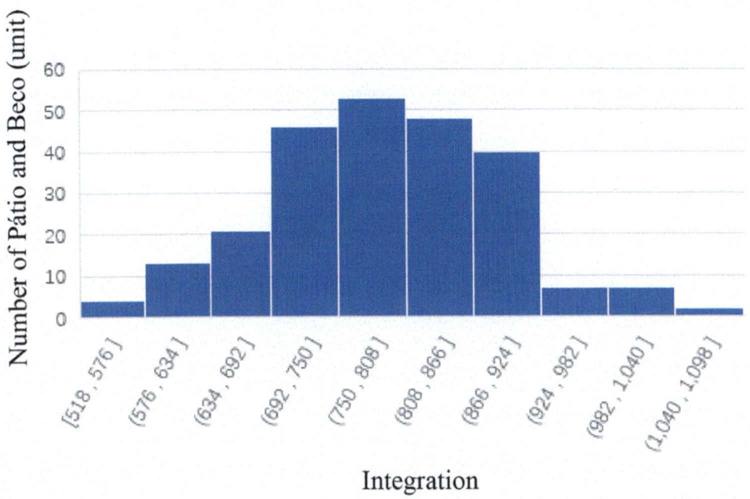

Fig. 2 Macau Peninsula "Pátio" and "Beco" global integration statistics histogram (*Image source* drawn by the author)

Table 1 Macau Peninsula "Pátio" and "Beco" global integration statistics

Macau Peninsula "Pátio" and "Beco" global integration statistics					
Count	Minimum	Maximum	Sum	Mean	Standard deviation
245	517.91	1073.07	193,704.73	790.88	102.13

Source Author based on space syntax statistics

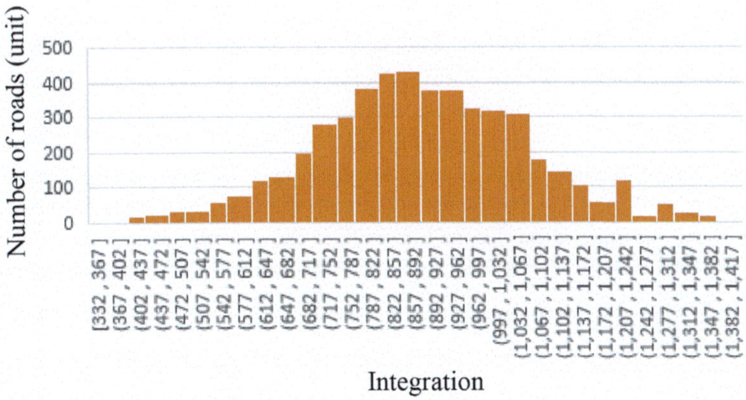

Fig. 3 Statistical histogram of the global integration of the street network in the Macau Peninsula. (*Image source* drawn by the author)

Table 2 Macau Peninsula street network global integration statistics

Macau Peninsula street network global integration statistics					
Count	Minimum	Maximum	Sum	Mean	Standard deviation
4954.00	332.11	1386.10	4,425,668.78	893.35	172.67

Source Author based on space syntax statistics

the background network in an urban network [4]. The so-called background network is relative to the theory of universal cities. The space syntax theory believes that all cities have a universal concept, that is, they are all composed of a foreground network and a background network. The background network can be simply explained as a space that is formally short and functionally used for residence, while the foreground network is the opposite [5]. Relevant studies also summarize the foreground network and the background network from a quantitative perspective. The common definition is that the road segments with the top 20% of integration and choice are the foreground network, and the rest are the background network [6]. If we look at Pátio and Beco from this meaning, it can be explained that even in the residential blocks of the Macau Peninsula today, the low-level Pátio and Beco streets are more private, less noticeable, and more difficult to reach spaces. Therefore, from the scale of the Macau Peninsula today, due to the changes of the times and the development of transportation technology, Pátio and Beco streets have indeed become a kind of street space hidden in the city and not easy to be found. This characteristic of being difficult to reach is more obvious for some higher-level and large-scale vehicle flows.

At the same time, the global integration degree of Pátio and Beco in Macau also fluctuates significantly. Among them, the global integration degree of Pátio de Tomé Pires reaches approximately 1,073 at the highest, which is incorporated into the street structure of the foreground network in the overall network structure of the Macau Peninsula. The lowest is Pátio da Pomba, with only about 518, belonging to a relatively isolated area in the overall structural network of the Macau Peninsula.

By sorting and summarizing, the Pátio and Beco streets in the Macau Peninsula according to the average value of integration degree into analysis diagrams (such as Figs. 4, 5, and 6), it can be found that among the Pátio in the Macau Peninsula, 34 Pátio have an integration degree higher than the overall average level of urban streets and alleys in the Macau Peninsula. 85 Pátio have an integration degree higher than the overall average level of Pátio and Beco spaces in the Macau Peninsula. The remaining 122 Pátio are those with relatively low integration degree even for Pátio and Beco streets. Therefore, according to the average value, the Pátio and Beco streets in the Macau Peninsula can be divided into three categories: InA, InB, and InC. Among them, the InA category belongs to the Pátio and Beco streets with the highest global accessibility, and in the entire street network of the Macau Peninsula, their accessibility potential at the overall urban scale is considered to be relatively high. The InB category belongs to the Pátio and Beco streets with general accessibility. Although their accessibility is not prominent in the street system of the Macau

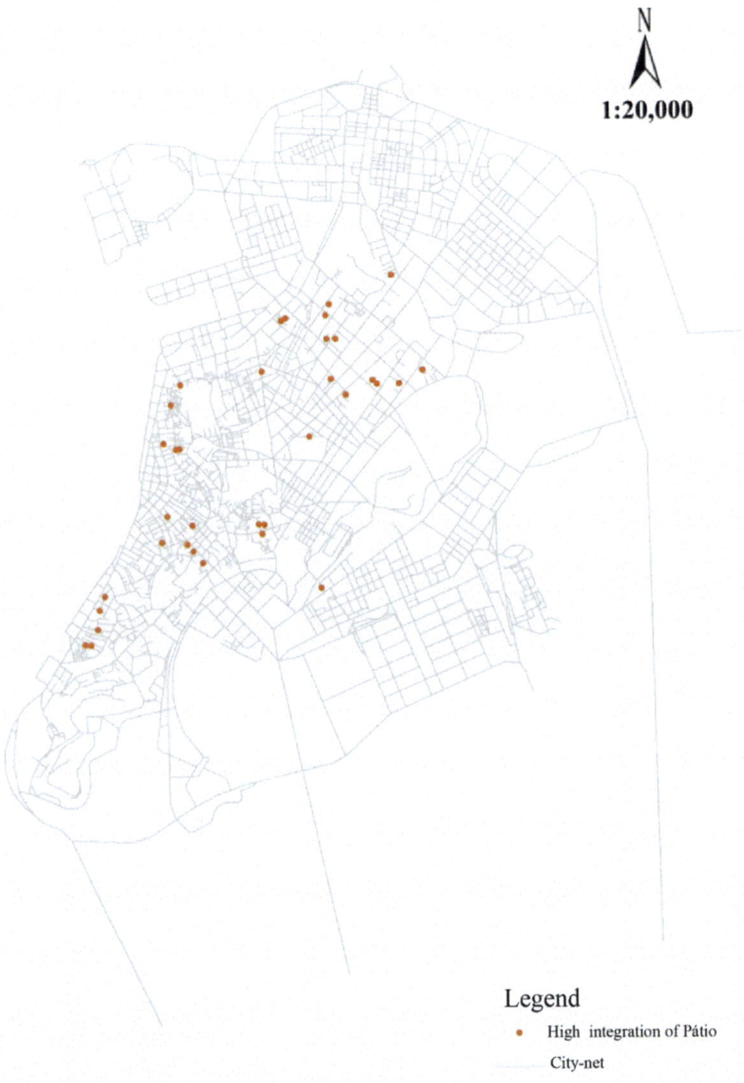

Fig. 4 Distribution map of Pátio with high integration

Peninsula, their accessibility potential is relatively high among these low-level Pátio and Beco streets. The InC category is the one with the lowest accessibility. From the perspective of the overall urban scale, most of the Pátio and Beco in Macau have poor spatial accessibility potential, but there are still some Pátio streets with relatively prominent accessibility potential.

Convert the point data of these three types of Pátio and Beco into kernel density analysis maps, and then sort out the foreground network, which is the top 20% of the road network in terms of integration degree calculated by the space syntax road

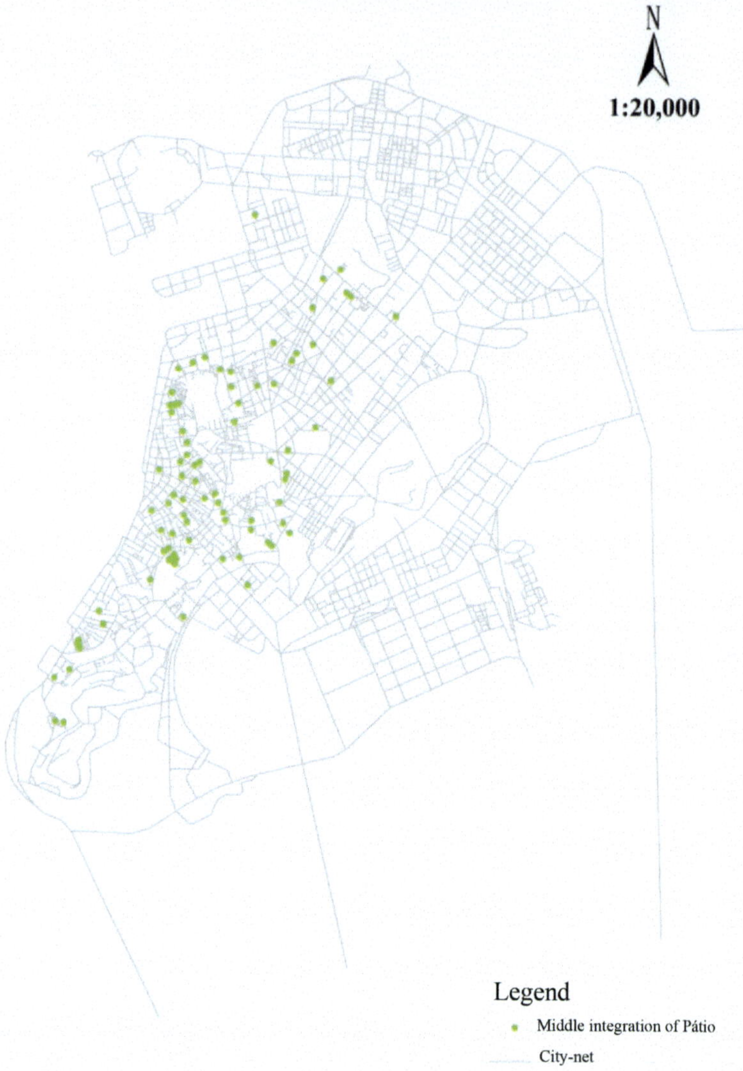

Fig. 5 Distribution map of Pátio with middle integration. (*Image source* drawn by the author)

network, and overlay the two into one map (such as Figs. 7, 8, and 9). These analysis maps show that the better the accessibility of the Pátio type, the closer the relationship between its distribution core and the foreground network, and even some Pátio streets are included in the foreground network, including Pátio de Tomé Pires, Beco da Concha, and the partial connection sections of Beco de Tomé Pires and Avenida do Almirante Lacerda. The poorer the accessibility of the Pátio type, the more distant the relationship between its distribution core and the foreground network. Based on this phenomenon, a simple conjecture can be drawn from the urban scale, the reason why

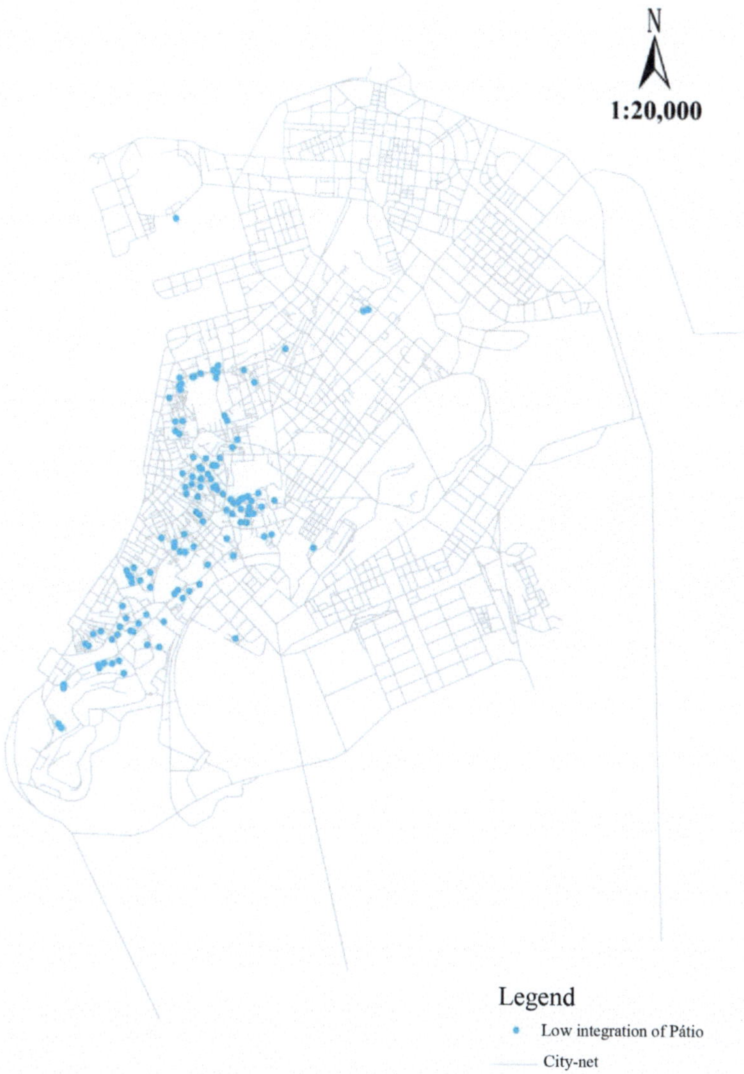

Fig. 6 Distribution map of Pátio with low integration. (*Image source* drawn by the author)

Pátio with high accessibility is easy to reach is that it is more directly connected to the foreground network, while Pátio with poor accessibility is often distributed in the gaps surrounded by the foreground network. Corresponding statistical findings can also prove this conjecture: by sorting out the number of direct connections between various types of Pátio and the foreground network and making an intuitive distribution map (such as Fig. 10), it can be found that the better the global accessibility of the type, the larger the proportion of Pátio directly connected to the foreground network, and there is not even one Pátio street directly connected to the foreground network

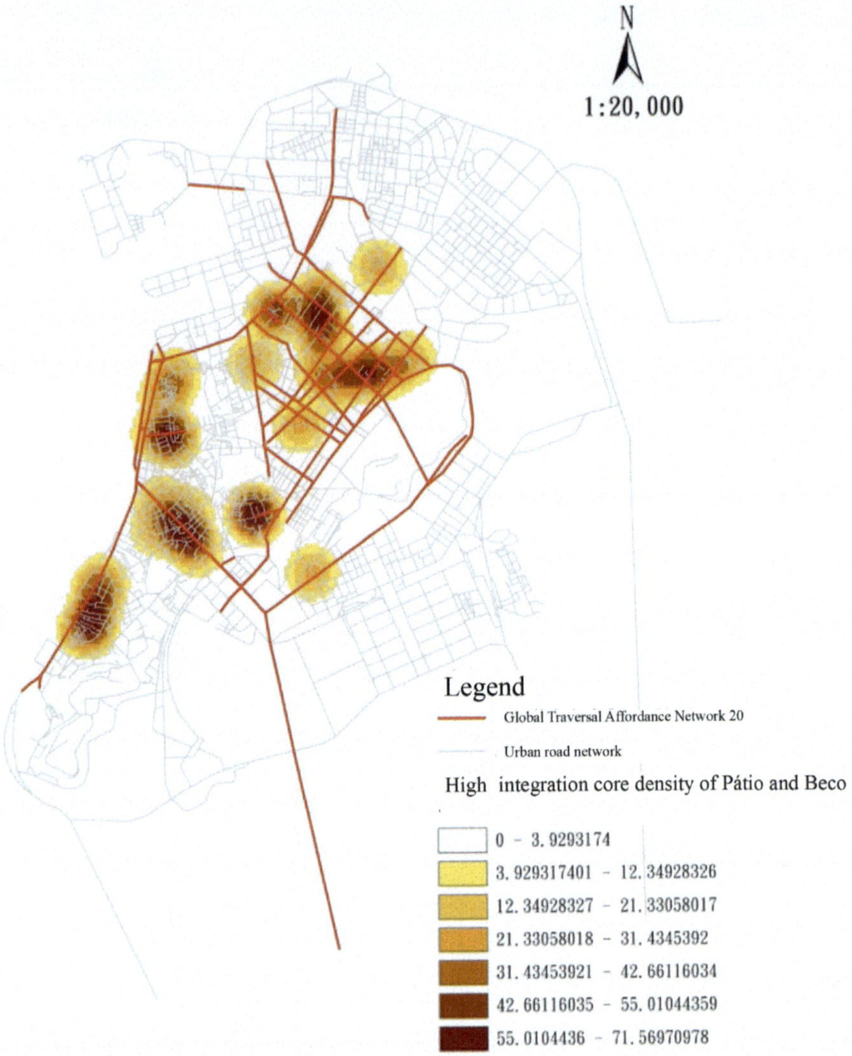

Fig. 7 Distribution characteristics of Pátio with high integration. (*Image source* drawn by the author)

in the low-accessibility category. This distribution characteristic reflects that the accessibility characteristics of urban street morphology show a clustering feature: the closer the street is to the foreground network, the higher its accessibility (the average change rate of direct connection proportion: 49.6%). For low-level Pátio and Beco streets, due to their smaller scale and fewer street openings, this morphological feature more reflects a dependence on the accessibility of external higher-level streets.

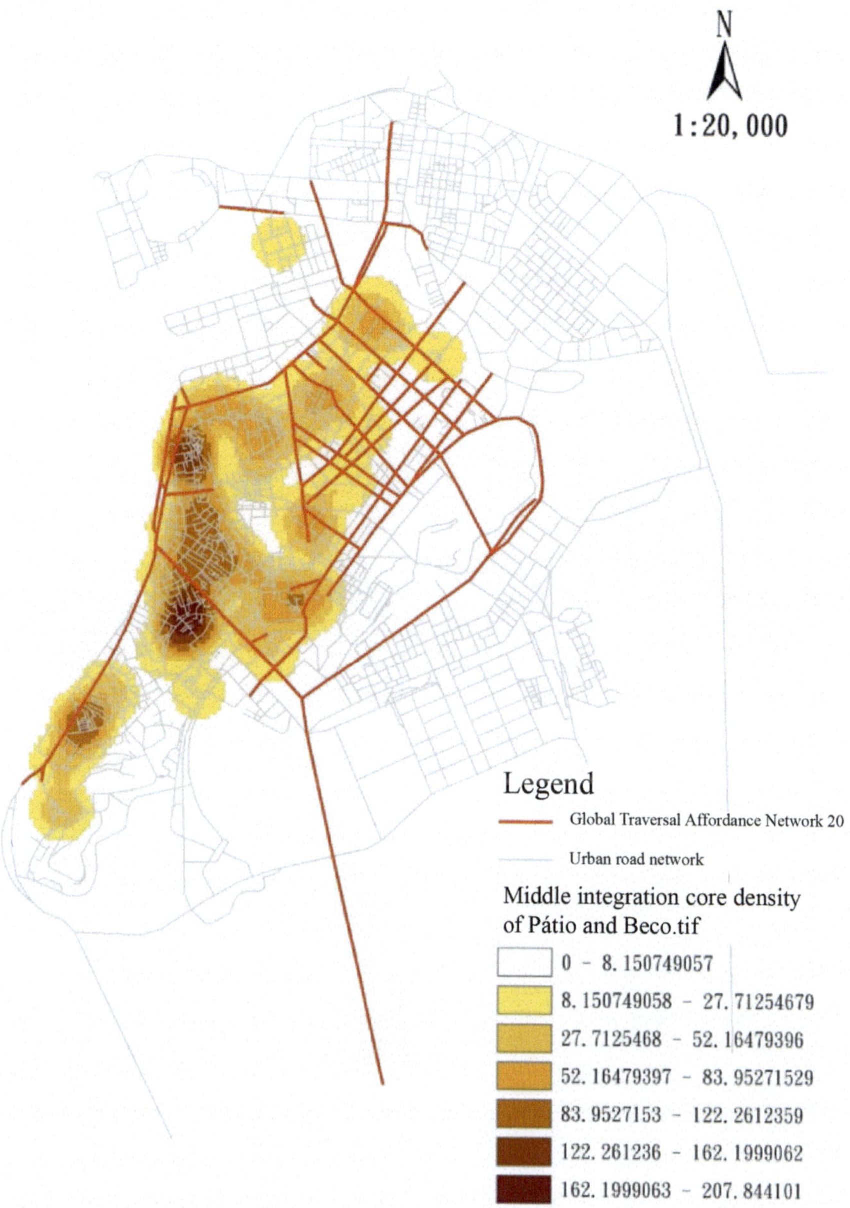

Fig. 8 Distribution characteristics of Pátio with middle integration. (*Image source* drawn by the author)

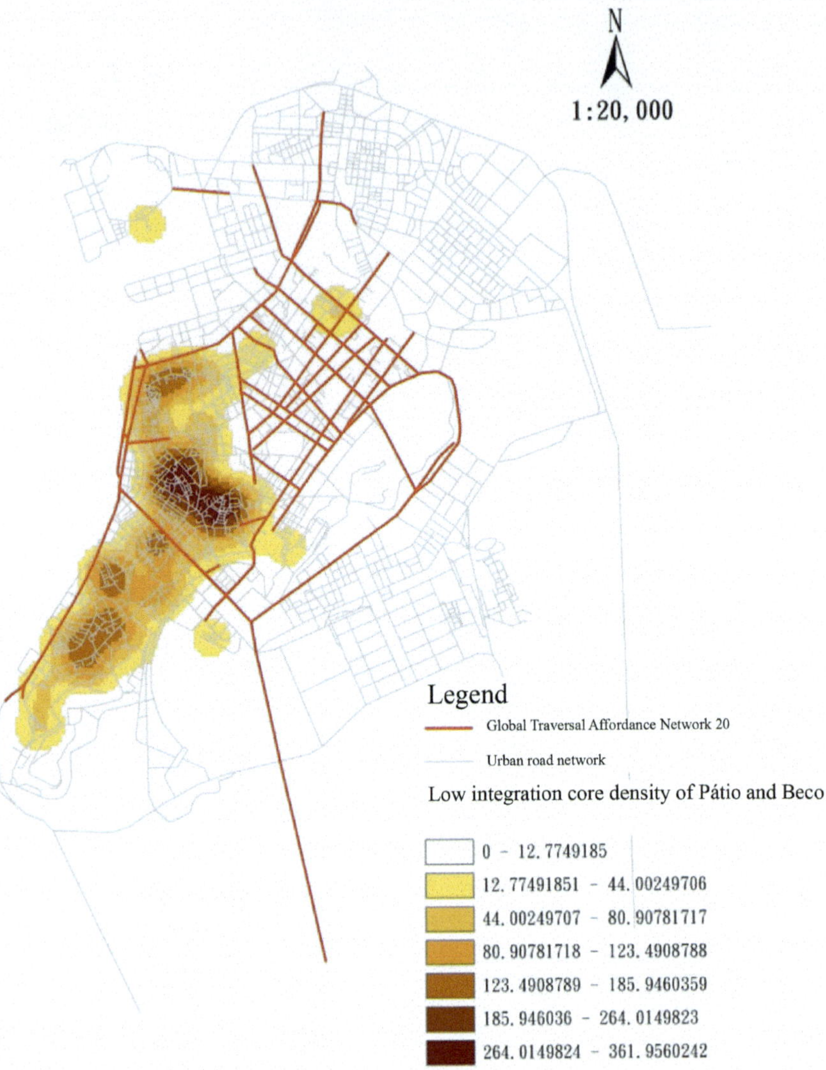

Fig. 9 Distribution characteristics of Pátio with low integration. (*Image source* drawn by the author)

In specific terms, the accessibility of Pátio and Beco refers to a spatial potential as a destination at the overall urban scale. This spatial potential serves as a prerequisite for the current development of Pátio streets, making the space itself easily reachable for pedestrian flow. Such spatial prerequisites also provide a basis for urban functional development, with the most typical example being commercial service development. Due to their profit-driven objectives, commercial projects are often more sensitive to actual urban traffic flows [5].

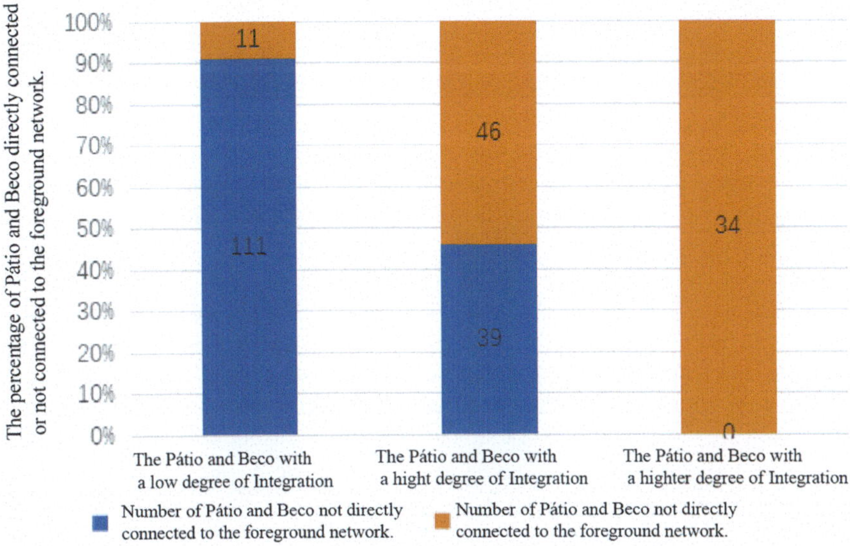

Fig. 10 Connection degree between each Pátio and the accessible foreground network. (*Image source* drawn by the author)

From the perspective of overall urban accessibility, Pátio and Beco streets with better accessibility should be prioritized for corresponding public commercial development. This is because their spatial morphological characteristics bring more opportunities for cultural and commercial activities in the city. Such spatial conditions can better support public activities with various pedestrian flows, bringing more positive impacts on urban economic development and cultural promotion. In contrast, Pátio and Beco streets with low accessibility do not possess such spatial potential at the urban scale, as they are more difficult to reach for traffic flows at the global scale. This spatial feature restricts their own vitality, so such Pátio and Beco streets should be considered later and serve the community life of residents.

And the clustered distribution characteristics of accessibility also strengthen such development priorities, because it indicates that the accessibility of Pátio and Beco streets is closely related to that of external streets. This means that highly accessible Pátio and Beco often distribute around more reachable streets, which in fact brings spatial advantages to the public activity carrying capacity of Pátio and Beco themselves, injecting more dynamic vitality into accessible Pátio and Beco and contributing to the overall vitality improvement and development of urban streets. In contrast, low-accessibility Pátio and Beco are usually located in hard-to-reach surrounding areas, where both internal and external spaces restrict their vitality development.

3.2 Openness Analysis

The choice was proposed at the International Space Syntax Symposium in 2012 [3]. In essence, it is the standardization of the choice degree, and its advantage lies in enabling cross-scale and cross-time–space comparisons. What it reflects is the possibility that people choose a certain road segment to pass through when traveling from any area to another arbitrary area in the city. In this paper, the quantitative index of openness analysis is reflected by the calculation results of choice. Because this parameter can also directly indicate the probability of space being disturbed by outsiders. If there is no situation of outsiders passing through and disturbing a Pátio space, it can be explained that the space is a completely private and inward-looking space. On the contrary, it is more like a public place with a high degree of openness.

Extract the choice data of Pátio and Beco in the Macau Peninsula and the global choice data of the street network in the Macau Peninsula and draw them into histograms and statistical tables (such as Fig. 11, Table 3, Fig. 12, and Table 4). These analysis diagrams show that from the average level, the average value of the global choice of Pátio and Beco in the Macau Peninsula (0.36) is significantly lower than the average level of the choice of the street network in the Macau Peninsula (0.91). This indicates that in the overall urban street network, few people pass through these Pátio and Beco spaces during travel. Similar to accessibility, they belong to the more inward-looking type of space even in the background network. At the same time, the global choice of Pátio and Beco in Macau also fluctuates greatly. Among them, the global choice of Pátio de Hó Chin Sin Tong (柯傳善堂圍 / 柯傳善堂里) is the highest, reaching about 1.12, which belongs to the area where it is relatively easy to pass through in the overall network structure of the Macau Peninsula. However, the choice of many Pátio is 0, indicating that there is no possibility of being disturbed from the perspective of crossing pedestrian flow.

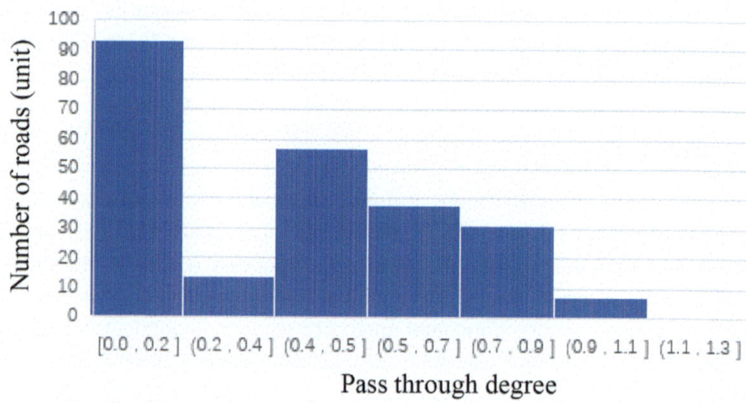

Fig. 11 Histogram of global pass-through statistics for Pátio and Beco in Macau Peninsula. (*Image source* drawn by the author)

Table 3 Macau Peninsula "Pátio" and "Beco" global choice statistics

Macau Peninsula "Pátio" and "Beco" global choice statistics						
Count	Minimum	Maximum	Sum	Mean	Standard deviation	Variance
245	0	1.12	88.57	0.36	0.32	0.10

Source Author based on space syntax statistics

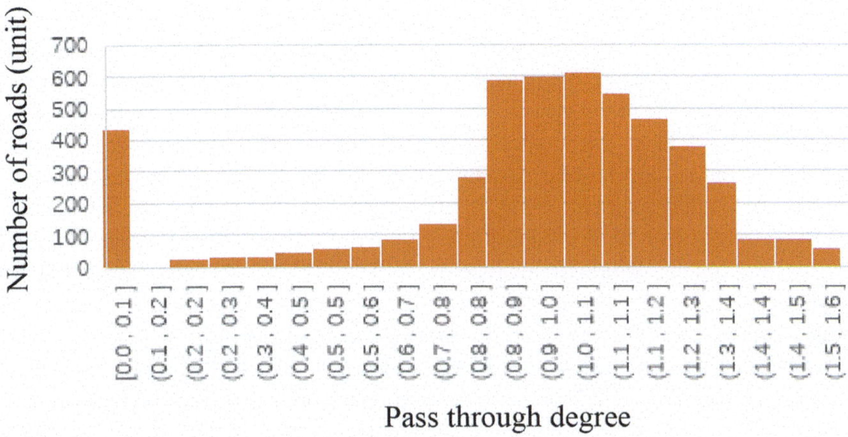

Fig. 12 Histogram of global pass-through statistics for street network in Macau Peninsula. (*Image source* drawn by the author)

Table 4 Macau Peninsula street network global pass-through statistics

Macau Peninsula street network global pass-through statistics						
Count	Minimum	Maximum	Sum	Mean	Standard deviation	Variance
4954	0	1.56	4510.60	0.91	0.37	0.14

Source Author based on space syntax statistics

Different from the classification of accessibility, the classification of Pátio and Beco openness not only takes the average value as the division standard but also should consider the number of entrances and exits of Pátio and Beco. Through the above statistical charts (such as Tables 3 and 4), a phenomenon can be found: the global choice values of many street segments are 0, because most of these street segments are "linear" dead-end roads. Therefore, the movement on these street segments only involves destination-oriented pedestrian flow, rather than crossing pedestrian flow. At the same time, since the space syntax model calculates by taking the corner angle of street segments as the weight, many Pátio and Beco still have choice values even with only one entrance and exit. For these Pátio and Beco, public crossing activities still occur in the space outside the Pátio and Beco unless the travel destination or origin is inside the Pátio and Beco. Therefore, in the classification of passable Pátio and Beco, the worst category should simultaneously consider whether the Pátio and Beco have a single entrance and exit or whether their choice value is 0. When there is only one entrance and exit or the choice value is 0, they are inward-looking and private Pátio and Beco. In terms of quantity, such Pátio and Beco account for the majority, with a total of 148, indicating that from the scale of the entire Macau Peninsula, Pátio and Beco spaces are mainly inward-looking streets.

When the Pátio and Beco spaces have multiple entrances and exits, the remaining Pátio and Beco streets can be divided into two categories based on the average choice value of the urban system. One category is the Pátio and Beco streets with choice values higher than the urban system average (such as Fig. 13), which are the smallest in number, only 8. From the perspective of public openness, these streets have strong spatial potential to carry public crossing pedestrian flows, so these spaces are often suitable for the development of public place types. The other category is the Pátio and Beco spaces with choice values lower than the urban system average (such as Fig. 14), totaling 85. Although this type of Pátio and Beco has a certain potential for outward orientation, it is lower than the urban average level. Therefore, in terms of public openness, this type can only be classified as semi-public.

To sum up, based on passability, Pátio and Beco streets can be divided into three categories: NnA, NnB, and NnC. Among them, NnA refers to Pátio and Beco streets with overall public openness; NnB refers to semi-open Pátio and Beco streets, which have the possibility of being passed through, but the possibility is low; NnC refers to private Pátio and Beco streets (such as Fig. 15).

First, analyze the distribution characteristics of Pátio with superior openness. Due to the extremely small number of open Pátio, the study chooses to directly analyze their distribution characteristics. It can be seen from the analysis diagram (such as Fig. 16) that these eight Pátio with high choice have the following characteristics: on the one hand, there are no dead-end roads inside, and they all have multiple entrances and exits; on the other hand, most of the Pátio with high choice are directly connected to the foreground network of choice. This indicates that the distribution characteristics of the openness of Pátio and Beco streets may be like accessibility and also have a spatial dependency characteristic on external high-level streets.

Next, analyze the openness distribution characteristics of all other Pátio and Beco streets except these eight publicly open ones. First, it is necessary to exclude Pátio and Beco spaces with single entrances and exits and no turns in internal streets, because regardless of the openness of external streets, the choice values of these streets always remain 0, and they fundamentally lack passability in calculations. However, single-entrance Pátio and Beco with non-zero choice values still need to be considered, as although they are as impassable as the former in terms of calculation, they can still reflect the influence of external choice values. Based on the average choice value (0.55) of the remaining Pátio and Beco, the remaining ones are quantitatively divided into two categories: those with better openness and those with poor openness. The kernel density of these two categories is calculated, respectively, and overlaid with the top 20% foreground network (such as Figs. 17 and 18). The analysis diagrams show that both the lower and higher categories have seemingly close connections

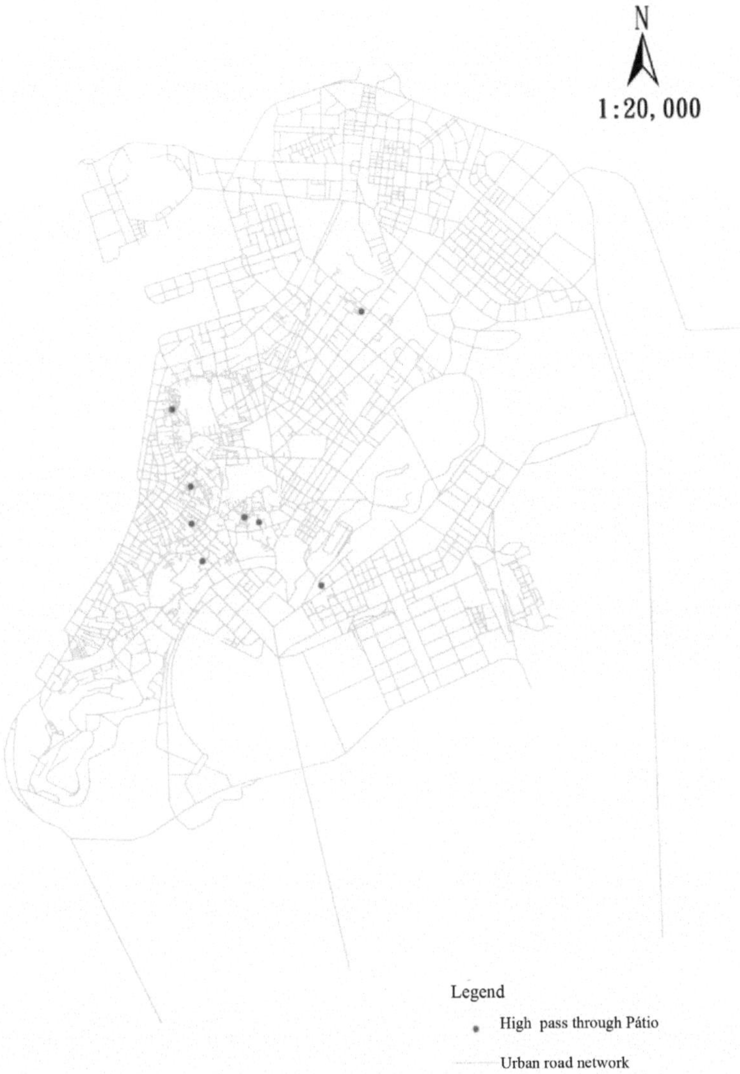

Fig. 13 Distribution of public and open Pátio and Beco. (*Image source* drawn by the author)

between their distribution cores and the foreground network. Using the same method as Figs. 2, 3, 4, 5, 6, 7, 8 and 9 to sort out the statistical chart (Fig. 19), it can be found that although the proportion change rule of Pátio and Beco directly connected to the foreground network is the same as that of accessibility—i.e., the better the openness type, the higher the proportion of Pátio and Beco directly connected to the foreground network—the degree of this change is not obvious compared with accessibility (the average change rate of direct connection proportion: 16.3%). Therefore, based on this finding, the conclusion is drawn that the passability within Pátio and Beco will

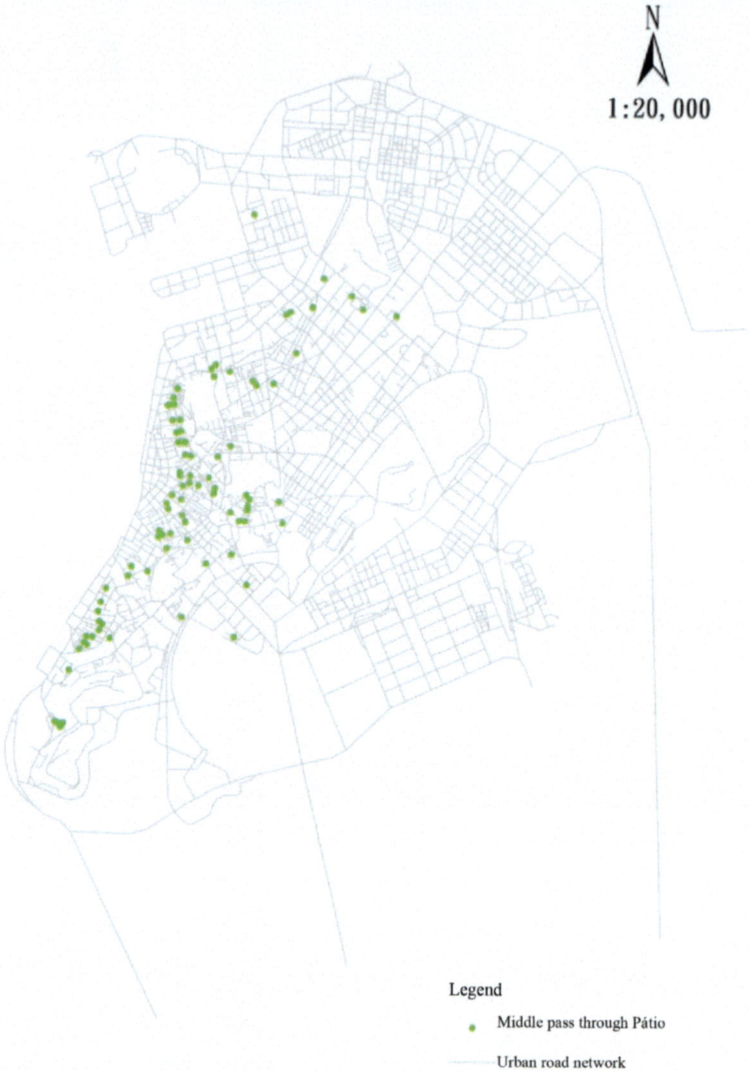

Fig. 14 Distribution map of semi-open Pátio and Beco. (*Image source* drawn by the author)

be affected by the passability of external streets, but the degree of influence is weak, and its openness potential has no strong dependency on the outside.

Therefore, different from the distribution characteristics of accessibility, although the openness distribution also mainly presents an aggregated feature, it is more dispersed compared to accessibility. As a result, the dependency of low-level Pátio and Beco streets' open vitality on the openness of external streets is weaker than that of accessibility. The probability of being chosen for passage by traffic flow is more

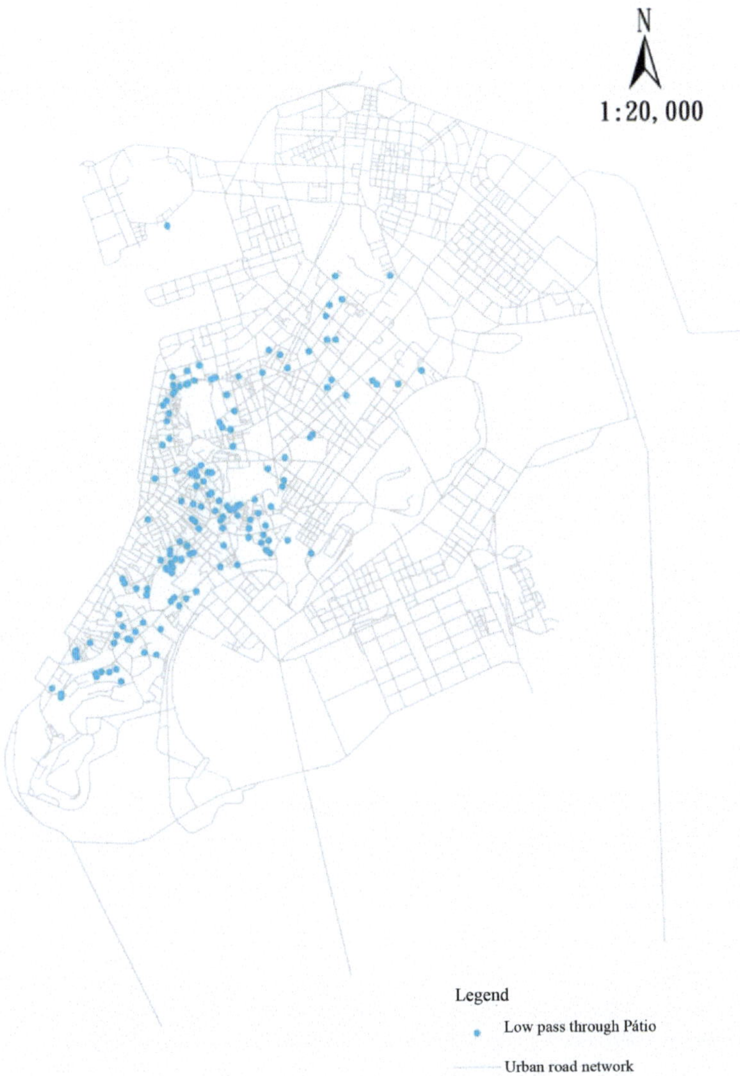

Fig. 15 Distribution map of privacy Pátio and Beco. (*Image source* drawn by the author)

related to their own street forms (such as the number of openings, connection corner angles, etc.).

To sum up, the choice value of a Pátio reflects the public openness of its streets, which is manifested in whether the street is frequently traversed. Pátio streets with high global choice indicate that at the urban scale, these areas are also suitable for use as public passages. The benefit of this characteristic is that although the space itself may not be an easily accessible destination, in the complex movement of urban traffic flow, these spaces often carry a large traffic volume. Therefore, such Pátio

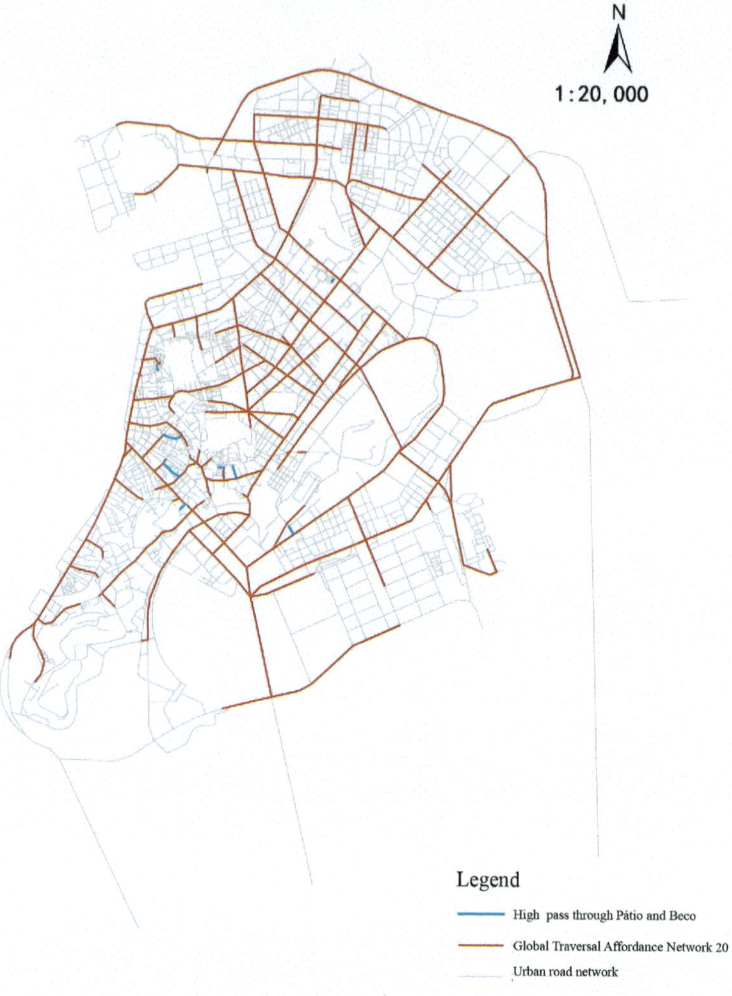

Fig. 16 Analysis of high pass through Pátio and Beco. (*Source* drawn by the author)

streets also have high potential spatial vitality, and their surrounding plots are suitable for development and use as general living services and lower-grade shops (such as convenience stores). In contrast, Pátio spaces with poor passability or single entrances and exits reflect the inwardness of the space, which are more private and suitable for development and activation as local residential communities. From the perspective of priority, the former is more important for the public use and commercial openness of urban spaces and has a greater impact on the urban collective public life and traffic flow. Therefore, they should be considered first.

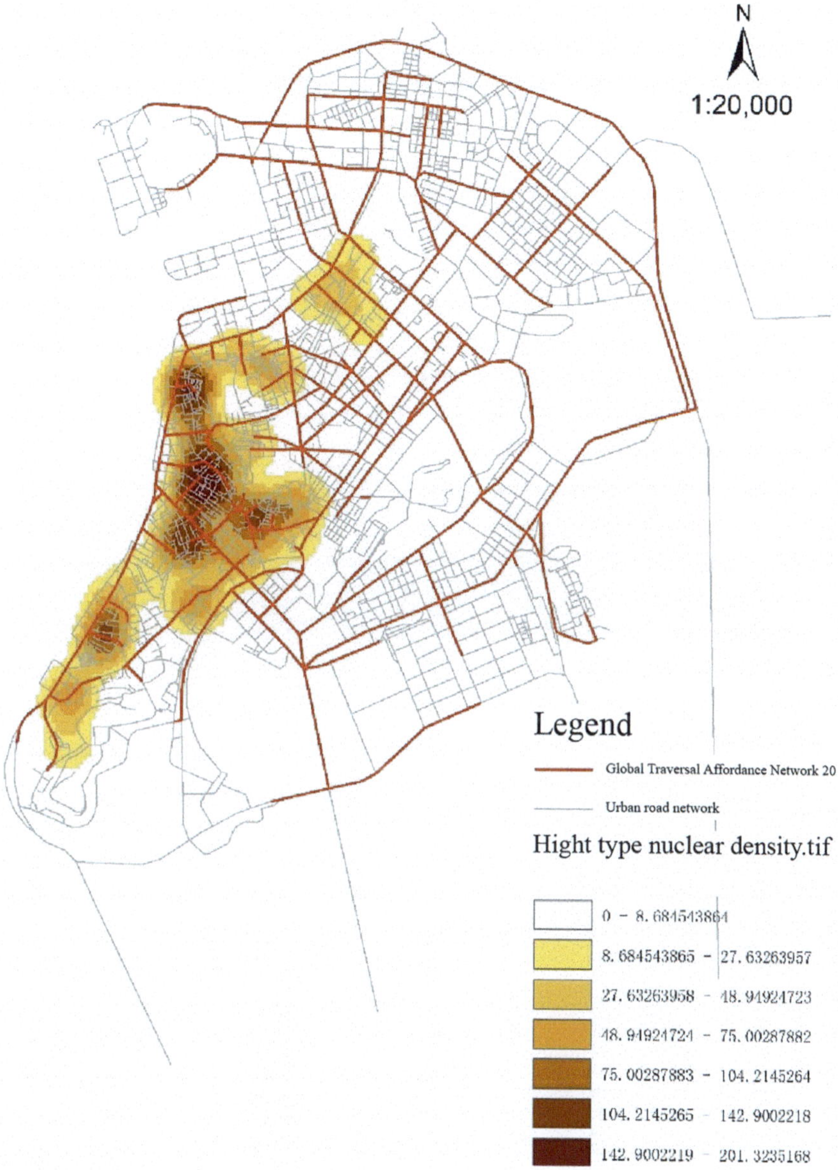

Fig. 17 Distribution map of relatively high pass through in Pátio and Beco. (*Image source* drawn by the author)

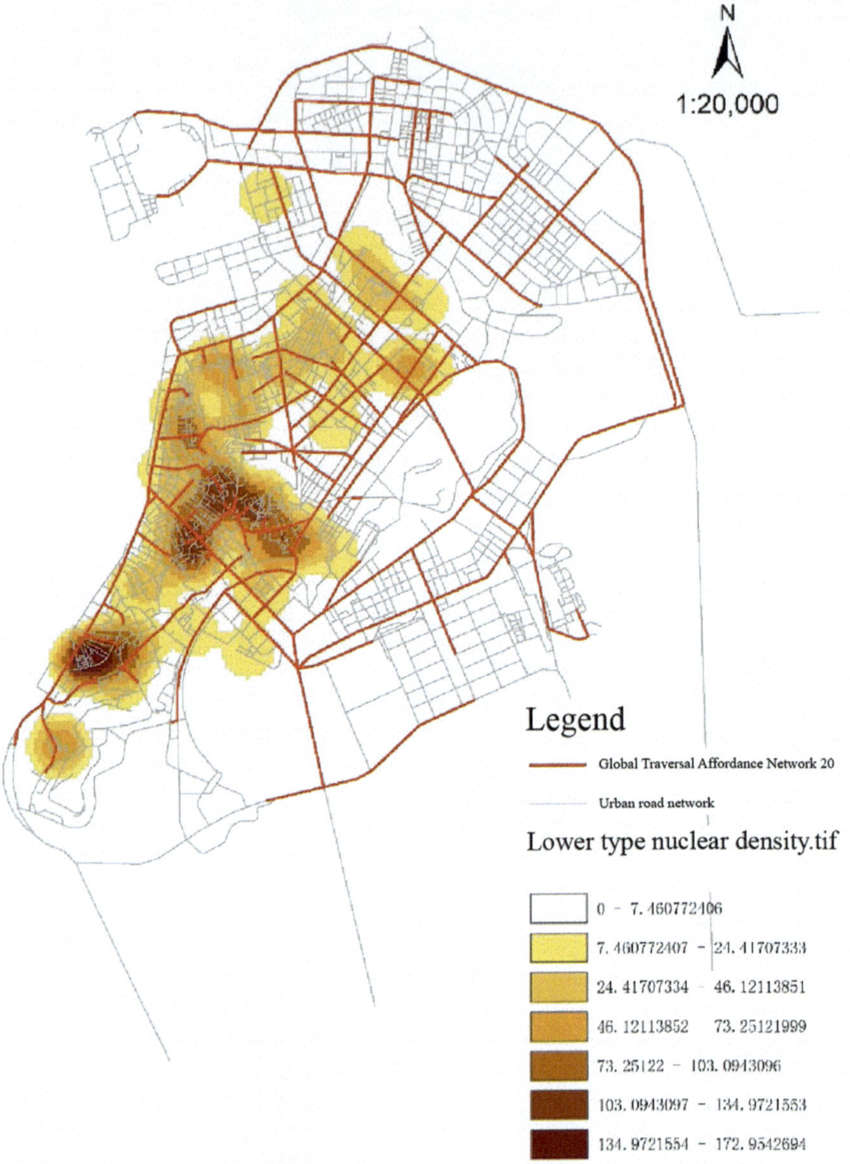

Fig. 18 Distribution map of relatively low pass through in Pátio and Beco. (*Image source* drawn by the author)

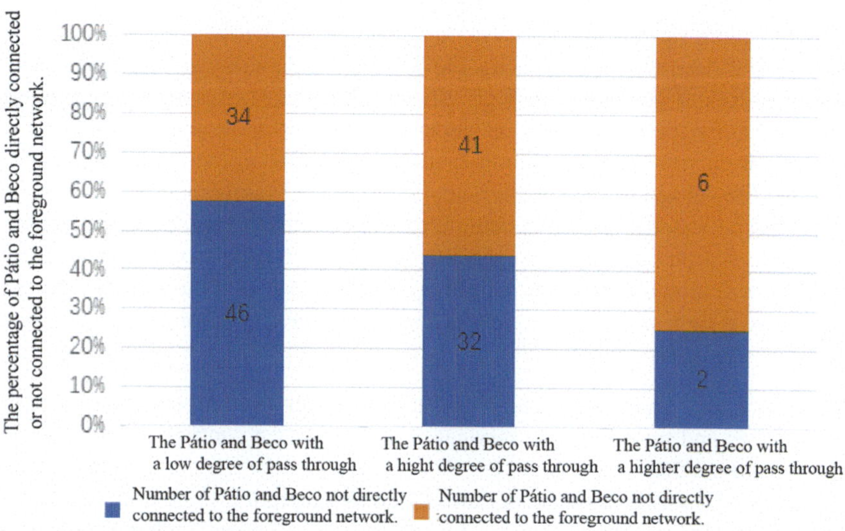

Fig. 19 Connection degrees of various Pátio and Beco with the passable foreground network. (*Image source* drawn by the author)

4 Morphological Characteristics of "Pátio" and "Beco" at the Walking Scale

4.1 Reachability Analysis

In spatial syntax analysis, the verification of the model is of great importance. The morphological characteristics of space vary significantly at different scales, so it is necessary to introduce actual urban big data to assist in selecting the optimal scale. On the one hand, this can enhance the accuracy of the syntax model, and on the other hand, it can better explore the characteristics of urban centers and peripheries across multiple scales. Early spatial syntax models were verified using pedestrian flow data, but such data is costly to obtain. Therefore, many studies have used urban point of interest (POI) data for model verification [7, 8]. Reference data includes commercial shopping data, life service data, and transportation service data, as they often show a high consistency with integration, while transportation service data tends to align with choice [9, 10]. Building on these prior experiences, this study uses Gaode (Amap) POI data to test and verify the correlation of syntax models at various scales. For the selection of empirical scales, this study refers to the research by Feng Chen et al. [11]. Although there are significant differences in data acquisition methods and the selected street network data compared to this study, their research on scale correlation has already drawn some conclusions about high correlation.

By obtaining various urban data through the Gaode (Amap) API, including four types of POI data: catering services, life services, commodity services, and transportation services (as shown in Figs. 20, 21, 22, and 23). Among them, catering services mainly include various restaurants, cafes, pastry shops, and other food-related places; life services mainly include real estate agencies, beauty salons, and other life-related service points; shopping services mainly include convenience stores, specialty stores, and other commercial places; transportation services mainly include gas stations, bus stops, parking lots, and other transportation-related service places. Correlation analysis was carried out on these data and the syntactic parameter results at each scale, and the results are shown in Fig. 24, Table 5, Table 6, and Fig. 25. It can be found that:

(1) Both choice value and integration value show a high correlation with commercial services and transportation services, with correlation coefficients at all scales exceeding 75%.
(2) In terms of variation range, the correlation of integration value is more affected by scale changes, while choice value is more stable, indicating that the influence of spatial openness reflected by choice value seems to be unaffected by scale.
(3) From an overall correlation perspective, the characteristics of integration and choice values in Macau are consistent with previous studies, that is, commercial and life service functions are more correlated with integration, while transportation services are more correlated with choice value.
(4) However, when comparing different scales, some abnormal values contradicting the hypothesis can be found. For example, the correlation coefficient between integration value and life services at the 500m scale is lower than that of choice value; the correlation coefficients between choice value and transportation services at 2500, 3000, 4000 m, and global scales are lower than those between integration value and transportation services. Of course, the differences in these abnormal values are very small, with the maximum only 0.2%, so the two can be approximately considered equivalent.

As the degree of correlation between various urban data and spatial openness varies slightly with scale, this study primarily refers to the correlation between spatial accessibility and urban commercial/life service POI data as the basis for selecting the optimal scale. This is because functional data of this type is more significant for exploring the activation potential of Macau's Pátio spaces. By calculating the average correlation coefficient across scales, this study selects 750 m as the optimal radius for the local scale. At this scale, the correlation between accessibility and the development of actual urban functions such as commercial and life services is the highest. According to different studies [9], 750 m falls within the pedestrian scale, which can reflect the urban spatial characteristics of Macau based on human walking distance. For pedestrians, the time spent walking 750 m should be less than 10 min. Therefore, this scale is a familiar living area for people residing in Pátio, and for those outside, it is also a local scale that can be directly experienced through walking perception.

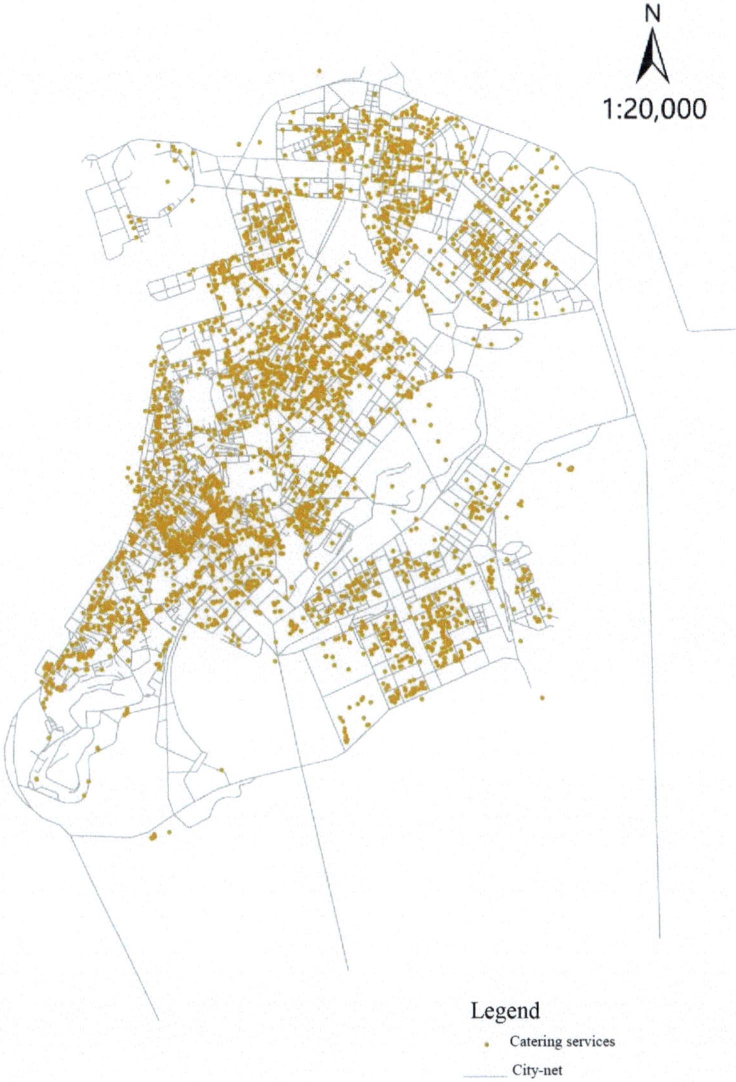

Fig. 20 Distribution map of catering service POIs. (*Image source* drawn by the author)

Calculate the integration value of each segment of the Macau urban street network with a search radius of 750 m, as shown in Figs. p3, 4, 5, 6 and 7. Meanwhile, filter out the 750 m integration value reflecting the pedestrian accessibility of Pátio, and compare it with the pedestrian integration value of the street network in the Macau Peninsula (as shown in Figs. 26, 27, and 28).

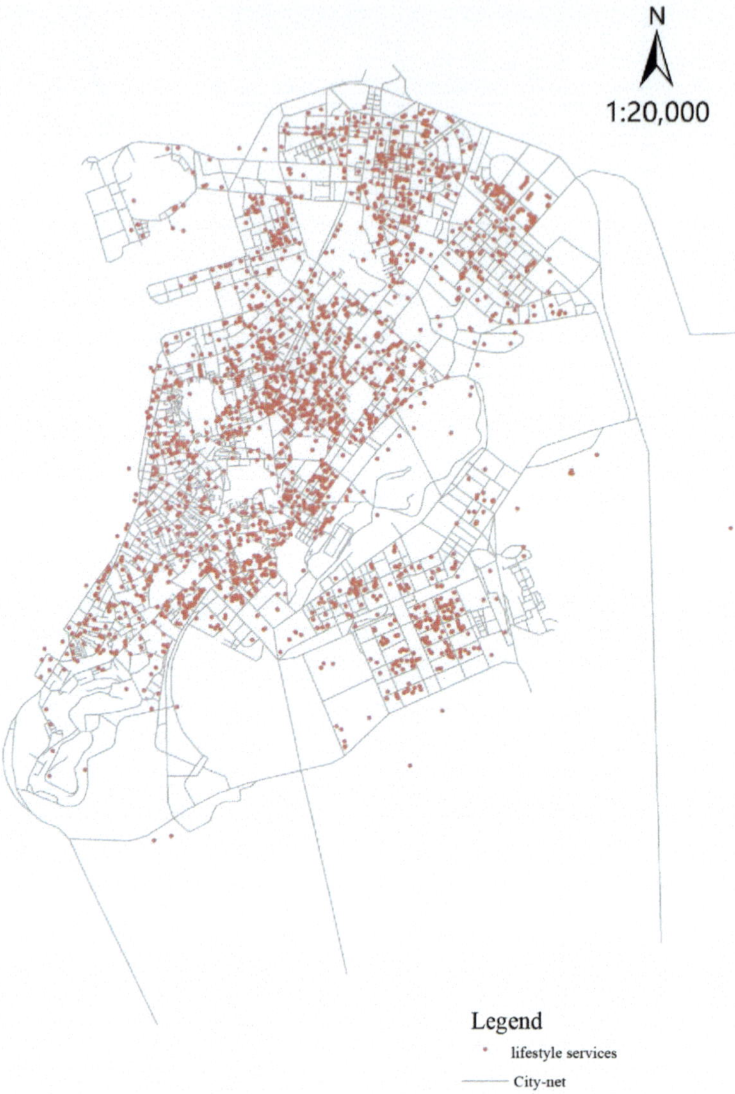

Fig. 21 Distribution map of lifestyle service POIs. (*Image source* drawn by the author)

From an average perspective, although the pedestrian accessibility of Macau's Pátio (176.92) is still slightly lower than that of the street network in Macau Peninsula (177.07), the status of Pátio's local accessibility has significantly risen within the urban network system when compared to their overall spatial accessibility characteristics. This is because quantitative results show minimal difference between the two averages. Therefore, at the pedestrian scale, Pátio are more accessible than at larger urban scales. This "pedestrian-friendly" feature enables different groups

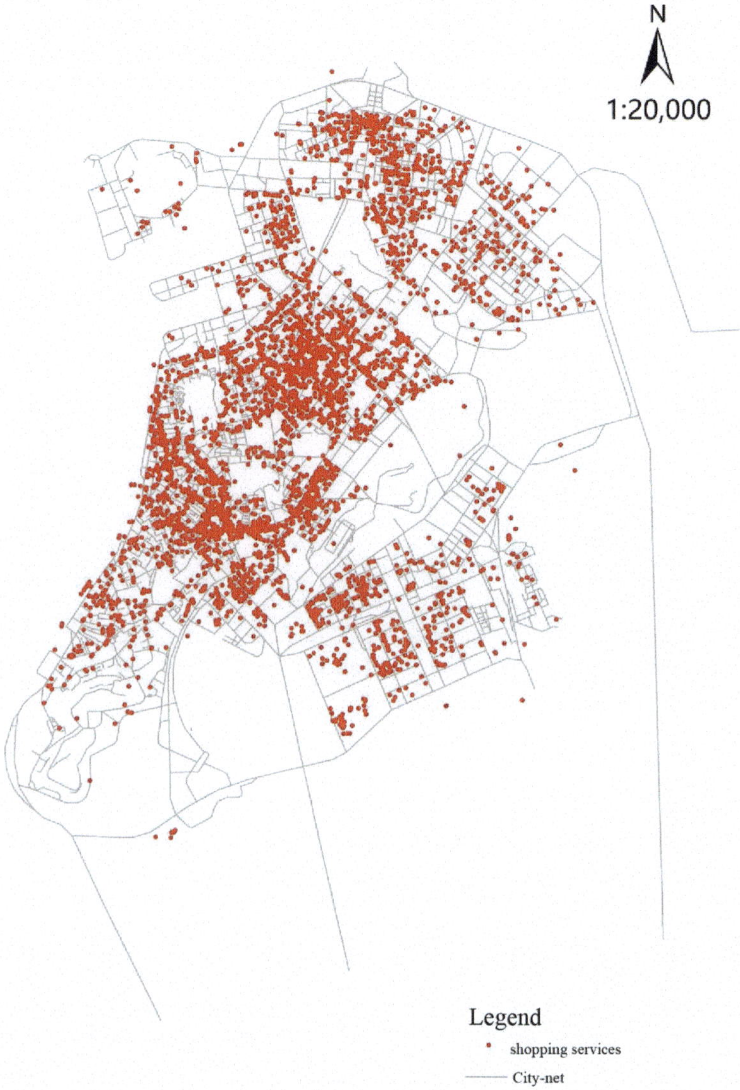

Fig. 22 Distribution map of shopping service POIs. (*Image source* drawn by the author)

walking in Macau Peninsula to more easily reach and experience the inner streets of Pátio directly. As such, they are not merely forgotten old street spaces as they currently appear. Furthermore, the pedestrian scale endows Pátio spaces with more significance than the urban scale, providing specific directions for how people can use.

Similar to their accessibility characteristics at the urban scale, the pedestrian accessibility of Pátio also varies significantly. The lowest pedestrian accessibility is

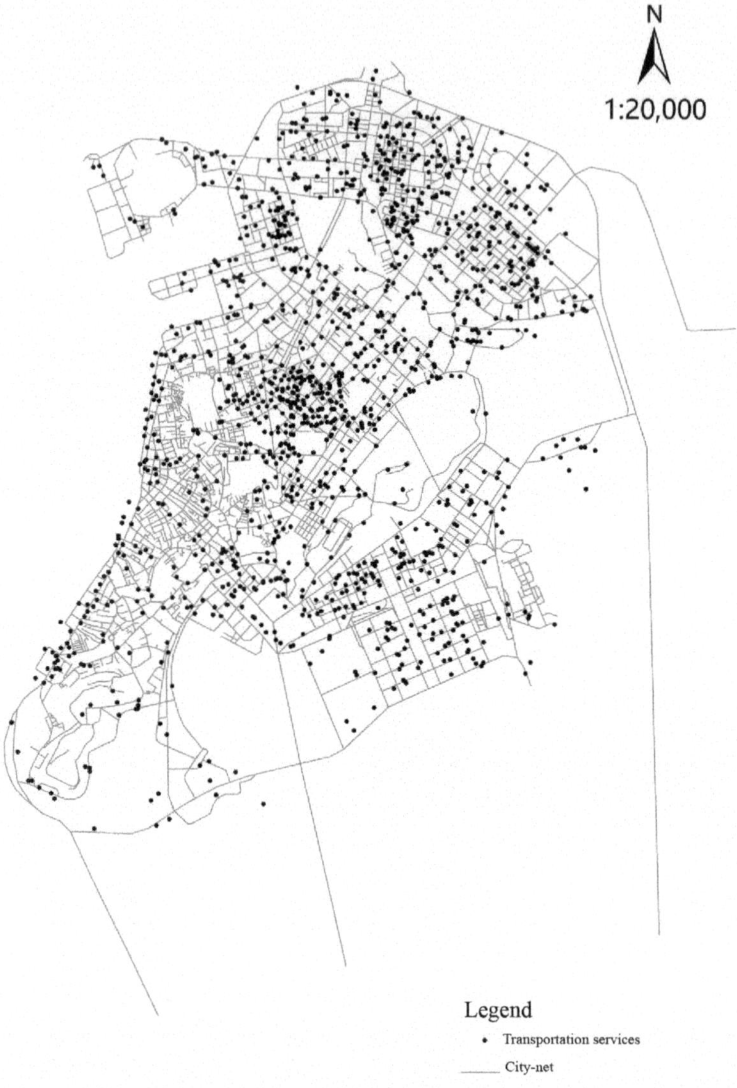

Fig. 23 Distribution map of transportation services POIs. (*Image source* drawn by the author)

found in Beco da Ilha Verde (青州里), with a value of only 49.66. In contrast, Pátio da Palha (乾草圍) has the highest accessibility, with an integration value of 295.24, making it part of the foreground network.

In GIS, the calculated 750 m integration values were averaged and input into the point data of Pátio and Beco. Due to the significant increase in the average accessibility of Pátio and Beco at the pedestrian scale, the mean integration value of Pátio and Beco is nearly identical to that of the Macau Peninsula street network. As a

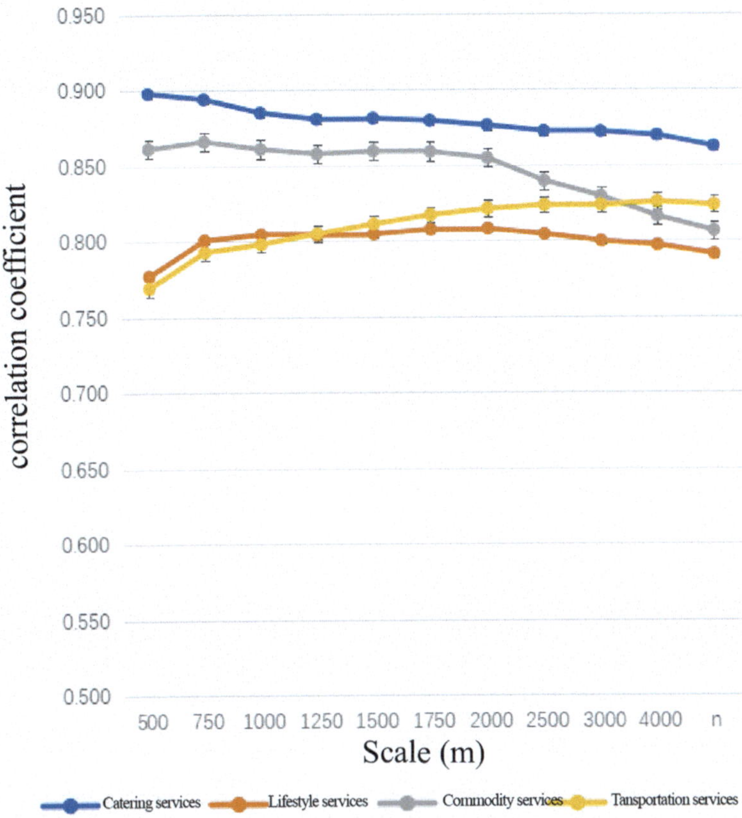

Fig. 24 Statistical chart of multi-scale Pátio and Beco accessibility verification. (*Image source* drawn by the author)

result, the method used in the previous chapter—classifying accessibility into low, medium, and high categories based on the overall average of the Macau Peninsula network and the individual average of Pátio and Beco—can no longer be applied. However, the universal improvement in pedestrian accessibility of Macau's Pátio and Beco reveals that many Pátio streets are partially or fully integrated into the 750 m accessibility foreground network of Macau Peninsula. Therefore, this study designates Pátio interacting with the foreground network as high pedestrian accessibility PnA (as shown in Fig. 29), totaling 26 cases. The remaining Pátio and Beco spaces are divided into moderate pedestrian accessibility PnB and poor pedestrian accessibility PnC based on the average pedestrian integration value of Macau Peninsula (as shown in Figs. 30 and 31). The former type indicates that although Pátio streets lack outstanding accessibility, they still exceed the average accessibility of Pátio and Beco or the urban network, with 99 cases in total. PnC refers to Pátio with accessibility below the average level of Pátio and Beco or the urban network, comprising 116 cases.

Table 5 Correlation statistics between multi-scale "Pátio" and "Beco" accessibility and urban functions

Integration verification scale (m)	Catering services	Life services	Commodity services	Transportation services
500	0.898	0.777	0.861	0.769
750	0.894	0.801	0.866	0.793
1000	0.885	0.804	0.861	0.798
1250	0.880	0.804	0.857	0.805
1500	0.881	0.805	0.859	0.811
1750	0.879	0.807	0.859	0.817
2000	0.876	0.808	0.854	0.821
2500	0.872	0.804	0.839	0.823
3000	0.872	0.800	0.829	0.824
4000	0.869	0.797	0.816	0.826
n	0.862	0.791	0.806	0.824
Average	0.879	0.800	0.846	0.810

Source Author's statistics

Table 6 Correlation statistics between multi-scale "Pátio" and "Beco" openness and urban functions

Integration verification scale (m)	Catering services	Life services	Commodity services	Transportation services
500	0.850	0.778	0.783	0.821
750	0.849	0.778	0.783	0.823
1000	0.848	0.779	0.782	0.822
1250	0.848	0.779	0.782	0.823
1500	0.848	0.779	0.783	0.823
1750	0.849	0.780	0.785	0.824
2000	0.849	0.781	0.786	0.823
2500	0.849	0.782	0.788	0.823
3000	0.849	0.782	0.789	0.823
4000	0.848	0.783	0.788	0.824
n	0.849	0.783	0.790	0.824
Average	0.849	0.780	0.785	0.823

Source Author's statistics

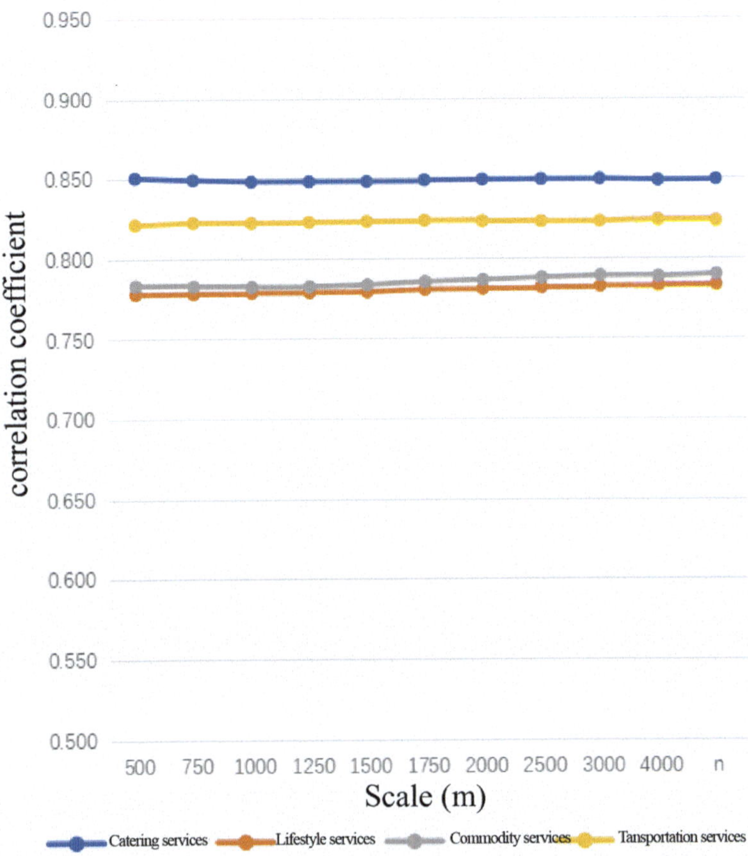

Fig. 25 Statistical chart of multi-scale Pátio and Beco openness verification. (*Image source* drawn by the author)

Overlay analysis was conducted by superimposing the top 20% foreground network with the kernel density of various Pátio and Beco (as shown in Figs. 32, 33, and 34). It can be found that the same rule as the urban scale accessibility of Pátio applies: Pátio and Beco streets with high accessibility often maintain a close connection with the foreground network, while those with poor accessibility are often distributed in the interstices of the foreground network. This indicates that even when the scale is reduced to the pedestrian scale, the spatial accessibility of Pátio and Beco still shows obvious dependency. Through statistics on the degree of connection between Pátio and Beco spaces and the external foreground network (as shown in Fig. 35, this dependency is very significant. With the increase in accessibility, the proportion of direct connections between Pátio and Beco spaces and the foreground network significantly rises (the average change rate of direct connection proportion: 49.9%). Therefore, although the scale has changed, the accessibility distribution characteristics of Pátio and Beco streets still show obvious "dependency". Low-level

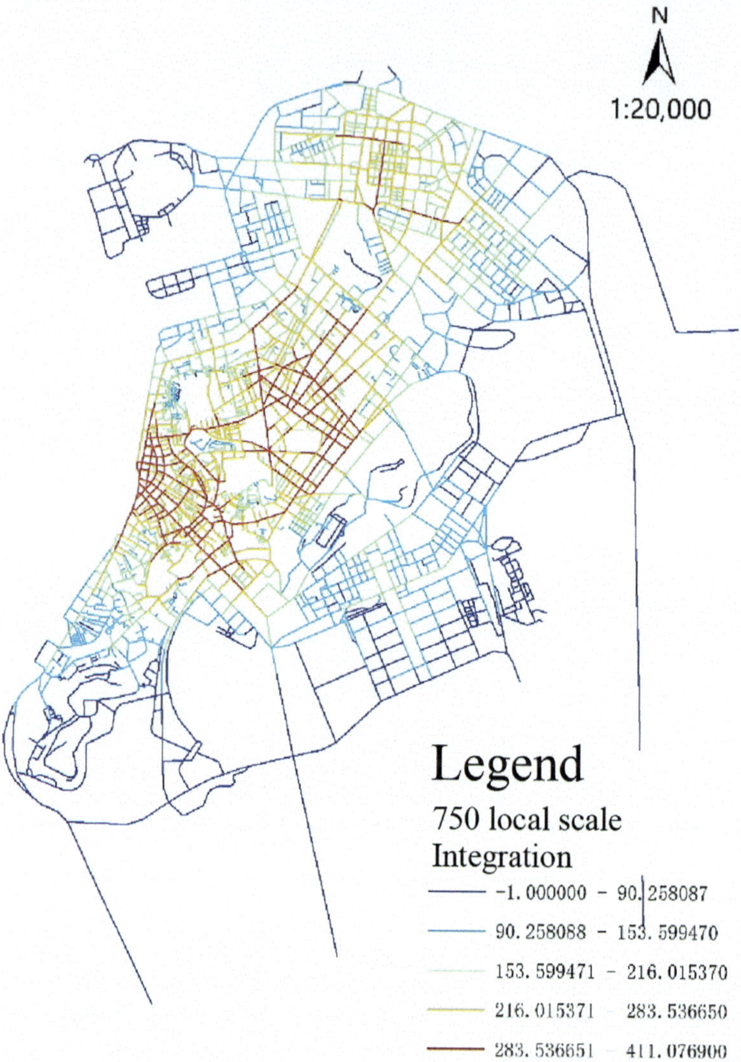

Fig. 26 750 m local integration analysis map of Macau road network. (*Image source* drawn by the author)

Pátio and Beco with high accessibility gather around high-accessibility high-level streets, while on the contrary, low-level Pátio and Beco with low accessibility gather around low-accessibility high-level streets.

In summary, compared with the accessibility of Pátio and Beco at the urban scale, within the scale of actual pedestrian experience, the accessibility of Pátio and Beco streets has seen an increase in average level on the one hand, and this spatial potential has become more influential on the other hand, as it shows a stronger correlation with

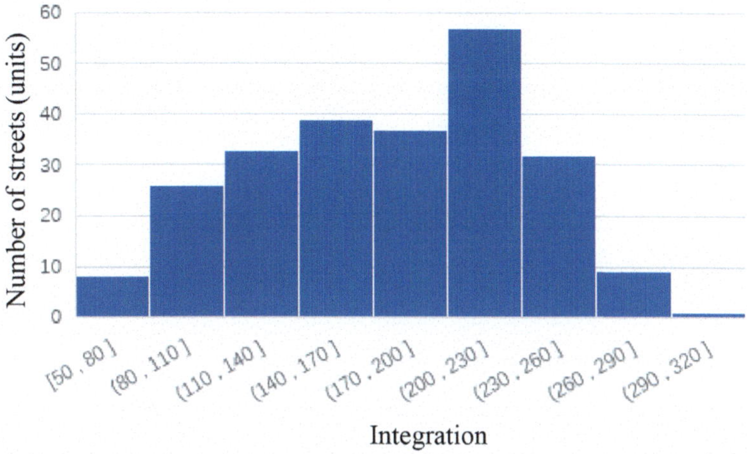

Fig. 27 Histogram of 750 m local integration statistics for Pátio and Beco in Macau Peninsula. (*Image source* drawn by the author)

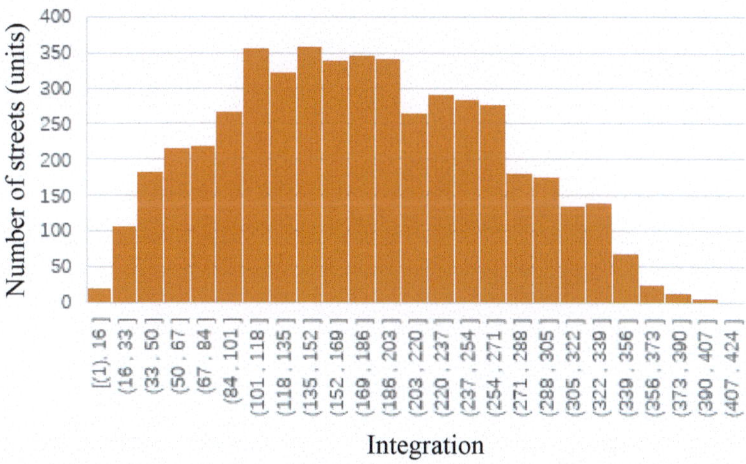

Fig. 28 Histogram of 750 m local integration statistics for street networks in Macau Peninsula. (*Image source* drawn by the author)

commercial service data in urban POI datasets. This indicates that the accessibility of Pátio and Beco streets at the pedestrian scale is a key factor affecting their activation potential. However, this does not negate the significance of Pátio and Beco street accessibility at the urban scale, as pedestrian scale accessibility reflects the vitality of pedestrian flow within walking distance, while urban-scale accessibility reflects the movement of pedestrian and vehicular flows over a larger area. The superposition of movement flows at different scales or levels can further enhance the activation potential of these streets, making both factors influencing activation potential.

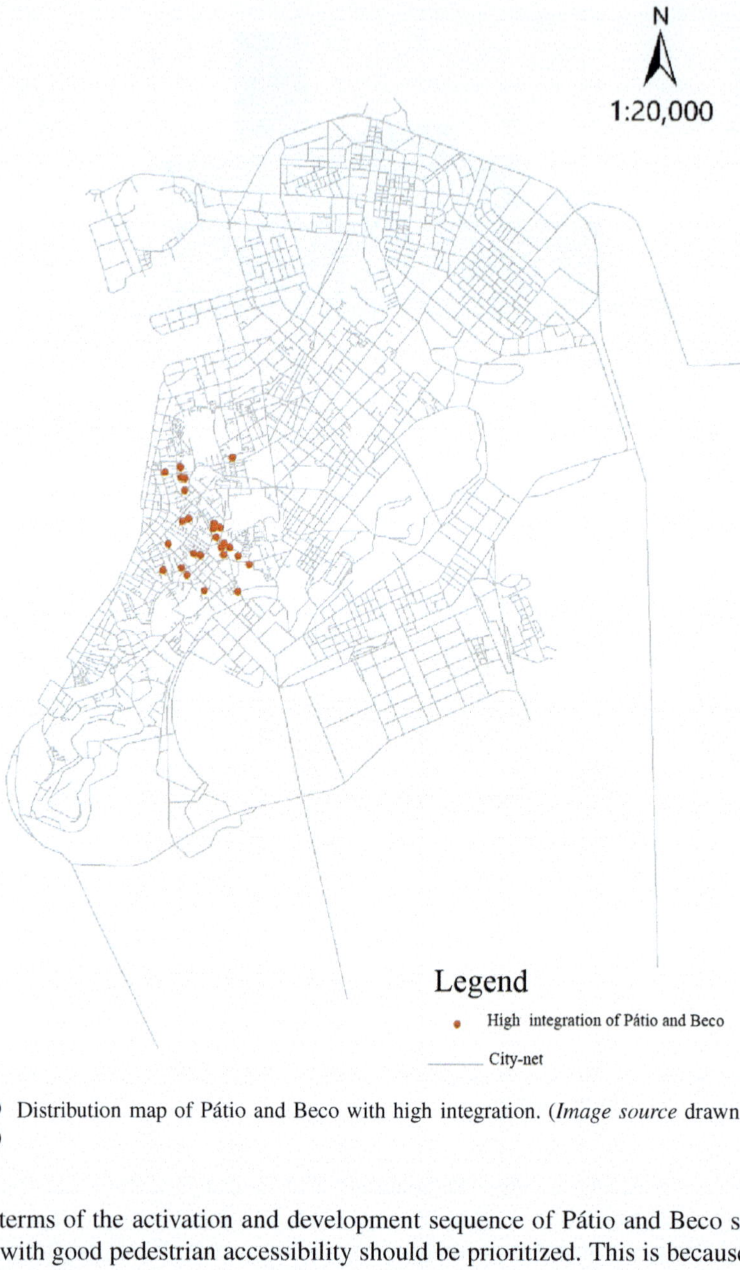

Fig. 29 Distribution map of Pátio and Beco with high integration. (*Image source* drawn by the author)

In terms of the activation and development sequence of Pátio and Beco streets, those with good pedestrian accessibility should be prioritized. This is because such Pátio and Beco are more accessible to pedestrian flows, and their favorable accessibility spatial potential makes them suitable for gathering people, thus providing a spatial basis for subsequent commercial site selection, development, and activation. Meanwhile, the distribution characteristics of accessibility also indicate that these Pátio and Beco streets are located around more accessible external streets, which

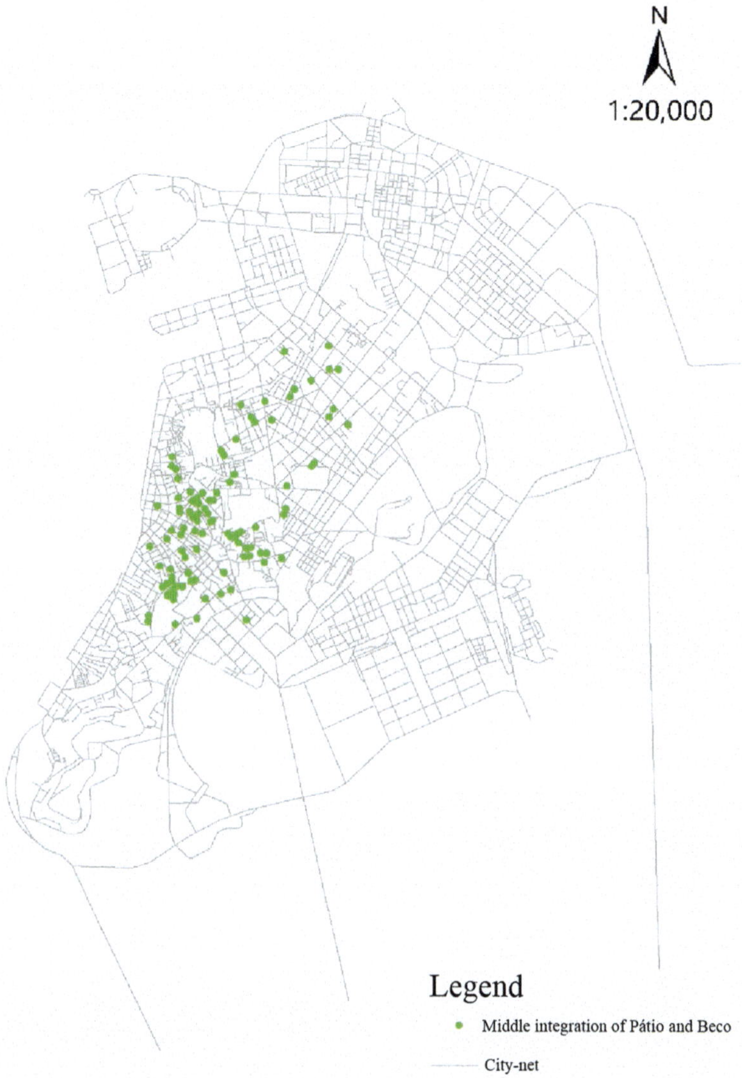

Fig. 30 Distribution map of Pátio and Beco with middle integration. (*Image source* drawn by the author)

brings more development opportunities to their interiors. This is because their activation can be considered in integration with the high-level streets they are closely connected to.

Conversely, Pátio and Beco with poor pedestrian accessibility are difficult for pedestrians to reach. Generally, they are distributed around low-accessibility high-level streets, and their spatial morphology itself restricts their own spatial vitality.

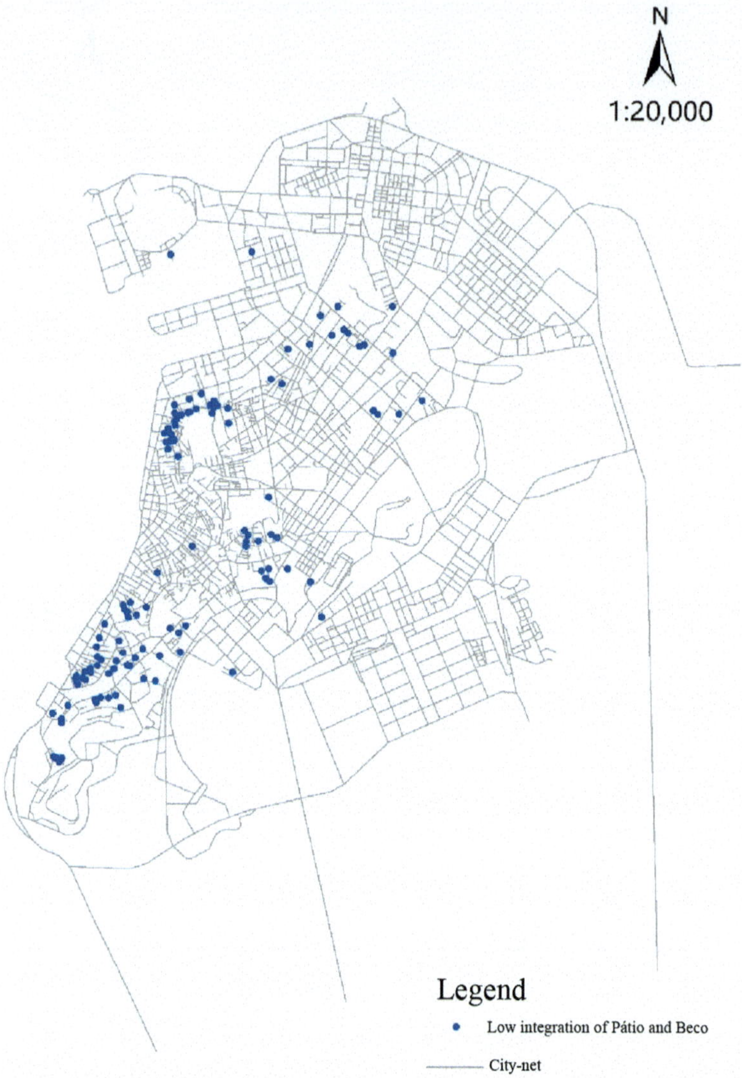

Fig. 31 Distribution map of Pátio and Beco with low integration. (*Image source* drawn by the author)

Therefore, they should be considered later and are suitable for use as living communities for residents.

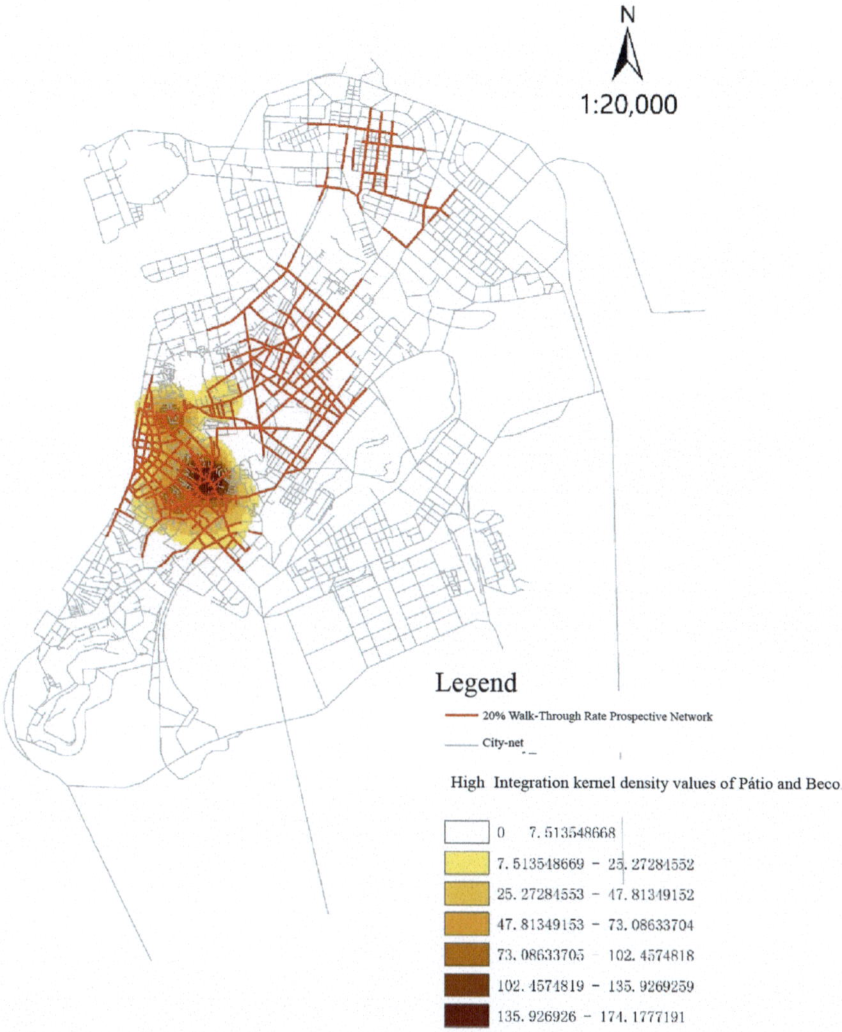

Fig. 32 Distribution characteristics of Pátio and Beco with high walk through. (*Image source* drawn by the author)

4.2 Openness Analysis

Calculate the choice value data for each segment of the Macau urban street network with a search radius of 750 m (as shown in Fig. 36). Filter out the 750 m choice value data reflecting the pedestrian openness of Pátio and Beco, and compare it with the pedestrian choice value of the street network in the Macau Peninsula (as shown in Figs. 37 and 38).

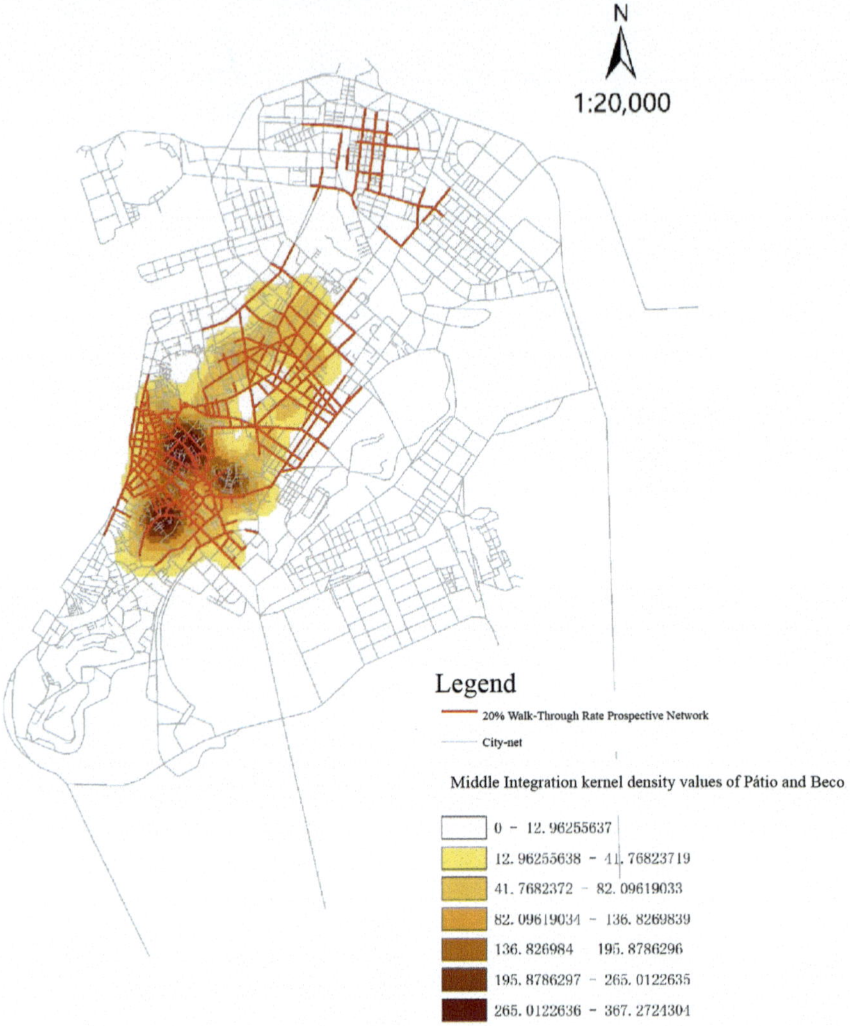

Fig. 33 Distribution characteristics of Pátio and Beco with middle walk through. (*Image source* drawn by the author)

In terms of average levels, the pedestrian choice value of Macau's Pátio and Beco (0.40) is significantly lower than that of the overall street network in Macau Peninsula (0.97). This indicates that even at the pedestrian scale, Pátio and Beco streets in Macau Peninsula remain generally introverted, with residents having a low probability of choosing Pátio streets for traversal during urban movement.

Since choice values allow cross-scale comparisons, this study compares choice values at the pedestrian scale and urban scale. The results show that both the average choice value of Macau's street network and that of Pátio and Beco streets are higher

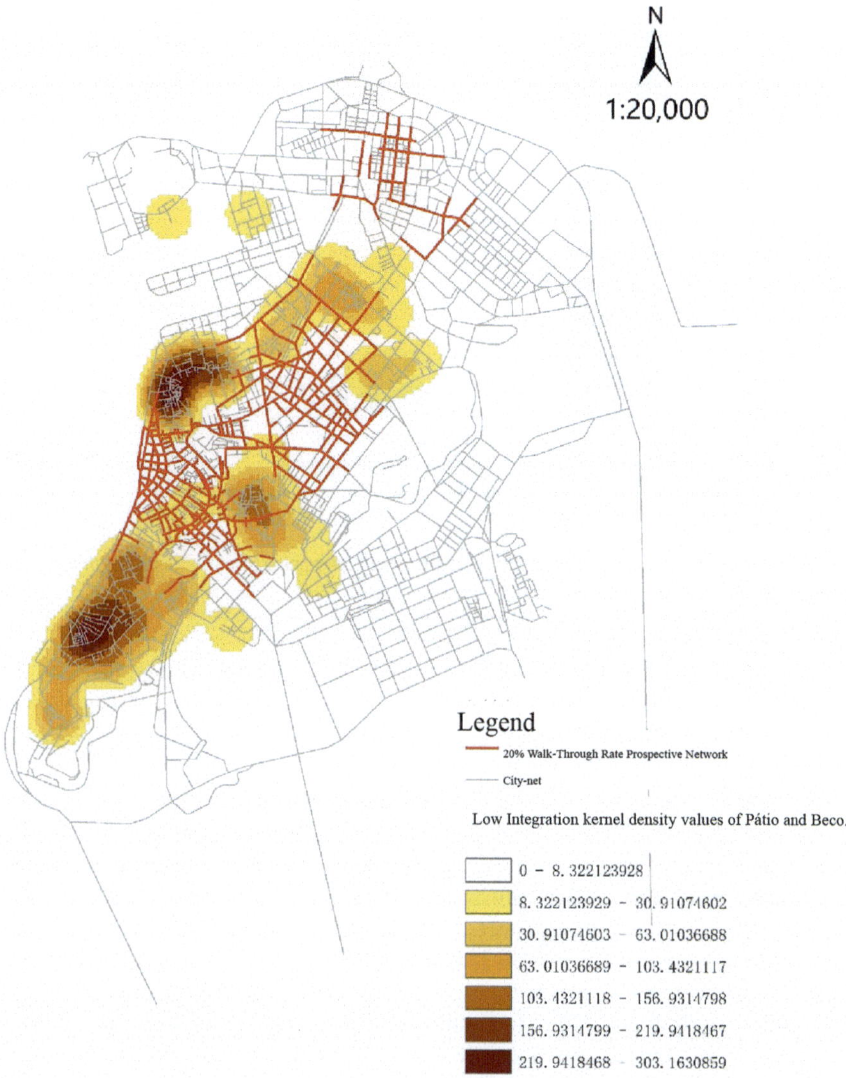

Fig. 34 Distribution characteristics of Pátio and Beco with low walk through. (*Image source* drawn by the author)

at the pedestrian scale than at the overall urban scale. This indicates that both Pátio and Beco spaces and the street system of Macau Peninsula are more "friendly" to pedestrian traversal activities, as pedestrian movement through urban streets in Macau Peninsula is more accessible at the pedestrian scale than at the urban scale. However, significantly different from accessibility, the public openness status of Pátio and Beco streets does not rise with the reduction of scale. The comparison between

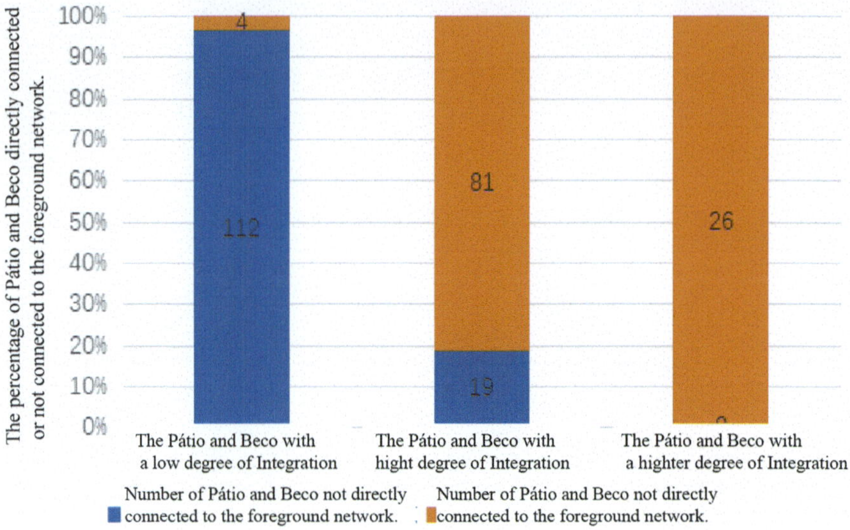

Fig. 35 Connection degree of each Pátio and Beco with the pedestrian accessibility foreground network. (*Image source* drawn by the author)

overall choice values and pedestrian choice values reveals that the gap between the average values of Pátio and Beco and the street system has not significantly narrowed. Therefore, even at a smaller scale, Pátio and Beco remain introverted private spaces relative to the entire urban street hierarchy.

Meanwhile, compared with the choice value at the urban scale, the degree of openness of Pátio and Beco spaces at the pedestrian scale also varies significantly. On the one hand, many Pátio and Beco spaces are dead-end roads, leading to numerous pedestrian choice values of 0 in Pátio spaces. On the other hand, many streets in Pátio and Beco spaces still exhibit extroversion at the pedestrian scale. The most open one is Pátio de Hó Chin Sin Tong (柯傳善堂圍 / 柯傳善堂里), with a pedestrian choice value of 1.12, which is significantly higher than the average level of streets.

In GIS, the calculated 750 m integration data is averaged and then input into the point data of Pátio and Beco in Macau Peninsula. The Pátio spaces in Macau Peninsula are divided into three categories: PaA, PaB, and PaC (as shown in Figs. 39, 40, and 41). The PaA type indicates that these Pátio and Beco spaces have good openness. At the pedestrian scale, residents tend to choose these spaces for traversal in their natural travel, so such Pátio and Beco more reflect a sense of publicity in spatial organization. This category is relatively special, including only only Pátio and Beco spaces. The extroversion degree of the PaB type is lower than the average level of Macau Peninsula, but because they have multiple entrances and exits and non-zero choice values, it shows that although these spaces may be disturbed by outsiders' traversal, their public openness is not high. There are 92 Pátio and Beco in this type. PaC belongs to the completely introverted Pátio and Beco type because they are all dead-end roads. Since this type is mostly determined by the number of entrances

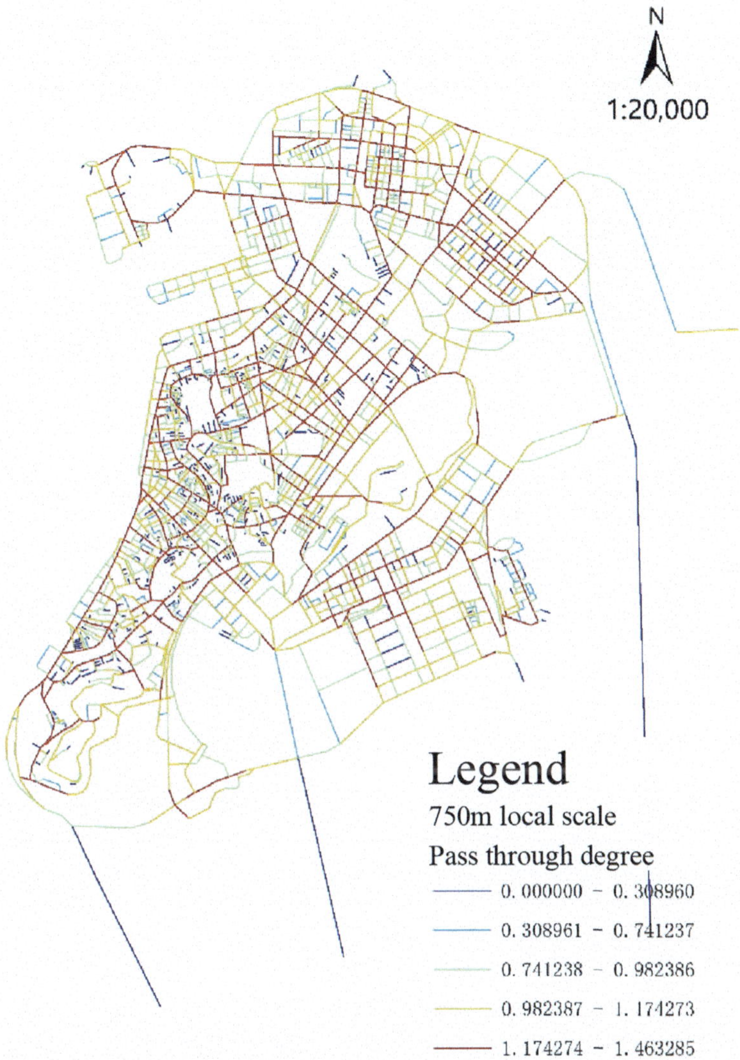

Fig. 36 750 m local pass-through degree value analysis map of Macau road network

and exits of the streets, regardless of the scale, their number is more stable compared with accessibility, and there are 144 Pátio in this type.

Further collation and analysis of the openness characteristics of Pátio and Beco streets, as presented in the analysis diagrams (such as Figs. 42, 43, and 44), show that although most Pátio and Beco streets with high choice values are directly connected to the external foreground network, the connection distribution between semi-open and introverted Pátio and Beco streets and the foreground network does not appear to have obvious differences. Therefore, similar to the openness at the urban scale,

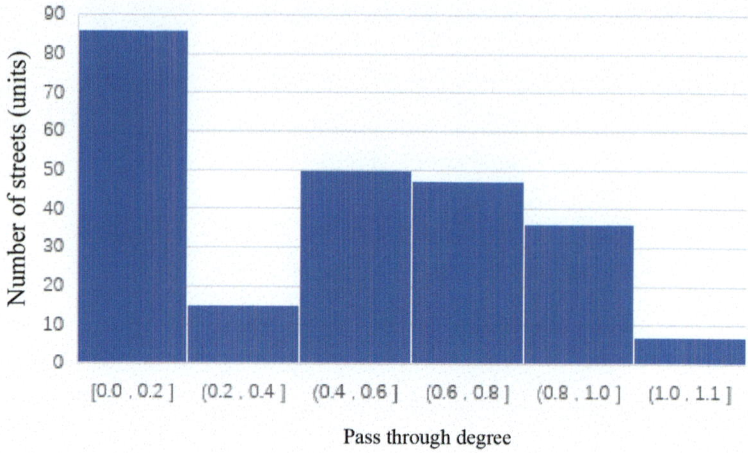

Fig. 37 Histogram of 750 m local pass-through degree value statistics for Pátio and Beco in Macau Peninsula. (*Image source* drawn by the author)

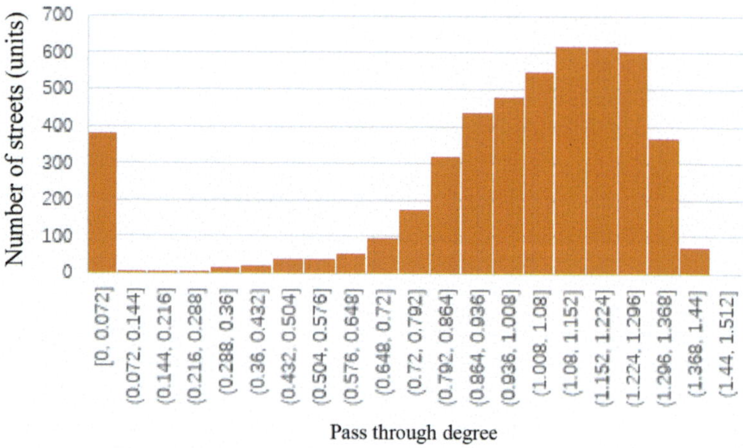

Fig. 38 Histogram of 750 m Local pass-through degree value statistics for street networks in Macau Peninsula. (*Image source* drawn by the author)

while the openness of Pátio and Beco streets at the pedestrian scale is also influenced by the openness of external streets, this influence is not as significant as that reflected in accessibility (as shown in Fig. 45, the average change rate of direct connection is approximately 15.7%). Thus, analogous to the openness at the Macau Peninsula scale, the public openness of Pátio is more related to their own formal characteristics, such as the number of entrances and exits, and the angle between roads and external streets. Although the openness of external streets still slightly affects the accessibility

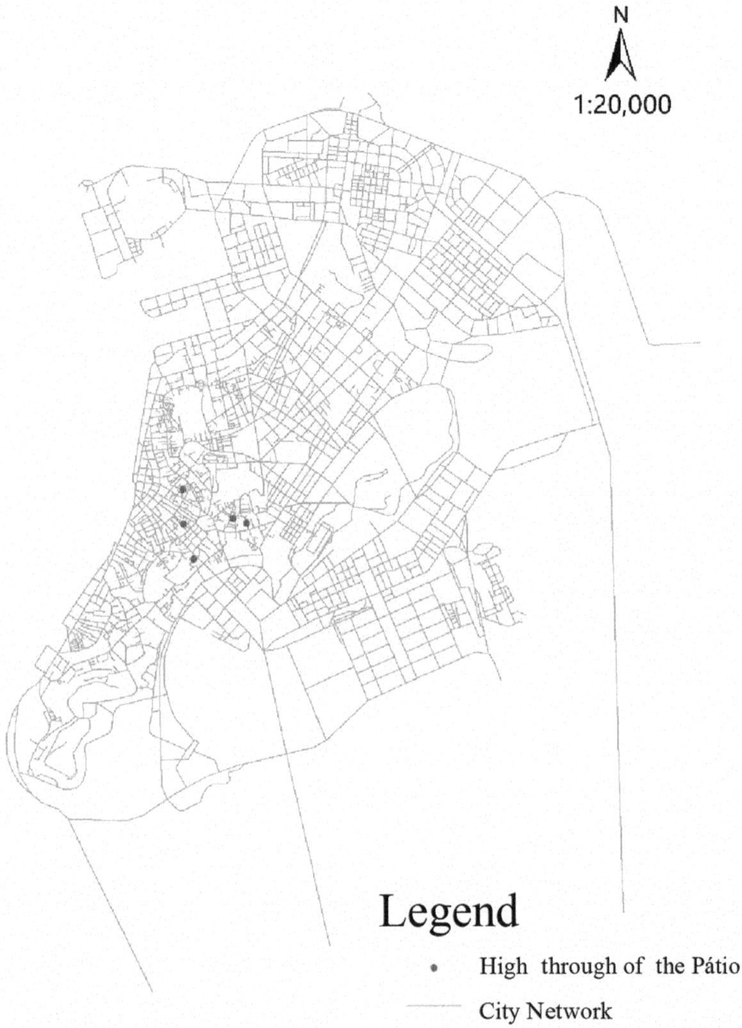

Fig. 39 Distribution map of public high pass-through degree in Pátio and Beco. (*Image source* drawn by the author)

within Pátio and Beco in practical development, this type of accessibility supporting traversing pedestrian flow is more independent compared to accessibility.

In conclusion, the choice value at pedestrian scale reflects public openness based on pedestrian flow—manifested in whether people's urban walking activities frequently traverse Pátio or Beco streets spaces. Compared to the urban scale, Pátio and Beco streets are indeed more accessible to pedestrian flow, yet this has not altered their inherently introverted status within the street hierarchy.

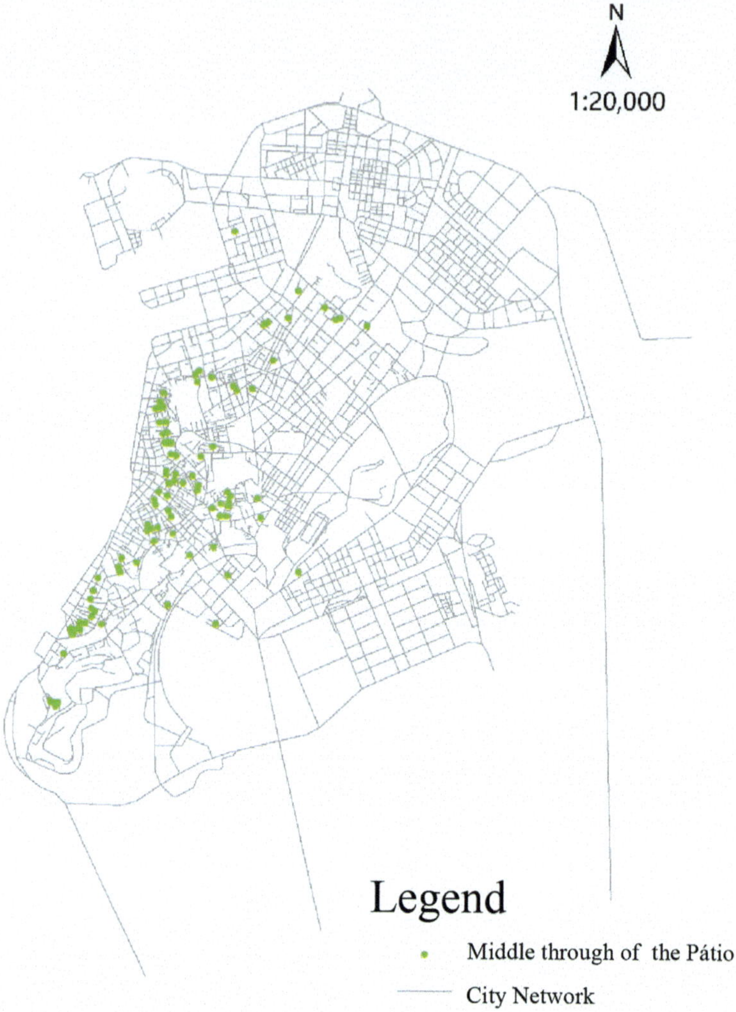

Fig. 40 Distribution map of middle pass through Pátio and Beco spaces. (*Image source* drawn by the author)

That said, this does not rule out individual Pátio or Beco streets from serving as public thoroughfares. Such spaces exert stronger influence on low-level commercial development, public activity organization, and traffic flow accommodation. Thus, Pátio and Beco with high openness should be prioritized, as they support intensive pedestrian traversal. In contrast, poorly open Pátio and Beco are better suited for private, introverted residential communities and should be scheduled later in the development sequence.

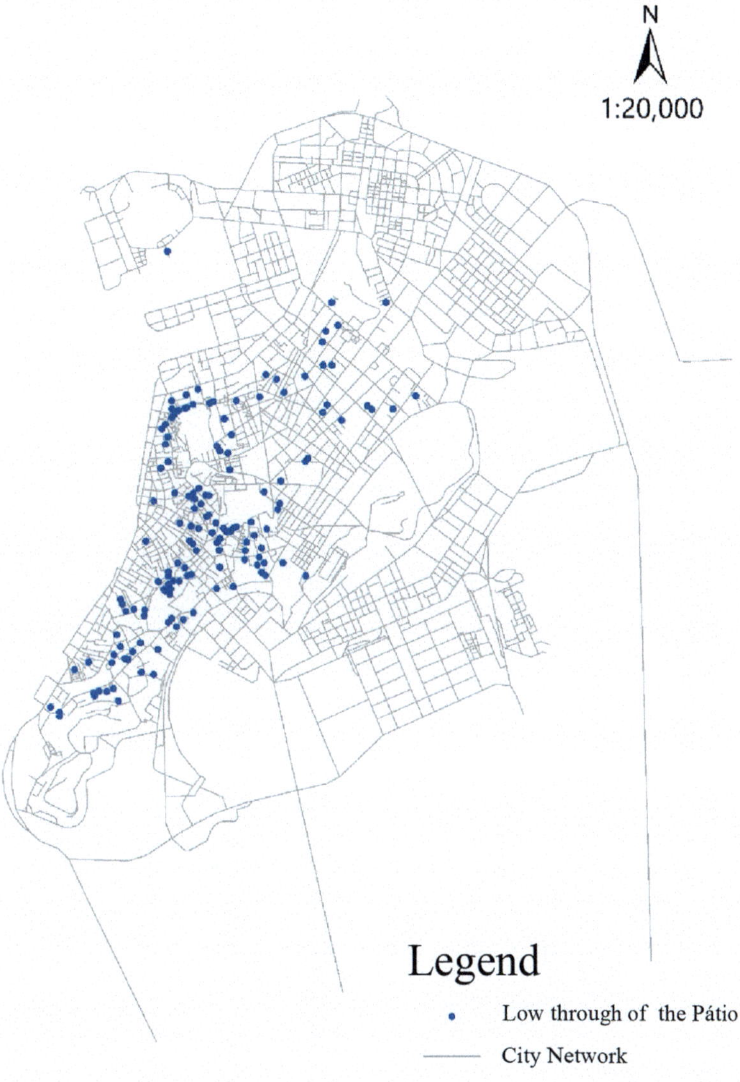

Fig. 41 Distribution map of low pass through Pátio and Beco spaces. (*Image source* drawn by the author)

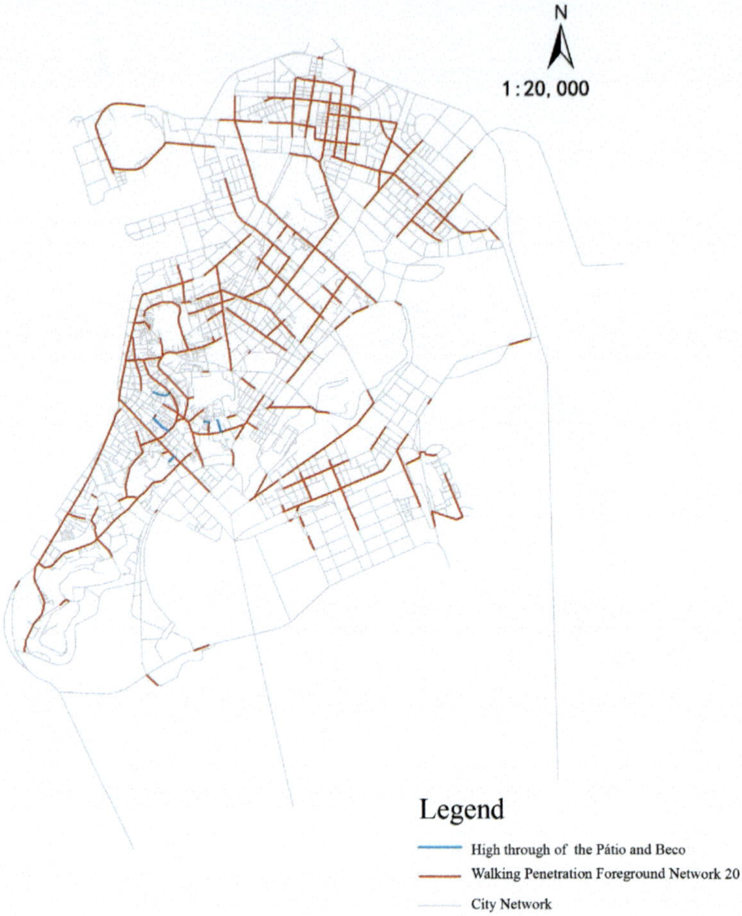

Fig. 42 Distribution characteristics of Pátio and Beco with high pass through. (*Image source* drawn by the author)

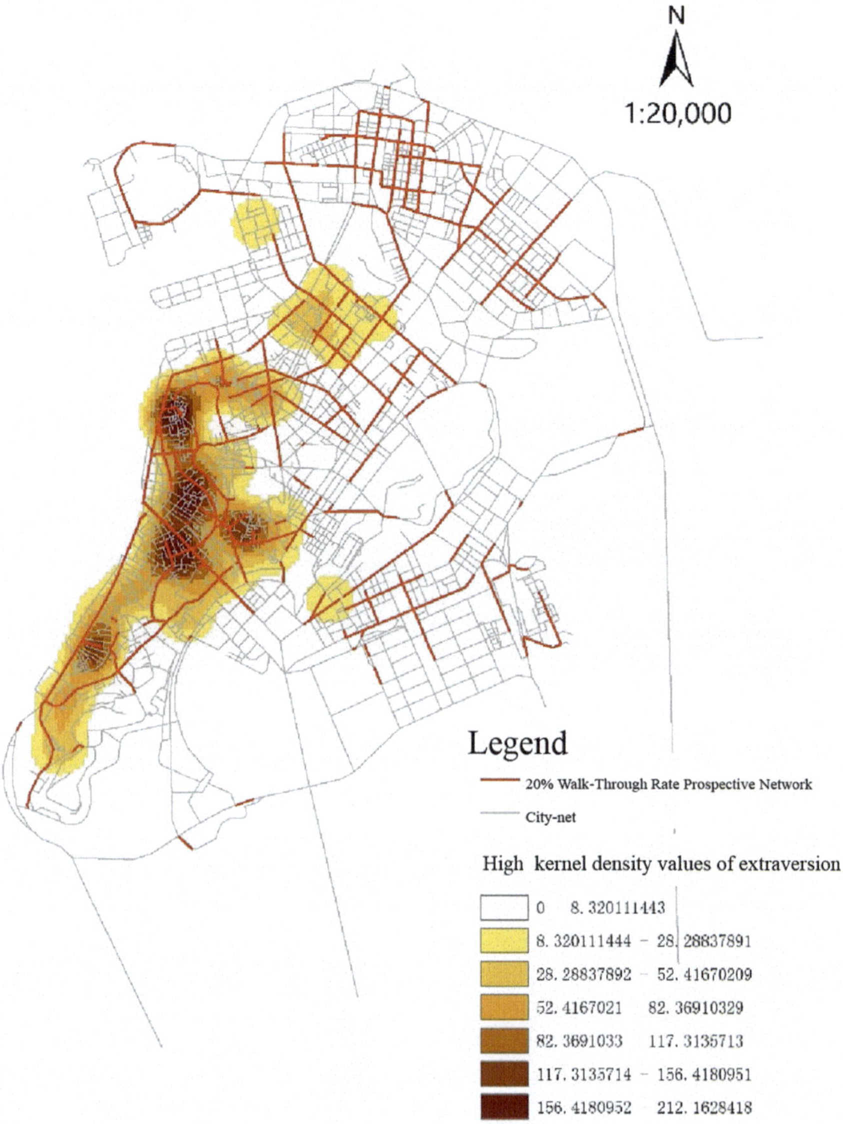

Fig. 43 Distribution characteristics of Pátio and Beco with high pass through. (*Image source* drawn by the author)

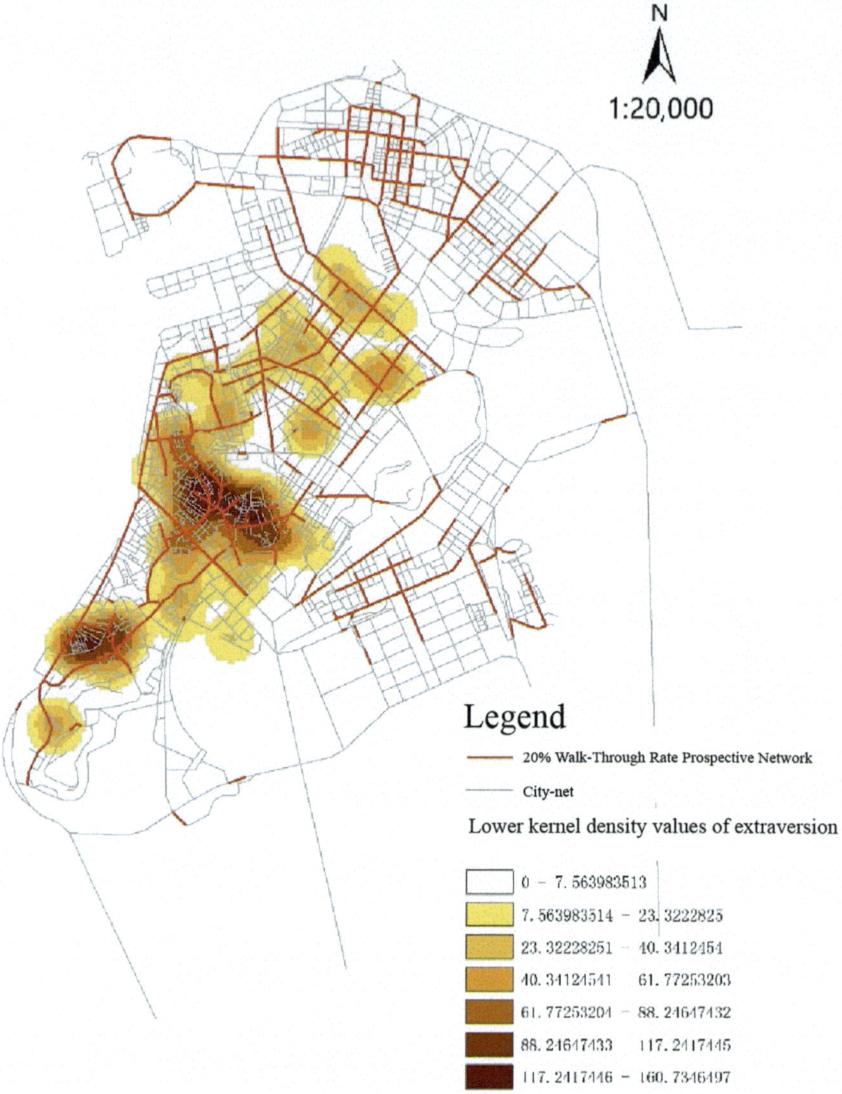

Fig. 44 Distribution characteristics of Pátio and Beco with low pass through. (*Image source* drawn by the author)

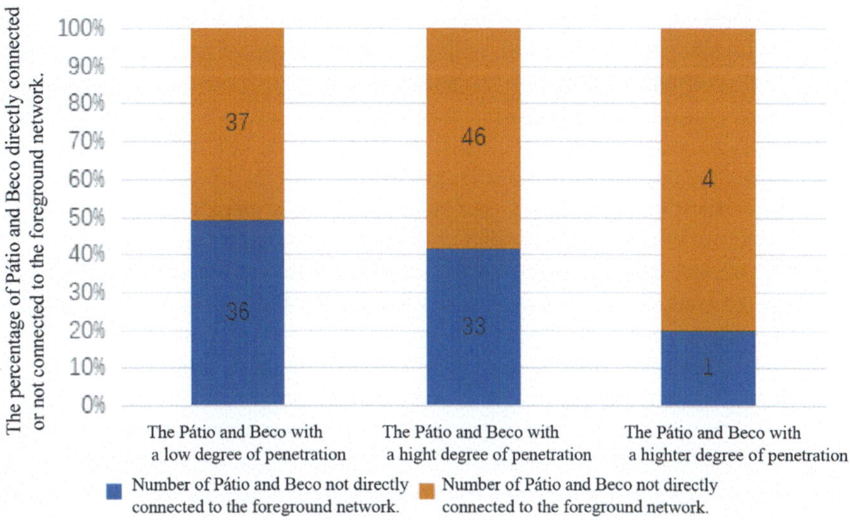

Fig. 45 Connection degree of each Pátio and Beco with the pedestrian openness foreground network. (*Image source* drawn by the author)

5 Conclusion

This paper combines space syntax, geographic information system (GIS) to analyze the multi-scale spatial characteristics of Pátio and Beco streets spaces in Macau from a spatial quantification perspective.

From a spatial perspective, based on the space syntax theory, this paper comprehensively analyzes the potential spatial characteristics of Pátio and Beco streets in the Macau Peninsula from three aspects: accessibility, public openness, and block harmony, through integration, choice, and intelligibility. These potential spatial characteristics are reflected in the following three aspects:

In terms of accessibility, Pátio and Beco streets show obvious dependence on external high-level streets. Therefore, if Pátio and Beco are to be activated as destinations for corresponding functions, it is necessary to comprehensively consider the streets directly connected to them to jointly promote the vitality improvement of multiple streets. At the same time, Pátio and Beco are more accessible at the pedestrian scale than at the overall urban scale, which also reflects that the activation of Pátio and Beco streets should mainly consider people's walking activities.

In terms of openness, although Pátio and Beco streets are influenced by external high-level streets, they are more "independent" compared to accessibility. Although their openness increases with the reduction of scale, their hierarchical status in the overall urban system has not improved, which is related to the fact that most Pátio and Beco streets are dead-end roads. Therefore, when Pátio and Beco serve as a passage, they always maintain an introverted characteristic.

In terms of block harmony, from the perspective of human cognition, small-scale block activities are obviously easier to grasp and understand than large-scale urban activities. Regarding the distribution characteristics of harmony in each block, the southeastern blocks of the St. Lawrence District are clearly unfriendly to external pedestrian flows. The morphological characteristics of the road network within the blocks hinder the cognition of external pedestrians towards this area, causing great difficulties in their entry, exit, and guidance. This area is more suitable for serving local Macau residents who are familiar with the area during actual activation.

From a cultural value perspective, this paper argues that this layer of vitality more embodies a key influencing factor of vitality that can be directly perceived by people. Its significance lies in supplementing the limitations of space syntax theory research. Historical buildings, historical features, and related humanistic stories within streets can obviously make Pátio and Beco become a cultural attraction for different pedestrian flows, which cannot be explained or measured by space syntax theory. Therefore, in the actual analysis, this paper mainly refers to previous studies on the cultural values of Pátio and Beco, whose results show that only a small number of Pátio and Beco streets have preserved intact historical features.

Finally, this paper comprehensively quantifies the activation potential indicators of all Pátio and Beco streets in Macau Peninsula by integrating the above two aspects. The significance of this scale lies in its explanation of the current spatial vitality status of each Pátio and Beco streets from both spatial and cultural perspectives. On the one hand, the magnitude of the indicator can assist in the orderly activation of Pátio spaces—the higher the vitality, the higher the priority for activation. On the other hand, the level of the activation potential scale also provides a basis for determining their development direction—a higher activation potential indicates that the street space tends to be more public, while a lower activation potential suggests a more private spatial characteristic. Therefore, in terms of development orientation, it also explores the suitability of development directions for Pátio and Beco streets.

The innovations of this paper first lie in integrating the spatial morphological characteristics of Pátio and Beco (factors imperceptible to people) with cultural values (factors directly perceivable by people). Through quantitative methods such as space syntax analysis, it objectively measures the spatial activation potential of Pátio and Beco streets, so as to provide references for activation sequences and directions in subsequent development and activation. Secondly, in terms of specific methods, this paper also refers to previous theories and makes some innovations in methods around the special research object of Pátio and Beco. For example, this paper comprehensively defines the "blocks" divided by public streets through bottom-up network analysis and top-down public service technology analysis, and measures the intelligibility of Pátio and Beco streets within this area. In the application of space syntax quantitative parameters, some re-interpretations are also made to endow the research with more planning significance.

Finally, this paper still has some shortcomings: First, regarding the potential activation scale. In fact, although the paper attempts to comprehensively quantify the activation potential of Pátio and Beco streets from multiple angles based on the influence of the space itself and the usage meaning of the space, there are still more influencing factors to be considered, such as the land property rights within the streets, the social background of residents, and the damage status of buildings, which will also affect the development urgency order of Pátio and Beco streets. Therefore, theoretically, more influencing factors need to be supplemented to this activation potential scale, which can only be said to be a reference factor rather than a decisive one. Second, there are many subjective considerations in the construction of the urban model. For example, some road networks are not drawn or not fully and finely drawn in the actual model construction, such as special spaces with obvious boundaries like parks or commercial streets inside buildings. This paper believes that the roads within such places are more special compared to the external open street network. In terms of the meaning of Macau streets, they do not belong to the traditional road hierarchy in Macau. On the other hand, they are often under special management with obvious internal and external boundaries, which also control the actual use of people, making them different from the free traffic flow outside, but these places do affect the movement of traffic flow. Finally, there are technical problems in this paper. Since the traditional DepthmapX software of space syntax is used, which cannot perform three-dimensional modeling of urban streets, the quantification of actual streets by this model will also have certain errors in terms of human spatial cognition.

Funding: This research received no external funding.

Institutional Review Board Statement: Not applicable for studies not involving humans or animals.

Informed Consent Statement: Not applicable for studies not involving humans.

Conflicts of Interest: The authors declare that the research was conducted in the absence of any commercial or financial relationships that could be construed as a potential conflict of interest.

References

1. Wang Weijen Architectural Design and Research Studio. (2011). Conservation of Cilinwei: St. Paul's Church, Urban Village and Pastoral Texture in Macau. *Time Architecture,* (05), 68–69. https://chn.oversea.cnki.net/KCMS/detail/detail.aspx?dbcode=CJFD&dbname=CJFD2011&filename=SDJZ201105023&uniplatform=OVERSEA&v=qajTTGj4ML8yoeskeyGi74bk9GiuHiddgVrjAEW8gdqKwIgTJlYqJ64z7_LBbnQU
2. Hillier, B. (2008). *Space is a machine: A theory of architectural composition.* Beijing: China Architecture & Building Press. ISBN: 9787112099528
3. Hillier, W. R. G., Yang, T., & Turner, A. (2012). Normalising least angle choice in Depthmap-and how it opens up new perspectives on the global and local analysis of city space. *Journal of Space Syntax, 3*(2), 155–193. https://discovery.ucl.ac.uk/id/eprint/1389938

4. Sayid, J., Alasdair, T., Hillier, B., Iida, S., Payne, A., Shibo, G., & Tao, Y. (2016). Line segment analysis and advanced axis and line segment analysis: From Chapters 5 and 6 of space syntax methods: A teaching guide. *Urban Design, 01*, 32–55. https://doi.org/10.16513/j.urbandesign.2016.01.005
5. Jin, D., Tao, Y., Qiang, S., Haofeng, W., & Xiaoling, D. (2019). *Space syntax course*. Beijing: China Architecture & Building Press. ISBN: 9787112245703.
6. Lin, Z., Shan, J., Keshi, C., et al. (2015). Study on the spatial morphological evolution of the old city center based on space syntax: A case study of the Xiaogongyuan opening port area in Shantou City. *Modern Urban Research*, (7), 68–76. https://chn.oversea.cnki.net/KCMS/detail/detail.aspx?dbcode=CJFD&dbname=CJFDLAST2015&filename=XDCS201507014&uniplatform=OVERSEA&v=1MvZXnpLdwYVgZkAWBAx5oR562BujorhHRHR9oaOISObzZDxzarN6GUGcJmmsllE
7. Zheng, L., & Chen, Y. (2022). Research on the spatial characteristics of high-density Urban road network and functional agglomeration taking the northern part of Macau as an example. In *International Conference on Spatial Data and Intelligence* (pp. 189–206). Cham: Springer Nature Switzerland. https://doi.org/10.1007/978-3-031-24521-3_14
8. Huang, Y., Chen, Y., Song, J., Zheng, L., Yang, S., Gao, Y., Li, R., & Huang, L. (2025). Explore the ultra-high density Urban waterfront space form: An Investigation of Macau Peninsula Pier District via point of interest (POI) and space syntax. *Buildings, 15*(10), 1735. https://doi.org/10.3390/buildings15101735
9. Duo, H., Hengyu, G., & Hongqing, J. (2018). Construction of urban design method and process integration mechanisms based on space syntax. *Planner, 34*(03), 59–65. https://chn.oversea.cnki.net/KCMS/detail/detail.aspx?dbcode=CJFD&dbname=CJFDLAST2018&filename=GHSI201803011&uniplatform=OVERSEA&v=0dES7YDBoFam4NVrSSXDQ0-INITj_LupJXRRUpyF44KoeCj4HOAJ9SUK4iQQUCwW
10. Hengyu, G., Xin, M., Tiyan, S., Huiling, C., Fan, X. (2018). Research on the impact of road network morphology on housing prices in Guangzhou based on sDNA model. *Modern Urban Studies*, (06), 2–8. https://chn.oversea.cnki.net/KCMS/detail/detail.aspx?dbcode=CJFD&dbname=CJFDLAST2018&filename=XDCS201806003&uniplatform=OVERSEA&v=ADq0OU3JRtt7mZyN_yNFguXIIuDrPqtGf277cYQuZ7ozl-O3D69-wuFVOX6-4vND
11. Chen, F., Haofeng, W., & Xiaojun, R. (2016). Study on the street network structure and commercial density distribution of Macau Peninsula. *Southern Architecture*, (03), 122–128. https://chn.oversea.cnki.net/KCMS/detail/detail.aspx?dbcode=CJFD&dbname=CJFDLAST2016&filename=NFJZ201603023&uniplatform=OVERSEA&v=k6Ti0xAxXHlNhbs6n3meyDz_Jk5BNAfPSao0QpajdvPzfcB5Qy9ahxMgyIuJSALb

Route Optimization for the Activation of Historical Urban Areas: Macau Route

Song Yang

Abstract Urban heritage, as a carrier of a city's history, spirit, culture, and collective memory, is increasingly valued. Establishing urban heritage trails enables the threading of heritage sites together, integrating them more broadly into people's lives. The primary task in creating urban heritage trails lies in selecting optimal routes. As theories of urban heritage conservation evolve, the concept of the Historic Urban Landscape (HUL) has gained wider acceptance. This theory posits that urban heritage exhibits historical layering, which visually manifests the imprints of a city's development and evolution across different historical periods. This study adopts the HUL perspective to systematically review the historical layering of the Macau Peninsula, examining the characteristics and shifting focal points of urban development throughout its history. Subsequently, an evaluation system for the layering richness of urban cultural heritage trails is developed based on the concept of heritage layering. This framework is applied to analyze and assess existing routes within the core zone of the Historic Centre of Macau. Relevant tools are then applied to optimize these routes based on evaluation results. Finally, the optimized outcomes are compared and validated. The results indicate a measurable improvement in layering richness following optimization. Using the Macau Peninsula as a case study, this research explores an evaluation system and optimization methodology for heritage trails based on layering richness. It aims to provide innovative insights for route selection in themed heritage trail development while enhancing public understanding of urban heritage layering and strengthening awareness of heritage conservation.

Keywords Historic Urban Landscape (HUL) · Historical layering · Urban cultural heritage trail · Historic Centre of Macau

S. Yang (✉)
Faculty of Humanities and Arts, Macau University of Science and Technology, Tapai, Macau, China
e-mail: 694217541@qq.com

Y. Chen et al. (eds.), *City-Sea Symbiosis*, Springer Geography,
https://doi.org/10.1007/978-3-032-13725-8_12

1 Research Background

1.1 Development of Themed Heritage Trail Initiatives

In recent years, numerous countries have achieved substantial success in developing heritage trails that reflect their unique cultural identities. For instance, the United States established 19 "National Historic Trails" to showcase its multifaceted historical evolution. Similarly, the UK's Hadrian's Wall Path National Trail runs primarily along Hadrian's Wall, allowing visitors to traverse ancient fortifications while experiencing the profound weight of history. In May 2023, Chinese national ministries issued directives initiating preparatory work for developing themed heritage trails across the country [1]. Constructing urban cultural heritage trails interconnects and integrates scattered heritage sites within urban spaces, embedding them into daily life. This approach not only facilitates the preservation and revitalization of urban cultural heritage—infusing it with contemporary vitality—but also promotes the synergistic development of cultural heritage and tourism. Consequently, it enhances public cultural literacy and strengthens national pride.

1.2 Conceptual Framework of the Historic Urban Landscape

The concept of the "Historic Urban Landscape" (HUL) was formally introduced in the 2005 *Vienna Memorandum on the Conservation of the Historic Urban Landscape* (hereinafter referred to as the "Memorandum"). HUL represents a holistic and integrated theoretical and methodological framework for urban heritage conservation. Its core principle emphasizes contextualizing heritage within its broader urban environment, underscoring the interconnectedness between urban heritage assets [2, 3]. This conceptual foundation offers critical guidance for designing urban cultural heritage trails.

1.3 Current Status of Heritage Trails in the Historic Centre of Macau

In July 2005, the "Historic Centre of Macau" was inscribed on the *UNESCO World Heritage List*. The core zone of this Centre (hereinafter referred to as the "Core Zone") encompasses 22 historic monuments, 8 urban spaces, and the streets connecting its squares and architectural landmarks. These elements include numerous historic urban spaces of Portuguese and Chinese heritage, such as streets, "largo" (public squares), residences, public buildings, churches, and temples. The World Heritage Committee recognizes the Historic Centre of Macau as "the oldest, most complete, and longest-lasting fusion of Chinese-Western cultural exchange, rooted in the flourishing development of international trade and economy" [4].

Based on the *Map of the Historic Centre of Macau* (Fig. 1) included in the Annex to Macau SAR's *Cultural Heritage Protection Law* promulgated in 2013, the current delimitation of the Core Zone features a "de facto" path connecting World Heritage attractions on the western side of the Macau Peninsula. However, this route lacks formal designation as a structured tourist trail. Moreover, "Farol de Guia" and its adjacent World Heritage sites remain isolated on the eastern periphery, with no clearly defined access route linking them to other heritage assets.

Beyond the Core Zone, numerous urban heritage sites embody rich historical layering yet receive minimal visitor engagement due to limited public awareness. Consequently, tourist flows concentrate disproportionately at a few high-profile attractions, causing congestion, while historically significant spaces elsewhere remain underutilized. Establishing urban cultural heritage trails would not only redistribute visitor traffic to alleviate overcrowding but also revitalize neglected heritage sites.

This chapter adopts the HUL theoretical framework to analyze and evaluate existing routes within the Core Zone. Building on this assessment, we propose an optimized urban cultural heritage trail for the Historic Centre of Macau that more effectively embodies HUL principles.

2 Research Framework and Methodology

2.1 Research Framework

Optimizing heritage trails to revitalize historic districts requires first identifying deficiencies in existing routes through systematic evaluation. Anchored in the HUL theoretical perspective [5–16], this study assesses current trails in the Core Zone using the "layering richness" of heritage attraction nodes (hereafter "attraction nodes") along routes as the core evaluation criterion. Subsequent optimization yields an enhanced trail with superior layering richness. The research methodology comprises three phases (Fig. 2):

(1) Delineating the Historical Layering of the Macau Peninsula. As optimization may involve modifying or adding attraction nodes, the study scope extends to the entire Macau Peninsula. Historical records are analyzed to reconstruct the peninsula's evolving physiographic conditions, urban morphology, and socio-cultural transformations. Through an HUL lens [5–10], we catalog historical layering across the peninsula and its attraction nodes. These nodes are then georeferenced onto period-specific historical maps based on their construction eras to construct a "spatiotemporal heritage model."
(2) Evaluation of Existing Core Zone Trails. A quantitative methodology for evaluating trail "layering richness" is established [17–20]. This framework is

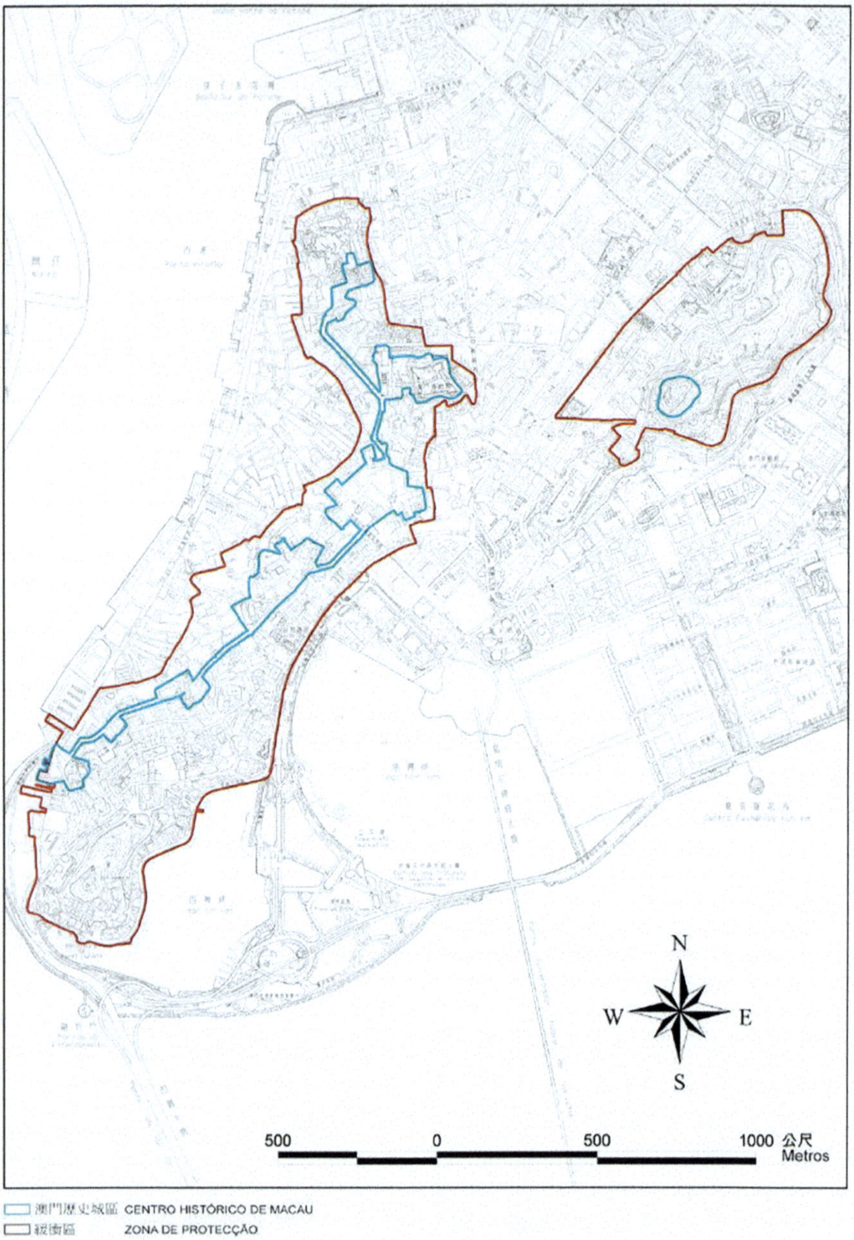

Fig. 1 Map of the Historic Center of Macau (*Data source* https://bo.io.gov.mo/bo/i/2013/36/lei11_cn.asp)

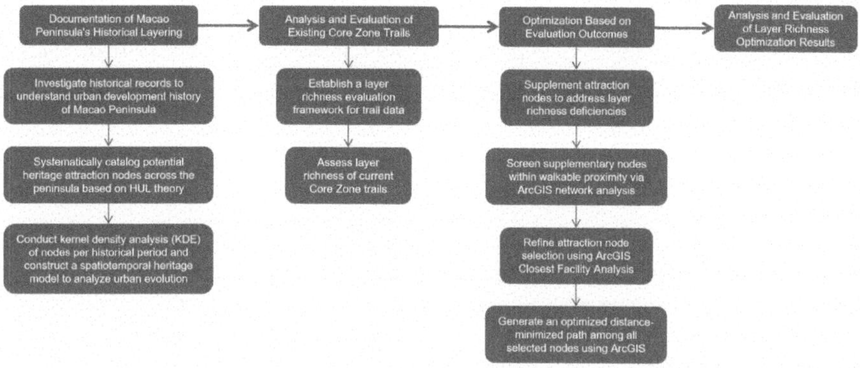

Fig. 2 Research framework. (*Image source* drawn by the author)

applied to assess current Core Zone routes, identifying deficiencies to inform optimization strategies.

(3) Optimization Based on Evaluation Outcomes. After obtaining the evaluation outcomes of the existing route layering richness in the Core Zone, it needs to be optimized. The optimization process will involve the addition and modification of attraction nodes and the regeneration of the optimal route and finally get an upgraded heritage trail based on the existing route in the Core Zone but with better layering richness.

2.2 *Research Methodology*

2.2.1 Literature Analysis

This study employs documentary research. Primary historical archives and cartographic sources are analyzed to reconstruct the Macau Peninsula's physiographic evolution, urban morphology, and socio-cultural transformations. Concurrently, HUL theoretical literature is interrogated to derive methodologies for examining historical layering and assessing heritage values. This informs the criteria for selecting attraction nodes and establishing the layering richness evaluation system. Statutory instruments and management frameworks governing Macau's cultural heritage are also reviewed to identify attraction nodes and construct the peninsula's spatiotemporal heritage model.

2.2.2 ArcGIS Kernel Density Estimation (KDE)

KDE—a widely used spatial analysis technique in ArcGIS—calculates feature density per unit area within a defined radius, generating continuous "raster density surfaces." This study applies KDE to quantify spatial clustering patterns of heritage attraction nodes across historical periods. Resultant density surfaces provide spatial proxies for shifting urban development foci over time.

2.2.3 ArcGIS Network Analysis

ArcGIS network analysis tools can perform service area analysis, closest facility analysis, and optimal route solving by constructing network datasets. The establishment of a service area is used to conduct accessibility analysis on the location of existing World Heritage sites, to screen the range of attractive nodes for route optimization. Closest facility analysis can identify points closer to the Core Zone from near to far, and further screen for attractive nodes. Optimal route solving can search for the optimal route between multiple stopping points and is used to find an optimized route between all selected attraction nodes.

3 Documentation of Historical Layering in Macau Peninsula

3.1 Analysis of Historical Records

According to established scholarship [21], urban development in the Macau Peninsula evolved through four distinct historical phases. Representative historical maps from each period were comparatively analyzed to trace urban morphological transformations:

(1) Initial Urbanization Period (1557–1586): Following Macau's establishment as a trading port in 1557, Chinese communities clustered in villages like "Templo de A-Má" and "Mong-Há," forming traditional settlements. Portuguese settlers occupied central and southern areas, establishing church-centered parishes such as "Igreja de St.º António," "Igreja de S. Lázaro," and "Igreja de S. Lourenço." A fishbone street pattern gradually emerged. No extant historical maps survive this phase.
(2) Urban Expansion Period (1586–1840): After gaining municipal autonomy under Portuguese India in 1586, defensive infrastructure (city walls, fortresses) was constructed in the peninsula's core by the Portuguese. This physically demarcated Portuguese settlements within the walls from Chinese communities outside, creating a binary spatial structure (Fig. 3).
(3) Colonial Expansion Period (1840–1974): Following the Opium War (1840), Portuguese colonial administration-initiated dismantling of city walls (1863) and territorial expansion. Street networks consolidated while large-scale sea reclamation fueled diversified urban development—commercial, residential, and institutional functions proliferated (Fig. 4).

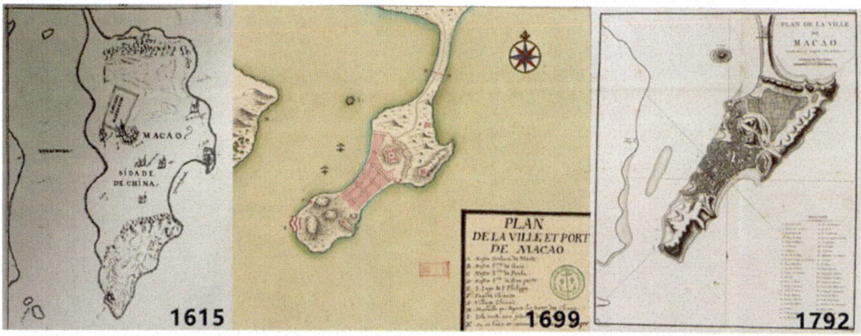

Fig. 3 Historical map of the Macau Peninsula during the urban growth period (1586–1840). (Macau University of Science and Technology Library Collection, https://gmom.must.edu.mo/)

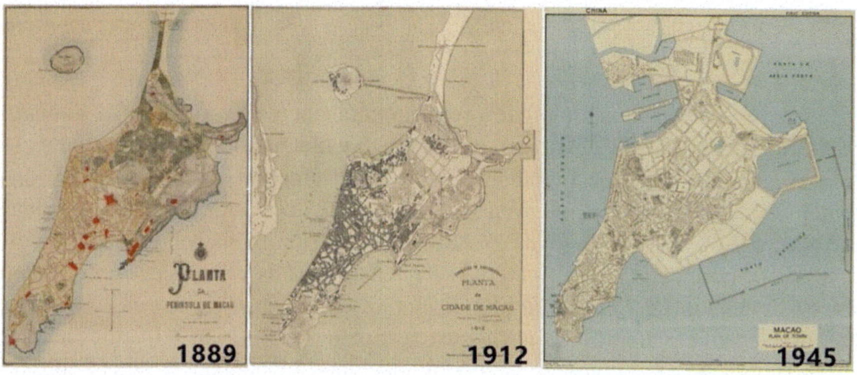

Fig. 4 Historical map of the Macau Peninsula during the period of urban expansion (1840–1975). (*Image source* Macau University of Science and Technology Library Collection, https://gmom.must.edu.mo/)

(4) Urban Development Period (1974–Present): Portugal's Carnation Revolution (1974) ended colonial policies, affirming Macau as Chinese territory. Transitioning from autonomy to handover (1999), the city entered stable development with continuous reclamation at Outer Harbour, "Areia Preta," and "NAPE" (Zona Nova de Aterros do Porto Exterior), expanding urban boundary (Fig. 5).

3.2 *Value Carriers of Urban Heritage from a HUL Perspective*

Research and summary of HUL theory indicate that urban heritage exhibits "layering," "dynamicity," "interconnectedness," "sustainability," and "holism." From this perspective, heritage values are primarily manifested through material carriers of

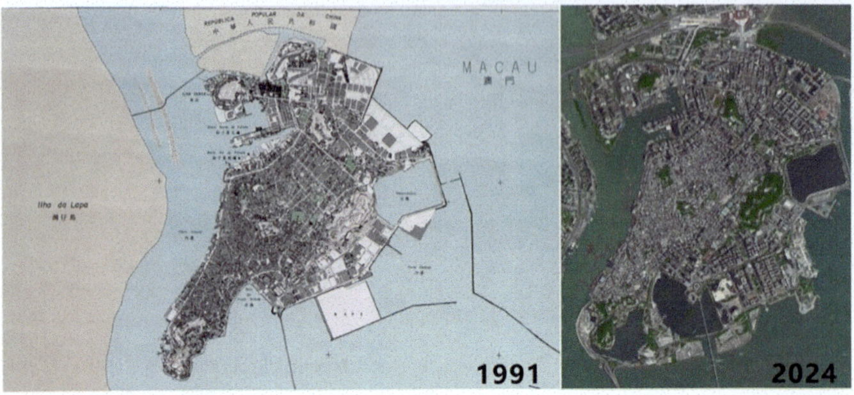

Fig. 5 Map of the Macau Peninsula during the Urban Development Period (1974-present). (*Image source* Macau University of Science and Technology Library Collection, https://gmom.must.edu.mo/, On the right is a satellite image from Baidu Maps)

spatial layering and intangible carriers of cultural memory. Analysis of material spatial layering typically involves constructing spatiotemporal heritage models to enhance research clarity, encompassing both vertical temporal periodization and horizontal spatial element analysis. Vertical temporal periodization primarily employs two approaches: division by dynastic shifts and division by morphological phases. Analysis of horizontal spatial-material elements generally references conservation objects defined in UNESCO's *Recommendation on the Historic Urban Landscape* (2011), adapted to specific research contexts, with the "Anchoring-Layering" analytical model serving as a common methodological tool. Intangible carriers of cultural memory include collective memory, belief systems, folk rituals, etc. (Fig. 6).

3.3 Construction of Macau Peninsula's Spatiotemporal Heritage Model

Constructing the spatiotemporal heritage model of Macau Peninsula requires first identifying urban spaces that function as attraction nodes—locations embodying the value propositions of urban heritage conveyed through HUL theory. Per HUL principles, such attraction nodes include but are not limited to formally designated tangible cultural heritage, urban spaces bearing intangible historical-cultural significance, and urban landmarks reflecting developmental characteristics across distinct historical periods. Designated tangible cultural heritage is identified through the *Cultural Heritage Protection Law* (Law No. 11/2013) and the *List of Classified Immovable Properties of Macau* (Revised Edition, 2023). Supplementary attraction nodes with heritage value—though not formally listed—are sourced through auxiliary channels such as tourism applications, including Phase IV urban landmarks,

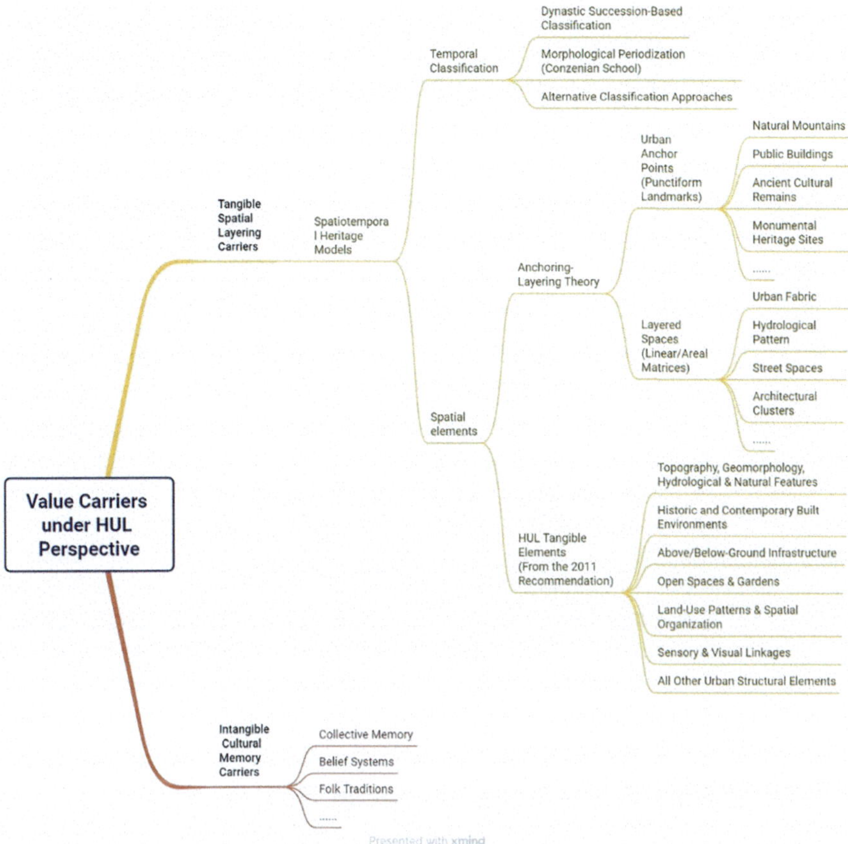

Fig. 6 Analysis of heritage value carriers from the perspective of HUL. (*Image source* drawn by the author)

memorials to historical figures or events, and spaces manifesting intangible spiritual culture. Legally protected heritage encompasses classified immovable properties and the core zone of the UNESCO World Heritage property.

World Heritage attraction nodes primarily comprise 22 historic monuments, 8 urban spaces, and connecting streets between squares and architectural landmarks (Fig. 1). Classified immovable properties fall into four categories: monuments, buildings of architectural-artistic value, architectural ensembles, and sites (Fig. 7). Data compilation also referenced key landscape view corridors from the *Conservation and Management Plan for the Historic Centre of Macau* (2018), along with Macau's designated "Eight Scenes," "New Eight Scenes," and venues sustaining intangible cultural heritage. Attraction nodes were categorized according to Zhang Jie's four cultural landscape systems: Human Habitat, Functional, Historical, and Spiritual. The Human Habitat System encompasses daily living spaces, including residential buildings, public structures, parks, and squares linked to communal life. The Functional System covers port facilities, defense structures, and traditional industrial remnants, such as piers, city walls, and fortresses. The Historical System incorporates commemorative spaces honoring figures or events, such as memorial halls and monuments. The Spiritual System includes religious sites and venues embodying folk culture. When classifying nodes, overlapping designations may occur (e.g., a historic residence may simultaneously belong to both Human Habitat and Historical Systems, requiring dual listing). This process yielded a categorized heritage inventory (see Appendix 1).

The existing attraction nodes were georeferenced onto period-specific peninsula boundary maps based on their construction eras, establishing Macau Peninsula's Spatiotemporal Heritage Model (Fig. 8). Kernel Density Estimation (KDE) in ArcGIS was applied to analyze node clustering patterns per period, revealing shifts in urban development cores. Linear features (like streets) and areal features (like architectural ensembles and sites with larger areas) were abstracted as point features for analytical compatibility. Cross-verification with the four historical phases shows: Phase I (1557–1586) focused on "Templo de A-Má," "Igreja de S. Lázaro," and "Mong Há" areas; Phase II (1586–1840) centered around "Igreja de St.º António," "Ruínas do Colégio de S. Paulo," "Largo do Senado," and "Colina da Guia"; Phase III (1840–1974) developed near "Igreja de S. Lázaro e adro" and the western slope of "Colina da Guia"; Phase IV (1974–present) expanded to reclaimed areas at "Praça de Ferreira do Amaral" and Macau Cultural Centre.

4 Layer Richness Assessment Based on HUL

4.1 Establishing the Layering Evaluation System for Trail Attraction Nodes

Urban heritage embodies historical layering—a cumulative record of urban development and transformation across different periods. This layering represents the sedimentation of historical changes and integrating it with urban cultural heritage trails allows visitors to experience socio-spatial imprints of human activities under evolving historical contexts. As heritage attraction nodes serve as primary navigational targets during visits, this study evaluates existing Core Zone trails through the aggregate layer richness of their constituent nodes. The current trail network and accessible attraction nodes are mapped (Fig. 9).

This study's layer richness assessment, grounded in HUL theory, evaluates three dimensions [22, 23]: construction eras, functional typologies, and heritage classifications of attraction nodes (Table 1). Construction eras follow the four historical phases of Macau Peninsula's urban development. Functional typologies adopt Zhang Jie et al.'s cultural landscape framework, categorizing systems as residential, institutional, historical, and spiritual. Heritage classifications align with Macau's statutory

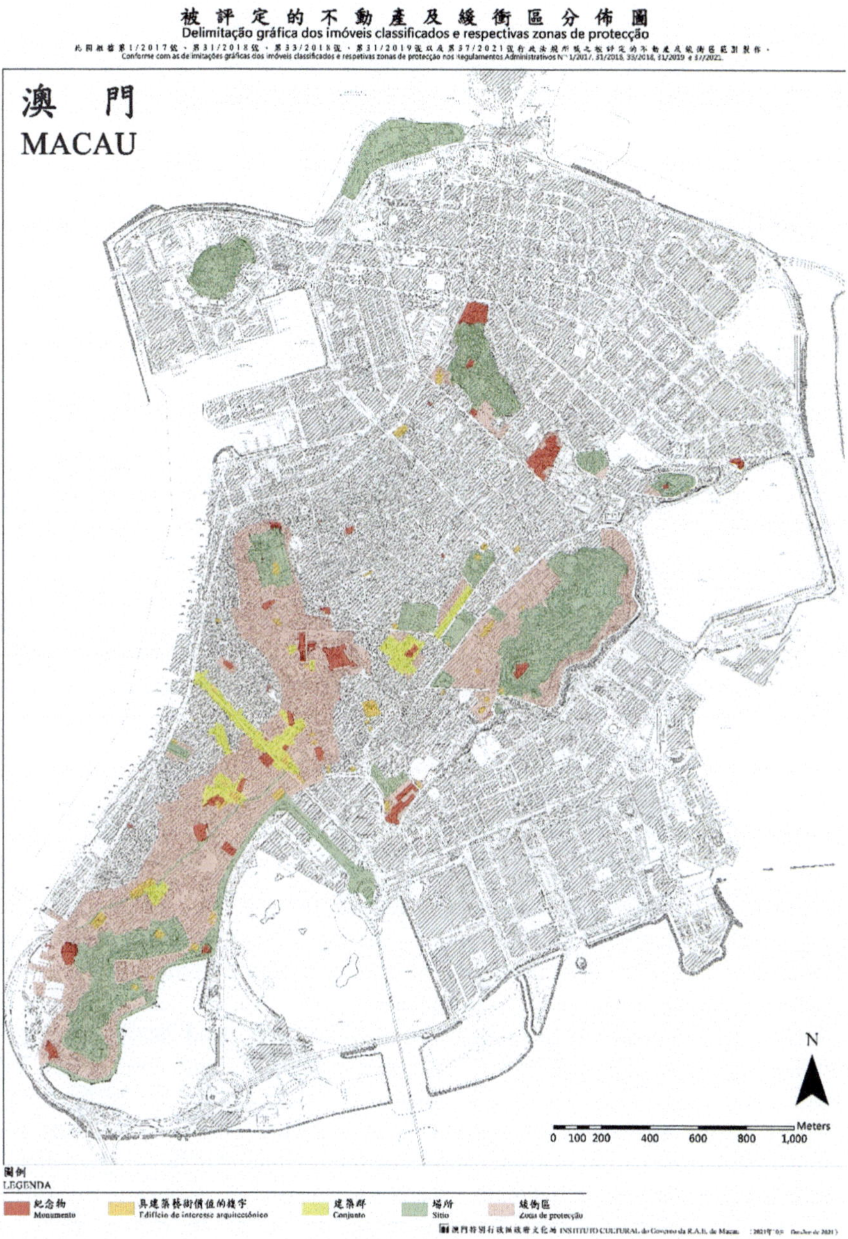

Fig. 7 Demonstration map of Classified Immovable Properties and its buffer zone. (*Image source* http://www.culturalheritage.mo/_ref/site/images/macau2021F.jpg?AspxAutoDetectCookieSupport=1)

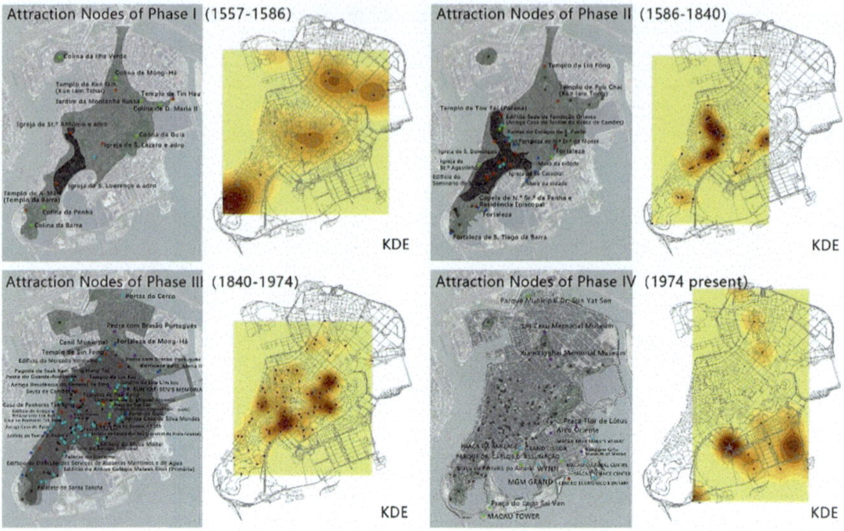

Fig. 8 Macau Peninsula's Spatiotemporal Heritage Model. (drawn by the author)

immovable heritage categories: monuments, buildings of architectural merit, architectural ensembles, and sites. Consequently, the layer richness evaluation system comprises 3 dimensions and 12 subcategories to assess trail heritage layering.

4.2 Layer Richness Evaluation of Existing Core Zone Trails

The existing Core Zone trails were evaluated per the aforementioned system by quantifying layer richness types across their attraction nodes. These nodes encompass all listed sites accessible along the trails—not limited to World Heritage properties—with relevant information cataloged in Appendix 2.

Tabular data for node subtypes along current trails (Table 2) indicate: construction eras show strong coverage of Phases II–III, minimal Phase I, and absent Phase IV nodes; functional typologies reveal dominance of residential and spiritual systems against underrepresented institutional and historical systems; heritage classifications exhibit predominance of monuments, moderate representation of buildings of architectural merit, and scarce architectural ensembles and sites. Consequently, attraction nodes will be supplemented and trails optimized to address these layer richness deficiencies.

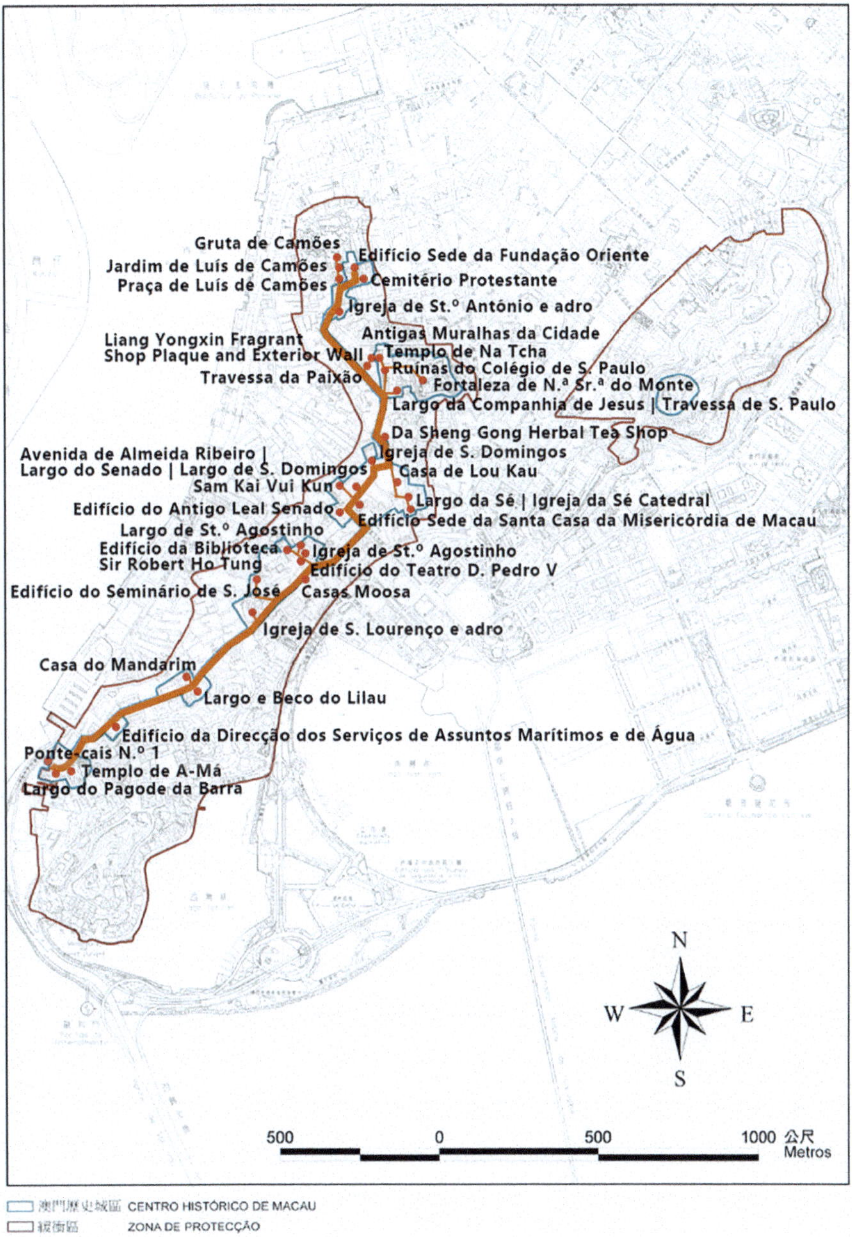

Fig. 9 Current trail network and accessible attraction nodes. (drawn by the author)

Table 1 Evaluation content of layer richness

Construction eras	Phase I (1557–1586)	Phase II (1586–1840)	Phase III (1840–1974)	Phase IV (1974–present)
Functional typologies	Residential system	Institutional system	Historical system	Spiritual system
Heritage classifications	Monuments	Buildings of architectural merit	Architectural ensembles	Sites

Table 2 Counts of attraction node subtypes along existing Core Zone trails

Phase I	Phase II	Phase III	Phase IV
5	16	16	0
Residential system	Institutional system	Historical system	Spiritual system
23	4	4	14
Monuments	Buildings of architectural merit	Architectural ensembles	Sites
17	9	5	3

5 Optimization of Existing Core Zone Trails

The specific approach for optimizing existing Core Zone trails is illustrated (Fig. 10). To address insufficient layer richness caused by underrepresented node subtypes along current trails, supplementary nodes of deficient subtypes will be added to regenerate trails, thereby enhancing trail layer richness. This involves resolving two key issues: First, determining how to select supplementary nodes. Since this study assumes walking as the trail mode, we identify attraction nodes of each underrepresented subtype within walkable proximity to the Core Zone, prioritized from near to far. Second, determining how to generate trails. After node selection, an optimal path (i.e., the shortest walking distance path) must be generated among all selected nodes. ArcGIS Network Analyst tools facilitate finding the optimal path among multiple stops. Following optimized path generation, layer richness before and after optimization will be compared to analyzing and evaluating optimization outcomes.

Optimization of existing Core Zone trails proceeds through sequential steps:

(1) Identifying layer types requiring supplementation and target node quantities per type based on layer richness evaluation results.
(2) Conducting distance-based accessibility analysis at walkable scales to preliminarily screen candidate nodes for underrepresented subtypes.
(3) Selecting all accessible nodes for subtypes with insufficient candidates, while for subtypes with adequate candidates, performing closest facility analysis from Core Zone nodes to progressively incorporate nearest candidates until reaching the mean node count across all subtypes.

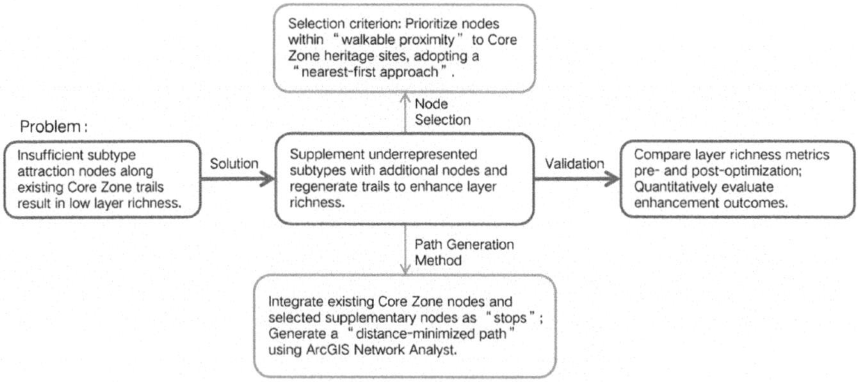

Fig. 10 Optimization approach diagram. (*Image source* drawn by the author)

(4) Integrating Core Zone nodes and selected supplementary nodes to generate the optimized urban cultural heritage trail based on Macau's World Heritage core.

5.1 *Identifying Layer Types Requiring Optimization*

First, the average number of attraction nodes for all subtypes along the trail was calculated using prior data: $(5 + 16 + 16 + 0 + 23 + 4 + 4 + 14 + 17 + 9 + 5 + 3)/12 \approx 9$. Subtypes with counts below 9 require node supplementation. Construction eras dimension, Phase I and Phase IV nodes need supplementation; Functional typologies dimension, institutional and historical systems require supplementation; Heritage classifications dimension, architectural ensembles and sites need supplementation (Fig. 11).

5.2 *Walkable Proximity Screening Beyond the Core Zone*

To ensure walking comfort on the new trail, candidate attraction nodes must undergo location screening before supplementing underrepresented layer types. This entails selecting nodes beyond the Core Zone within walkable proximity, guaranteeing suitable walking distances between each pair of navigational nodes. Wang and Du [24] Revealed Preference survey in Shanghai residential areas established that a higher proportion of residents choose to walk within a walking distance of 787 m, thus defining this distance threshold [24]. Consequently, ArcGIS Network Analyst's "Service Area Analysis" was applied to World Heritage sites—notably including "Farol de Guia" and its adjacent World Heritage sites, previously isolated due to its exclusion from existing Core Zone trails but now integrated as mandatory routing points—to delineate accessible zones for supplementary nodes (Fig. 11).

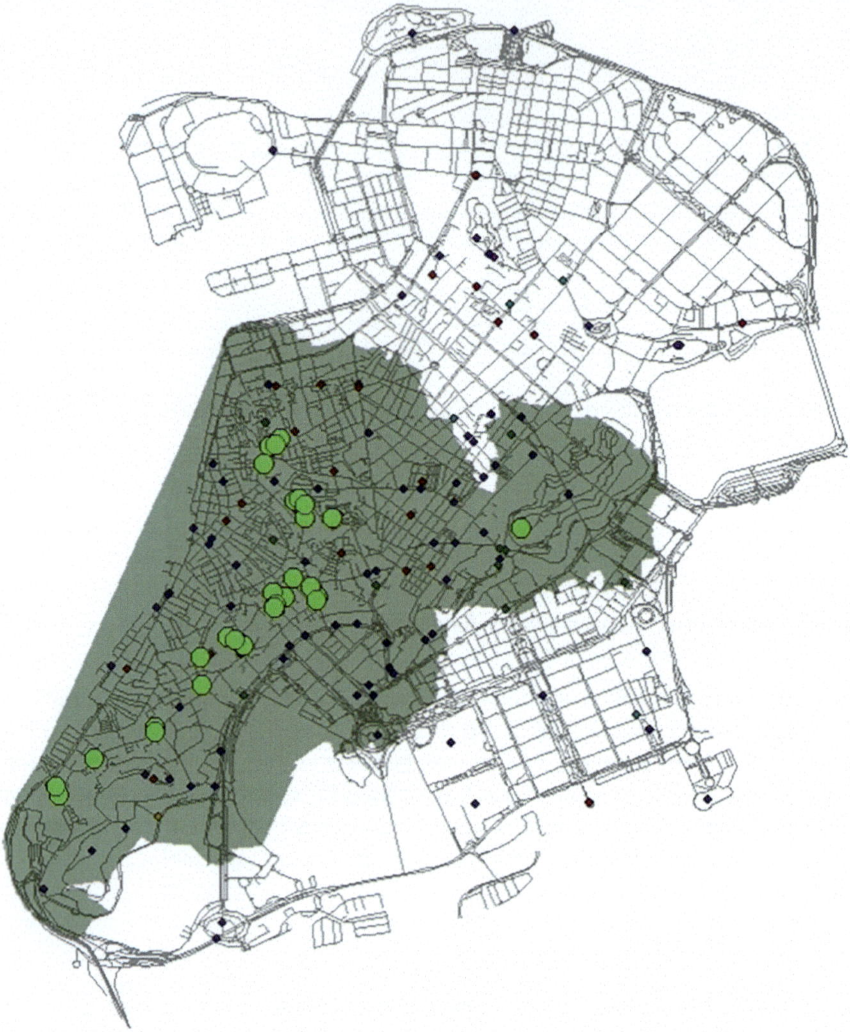

Fig. 11 Accessibility Analysis within Walkable Proximity of the World Heritage Core Zone. Light green circles denote Core Zone attraction nodes; dark green zones indicate walkable proximity buffers from these nodes; dark dots represent other attraction node locations. (*Image source* drawn by the author)

5.3 Supplementing Attraction Nodes for Deficient Layer Types

After delineating the search area, attraction nodes of underrepresented layer types within walkable proximity were identified: from the perspective of construction eras, Phase I and IV nodes; from the perspective of functional typologies, institutional and historical systems; from the perspective of Heritage classifications, architectural ensembles and sites. Metadata for these six subtypes were cataloged (Appendix 3).

Supplementation quantities were then determined by accounting for new mandatory routing points at "Farol de Guia" and its adjacent World Heritage sites. This added one Phase I node (Colina da Guia), two institutional system nodes (Fortaleza de N.ª Sr.ª da Guia e Farol) and one site node (Colina da Guia) (Fig. 12).

Using the mean value of 9 attraction nodes per subtype along existing trails as the benchmark, supplementation targets were quantified: Phase I requires three additional nodes; Phase IV, nine nodes; institutional systems, three nodes; historical systems, five nodes; architectural ensembles, four nodes; and sites, five nodes. Within the walkable proximity zone, only five Phase IV nodes and three architectural ensemble nodes were available—both fewer than required quantities—thus all were selected. Their spatial distribution is mapped (Fig. 13).

Subsequently, ArcGIS Network Analyst's Closest Facility tool processed Core Zone nodes alongside Phase I, institutional systems, historical systems, and site nodes—excluding subtype nodes spatially coincident with Core Zone locations. The resultant outputs were integrated with previously selected Phase IV and architectural ensemble nodes (Fig. 14). Candidate nodes within accessible zones were then selected by nearest-to-farthest walking distance from Core Zone nodes, compiling all supplementary selections (Table 3).

5.4 Optimized Path Solving

Supplementary node locations and Core Zone attraction nodes were loaded as stops in ArcGIS. The optimal path was solved using network analysis tools, adjusted according to field-verified urban network conditions, and finalized as shown (Fig. 15).

6 Analysis and Evaluation of Optimization Results

6.1 Comparison of Layer Richness Pre- and Post-optimization

Statistical results indicate the optimized trail traverses 70 attraction nodes (Appendix 4). Node counts per subtype along the new trail are tabulated (Table 4).

To quantify changes in layer richness along the trail before and after optimization, the proportion of nodes per subtype relative to total nodes per dimension was calculated and tabulated, with standard deviations (SD) computed for each dimension (Table 5).

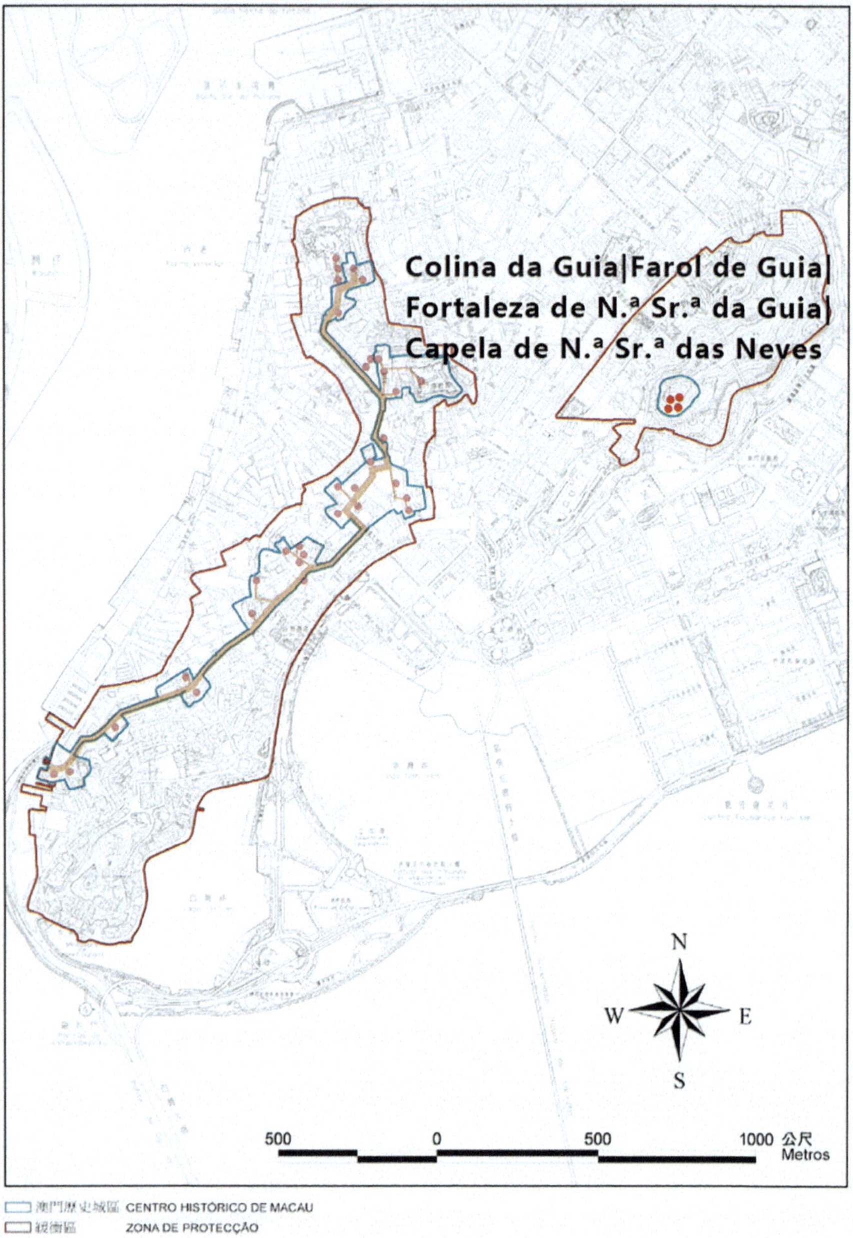

Fig. 12 Mandatory attraction nodes increase before the optimization. (*Image source* drawn by the author)

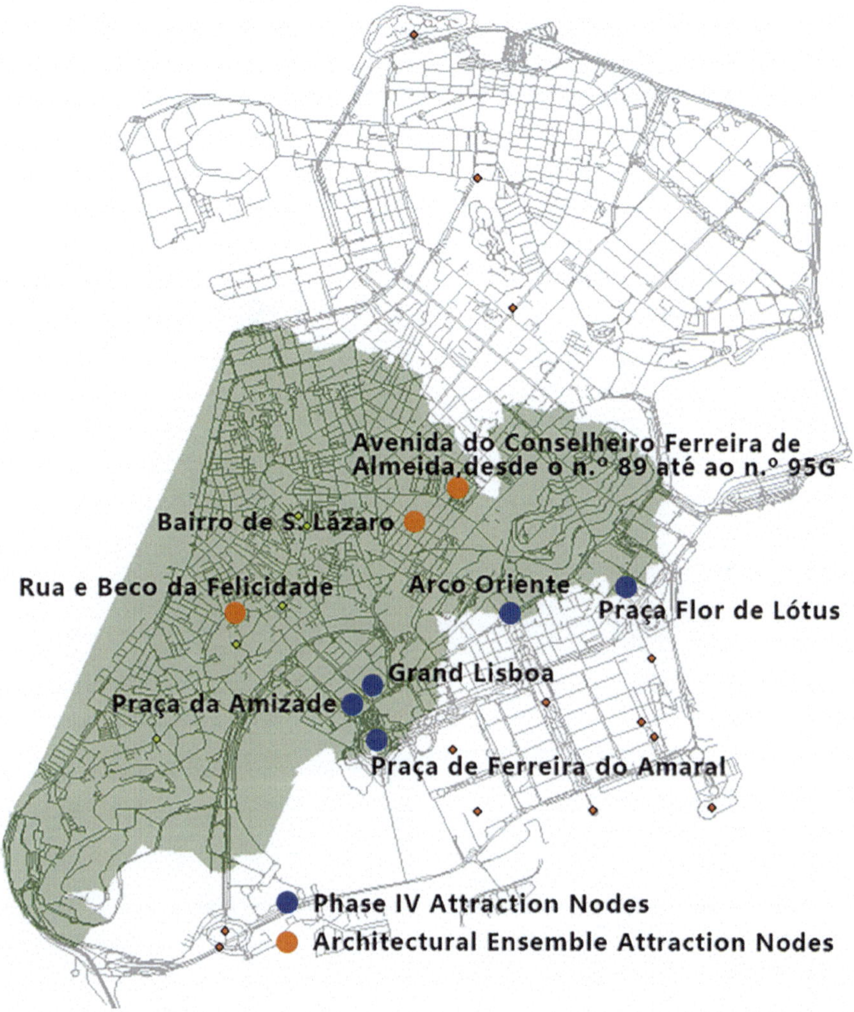

Fig. 13 Location of Attraction Nodes for Phase IV and Architectural Ensemble within the Walkable Proximity Zone. (*Image source* drawn by the author)

From the ratio of the number of attraction nodes for each subtype to the total number of attraction nodes in its dimension before and after optimization, subtypes with lower pre-optimization ratios—Phase IV, institutional systems, historical systems, and sites—exhibit increased post-optimization proportions. Phase I and architectural ensembles show marginal decreases due to smaller absolute quantities. Standard deviation comparisons demonstrate significant reductions in temporal and typological dimensions, indicating diminished dispersion and more balanced node count distributions, thereby enhancing layer richness in these dimensions. The functional dimension exhibits unchanged standard deviation because the optimized path

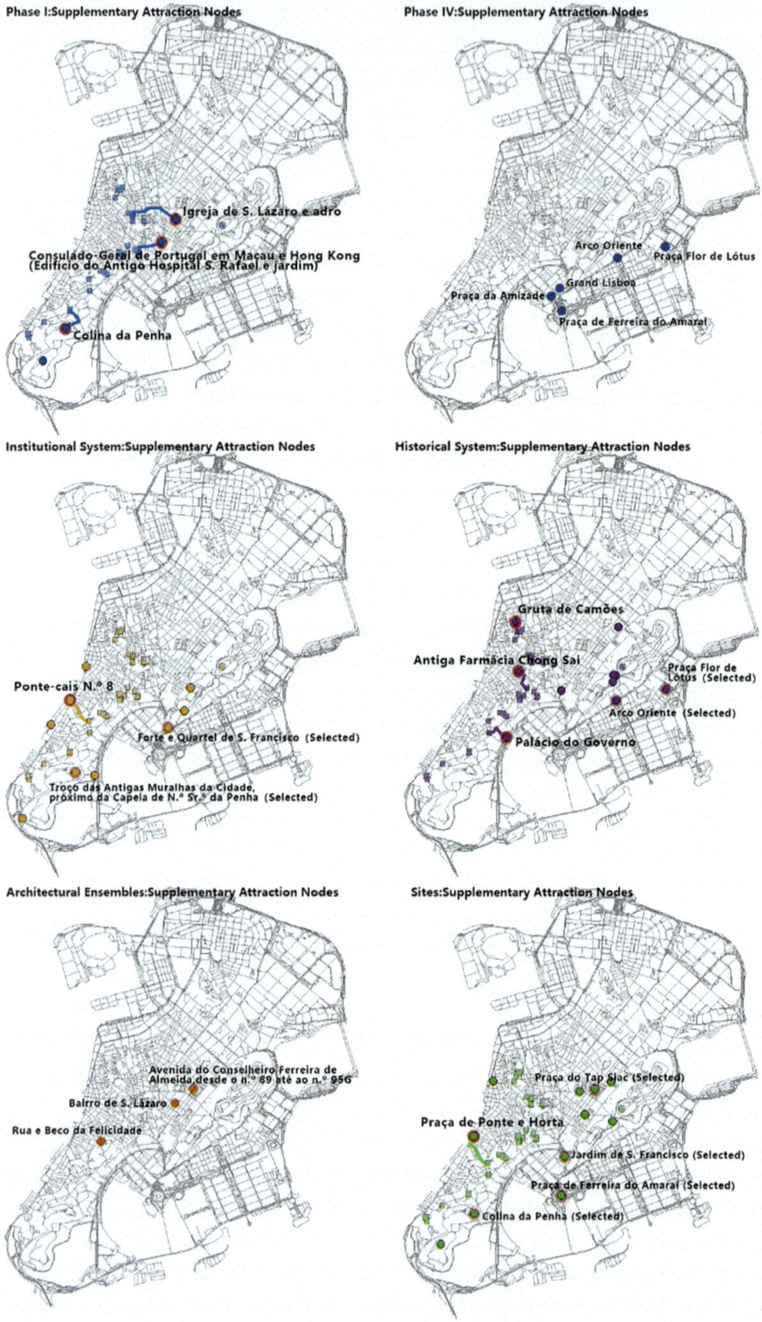

Fig. 14 Aggregation of Attraction Nodes from Closest Facility Analysis and Pre-Supplemented Nodes. Square markers denote Core Zone attraction nodes; red circles connected by trail paths represent attraction nodes retained after Closest Facility analysis (eliminating distant paths); unconnected red circles indicate previously selected nodes. (*Image source* drawn by the author)

Table 3 Selected supplementary attraction nodes list

Selected supplementary attraction nodes (Portuguese or English)	Construction eras	Heritage classifications	Functional typologies
Igreja de S. Lázaro e adro	Phase I	Monuments	Spiritual system
Consulado-Geral de Portugal em Macau e Hong Kong (Edifício do Antigo Hospital S. Rafael e jardim)	Phase I	Buildings of architectural merit	Residential system
Colina da Penha	Phase I	Sites	Residential system
Troço das Antigas Muralhas da Cidade, próximo da Capela de N.ª Sr.ª da Penha	Phase II	Monuments	Institutional system
Praça de Ponte e Horta	Phase II	Sites	Residential system
Ponte-cais N.º 8	Phase III		Institutional system
Gruta de Camões	Phase III	Monuments	Historical system
Antiga Farmácia Chong Sai	Phase III	Monuments	Historical system
Palácio do Governo	Phase III	Monuments	Historical system, Residential system
Rua e Beco da Felicidade	Phase III	Architectural ensembles	Residential system
Bairro de S. Lázaro	Phase III	Architectural ensembles	Residential system
Avenida do Conselheiro Ferreira de Almeida,desde o N.º 89 até ao N.º 95G	Phase III	Architectural ensembles	Residential system
Antigo Mercado do Tarrafeiro	Phase III	Sites	Residential system
Praça do Tap Siac (Antigo Campo Coronel Mes quita)	Phase III	Sites	Residential system
Grand Lisboa	Phase IV		Residential system
Praça de Ferreira do Amaral	Phase IV		Residential system
Praça da Amizade	Phase IV		Residential system
Arco Oriente	Phase IV		Historical system
Praça Flor de Lótus	Phase IV		Historical system

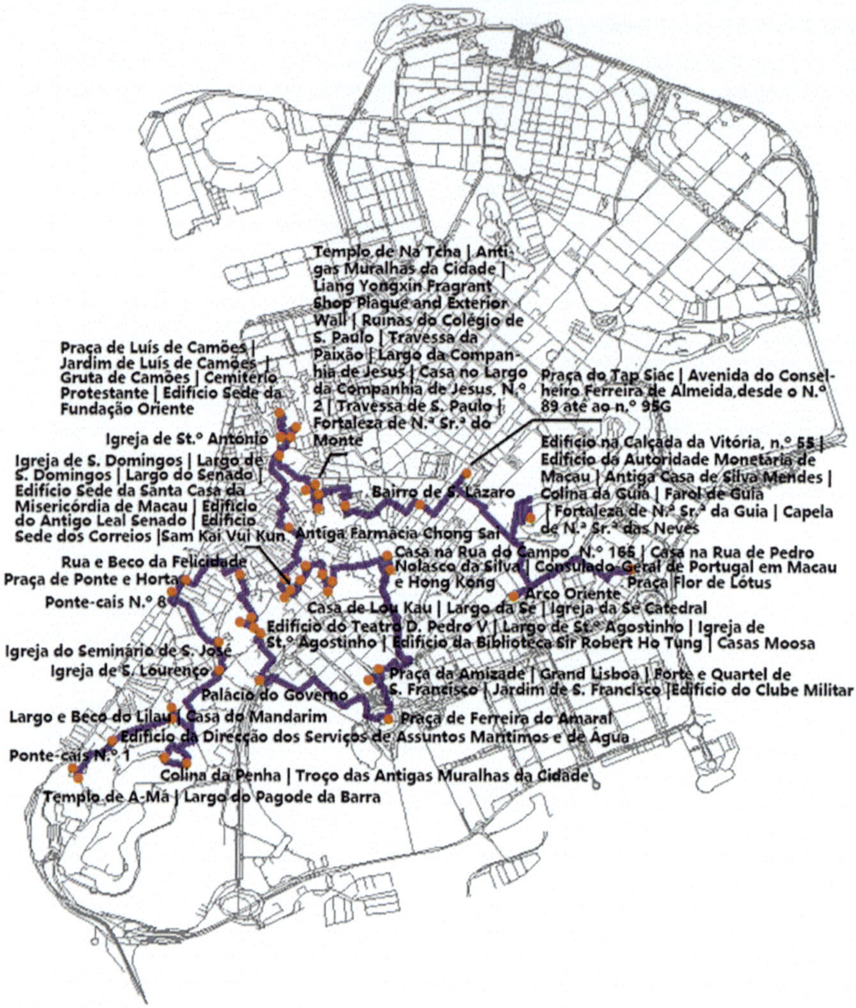

Fig. 15 Optimized path generated based on attraction nodes' location. (*Image source* drawn by the author)

Table 4 Number of attraction nodes for each subtype of optimized path

Phase I	Phase II	Phase III	Phase IV
9	23	33	5
Residential system	Institutional system	Historical system	Spiritual system
45	9	12	16
Monuments	Buildings of architectural merit	Architectural ensembles	Sites
26	19	8	10

Table 5 Comparison of data related to path attraction nodes before and after optimization

Before optimization				
Phase I	Phase II	Phase III	Phase IV	SD
0.14	0.43	0.43	0	0.19
Residential system	Institutional system	Historical system	Spiritual system	
0.51	0.09	0.09	0.31	0.17
Monuments	Buildings of architectural merit	Architectural ensembles	Sites	
0.50	0.26	0.15	0.09	0.16
After optimization				
Phase I	Phase II	Phase III	Phase IV	SD
0.13	0.33	0.47	0.07	0.16
Residential system	Institutional system	Historical system	Spiritual system	
0.55	0.11	0.15	0.20	0.17
Monuments	Buildings of architectural merit	Architectural ensembles	Sites	
0.41	0.30	0.13	0.16	0.11

traverse's additional residential system nodes and partially overlaps with historical/spiritual systems, amplifying residential proportions. Nevertheless, institutional and historical systems show measurable ratio increases, confirming enhanced richness.

In summary, optimized trails exhibit substantially improved layer richness in construction eras and heritage classifications dimensions, while functional typologies dimension improvement remains limited primarily due to: (1) The functional typologies of attraction nodes have overlapping subtypes; (2) Abundant residential heritage resources increasing residential node ratios, introducing moderate bias in functional typologies dimension analysis.

6.2 *Analysis of Optimized Trail Attributes*

The optimized trail spans approximately 10,880 m with an estimated walking duration of 136 min. Originating at "Templo de A-Má" and terminating at "Praça Flor de Lótus," it enables visitors to experience heritage across distinct historical periods, functional typologies, and heritage classifications while traversing all World Heritage nodes. This suggests that this path possesses a relatively high degree of layer richness. Create a route map by sequentially taking photos of all attraction nodes (Appendix 5).

Table 6 10 of New Zealand's best hiking trails

Serial number	Trail name	Total length of the walking trail/km	Completion days/d
1	Lake Waikaremoana	46	3–4
2	Tongariro Northern Circuit	43	3–4
3	Whanganui Journey	145	3–5
4	Abel Tasman Coast Track	60	3–5
5	Heaphy Track	78.4	4–6
6	Paparoa Track	55	2–3
7	Routeburn Track	32	2–4
8	Milford Track	53.5	4
9	Kepler Track	60	3–4
10	Rakiura Track	32	3

Regarding the reasonableness of the trail's total length, New Zealand's ten premier hiking trails serve as comparative benchmarks for their lengths and recommended completion days [25] (Table 6). It can be observed that all these hiking trails exceed 30 km in total length, and based on their recommended completion days, their average daily walking distances surpass 10 km. These trails thus provide actual comparative cases for this study's optimized path, validating the reasonableness of its total length.

7 Research Conclusions

This study, grounded in the Historic Urban Landscape (HUL) perspective [26, 27], systematically documented historical layering in the Macau Peninsula and constructed a spatiotemporal heritage model to visually demonstrate urban morphological evolution across periods. The core concept of heritage layering in HUL theory was introduced into urban cultural heritage trail route selection and evaluation. A layer richness evaluation system was developed to assess existing Core Zone trails, analyzing their layer richness as heritage routes. Subsequently, attraction nodes were supplemented to address identified layer richness deficiencies, thereby optimizing the trail.

Results confirm enhanced layer richness in the optimized trail with more balanced distribution of attraction nodes across various subtypes. For future trail planning and management in the peninsula, these outcomes provide actionable references: redirecting overloaded tourist flows from core World Heritage sites; alleviating visitor pressure while diversifying heritage experiences across eras and typologies; and deepening public understanding of the peninsula's holistic urban development.

Methodological innovations include introducing HUL's heritage layering concept into trail route selection and establishing a layer richness evaluation system for heritage trails. This enables multidimensional interpretation of urban development through route design, enriching visitor experiences and providing novel perspectives for themed heritage trail development. Concurrently, establishing such trails extends cultural spaces, integrates heritage into community life, and catalyzes community revitalization.

8 Limitations and Future Research

This study's limitations and corresponding research prospects are outlined below:

(1) The evaluation focused solely on the "layering" dimension of heritage trails, whereas HUL theory encompasses broader aspects—"integrity," "dynamic nature," "interconnectedness," and "sustainability." Future studies may integrate these multidimensional criteria into trail route selection.
(2) The layer richness evaluation system considered construction eras, functional typologies, and heritage classifications. Future frameworks could incorporate additional layering dimensions.
(3) ArcGIS Network Analyst's optimal path-solving function relies on point-based stops, potentially inadequately representing linear features (streets) or areal elements (green spaces, building complexes). Future methodologies should explore integrating linear and areal heritage into routing algorithms.
(4) Supplementary node selection prioritized spatial proximity, neglecting heritage value significance. Future optimizations should balance accessibility with value-based criteria (e.g., historical importance or conservation urgency).

Appendix 1: Summary List of Attraction Nodes for Macau Peninsula Urban Heritage

Residential and public buildings			
Name (Portuguese or English)	Construction eras	Heritage classifications	Notes
Consulado-Geral de Portugal em Macau e Hong Kong (Edifício do Antigo Hospital S. Rafael e jardim)	Phase I	Buildings of architectural merit	

(continued)

(continued)

Residential and public buildings			
Name (Portuguese or English)	Construction eras	Heritage classifications	Notes
Ruínas do Colégio de S. Paulo (Antigo Muro, troço na Rua de D. Belchior Carneiro)	Phase II	Monuments	
Vestígios históricos encontrados em fosso aberto no subs-trato rochoso, na Rua de D. Belchior Carneiro	Phase II	Monuments	
Ruínas do Colégio de S. Paulo (Antigo Muro: dois troços no Pátio do Espinho) (Antigo Muro: troço no Beco do Craveiro)	Phase II	Monuments	
Edifício Sede da Santa Casa da Misericórdia de Macau	Phase II	Monuments	
Edifício Sede da Fundação Oriente (Antiga Casa do Jardim da Gruta de Camões)	Phase II	Buildings of architectural merit	
Edifício Sede do Instituto para os Assuntos Municipais (Edifício do Antigo Leal Senado)	Phase II	Monuments	
Casa de Penhores Tak Sang	Phase III	Buildings of architectural merit	
Casa de Penhores Kou Seng	Phase III	Buildings of architectural merit	
Palácio do Governo	Phase III	Monuments	
Casa no Largo da Companhia de Jesus, N.º 2	Phase III	Buildings of architectural merit	
Edifício do Teatro D. Pedro V	Phase III	Buildings of architectural merit	
Palacete de Santa Sancha	Phase III	Buildings of architectural merit	
Edifício do Antigo Colégio Mateus Ricci (Primária)	Phase III	Buildings of architectural merit	
Casa do Mandarim	Phase III	Buildings of architectural merit	Intangible Cultural Heritage "Nanyin Rap" and "Taoist Ritual Music"

(continued)

(continued)

Residential and public buildings

Name (Portuguese or English)	Construction eras	Heritage classifications	Notes
Residência Oficial do Cônsul de Portugal em Macau e Hong Kong (Edifício do Antigo Hotel Bela Vista)	Phase III	Buildings of architectural merit	
Edifício do Clube Militar	Phase III	Buildings of architectural merit	
Edifício da Direcção dos Serviços de Assuntos Marítimos e de Água (Antigo Quartel dos Mou ros)	Phase III	Buildings of architectural merit	
Antiga Casa do Ópio	Phase III	Buildings of architectural merit	
Casas Moosa	Phase III	Buildings of architectural merit	
Casa de Lou Kau	Phase III	Buildings of architectural merit	
Edifício da Biblioteca Sir Robert Ho Tung	Phase III	Buildings of architectural merit	
Edifício da Antiga Farmácia Chinesa Tai Neng	Phase III	Buildings of architectural merit	
Casa no Largo da Sé, N.º 1	Phase III	Buildings of architectural merit	
Casa na Avenida da República, N.º 6	Phase III	Buildings of architectural merit	
Da Sheng Gong Herbal Tea Shop	Phase III		Intangible Cultural Heritage "Herbal Tea Preparation"
Edifício Administrativo da Escola Secundária Pui Ching (Antiga Residência de Lou Lim Ioc)	Phase III	Buildings of architectural merit	
Pavilhão da Primavera (no Jardim de Lou Lim Ioc)	Phase III	Buildings of architectural merit	
Antiga Casa de Silva Mendes	Phase III	Buildings of architectural merit	
Edifício da Escola Pui Tou (Sucursal da Praia Grande)	Phase III	Buildings of architectural merit	
Antigo Estábulo Municipal de Gado Bovino e Canil Municipal	Phase III	Buildings of architectural merit	

(continued)

(continued)

Residential and public buildings			
Name (Portuguese or English)	Construction eras	Heritage classifications	Notes
Casa na Rua do Campo, N.º 165	Phase III	Buildings of architectural merit	
Casa de Penhores Tak Seng	Phase III	Buildings of architectural merit	
Edifício da Autoridade Monetária de Macau (Edifício do Antigo Convento do Precioso Sangue)	Phase III	Buildings of architectural merit	
Edifício na Estrada Nova, N.º 2	Phase III	Buildings of architectural merit	
Edifício da Escola Ling Nam (Antiga Vila Alegre)	Phase III	Buildings of architectural merit	
Casa na Avenida da Praia Grande, N.º 679	Phase III	Buildings of architectural merit	
Edifício do Corpo de Bombeiros	Phase III	Buildings of architectural merit	
Pavilhão de Exposições e Espectáculos Artísticos para Jovens (Edifício da Antiga Caixa Escolar)	Phase III	Buildings of architectural merit	
Lar de Nossa Senhora da Misericórdia	Phase III	Buildings of architectural merit	
Edifício Sede do Banco Nacional Ultramarino	Phase III	Buildings of architectural merit	
Edifício na Calçada da Vitória, N.º 55	Phase III	Buildings of architectural merit	
Casa na Rua de Pedro Nolasco da Silva, N.os 24A, 26, 28 e 28A	Phase III	Buildings of architectural merit	
Edifício na Calçada do Gaio N.º 6	Phase III	Buildings of architectural merit	
Edifício na Avenida de Horta e Costa, N.os 14 e 16	Phase III	Buildings of architectural merit	
Edifício na Estrada da Vitória N.º 30	Phase III	Buildings of architectural merit	
Edifício Sede dos Correios	Phase III	Buildings of architectural merit	
Edifício do Mercado Vermelho	Phase III	Buildings of architectural merit	

(continued)

(continued)

Residential and public buildings			
Name (Portuguese or English)	Construction eras	Heritage classifications	Notes
Edifício do Antigo Restaurante Lok Kok	Phase III	Buildings of architectural merit	
Edifício da Escola Secundária Luso-Chinesa de Luís Gonzaga Gomes (Edifício da Antiga Escola Primária Oficial Pedro Nolasco da Silva)	Phase III	Buildings of architectural merit	
Edifício na Estrada do Cemitério, N.º 6 (Casa Azul)	Phase III	Buildings of architectural merit	
Posto do Guarda-Nocturno (Patane)	Phase III	Monuments	
Edifício do Antigo Tribunal	Phase III	Buildings of architectural merit	
Cinema Alegria	Phase III		Intangible Cultural Heritage "Cantonese Opera"
Hotel Lisboa Macau	Phase III		
Macau Tower	Phase IV		
Macau Cultural Centre	Phase IV		
Macau Fisherman's Wharf	Phase IV		
WYNN	Phase IV		
MGM Grand	Phase IV		
Grand Lisboa	Phase IV		
Macau Science Center	Phase IV		
Streets, architectural ensembles, urban texture			
Rua e Beco da Felicidade	Phase III	Architectural ensembles	One of the New Eight Scenic Wonders of Macau
Travessa da Paixão	Phase III	Architectural ensembles	One of the New Eight Scenic Wonders of Macau, Style Street
Bairro de S. Lázaro	Phase III	Architectural ensembles	
Avenida do Conselheiro Ferreira de Almeida,desde o N.º 89 até ao N.º 95G	Phase III	Architectural ensembles	
Largo e Beco do Lilau	Phase III	Architectural ensembles	
Largo de St.º Agostinho	Phase III	Architectural ensembles	

(continued)

(continued)

Residential and public buildings			
Name (Portuguese or English)	Construction eras	Heritage classifications	Notes
Avenida de Almeida Ribeiro I Largo do Senado I Largo de S. Domingos	Phase III	Architectural ensembles	One of the New Eight Scenic Wonders of Macau, Style Street, Critical Visual Corridors I
Travessa de S. Paulo	Phase III	Architectural ensembles	
Public Spaces (Parks, Squares, Gardens)			
Colina da Ilha Verde	Phase I	Sites	
Colina de Mong-Há	Phase I	Sites	
Colina de D. Maria II	Phase I	Sites	
Jardim da Montanha Russa	Phase I	Sites	
Marginal, desde a Ponte Governador Nobre de Carvalho até à Fortaleza de S. Tiago da Barra	Phase I	Sites	One of the Eight Scenic Wonders of Macau, One of the New Eight Scenic Wonders of Macau
Percurso entre a Rua Central, a Rua de S. Lou renço, a Rua do Padre António, a Rua da Barra e a Calçada da Barra	Phase I	Sites	
Colina da Barra	Phase I	Sites	
Colina da Penha	Phase I	Sites	
Colina da Guia	Phase I	Sites	
Largo do Pagode da Barra	Phase I	Monuments	
Praça de Ponte e Horta	Phase II	Sites	
Jardim de Luís de Camões	Phase II	Sites	Style Street, Critical Visual Corridors F
Praça de Luís de Camões	Phase II	Sites	
Cemitério Protestante	Phase II	Sites	
Cemitério dos Parses	Phase II	Sites	
Largo da Sé	Phase II		
Largo da Companhia de Jesus	Phase II	Monuments	
Jardim de Lou Lim Ioc	Phase III	Sites	One of the Eight Scenic Wonders of Macau
Praça do Tap Siac (Antigo Campo Coronel Mes quita)	Phase III	Sites	Critical Visual Corridors H
Jardim de S. Francisco	Phase III	Sites	
Jardim da Vitória	Phase III	Sites	

(continued)

(continued)

Name (Portuguese or English)	Construction eras	Heritage classifications	Notes
Residential and public buildings			
Jardim de Vasco da Gama	Phase III	Sites	
Cemitério de S. Miguel Arcanjo	Phase III	Sites	
Antigo Mercado do Tarrafeiro	Phase III	Sites	
Parque Municipal Dr. Sun Yat-Sen	Phase IV	Sites	
Praça do Lago Sai Van	Phase IV		
Praça da Amizade	Phase IV		
Praça de Ferreira do Amaral	Phase IV	Sites	
Parque Dr. Carlos D' Assumpção	Phase IV		
Institutional system			
Fortaleza de N.ª Sr.ª do Monte	Phase II	Monuments	Critical Visual Corridors E, F, G
Fortaleza de N.ª Sr.ª do Bom Parto	Phase II	Monuments	
Fortaleza de N.ª Sr.ª da Guia	Phase II	Monuments	Critical Visual Corridors E, H, J
Forte e Quartel de S. Francisco	Phase II	Monuments	
Fortaleza de S. Tiago da Barra	Phase II	Monuments	Critical Visual Corridors C
Antigas Muralhas da Cidade (Troço na Calçada de S. Francisco Xavier)	Phase II	Monuments	
Fortaleza de Mong-Há	Phase III	Monuments	
Fortaleza de D. Maria II	Phase III	Monuments	
Farol de Guia	Phase III	Monuments	One of the Eight Scenic Wonders of Macau
Ponte-cais N.º 8	Phase III		
Ponte 16	Phase III		
Troço das Antigas Muralhas da Cidade, próximo da Estrada de S. Francisco	Phase II	Monuments	

(continued)

(continued)

Residential and public buildings			
Name (Portuguese or English)	Construction eras	Heritage classifications	Notes
Troço das Antigas Muralhas da Cidade, próximo da Estrada do Visconde de S. Januário	Phase II	Monuments	
Troço das Antigas Muralhas da Cidade, próximo da Capela de N.ª Sr.ª da Penha	Phase II	Monuments	
Ponte-cais N.º 1	Phase I	Monuments	
Ponte-cais N.º 5B	Phase III		
Liang Yongxin Fragrant Shop Plaque and Exterior Wall	Phase III		
Portas do Cerco	Phase III	Monuments	
Historical system			
Edifício Sede da Fundação Oriente (Antiga Casa do Jardim da Gruta de Camões)	Phase II	Buildings of architectural merit	
Gruta de Camões	Phase III	Monuments	
Pedra com Brasão Português (junto à Rampa dos Cavaleiros)	Phase III	Monuments	
Pedra com Brasão Português (junto ao Templo de Lin Fong)	Phase III	Monuments	
Palácio do Governo	Phase III	Monuments	
Casa do Mandarim	Phase III	Buildings of architectural merit	Intangible cultural heritage “Taoist Ritual Music”
Casa de Lou Kau	Phase III	Buildings of architectural merit	
Antiga Farmácia Chong Sai	Phase III	Monuments	
Edifício da Biblioteca Sir Robert Ho Tung	Phase III	Buildings of architectural merit	
Antiga Residência do General Ye Ting	Phase III	Monuments	
Antiga Casa de Silva Mendes	Phase III	Buildings of architectural merit	
Casa na Rua do Campo, N.º 165	Phase III	Buildings of architectural merit	

(continued)

(continued)

Name (Portuguese or English)	Construction eras	Heritage classifications	Notes
Residential and public buildings			
Dr. Sun Yat-Sen's Memorial House	Phase III		
Edifício da Autoridade Monetária de Macau (Edifício do Antigo Convento do Precioso Sangue)	Phase III	Buildings of architectural merit	
Edifício da Escola Ling Nam (Antiga Vila Alegre)	Phase III	Buildings of architectural merit	
Arco Oriente	Phase IV		
Lin Zexu Memorial Museum	Phase IV		
Praça Flor de Lótus	Phase IV		
Handover Gifts Museum of Macau	Phase IV		
Xian Xinghai Memorial Museum	Phase IV		
Spiritual system			
Templo de A-Má (Templo da Barra)	Phase I	Monuments	One of the Eight Scenic Wonders of Macau, Intangible cultural heritage "Mazu believes in customs" and "Cantonese Opera," Critical Visual Corridors D
Templo de Tin Hau	Phase I	Monuments	
Templo de Kun Iam (Kun Iam Tchai)	Phase I	Monuments	
Templo de Lin Fong	Phase II	Monuments	Intangible cultural heritage "Mazu believes in customs"
Templo de Pou Chai (Kun Iam Tong)	Phase II	Monuments	One of the Eight Scenic Wonders of Macau
Templo de Na Tcha (Calçada das Verdades)	Phase II	Monuments	Intangible cultural heritage "Nezha believes in customs"
Sam Kai Vui Kun (Templo de Kuan Tai)	Phase II	Monuments	
Templo de Tou Tei (Patane)	Phase II	Monuments	Intangible cultural heritage "land customs," "Taoist Ritual Music"
Templo de Foc Tac (Rua do Teatro)	Phase II	Monuments	Intangible cultural heritage "land customs"
Templo de Lin Kai	Phase III	Monuments	

(continued)

(continued)

Residential and public buildings			
Name (Portuguese or English)	Construction eras	Heritage classifications	Notes
Templo de Foc Tac (Bairro da Horta da Mitra)	Phase III	Monuments	Intangible cultural heritage "land customs"
Ian Tak Tong Kong So (Templo de Lou Pan Si Fu)	Phase III	Monuments	
Templo de Sin Fong	Phase III	Monuments	
Templo de Hong Chan Kuan (Templo do Bazar)	Phase III	Monuments	
Templo de Foc Tac (Rua do Almirante Sérgio)	Phase III	Monuments	Intangible cultural heritage "land customs"
Templo de Foc Tac (Rua do Patane)	Phase III	Monuments	Intangible cultural heritage "land customs"
Pagode de Seak Kam Tong Hang Toi	Phase III	Monuments	
Templo de Na Tcha (junto às Ruínas do Colégio de S. Paulo)	Phase III	Monuments	Intangible cultural heritage "Nezha believes in customs," "Taoist Ritual Music," Style Street
Templos de Nam San, Pao Kong, I Leng e Loi Tsou	Phase III	Monuments	
Centro Ecuménico Kun Iam	Phase IV		
Igreja de St.º António e adro	Phase I	Monuments	Style Street
Igreja de S. Lázaro e adro	Phase I	Monuments	
Igreja de S. Lourenço e adro	Phase I	Monuments	
Igreja de S. Domingos	Phase II	Monuments	Intangible cultural heritage "Procissão de Nossa Senhora de Fátima"
Igreja de St.º Agostinho	Phase II	Monuments	Intangible cultural heritage "Procissão de Nosso Senhor Bom Jesus dos Passos"
Ruínas do Colégio de S. Paulo (antiga Igreja da Madre de Deus, adro e escadaria)	Phase II	Monuments	One of the Eight Scenic Wonders of Macau, Critical Visual Corridors K, Style Street
Igreja da Sé Catedral	Phase II	Monuments	
Capela de N.ª Sr.ª das Neves	Phase II	Monuments	
Capela de N.ª Sr.ª da Penha e Residência Episcopal	Phase II	Buildings of architectural merit	One of the New Eight Scenic Wonders of Macau, Critical Visual Corridors A, B

(continued)

(continued)

Residential and public buildings			
Name (Portuguese or English)	Construction eras	Heritage classifications	Notes
Edifício do Seminário de S. José	Phase II	Monuments	
St. Michael's Chapel	Phase III		
Igreja de São Francisco Xavier	Phase III		

Appendix 2: Attraction Nodes of the Existing Core Zone Trails

Name (Portuguese or English)	Construction eras	Heritage classifications	Functional typologies
Templo de A-Má (Templo da Barra)	Phase I	Monuments	Spiritual system
Igreja de St.º António e adro	Phase I	Monuments	Spiritual system
Igreja de S. Lourenço e adro	Phase I	Monuments	Spiritual system
Largo do Pagode da Barra	Phase I	Monuments	Residential system, Spiritual system
Ponte-cais N.º 1	Phase I	Monuments	Institutional system
Sam Kai Vui Kun (Templo de Kuan Tai)	Phase II	Monuments	Spiritual system
Igreja de S. Domingos	Phase II	Monuments	Spiritual system
Igreja de St.º Agostinho	Phase II	Monuments	Spiritual system
Ruínas do Colégio de S. Paulo (antiga Igreja da Madre de Deus, adro e escadaria)	Phase II	Monuments	Spiritual system
Igreja da Sé Catedral	Phase II	Monuments	Spiritual system
Edifício do Seminário de S. José	Phase II	Monuments	Spiritual system
Fortaleza de N.ª Sr.ª do Monte	Phase II	Monuments	Institutional system
Antigas Muralhas da Cidade (Troço na Calçada de S. Francisco Xavier)	Phase II	Monuments	Institutional system
Edifício Sede do Instituto para os Assuntos Municipais (Edifício do Antigo Leal Senado)	Phase II	Monuments	Residential system

(continued)

(continued)

Name (Portuguese or English)	Construction eras	Heritage classifications	Functional typologies
Edifício Sede da Fundação Oriente (Antiga Casa do Jardim da Gruta de Camões)	Phase II	Buildings of architectural merit	Residential system, Historical system
Edifício Sede da Santa Casa da Misericórdia de Macau	Phase II	Monuments	Residential system
Cemitério Protestante	Phase II	Sites	Residential system
Largo da Sé	Phase II		Residential system, Spiritual system
Largo da Companhia de Jesus	Phase II	Monuments	Residential system, Spiritual system
Jardim de Luís de Camões	Phase II	Sites	Residential system
Praça de Luís de Camões	Phase II	Sites	Residential system
Templo de Na Tcha (junto às Ruínas do Colégio de S. Paulo)	Phase III	Monuments	Spiritual system
Casa do Mandarim	Phase III	Buildings of architectural merit	Residential system, Historical system
Edifício da Direcção dos Serviços de Assuntos Marítimos e de Água (Antigo Quartel dos Mou ros)	Phase III	Buildings of architectural merit	Residential system
Casa de Lou Kau	Phase III	Buildings of architectural merit	Residential system, Historical system
Edifício do Teatro D. Pedro V	Phase III	Buildings of architectural merit	Residential system
Edifício da Biblioteca Sir Robert Ho Tung	Phase III	Buildings of architectural merit	Residential system, Historical system
Largo e Beco do Lilau	Phase III	Architectural ensembles	Residential system
Largo de St.º Agostinho	Phase III	Architectural ensembles	Residential system
Avenida de Almeida Ribeiro I Largo do Senado I Largo de S. Domingos	Phase III	Architectural ensembles	Residential system
Travessa da Paixão	Phase III	Architectural ensembles	Residential system
Travessa de S. Paulo	Phase III	Architectural ensembles	Residential system

(continued)

(continued)

Name (Portuguese or English)	Construction eras	Heritage classifications	Functional typologies
Liang Yongxin Fragrant Shop Plaque and Exterior Wall	Phase III		Institutional system
Edifício Sede dos Correios	Phase III	Buildings of architectural merit	Residential system
Da Sheng Gong Herbal Tea Shop	Phase III		Residential system
Casas Moosa	Phase III	Buildings of architectural merit	Residential system
Casa no Largo da Companhia de Jesus, N.º 2	Phase III	Buildings of architectural merit	Residential system

Appendix 3: Location of Supplementing Attraction Nodes within the Walkable Proximity Zone

Name (Portuguese or English)	Construction eras	Heritage classifications	Functional typologies
Igreja de S. Lázaro e adro	Phase I	Monuments	Spiritual system
Consulado-Geral de Portugal em Macau e Hong Kong (Edifício do Antigo Hospital S. Rafael e jardim)	Phase I	Buildings of architectural merit	Residential system
Colina da Barra	Phase I	Sites	Residential system
Colina da Penha	Phase I	Sites	Residential system
Grand Lisboa	Phase IV		Residential system
Praça de Ferreira do Amaral	Phase IV	Sites	Residential system
Praça da Amizade	Phase IV		Residential system
Arco Oriente	Phase IV		Historical system
Praça Flor de Lótus	Phase IV		Historical system
Fortaleza de N.ª Sr.ª do Bom Parto	Phase II	Monuments	Institutional system
Forte e Quartel de S. Francisco	Phase II	Monuments	Institutional system
Fortaleza de S. Tiago da Barra	Phase II	Monuments	Institutional system
Troço das Antigas Muralhas da Cidade, próximo da Estrada de S. Francisco	Phase II	Monuments	Institutional system

(continued)

(continued)

Name (Portuguese or English)	Construction eras	Heritage classifications	Functional typologies
Troço das Antigas Muralhas da Cidade, próximo da Estrada do Visconde de S. Januário	Phase II	Monuments	Institutional system
Troço das Antigas Muralhas da Cidade, próximo da Capela de N.ª Sr.ª da Penha	Phase II	Monuments	Institutional system
Ponte-cais N.º 5B	Phase III		Institutional system
Ponte-cais N.º 8	Phase III		Institutional system
Ponte 16	Phase III		Institutional system
Gruta de Camões	Phase III	Monuments	Historical system
Antiga Farmácia Chong Sai	Phase III	Monuments	Historical system
Dr. Sun Yat-Sen's Memorial House	Phase III		Historical system
Palácio do Governo	Phase III	Monuments	Residential system, Historical system
Antiga Casa de Silva Mendes	Phase III	Buildings of architectural merit	Residential system, Historical system
Casa na Rua do Campo, N.º 165	Phase III	Buildings of architectural merit	Residential system, Historical system
Edifício da Autoridade Monetária de Macau (Edifício do Antigo Convento do Precioso Sangue)	Phase III	Buildings of architectural merit	Residential system, Historical system
Edifício da Escola Ling Nam (Antiga Vila Alegre)	Phase III	Buildings of architectural merit	Residential system, Historical system
Rua e Beco da Felicidade	Phase III	Architectural ensembles	Residential system
Bairro de S. Lázaro	Phase III	Architectural ensembles	Residential system
Avenida do Conselheiro Ferreira de Almeida,desde o N.º 89 até ao N.º 95G	Phase III	Architectural ensembles	Residential system
Antigo Mercado do Tarrafeiro	Phase III	Sites	Residential system
Praça de Ponte e Horta	Phase II	Sites	Residential system
Cemitério dos Parses	Phase II	Sites	Residential system
Praça do Tap Siac (Antigo Campo Coronel Mes quita)	Phase III	Sites	Residential system

(continued)

(continued)

Name (Portuguese or English)	Construction eras	Heritage classifications	Functional typologies
Jardim de S. Francisco	Phase III	Sites	Residential system
Jardim da Vitória	Phase III	Sites	Residential system
Jardim de Vasco da Gama	Phase III	Sites	Residential system
Cemitério de S. Miguel Arcanjo	Phase III	Sites	Residential system

Appendix 4: List of Attractive Nodes of the Optimized Trail

Name (Portuguese or English)	Construction eras	Heritage classifications	Functional typologies
Templo de A-Má (Templo da Barra)	Phase I	Monuments	Spiritual system
Igreja de St.º António e adro	Phase I	Monuments	Spiritual system
Igreja de S. Lourenço e adro	Phase I	Monuments	Spiritual system
Largo do Pagode da Barra	Phase I	Monuments	Residential system, Spiritual system
Ponte-cais N.º 1	Phase I	Monuments	Institutional system
Igreja de S. Lázaro e adro	Phase I	Monuments	Spiritual system
Consulado-Geral de Portugal em Macau e Hong Kong (Edifício do Antigo Hospital S. Rafael e jardim)	Phase I	Buildings of architectural merit	Residential system
Colina da Penha	Phase I	Sites	Residential system
Colina da Guia	Phase I	Sites	Residential system
Sam Kai Vui Kun (Templo de Kuan Tai)	Phase II	Monuments	Spiritual system
Igreja de S. Domingos	Phase II	Monuments	Spiritual system
Igreja de St.º Agostinho	Phase II	Monuments	Spiritual system
Ruínas do Colégio de S. Paulo (antiga Igreja da Madre de Deus, adro e escadaria)	Phase II	Monuments	Spiritual system
Igreja da Sé Catedral	Phase II	Monuments	Spiritual system
Igreja do Seminário de S. José	Phase II	Monuments	Spiritual system

(continued)

(continued)

Name (Portuguese or English)	Construction eras	Heritage classifications	Functional typologies
Fortaleza de N.ª Sr.ª do Monte	Phase II	Monuments	Institutional system
Antigas Muralhas da Cidade (Troço na Calçada de S. Francisco Xavier)	Phase II	Monuments	Institutional system
Edifício Sede do Instituto para os Assuntos Municipais (Edifício do Antigo Leal Senado)	Phase II	Monuments	Residential system
Edifício Sede da Fundação Oriente (Antiga Casa do Jardim da Gruta de Camões)	Phase II	Buildings of architectural merit	Residential system, Historical system
Edifício Sede da Santa Casa da Misericórdia de Macau	Phase II	Monuments	Residential system
Cemitério Protestante	Phase II	Sites	Residential system
Praça de Luís de Camões	Phase II	Sites	Residential system
Jardim de Luís de Camões	Phase II	Sites	Residential system
Largo da Sé	Phase II		Residential system, Spiritual system
Largo da Companhia de Jesus	Phase II	Monuments	Residential system, Spiritual system
Troço das Antigas Muralhas da Cidade, próximo da Capela de N.ª Sr.ª da Penha	Phase II	Monuments	Institutional system
Praça de Ponte e Horta	Phase II	Sites	Residential system
Capela de N.ª Sr.ª da Penha e Residência Episcopal	Phase II	Buildings of architectural merit	Spiritual system
Forte e Quartel de S. Francisco	Phase II	Monuments	Institutional system
Fortaleza de N.ª Sr.ª da Guia	Phase II	Monuments	Institutional system
Capela de N.ª Sr.ª das Neves	Phase II	Monuments	Spiritual system
Cemitério dos Parses	Phase II	Sites	Residential system
Templo de Na Tcha (junto às Ruínas do Colégio de S. Paulo)	Phase III	Monuments	Spiritual system
Casa do Mandarim	Phase III	Buildings of architectural merit	Residential system, Historical system
Edifício da Direcção dos Serviços de Assuntos Marítimos e de Água (Antigo Quartel dos Mou ros)	Phase III	Buildings of architectural merit	Residential system

(continued)

(continued)

Name (Portuguese or English)	Construction eras	Heritage classifications	Functional typologies
Casa de Lou Kau	Phase III	Buildings of architectural merit	Residential system, Historical system
Edifício do Teatro D. Pedro V	Phase III	Buildings of architectural merit	Residential system
Edifício da Biblioteca Sir Robert Ho Tung	Phase III	Buildings of architectural merit	Residential system, Historical system
Largo e Beco do Lilau	Phase III	Architectural ensembles	Residential system
Largo de St.º Agostinho	Phase III	Architectural ensembles	Residential system
Avenida de Almeida Ribeiro|Largo do Senado|Largo de S. Domingos	Phase III	Architectural ensembles	Residential system
Travessa da Paixão	Phase III	Architectural ensembles	Residential system
Travessa de S. Paulo	Phase III	Architectural ensembles	Residential system
Liang Yongxin Fragrant Shop Plaque and Exterior Wall	Phase III		Institutional system
Edifício Sede dos Correios	Phase III	Buildings of architectural merit	Residential system
Casas Moosa	Phase III	Buildings of architectural merit	Residential system
Casa no Largo da Companhia de Jesus, N.º 2	Phase III	Buildings of architectural merit	Residential system
Ponte-Cais N.º 8	Phase III		Institutional system
Gruta de Camões	Phase III	Monuments	Historical system
Antiga Farmácia Chong Sai	Phase III	Monuments	Historical system
Palácio do Governo	Phase III	Monuments	Residential system, Historical system
Rua e Beco da Felicidade	Phase III	Architectural ensembles	Residential system
Bairro de S. Lázaro	Phase III	Architectural ensembles	Residential system
Avenida do Conselheiro Ferreira de Almeida,desde o N.º 89 até ao N.º 95G	Phase III	Architectural ensembles	Residential system

(continued)

(continued)

Name (Portuguese or English)	Construction eras	Heritage classifications	Functional typologies
Praça do Tap Siac (Antigo Campo Coronel Mes quita)	Phase III	Sites	Residential system
Lar de Nossa Senhora da Misericórdia	Phase III	Buildings of architectural merit	Residential system
Farol de Guia	Phase III	Monuments	Institutional system
Jardim de S. Francisco	Phase III	Sites	Residential system
Edifício do Clube Militar	Phase III	Buildings of architectural merit	Residential system
Casa na Rua do Campo, N.º 165	Phase III	Buildings of architectural merit	Residential system, Historical system
Casa na Rua de Pedro Nolasco da Silva, N.os 24A, 26, 28 e 28A	Phase III	Buildings of architectural merit	Residential system
Antiga Casa de Silva Mendes	Phase III	Buildings of architectural merit	Residential system, Historical system
Edifício da Autoridade Monetária de Macau (Edifício do Antigo Convento do Precioso Sangue)	Phase III	Buildings of architectural merit	Residential system, Historical system
Pavilhão de Exposições e Espectáculos Artísticos para Jovens (Edifício da Antiga Caixa Escolar)	Phase III	Buildings of architectural merit	Residential system
Edifício na Calçada da Vitória, N.º 55	Phase III	Buildings of architectural merit	Residential system
Grand Lisboa	Phase IV		Residential system
Praça de Ferreira do Amaral	Phase IV	Sites	Residential system
Praça da Amizade	Phase IV		Residential system
Arco Oriente	Phase IV		Historical system
Praça Flor de Lótus	Phase IV		Residential system, Historical system

Appendix 5: A Route Map by Sequentially Taking Photos of All Attraction Nodes of the Optimized Trail

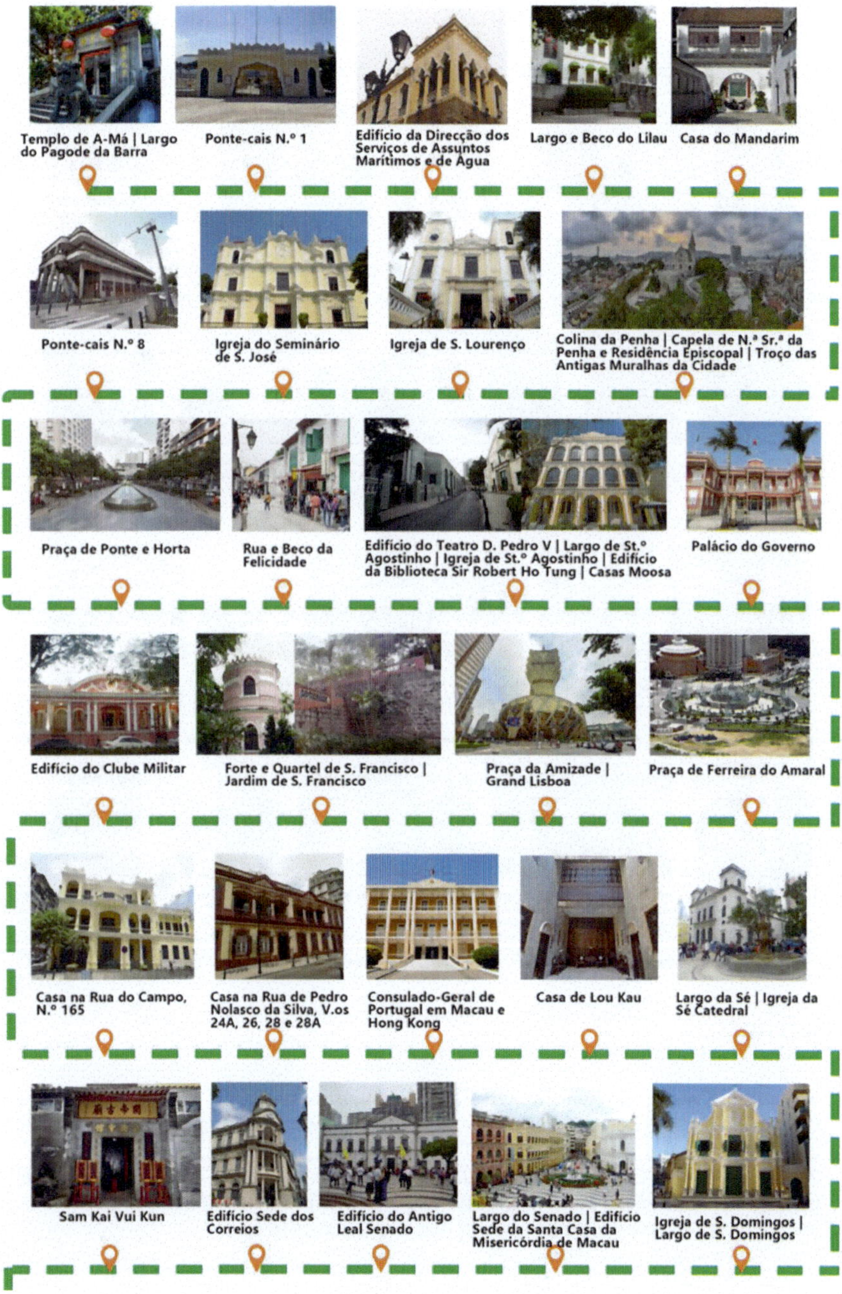

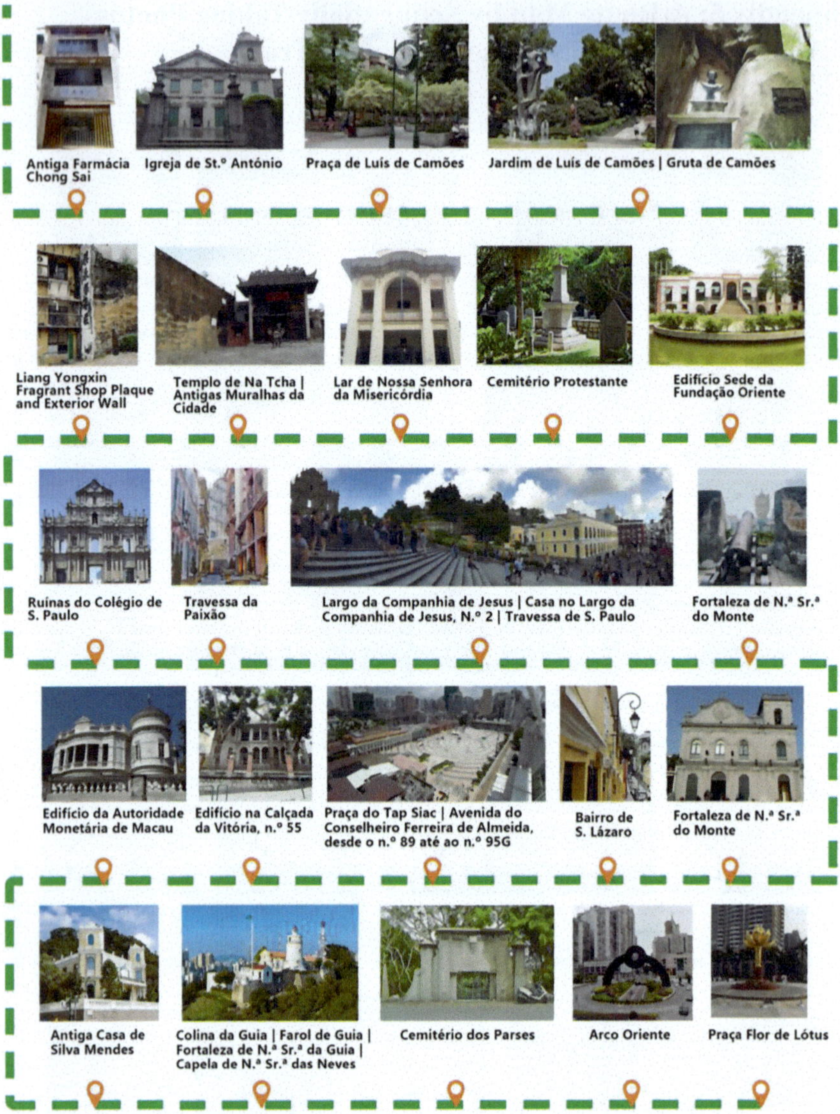

References

1. Ministry of Culture and Tourism. (2023). Three ministries issue the "Notice on Developing China's Themed Heritage Trails". Lvjie Media. https://www.lvjie.com.cn/brand/2023/0507/29172.html
2. Zhang, Y., & Guo, X. (2016). Debris and coupling: The exploration of construction of cultural heritage trails and integration of historical environment in metropolis. *Modern Urban Research, 31*(2), 105–113. https://doi.org/10.3969/j.issn.1009-6000.2016.02.017

3. Liu, W., Xiang, J., Liu, J., & Zhang, D. (2021). The protection and renewal of Luoyang Old City "Yujia Courtyard" is based on the "Historical Urban Land Scape" approach. *Architecture & Culture, 7*, 104–105. https://doi.org/10.19875/j.cnki.jzywh.2021.07.036
4. Cultural Affairs Bureau, Macau SAR Government. (2018). Conservation and management plan of the historic centre of Macau: Consultation document. Macau: Author.
5. Li, Y., & Yan, L. (2023). Review of domestic Historic Urban Landscape (HUL) research: Visualization analysis based on CNKI literature. *Architecture & Culture, 10*, 260–263.
6. Xiong, X., Dai, Y., Song, F., Lu, Y., & Zhu, D. (2017). Morphology-based value identification and geographical process analysis of Historic Urban Landscape heritage: Case study of Guling Town, Lushan. *Human Geography, 32*(3), 36–43.
7. Lei, Z., Luo, J., & Cao, L. (2021). Quantitative research on historic city landscape value assessment based on Ahp-gis—taking Taiyuan, a historic and cultural city, as an example. *Journal of Taiyuan University of Technology, 52*(4), 630–637. https://doi.org/10.16355/j.cnki.issn1007-9432tyut.2021.04.017
8. Tu, L., Zhao, P., Hu, Y., & Zhang, C. (2021). Theory of layering cognition and integrated conservation of traditonal villages: Introduction of historic urban landscape approach. *Urban Development Studies, 28*(11), 92–97. https://doi.org/10.3969/j.issn.1006-3862.2021.11.021
9. Pei, Y., Yang, R., & Dang, A. (2022). Historic layering process interpretation and value assessment of the Ancient City of Yulin based on historic urban landscape approach. *Urban Development Studies, 29*(7), 128–136. https://doi.org/10.3969/j.issn.1006-3862.2022.07.022
10. Bai, Y., Tang, X., Xu, Y., & Wang, X. (2023). A tentative discussion on heritage value from the perspective of historic urban landscape—taking Chenghuang temple in Jieyang as an example. *Chinese Landscape Architecture, 39*(9), 74–80. https://doi.org/10.19775/j.cla.2023.09.0074
11. Chen, H. (2022). Research on the urban feature enhancement of historical cities under HUL characteristic analysis: taking the historical district of Jining as an example. *Urban and Rural Planning, 4*, 49–60. https://doi.org/10.12049/j.urp.202204006
12. Cao, Y., & Li, H. (2019). Diachrony and synchronism in the protection of historical towns: Enlightenment and thought of "historic urban landscape." *Urban Development Studies, 26*(10), H13–H20. https://doi.org/10.3969/j.issn.1006-3862.2019.10.010
13. Xiao, J., & Cao, K. (2019). Characteristic extraction and layering interpretation of historic urban landscape in (U)-Tsang. *Chinese Landscape Architecture, 35*(10), 101–106. https://doi.org/10.19775/j.cla.2019.10.0101
14. Li, H., & Fu, P. (2020). The interpretation and preservation strategies of the layering laws of historic urban landscape. *Chinese Landscape Architecture, 36*(4), 49–54. https://doi.org/10.19775/j.cla.2020.04.0049
15. Xu, K., Qian, X., & Zeng, W. (2021). Enlightenment from the framework of historic cultural landscape elements in Germany——a case study of the State of Sachsen. *Chinese Landscape Architecture, 37*(2), 94–99. https://doi.org/10.19775/j.cla.2021.02.0094
16. Guo, Q., Xiao, L., & Huang, K. (2023). Research on the conservation of the historic site of Duanzhou Government Office in Zhaoqing from the perspective of historic urban landscape (HUL). *Chinese Landscape Architecture*, 99–105. https://doi.org/10.19775/j.cla.2023.03.0099
17. Tian, M., Zheng, W., Ai, Y., & Li, G. (2022). The interpretation of the spatial-temporal process of "anchoring-layering" structure of Guilin landscape city historical urban landscape. *Chinese Landscape Architecture, 38*(3), 26–31. https://doi.org/10.19775/j.cla.2022.03.0026
18. Peng, C., Zhang, H., & Wang, L. (2022). Visualization of trail system research in China based on CiteSpace. *Anhui Architecture, 29*(4), 74–78. https://doi.org/10.16330/j.cnki.1007-7359.2022.04.032
19. Zou, H., & Shi, L. (2014). A study on the conservation and revitalization approach of historic area based on heritage trail. *Huazhong Architecture, 32*(3), 154–157.
20. Guo, S. (2019). Route planning for urban heritage trails from a Historic Urban Landscape perspective [Master's thesis, Huazhong Agricultural University].
21. Zhang, S., & Zhen, X. (2014). Study on the Conservation Strategies for the Historic Urban Landscapes in Macau. *City Planning Review, 38*(1), 91–96. https://doi.org/10.11819/cpr20141318a

22. Liang, Z., Zhao, Y., Zhang, Y., & Feng, L. (2021). HUL-based conservation strategies for historic districts: Insights from Macau. *China Cultural Heritage, 3*, 4–12. https://doi.org/10.3969/j.issn.1672-7819.2021.03.001
23. Zhang, J., & Niu, Z. (2022). Cultural heritage protection of Macau from the perspective of cultural landscape. *Journal of Urban and Regional Planning*, z1, 118–129.
24. Wang, N., & Du, Y. (2015). Resident walking distance threshold of community. *Transportation Standardization, 2*(20–24), 30. https://doi.org/10.16503/j.cnki.2095-9931.2015.02.004
25. Li, Z., & Liu, M. (2021). "Education on site" of ecological civilization——enlightenment from building and using of the New Zealand walking and tramping tracks system. *Guangdong Landscape Architecture, 43*(3), 2–7. https://doi.org/10.12233/j.gdyl.2021.03.001
26. UNESCO. (2005). *Vienna memorandum*. UNESCO World Heritage Centre.
27. UNESCO. (2011). *Recommendation on the historic Urban landscape*. Paris.

Index

Y. Chen et al. (eds.), *City-Sea Symbiosis*, Springer Geography,
https://doi.org/10.1007/978-3-032-13725-8

The manufacturer's authorised representative in the EU is Springer Nature Customer Service Centre GmbH, Europaplatz 3, 69115 Heidelberg, Germany. If you have any concerns regarding our products, please contact ProductSafety@springernature.com

Printed and bound by CPI Group (UK) Ltd, Croydon, CR0 4YY

21/09/2026

02220013-0014